PRODUCED WATER

Volume 2:

Equipment
Process Configuration
Applications

John M. Walsh

Produced Water

Volume 1:
- Fundamentals
- Water Chemistry
- Emulsions
- Chemical Treatment

Volume 2:
- Equipment
- Process Configuration
- Applications

Authored by: John M. Walsh

Published By: Petro Water Technology, LLC

First Edition, Volume 1 ISBN: 978-1-7322736-0-3
Library of Congress Control Number:

First Edition, Volume 2 ISBN: 978-1-7322736-1-0
Library of Congress Control Number:

Table of Contents

Chapter Eight: Primary Separators 1

Chapter Nine: Interceptors CPI, TPI, PPI 33

Chapter Ten: Hydrocyclones 57

Chapter Eleven: Centrifuges 115

Chapter Twelve: Flotation 129

Chapter Thirteen: Tertiary Equipment 205

Chapter Fourteen: Process Engineering 281

Chapter Fifteen: Troubleshooting 305

Chapter Sixteen: Applications: Characterization 335

Chapter Seventeen: Shale Produced Water Treatment 371

Chapter Eighteen: Applications: Solids 423

Chapter Nineteen: Deepwater Best Practices 455

Chapter Twenty: Applications: GoM vs. North Sea 477

Chapter Twenty One: Applications: Dissolved & Water Soluble Organics 509

Appendix A: Units, Standard Conditions, Common Calculations 539

Index 547

Contributors:

I am extremely grateful to the following individuals who made major contributions to this book. They are listed in alphabetical order. In particular, Ron Bosch and Ted Frankiewicz were involved in the initial conception and planning of the book. But also many others provided valuable input who are not listed here, particularly those people affiliated with the Produced Water Society and the facilities discipline of the Society of Petroleum Engineers.

Munib Ahmad: Munib and I worked together in Oman. We shared a keen interest in using science to solve practical problems while at the same time gathering as much data we could on water chemistry and equipment performance in the field. Munib contributed very strongly to the quality of Chapter 12 on Flotation.

Kris Bansal: Shortly after Kris retired from ConocoPhillips, he accepted my offer to co-teach the SPE course. We worked on the course from the CETCO offices. During those months, he reviewed several chapters and gave me excellent feedback on the book.

Ron Bosch: Ron, Ted and I developed the outline for the book in the form of a detailed Table of Contents in November 2013. While we recognized the enormous ambition of the project, none of us could part with any subject in the outline. In the end, I am grateful to these two gentlemen for laying out an ambitious plan. In the several years that followed, Ron made intellectual contributions and provided ideas for Chapter 5 on Chemical Treatment.

Ted Frankiewicz: As mentioned in Ron's acknowledgement above, Ted, Ron and I developed the outline for the book. In addition, Ted provided copy editing for several chapters. Much of the Chapter on Primary Separation was written by Ted.

Morris Hoagland: Morris wrote a few sections and provided feedback on a number of subjects. The section on activated carbon in Chapter 13 is almost entirely his material. He is one of the world experts on that subject.

Dan Shannon: Dan and Ron Bosch provided the chemist perspective that was needed for Chapter 5 on production chemicals. Both gentlemen provided insightful material.

Ramesh Sharma: Ramesh and I worked together in New Mexico where we studied the high concentration of very fine particles associated with shale produced water. Ramesh provided significant intellectual insight that went into the chapter on Shale Produced Water.

Greg Simpson: Greg wrote the chapter on microbial biology and biological control. Also, Greg mentored me on the process of self-publishing. He is author of a multivolume series of texts on chlorine dioxide which he self-published. Greg is a true scholar. He wrote his chapter with minimal input from me and delivered it with an extensive set of references.

Colin Tyrie: There is a good chance that there would not have been a Produced Water Society if it were not for the efforts of Colin Tyrie and Dan Caudle. If there had not been a Produced Water Society, I and many others might not have become produced water specialists. Thus, much of the credit for the existence of this book goes to Colin. Also, I am grateful to him for helping me in my career, as he did for so many others.

Ming Yang: Ming helped write the chapter on sampling and analysis. It should also be mentioned that Gary Bartman provided a detailed review of that chapter. Between the two of these gentlemen, there is no one who knows the subject better.

Preface:

The material for this book originated in a course that I started teaching in Oman in 2007. I was an expatriate there working for a multinational oil company. Water treatment is an important issue in Oman since over 95 % of the fluid produced is water. As an expatriate, one of my duties was to mentor and coach the local staff so I began teaching a course in water treatment.

In 2010, I became the instructor for the SPE course on water treatment. I taught the course three or four times a year, across the globe, for more than ten years. Also, I chaired a series of SPE workshops each of which was held in an important area of produced water around the world. The SPE course and workshops provided a wealth of information.

For much of my career I spent a lot of time in the field both onshore and offshore. In addition to that practical experience, another valuable input for this book comes from the scientific literature. Eventually I amassed a collection of a couple thousand papers. One of the more gratifying experiences in writing this book has been the tying together of scientific literature with field observations. Understanding the chemical and physical fundamental mechanisms involved in produced water treatment provides a powerful tool in design and problem solving.

In 2013, Ron Bosch and Ted Frankiewicz had the idea to take the course material and write a book. Together we developed an inspiring table of contents. Kindly, John Occhipinti, my boss at CETCO gave me time to write. My experience at CETCO had a profound effect on the material in the book. The range and complexity of problems, the variety of systems and equipment, the ability to conduct lab studies expediently, and the team atmosphere made my time there a wonderful experience which is hopefully captured in the book.

In November 2018 I joined Jacobs, one of the largest and highly rated engineering firms in the world. The legacy Ch2m was still intact to a great extent within Jacobs. The contents of this book benefitted greatly by my interaction with this talented group of industrial water specialists including Ken Martins, Bruce Thomas-Benke, Michael Dunkel, and Derek Evans.

Looking back, I have been extremely fortunate to have made so many friendships around the subject of produced water. Produced water is a by-product of oil and gas production. As such it tends to be overlooked in the industry and is often considered of secondary importance. But it is also extremely complex, challenging and humbling. People who work in this area enjoy challenges and believe that we are doing something good for the environment. This book is my way of giving something to the many people who work in this area.

John
March 2019

Acknowledgement

This book would not have come into existence without the support and encouragement of my wife, Bawani. She helped to solve many publishing problems. I am extremely grateful for her help.

I would also like to thank my parents, Maryann and James for their constant encouragement.

CHAPTER EIGHT

Primary Separators

Chapter 8 Table of Contents

8.0 Introduction 5

8.1 Stokes Law Review 8

8.2 Separator Design Considerations 9

8.2.1 Fluid Phases in a Separator 9

8.2.2 Practical Considerations Included in Defining a Separator Size 10

8.2.3 Fluid Inlet Devices 14

8.2.4 Vortex Breakers 16

8.2.5 Sand Removal and Interface Drains 17

8.2.6 Process Control Philosophy and Instrumentation 19

8.3 Calculating Oil Droplet Retention Efficiency for a Horizontal Separator 23

8.4 Process (FWKO, Skim or Wash) Tank Design Considerations 27

8.4.1 Designing Skim Tanks for Horizontal Flow 28

8.4.2 Designing Skim Tanks for Vertical Flow 29

References to Chapter 8 31

8.0 Introduction

The objective of this Chapter is to define the parameters which can be utilized to size and design 3-phase or 2-phase liquid-liquid (as opposed to 2-phase gas-liquid) separation equipment. The type of equipment which is discussed here includes:

Primary Technologies:

- Upstream separators (gas/oil/water)
- Downstream Settling Tanks
- Free Water Knock-Outs (FWKO), a.k.a. Production Separators, Inlet Separators, etc.
- Skim Tanks (atmospheric) or Skim Vessels (pressurized)

The information presented in this chapter can also be utilized to diagnose performance limitations for existing equipment. This is an important capability since resolving a water treatment problem requires an understanding as to whether unacceptable water quality is related to equipment design or size, chemical issues, operational conditions such as severe gas/liquid slugging, or the process control philosophy.

In most product literature from vendors, and papers on the subject, the terminology "primary deoiling equipment" is generally agreed to include those processes and equipment which rely on Stokes Law variables to achieve separation. In these cases, there is a density difference between the water phase and the contaminants (oil droplets or solid particles). This density difference is what drives the motion of the contaminants and eventually results in oil/water separation. What is common to all primary separation processes is that the density difference is the driver. In addition, primary separation is enhanced for larger drops and higher temperature as a result of the lower water viscosity.

Primary, Secondary, and Tertiary Technologies: This definition of primary deoiling equipment helps to distinguish it from secondary and tertiary equipment. As explained in later chapters, secondary equipment, as we define it, refers to particle capture technology such as flotation. Tertiary refers to equipment which relies on the use of media of some kind, such as coalescing media, and filtration equipment. These broad definitions point the way to the different variables involved in separation performance. Primary separation, as noted, depends on density difference, contaminant (droplet and particle) size, and viscosity. Secondary (flotation) performance depends on contaminant size, interfacial adsorption (which depends on the contaminant chemistry and chemical treatment), bubble size, and total surface area of the bubbles (gas/water ratio). Tertiary performance depends on contaminant (floc) size, media surface properties (which depends on the contaminant chemistry and chemical treatment), media surface area, superficial velocity, and details regarding backwash or dirt holding capacity.

Obviously there is significant overlap between these definitions of primary, secondary and tertiary equipment. Flotation does after all depend on Stokes Law for bubble and drop relative velocities. Also, some vertical flotation cells have some form of coalescing media. However, we should not get too concerned with this classification scheme. We gratefully accept (and ignore) any criticism of this scheme. While there are some common variables between these technologies, there are also some important differences. There are also some important differences in the reject rate, coalescence of contaminants in the reject, and backwash details.

Different Types of Primary Separators: Ken Arnold has provided an excellent introduction and set of definitions for the terminology involved in primary separation [1, 2]:

1. A two phase separator separates gas and total liquids. The overall size of the separator (that is, its diameter and length) is either based on the gas capacity or the liquid capacity. Both have to be considered for the separator to work properly. If the gas capacity constraint governs the diameter and length the separator will be able to handle more liquid flow. If the liquid capacity governs, the separator will be able to handle more gas flow. Since we can't predict accurately the liquid loading and droplet size distribution to the mist extractor, trying to adjust the diameter and length of a separator to get 0.1 gal/MMSCF or 0.5 gal/MMSCF is a useless exercise. To get a better quality gas you need a better mist extractor. If the quality of gas is not important, do not use a mist extractor.

2. A three phase separator separates gas, oil and water. There are two kinds of three phase separators: those where the primary objective is to have a reasonably good quality oil stream with as low a BS&W content as possible, and those where the primary objective is to have as good a quality of water as possible. The former is generally referred to as a Free Water Knockout (FWKO), and the latter a Water Skimmer. In almost all cases the "oil" from the FWKO will still need further treatment to reach sales BS&W or salt specifications and the water from a Water Skimmer will still need further treatment for disposal or reuse. That is, the three phase separator is a bulk separation device.

3. A three phase separator cannot be both a FWKO and a Water Skimmer. The internals have to be set up differently if the goal is to get as good a quality of oil as is possible or if the goal is to get as good a quality water as is possible in this bulk separation.

4. A slug catcher is a vessel whose primary purpose is to catch and hold at inlet pressure a slug of liquid and release it at a controlled rate over time so as not to overwhelm the liquid handling capacity of the downstream separation equipment. A FWKO can handle only a very small slug without upsetting the oil-water interface that is crucial for its operation as a three phase separator. A two phase separator can handle a larger slug. However, because of the large momentum of a slug of liquid arriving at the very high velocities of the gas pushing the slug of liquid, an inlet diverter may not be possible and thus a mist extractor may flood during normal gas flow. For very large slugs a "finger type" slug catcher is needed to contain the large volume of liquid at pressure. Often the gas outlet of this type of slug catcher does not have a mist extractor and a two phase separator is required downstream of the slug catcher.

5. Typically downstream of a FWKO is equipment for treating the oil and emulsion from the FWKO. I call this vessel an "oil treater", an "electrostatic treater", an oil treating tank, or a gun barrel tank.

Generic Process Flow Diagrams (PFD) for offshore and for onshore production are shown in Figures 8.1, 8.2, and 8.3. The offshore PFD (Figure 8.1) utilizes hydrocyclones which are a compact separation technology. Also shown is a gas flotation unit which provides high performance removal of small oil droplets and oily solids. Since offshore produced water will be disposed of overboard, deep removal of contaminants is necessary. Figure 8.2 utilizes tanks and a flotation unit. Onshore contaminant removal may or may not require flotation depending on whether the treated effluent water will be disposed in a dipsal well or injected into the hydrocarbon reservoir for waterflood. The later usually requires cleaner water.

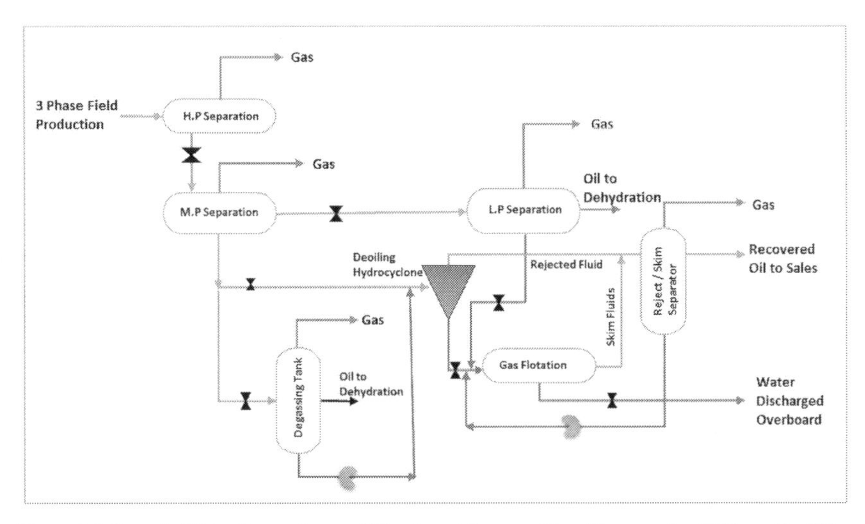

Figure 8.1 Generic Water Separation and Treatment Process Flow Diagram for Offshore Facilities

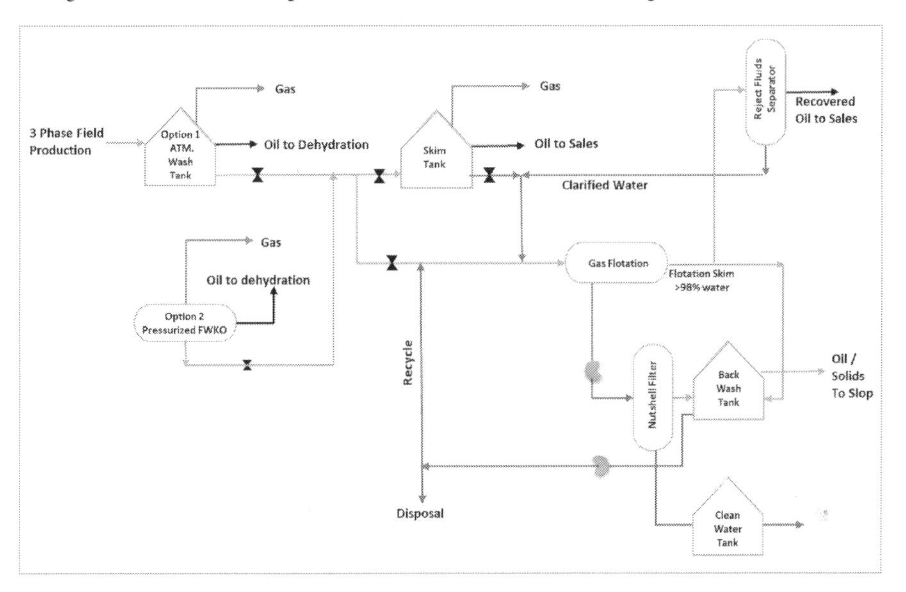

Figure 8.2 Generic Water Separation and Treatment Process Flow Diagram for Onshore Facilities

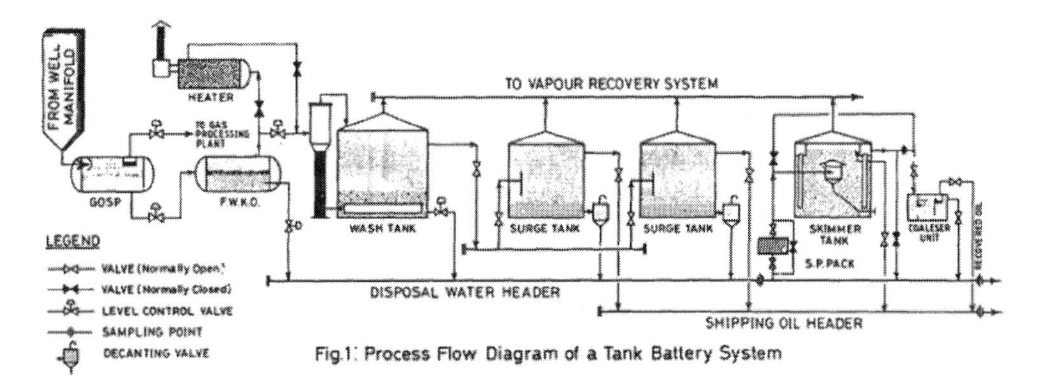

Figure 8.3 Generic Tank Battery Process Flow Diagram [3].

Figure 8.3 shows a slightly different onshore tank battery. This one is from Murti and Al-Nuaimi [3] who describe the modifications required to a facility as the water cut increases to relatively high values (20 % and higher). Often the primary design function of a tank battery is to provide storage volume to prevent shutting down the disposal pump if water from the field is temporarily curtailed, or to prevent shutdown of wells if the disposal pump is temporarily shutdown. The PFD shown above utilizes an SP-Pack which is a device for promoting oil droplet coalescence. Also shown is a so-called "coalescer unit." This is actually a corrugated plate interceptor. The use of the term "coalescer" is discussed in Chapter 9.

Besides settling rate, there are other significant issues in the selection and design of primary settling equipment. Theses issues are considered here:

- Is there a need for separating sand from the fluids either upstream of primary separation or within the primary separator itself?
- Are there fluid stability / compatibility issues which might lead to the formation of mineral scales?
- Is the primary separator the best location for injection of chemical in order to facility downstream separation? If so, what is the best way to inject the chemical? How should chemical selection be carried out?
- Are the vessel internals designed well enough to achieve a substantial portion of the superficial residence time?
- Optimum vessel design to achieve oil / water separation

8.1 Stokes Law Review

The primary equation of interest for gravity separation is Stokes Law. It allows prediction of the rise velocity of oil droplets in water, or the settling velocity of solid particles. Stokes Law has already been discussed in detail in Section 3.2.3 (Stokes Law Settling Rate). It is given here again for convenience. Also, the presentation here is from a slightly different perspective.

$$u = C\frac{(\rho_w - \rho_d)d^2}{\mu} \qquad \text{Eqn (8.1)}$$

Where:

u	= contaminant velocity
r_w	= density of water phase
r_o	= density of contaminant or dispersed phase
d	= diameter of oil drop or solid particle
μ	= viscosity of water phase

The value of the constant in the above equation depends on the units used for the dispersed droplet or particle, the density for each respective phase, and for fluid viscosity. The constant for often used sets of units are given in Table 8.1.

Table 8.1 Values of the Constant in Stokes Law for various sets of commonly used Units

V_{term}	Constant	ρ	δ	μ
cm/min	3.27×10^{-3}	g/cm^3	microns	cP
in/min	1.29×10^{-3}	g/cm^3	microns	cP
in/min	2.06×10^{-5}	lbs/ft^3	microns	cP

8.2 Separator Design Considerations

8.2.1 Fluid Phases in a Separator

As production fluids enter a separation vessel or tank, usually with an attendant decrease in pressure, the phases which form are illustrated in Figure 8.4. The concept that a vessel design only need consider the 3 primary phases of gas, oil, and water is simplistic and inhibits one's understanding of how phase separation actually takes place.

Foam forms in a separator due to gas entrainment in the liquid phase as the incoming fluid drops into the bulk liquid phase and also because dissolved gas evolves from the fluid for some time after the fluid has entered the vessel or tank. Gas bubbles rising in an oil phase can delay the dropout of water from the wet oil entering a separator. Experience indicates that the time required for an oil to fully degas and cease formation of gas bubbles gets longer as the viscosity of the oil increases. Since viscosity increases with a decrease in the API gravity of a crude, gas evolution for heavy oils can continue for a length of time which is significantly longer than the actual residence time in a separator.

Oil typically entrains water droplets which settle out slowly over time. These water droplets are often complex emulsions which contain smaller oil droplets within the water droplets. As these complex emulsion droplets enter the water phase further down the separator, they act to continuously introduce dispersed oil droplets into the bulk water phase and degrade the quality of water leaving the separation.

The oil/water interface forms as droplets of water settle out of the oil phase and droplets of oil migrate upward out of the water phase, together forming an interface which transitions from oil continuous emulsion in the upper portions of the interface to a water continuous "reverse emulsion" in the lower portion of the interface. Ideally, the thickness of the interface emulsion is small, less than 3 -4 inches, for example. In reality, the interface emulsion can be quite deep, even 1 - 3 feet in thickness. The control and resolution of interface emulsions is a key factor in optimizing the quality of water discharged from a separator because of the potential for an interface emulsion to

- Occupy significant volume in a separator and reduce available residence time for water clarification
- Trap solids which inhibit oil droplet and/or water droplet migration and coalescence
- Be stabilized by water treatment chemicals, especially polymers & surfactants, introduced either upstream of a separator or by a recycle stream from a downstream step in the water treatment process
- Interfere with oil/water interface instrumentation and control

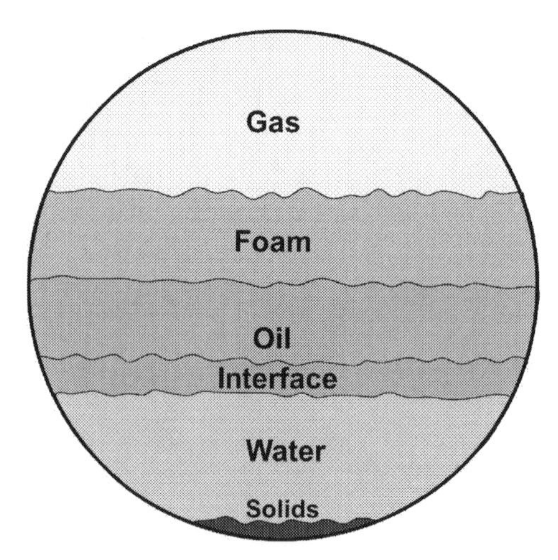

Figure 8.4 An illustration of phases which are present in a 3-phase separator.

Solids which accumulate in the bottom of a separator occupy volume which could be better used for improved water clarification, and serve as places where bacterial growth can either enhance vessel corrosion or provide stagnant volume where sulfate reducing bacteria (SRBs) can generate H2S and contribute to the formation of iron sulfides in the water. As a general rule, discharged water quality from a tank or vessel will be improved if both solids accumulation and oil/water interface accumulation are minimized. In fact one might even consider the First Law of Water Treatment to be:

"Keep all Vessels and Tanks Clean with Minimal Solids Accumulation Permitted."

Derivations of equations for sizing a vessel for 3-phase separation are presented in numerous sources, including the Surface Production Operations volumes by Ken Arnold and Maurice Stewart [2]. The geometric equations important for the design calculations of a separator can also be found in numerous sources, with Perry's Chemical Engineers Handbook [4] being an excellent starting point.

Arnold and Stewart [2] illustrate the use of vessel sizing equations for the case where 500 micron water droplets are to be separated from the oil phase and 100 micron oil droplets are to be separated from the water phase. They also recommend increasing the size of the separator by a factor of 1.8 in order to accommodate the fluid short circuiting which typically takes place in a vessel. The selection of the 1.8 factor is based solely on field experience. Based on numerous CFD studies, is likely that his factor should be further increased by an additional 2X – 3X for vessels and tanks which are not equipped with internals to effectively control the development of fluid flow path lines. As will be discussed below, the use of perforated plates to control fluid flow path lines through a separator (vessel or tank) can increase the effective utilization of a separator's volume to as high as 75 – 85%.

API 12J lists general guidelines for the recommended fluid residence time in a 3-phase separator. However, these guidelines are somewhat nebulous in their quantification and thus are not generally useful for establishing a recommendation for a vessel size. Residence time is defined in terms of time available in the vessel from the high level setting to the low level setting. Average residence time can be defined in terms of normal liquid level. Industry average residence time for water in a three phase separator is in the range of three to five minutes.

8.2.2 *Practical Considerations Included in Defining a Separator Size*

In addition to the sizing equations derived by Arnold and Stewart, several practical guidelines must also be considered when sizing and designing a separator. These include

Fluid Physical- Chemical Properties: The physical properties of water, oil, and gas at the specified operating conditions of the vessel should be used when determining size and/or anticipated performance. Using an API Gravity for the oil and an S.G. of the water phase is insufficient for good quality calculations. For separator design and sizing, the key physical properties of each fluid phase are density, viscosity, and composition at operating conditions.

Fluid Velocities in the Vessel or Tank: For 3-phase separation in a horizontal vessel, the velocity of the fluids down the vessel should not exceed about 7 ft/min. with 3 – 5 ft/min. preferred. This value range is derived largely from experience and not from a rigorous modeling of fluid flow. However, CFD studies support the use of 3 – 7 ft/min. as a desirable range for liquid velocity in a tank or vessel. Optimal oil- water separation occurs when the velocity for the oil phase and for the water phase down the vessel are similar (reference Lee & Phelps). To avoid liquid re-entrainment in the gas phase, the actual gas velocity down the vessel should be <12 ft/sec.

Setting Process Control Levels: Volume must be included in the separator to allow for the spacing of high & low level alarms and shutdowns. A good practice is to insure the selected vessel size allows for 30 to 60 seconds of fluid influx at the vessel's design conditions while the liquid outlet valves are all failed closed (a "Blocked Flow" condition) before the liquid or interface level in the vessel rises from one control level to the next For example, it should take 30 to 60 seconds for the liquid level in a separator to rise from the Normal Liquid Level (NLL) to the High Level Alarm (LAH) set point when fluid is entering the vessel at the Basis of Design Flow rates under a Blocked Flow Condition. Generally, >3 inches are recommended between any two control levels, more if the vessel is installed on a floating platform or FPSO where fluid sloshing is expected.

The levels to be set as part of the design and sizing exercise for a separator are:

LSHH	Level Safety High-High	High Liquid Level shut down for the vessel
LAH	Level Alarm High	The High liquid Level at which an alarm is initiated
NLL	Normal Liquid Level	The anticipated normal operating liquid level for the Vessel. For slugging conditions the NLL may be specified as a range rather than a fixed level. For example, 55+ 3" from BOV. Within this range the oil discharge control valve may open/close slowly over the specified range
LAL	Level Alarm Low	The Low liquid Level at which an alarm is initiated
LSLL	Level Safety Low-Low	The Low Liquid Level shut down for the vessel or for the oil level in the oil bucket if the separator is normally operating above the oil weir (a preferred condition for slug handling)

The above five levels must also be specified for the oil/water interface, giving a total of 10 levels which must be specified for effective and safe operation of the vessel. Special attention is generally given to setting the High Level and Low Level Interface shut-downs.

ILSHH	Interface Level Safety High-High	This safety shut down should be 3 to 12" below the top of the Oil Weir to prevent excessive water from getting into the oil dehydration train
ILSLL	Interface Level Safety Low-Low	This safety shut down needs to be sufficiently above the top of any installed vortex breaker that a vortex cannot drag significant oil into the water treatment system. In most instances, setting the ILSLL at 18 to 24" above the bottom of the vessel is sufficient.

The reader is encouraged to become familiar with API RP14C which describes the recommended practices for the installation of control systems on offshore separators.

Inlet Nozzle Elevation: For mechanical integrity and constructability reasons, the centerline for an inlet nozzle entering through the head of a pressure vessel will generally be no higher than 80% of the diameter of the vessel. For larger diameter vessels, the inlet nozzle centerline may be as high as 90% of the vessel diameter, but this higher inlet nozzle elevation should always be used with caution unless there has been consultation with a vessel fabricator. If the Inlet Nozzle Centerline is set at 80% of the vessel diameter, then the High Level Shut Down (LSHH) for a vessel should not exceed 80% of the diameter of the vessel minus ½ of the diameter of the inlet nozzle. In this way, the liquid level will never rise above the bottom of the inlet nozzle. Having a fluid enter a vessel tangentially near the top of a vessel's shell can sometimes allow the high level shutdown to exceed the 80% of diameter.

A key operational factor to recognize is that if the liquid level rises above the bottom of an inlet nozzle, then the probability of excessive foam, mist, or even bulk liquid carry-over into the gas system becomes high and, generally speaking, operators do not like to damage compressors unnecessarily by allowing excess liquids to reach the compressor suction.

An additional "best practice" recommendation is to allow for sufficient volume so that it will take 45 seconds of fluid influx at design conditions with all liquid outlets failed closed for the vessel level to rise from the LSHH (supposedly this is a US Coast Guard requirement but I do not have a reference for it) to a fully liquid packed condition.

Fluid Flow Path Control: Two points can be made which together provide the justification for carefully considering the need for well-engineered internals for controlling fluid flow paths in process vessels and tanks:

- The configuration of anything and everything installed in a separator will affect fluid flow paths within the vessel or tank
- The size of the water droplets to be removed from the oil and the size of the oil droplets to be removed from the water is process dependent. The identity and capability of equipment downstream of a FWKO or a skim tank will influence the size of oil droplets which need to be removed from water in the vessel or tank under consideration. Separators with well-designed internals will remove smaller sized contaminants with a significantly higher efficiency.

As the first step for controlling fluid flow paths, the installation of perforated plates is strongly recommended to reduce short circuiting of flow through a horizontal vessel or through a tank designed for horizontal flow. A minimum of 2 plates is recommended: one plate to isolate the inlet section of the vessel from the main body volume in the separator, and one plate to isolate the outlets of the separator from the main body volume in the separator. Perforated plate design considerations and calculations can be derived from equations presented in Perry's Chemical Engineers Handbook [4].

As a general guideline, a pressure drop on the order of 0.25 to 0.50 inches of water column across a perforated plate is required to effectively distribute flow over the cross sectional area available for liquid flow. The guideline pressure drop range is based on a value of < 5 for the Vmax/Vavg ratio for the velocity at which fluids approach the perforated plate. If a higher ratio is expected due to the design of the inlet nozzle, then the design flow ΔP across the plate should be increased.

Based on extensive (but unpublished) CFD studies and general field experience, the fraction of open area in a perforated plate will typically be in the range of 5 to 20% of the total available cross sectional area for fluid flow. The selection of hole size for the plate is a compromise between small diameters which would more effectively distribute the flow but which would impart more shear to fluids flowing through the plate and large diameters which would allow the formation of large, high velocity jets downstream of the plate. The former will reduce the oil droplet size distribution in the water phase while the latter will result in unintended short circuiting and the formation of stagnant recirculation zones in the separator. Again, based on CFD studies and field experience, perforated plates with holes less than 1" diameter or greater than 2" in diameter are not recommended. In this hole size range, flow jetting downstream of the perforated plate will dissipate completely within 24 to 36" downstream of the flow distributing plate.

As a general guideline, the hole size in and the pressure drop through a perforated plate should be selected to maintain the Reynolds Number for the fluid flowing through the plate in the mildly turbulent range, e.g., 10,000 to 25,000. The Shear at the circumference of the holes in the plate should be maintained at <100 sec-1. The equation for calculating the shear at the wall of a pipe can be applied to the shear through an orifice:

$$\text{Shear Rate} = (4\ Q) \div (\pi\ R3) \qquad \text{Eqn (8.2)}$$

where Q and R are defined (in consistent units) as follows:

Q = the flow rate through an individual hole in the perforated plate

R = the radius of the holes in the perforated plate

Since the Shear experienced by a fluid flowing through an orifice varies as 1/R3, the benefits from using a relatively large hole in the perforated plate become clear. This advantage is counter balanced by the formation of jets downstream of the plate as discussed above.

Operators will often request that space be left under the perforated plate for sand migration down the vessel. However, the height of this open area needs to be limited to avoid serious short circuiting. Typically, the open area for sand migration should not exceed 5% of the total open area in a perforated plate. A preferred practice would be to install and operate a sand jetting system in the vessel or tank which eliminates the need for the under-plate open area for sand migration.

Figure 8.5 illustrates the design and installation of a perforated plate. The height of each section of the perforated plate is limited by the need to install/remove plate sections through a manway. With the bolt-in construction, sections can be easily removed for access to all areas of the vessel during cleaning or maintenance activities. Either a square or a triangular hole pattern has proven to be acceptable.

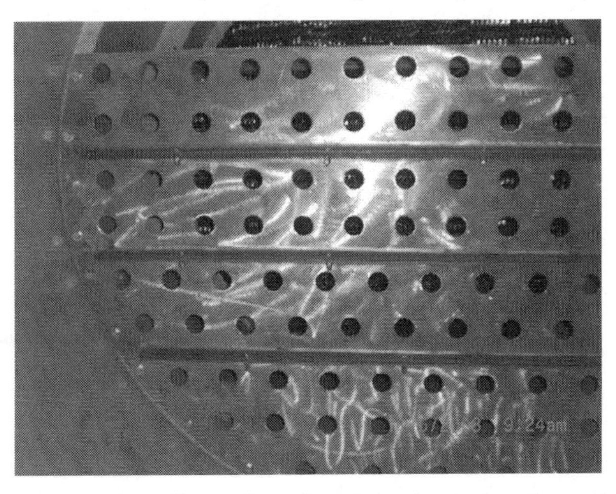

Figure 8.5 The design and installation of a perforated plate in a horizontal vessel is illustrated.

8.2.3 Fluid Inlet Devices

One of the most important design features for a vessel or tank is the design and selection of inlet devices and their configuration (placement) within the vessel. The main purpose of an inlet device is to arrest the momentum of the incoming fluid. To understand the importance of this design feature it is helpful to consider the geometries and fluid characteristics involved. Cylindrical pipes are almost always used to transport fluid into a vessel. The pipe diameter is selected (designed) on the basis of a superficial velocity criterion. Often, the operating company will have an internal document, such as a Design Engineering Practice (DEP) that provides this criterion for various types of fluids. For fluids containing sand and a high gas/liquid ratio, the superficial velocity must be low enough to prevent erosion. For fluids that do not contain sand or other solids, it is common to use an upper limit of about 2 to 3 m/sec. Once the fluid volumetric flow rate is specified, the pipe and vessel inlet nozzle diameter are selected in order to satisfy the design criteria.

By contrast, the fluid in the vessel has a much lower superficial velocity. The fluid in the vessel should move slowly, with a minimum of mixing, in order for gravity settling to occur. A well designed inlet device provides a gentle transition, in minimum space, from highly turbulent pipe flow to the quiescent laminar flow in the vessel.

Various criteria can be used to design the inlet device. In addition to superficial velocity, fluid momentum, energy or Reynolds number can be used. Chin [5] discusses the use of inlet momentum transport (ru^2) as a criterion. Inlet momentum transport has units of energy per unit volume of fluid. However, a better way to think about it is in terms of the velocity with which momentum is transported into the vessel (u x ru) per unit volume of fluid. Chin says that for inlet pipe having momentum transport of less than 1,000 Nm/m^3, a simple inlet diverter device is sufficient to arrest the forward momentum. An old-school conventional splash plate or impact plate "momentum breaker" would be sufficient although many water specialists object to such designs because they can shear oil droplets in water and water droplets in oil with both effects slowing oil/water separation. Using Chin's criteria and assuming an oil/water mixture having a density of 900 kg/m3, this corresponds to an inlet velocity of about 1 m/sec, which is relatively slow for pipe flow. Of course, this calculation assumes a liquid-packed system, which would seldom be the case. Nevertheless, the example provides a simple situation that can be easily understood.

At least three options can be considered for three-phase inlet configurations, all of which require the installation of a perforated plate to isolate the inlet section of the separator from the main body and confine or dissipate inlet turbulence. These options include:

- A Tangential inlet near the top of a horizontal vessel with the inlet nozzle being perpendicular to the vessel's access
- A Vane-type (Schoepentoeter^{TM}) inlet that gently distributes incoming flow over a wide area in the inlet section of the vessel
- A cyclonic inlet which effects rapid (1 – 5 seconds) gas/liquid separation and initiates oil/water separation and droplet coalescence within the cyclonic tubes.

Figure 8.6 illustrates the configuration for a cyclonic and vane-type inlet. It must be emphasized that fluid flow path control, e.g., by the installation of a well-engineered perforated plate, is required downstream of any and all inlet devices to mitigate short circuiting within the vessel or tank.

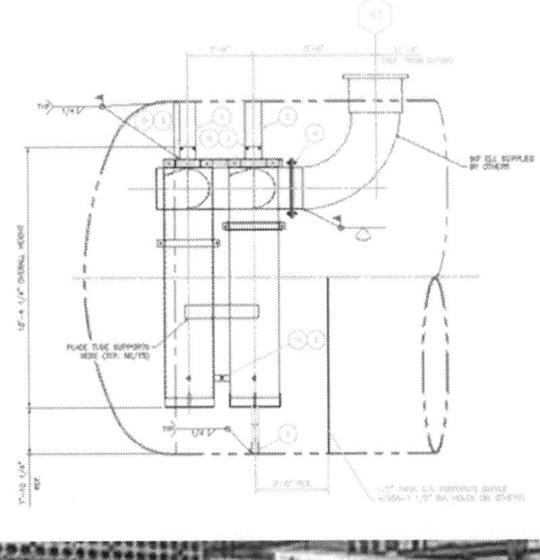

Figure 8.6 The configurations for a Cyclonic Inlet and for a Vane-type inlet for 3-phase flow coming into a separator are illustrated

The use of cyclonic inlets for introducing three-phase fluids into large tanks (wash tanks, skim tanks, or even FWKO tanks) with a vertical flow configuration can also be highly effective (see discussion below).

The design guidelines used by vendors who provide cyclonic inlets include consideration of

- The velocity at which multiphase fluids enter the cyclonic cluster manifold
- The velocity at which liquids travel vertically down the individual tubes
- The velocity at which separated gas travels upward in the individual tubes
- The velocity at which liquids emerge from the exit slot at the bottom of the tubes
- The minimum submergence required to prevent the gas vortex from exiting via the liquid exit slot at the bottom of the tubes
- Means to distribute or direct exiting liquid flow so as to prevent direct impingement of this flow onto a flow distributing plate
- Design flow rates that include the arriving velocity of liquid slugs

8.2.4 Vortex Breakers

The installation of well-designed vortex breakers on the water outlet of a vessel or tank is essential. Although the physics of vortex formation in a liquid phase is fairly well understood, these governing physical principals are seldom applied to oilfield vessels and tanks. The consequences of not having good vortex control are illustrated in Figure 8.7. This figure, a three-dimensional CFD simulation of fluid flowing down and then exiting a horizontal vessel is shown. Note from the fluid flow path lines that oil-laden water from the oil-water interface is drawn down into the water outlet nozzle. The installation of a well-designed vortex breaker would eliminate the discharge of this oily interface water and improve the overall quality of water leaving the vessel.

At a minimum, the design of a conventional crossed-plates vortex breaker for a water outlet nozzle should include the following properties:

- The cross vortex plates should extend down into the nozzle by > 1/2 of the nozzle diameter
- The cross vortex plates should extend upward into the vessel by 1X of the nozzle diameter. The length of the cross plates should be a minimum of 2X the nozzle diameter
- A cover plate should be installed on top of the cross plates which is a minimum of 3X the nozzle diameter
- The cross sectional area for flow under the vortex breaker cover plate should be a minimum of 4X the cross sectional area of the nozzle.

Other designs for vortex breakers are commonly used and they share the objective of reducing the velocity of flow into the periphery of the outlet nozzle. Several vendors, for example, install bar grating above an outlet nozzle at an elevation of 0.5 to 1.5 outlet nozzle diameters. This approach is effective at breaking vortex formation, but does not prevent the draw of water from higher up in the vessel (a situation which may be inconsequential in some cases). Other designs use a series of slots in a pipe or a sand pan configuration which have a number of rectangular slots with an aspect ratio of > 4:1 to draw distributed flow into an outlet nozzle. This approach works well as long as the velocity of flow into the slots remains sufficiently low.

Rochelle and Briscoe [6] review available literature which correlates vortex formation with the Froude Number and the ratio H/D where H is the height of fluid over a nozzle and D is the diameter of the nozzle. Based on their analysis, a conservative design to avoid vortex formation would be to operate a separator in such a manner that the Froude number for the exiting fluid remains below 0.5 and the H/D ratio for liquid in the separator exceeds 2.0 [6].

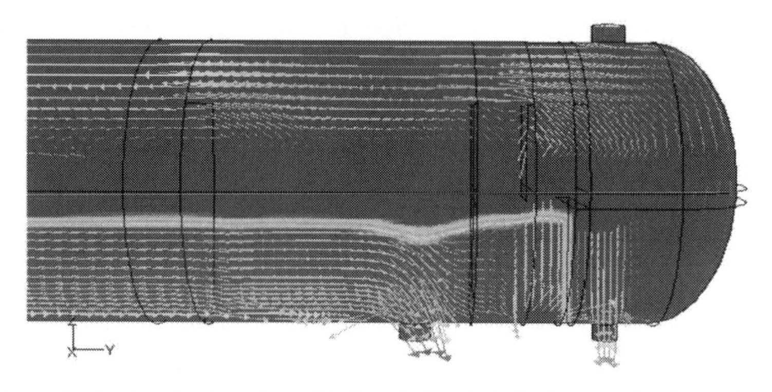

Figure 8.7 Water quality leaving a horizontal vessel is degraded by the lack of a vortex breaker and flow control into the outlet nozzle. Oily water from the oil/water interface is drawn down into the water outlet nozzle

8.2.5 Sand Removal and Interface Drains

As mentioned earlier, the First Law of water treatment is to keep vessels and tanks clean and free of accumulated solids and sludges. It is important to recognize that solids and sludges accumulate in vessels and tanks at two locations: (1) at and below the oil water interface, and (2) on the bottom of the vessel.

An essential part of preventing the accumulation of interface solids and sludge is to insure that contaminants removed from produced water by hydrocyclones, skim tanks, flotation cells, nutshell filter backwash water, etc. are not recycled to upstream process vessels and tanks. As a supplement to this philosophy, it is recommended to install interface drains on the vessel or tank which permit operators to draw off the problematic solids & chemical laden interface periodically without shutting down the vessel or "floating off" the interface to downstream equipment. A configuration for an interface drain is shown in Figure 8.8.

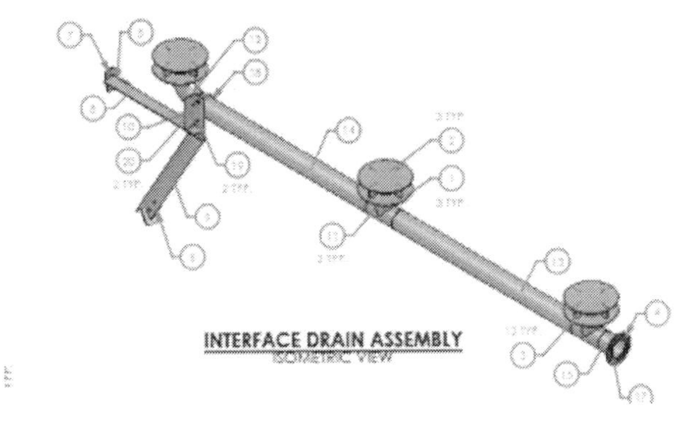

Figure 8.8 Interface Drain Assembly

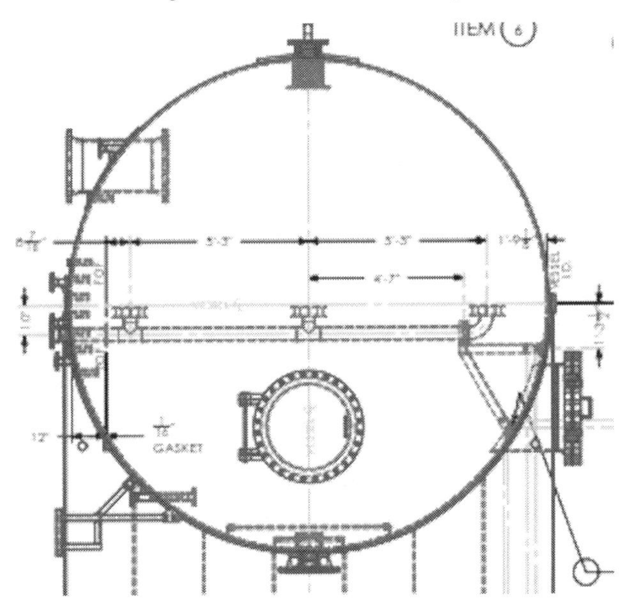

Figure 8.9 An example configuration for an interface drain installed on a horizontal vessel is shown.

To remove solids and fluid sludge from the bottom of a vessel, the installation of either on-line sand removal devices or a sand-jet plus sand-drain system is required. A generic sand drain and sand jet system is shown in Figure 8.9. For vessels larger than 6 – 8 ft. in diameter, two or more rows of sand jet nozzles are recommended. Considerations for the design of a sand pan and sand jet system include:

- The size and spacing of the V-notches in the sand pan should be such that water drained during jetting enters the V-notches at a velocity of 4 – 5 ft/sec. Note: To insure mechanical integrity, most sand pans are constructed of half-pipe sections rather than the 90o angle pans shown in the drawing
- The velocity under the sand pan should be maintained during a sand drain operation at 3 – 4 ft/sec. to insure entrained sand migrates to the drain nozzle
- Water with solids in the sand drain piping should maintain a velocity of 4 – 6 ft/sec. to prevent erosion from sand transport while insuring that solids do not deposit in the piping
- Sand jet water volume and sand drain water volume need to be similar in volume so the impact of sand jetting & draining on the oil/water interface level is minimal
- If scale is a potential issue, sand jet water should be treated with sufficient Scale Inhibitor (a recommended practice). Operators have had good success with having a scale inhibitor chemical pump activated whenever the valve or pump supplying jet water is active.
- Automated regular sand jet operation is recommended with each 5 ft section of vessel jetted and drained for 3 to 5 minutes (typical).

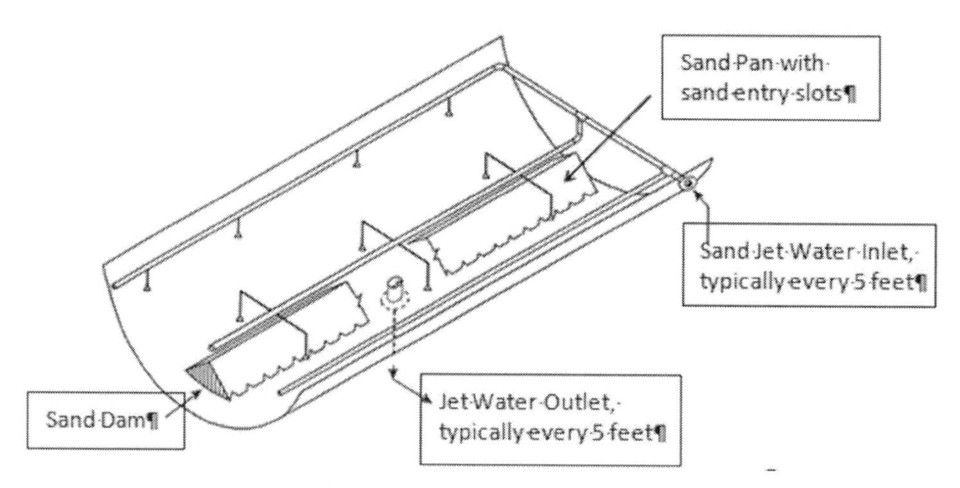

Figure 8.10 A generic sand jet and sand drain system is illustrated for a horizontal vessel

Several vendors offer devices which can be installed in a vessel in place of sand pans and sand jets. A generic version of this type of device is illustrated in Figure 8.10. Jet water is pumped into the devices to locally fluidize settled solids, generally in a cyclonic pattern. The fluidized solids are then withdrawn from the vessel through a separate outlet that is an integral part of the device. The advantage of these devices is that they are less disruptive to water quality and to the integrity of the oil/water interface than are sand jets/drains. Another advantage is that they remain functional even if covered with several feed of fluidic solids or sludge. Thus they can be operated less frequently than what is required for sand drains and sand jets. The disadvantage of the devices is that they do not fully scour all sand/solids from the vessel. Typically the devices will clear an area of sand within about a 2 ft diameter circle around the device, but the devices are typically installed at 4 to 6 ft intervals. Thus residual sand left in the vessel forms an egg-carton configuration with a peak sand level between devices that is several inches in height.

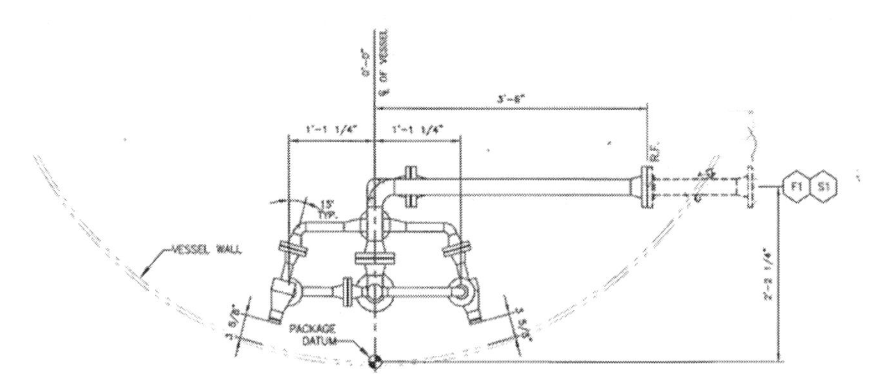

Figure 8.11 The installation of an on-line sand/solids fluidizing and removal device in a horizontal vessel is illustrated

8.2.6 Process Control Philosophy and Instrumentation

Fixed Level vs Floating Level Control: Historically, there has been an emphasis on controlling liquid and interface levels in separators and tanks as tightly as possible. This works well if the flow rate and the composition of the flow coming into the separator remains relatively constant. However, if the fluid arriving at the vessel or tank is multiphase (slugging) and/or includes significant variability in composition (e.g., oil/water ratio) over short periods of time , then an approach which allows both the liquid level and the interface level in a vessel or tank to vary over a range has process performance advantages. By operating a horizontal vessel at a liquid level which is above the top elevation of the oil weir, changes in both liquid level and interface level can be used to accommodate the arrival of slugs and dampen flow rate changes leaving the separator.

If the vessel were instead operated at the same level as the top of the oil weir, then the arrival of a slug would either force an increase in the interface level with a sudden excessive oil flow over the weir or, if the interface level were held constant, then a sudden reduction in water residence time in the vessel would result in a degradation of discharge water quality. Both situations can be problematic for the control of water and/or oil outlet valves. If slugging is moderate to severe, the PID control tuning of valve operations becomes a challenge and, in some cases, can be nearly impossible. However, if the adjustments of %-open for an outlet valve are left to the PLC based on a programmed curve for allowed variations in the liquid or interface level in the vessel or tank, then process control is simplified and becomes much more effective. For optimal water treatment performance, minimizing the variations in water discharge rates from a separator and maximizing the residence time for water in the separator is essential.

As mentioned, an alternative process control strategy is to allow both the liquid level and the interface level in the vessel to float over a prescribed range. The range of level variation will be a function of the vessel diameter and the character of slugging experienced. In larger diameter vessels, allowing levels to float over a range of 6" to 10" is often possible. To control the oil/water interface over the specified range requires that valve position transmitters be installed on the water outlet control valve and the position signal monitored by a PLC. Then the opening position for the valve is allowed to vary according to a curve which is defined by the Cv for the valve.

As an illustration of this strategy, assume that the maximum oil/water interface level to be allowed is defined by the Interface Level Alarm High (ILAH) and the minimum Interface level to be allowed is defined by the Interface Level Alarm Low (ILAL).

- When the interface level reaches the ILAH, the water outlet valve will be instructed by the PLC to be fully open.
- When the interface level reaches the ILAL, the water outlet valve will be instructed by the PLC to be fully closed.
- In between, the valve open position will be varied according to a polynomial curve which is defined by flow characteristics of the water outlet control valve.

For example, at an interface level half way between the ILAH and the ILAL, the valve may be opened to allow water to be discharged at 50% of the valve's maximum flow rate as defined by the Cv curve. If the interface level is at 80% of the range between the ILAL and the ILAH, then the valve may be set to open at a position which allows water to be discharged at 70% of the valve's maximum flow rate. Alternatively, the valve position can simply be controlled such that the rate of discharge varies linearly between 0% and 100% of the valve's maximum flow rate as the interface level moves in response to slug arrival between the ILAL and the ILAH.

With this approach the interface will normally migrate to a preferred level in the vessel, but this level will change as the short term average flow rate for arriving fluids changes.

Level & Interface Level Control Instrumentation: For liquid level control transmitters, the most common technologies in use as of 2015 are bridle-mounted floats and Guided Wave Radar. The use of vessel-internal large floats mechanically connected to a signal transmitter of some sort is used less frequently because of its limited ability to function over a large range of levels. Instead, floats are mounted in bridle columns and the position of the float is detected by magnetic sensors outside of the column. By the appropriate selection of float density, a single device can be used both for liquid level control and for interface level control. Figure 8.12 is an example of an externally mounted float-type level transmitter. A combination Liquid Level plus Interface level transmitter would require that one of the transmitter nozzles be located at an elevation on the tank or vessel which is always in the water phase, one of the transmitter nozzles be located at an elevation which is always in the oil phase, and the third transmitter nozzle be located at an elevation that is always in the gas phase. Two floats with their respective correct densities for detecting the liquid level and the interface level are installed in the combination transmitters.

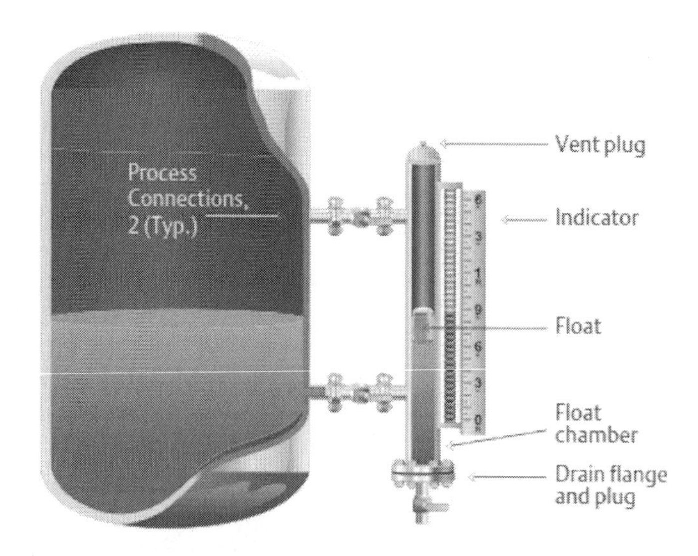

Figure 8.12 Courtesy of Emerson Process Management, Inc., a typical float-type level or interface Level transmitter is illustrated

The bridle-mounted magnetic float devices are versatile and can be equipped with external, easily repositioned magnetic switches to provide both level alarm and level shut-down signal capability. However, before using a single transmitter for level control, level alarms, and for level safety shut downs, the reader is encouraged to become familiar with the recommendations of API 14 E.

Although they are generally versatile and reliable, the use of bridle-mounted external magnetic level transmitter devices can become problematic if

- the crude viscosity is high at ambient temperature
- there are wax or asphaltenes colloids in the oil
- the oil/water contains significant solids, especially oil-wetted solids such as iron sulfides or scale mineral precipitates
- there are very rapid level changes that cause the float to hang-up in the transmitter column
- there are significant stable interface emulsions, especially solids or chemically (e.g., water clarifier chemical) stabilized emulsions

To maintain good level transmitter performance, the float column should be equipped with flush valves and a source of clean flush water so that the device can be cleaned on an as-needed basis with little or no interruption to separator operations.

Another often used option for liquid level control is guided wave radar. A guided wave radar transmitter can be mounted either internally to a vessel or tank, or externally in an external bridle similar to that used for float transmitters. Some manufacturers will claim that the guided wave radar technology is also suitable for determining the level of oil/water interfaces as long as there is a minimum oil pad thickness (typically 4 to 6") on the water. However, the ability of GWR technology to reliably detect oil/water interfaces is dependent on the interface being sharply defined. In most 3-phase separators this is not the case or may not be the case at all times. Thus the use of guided wave radar for interface level detection in most separators and process tanks (e.g., skimmers or wash tanks) is not recommended.

Excessive foam can also be a problem for guided wave radar, but severe and persistent foaming is not common in high water-cut production. Nevertheless, an operator should be aware of the effect of foam on the performance of guided wave radar level transmitters. Because guided wave radar can be easily installed to directly detect liquid levels inside a vessel without the use of an external bridle, their capability is not impacted by high oil viscosity, the presence of solids in the oil or water, or changes in liquid density.

Capacitance transmitters represent another technology which is viable in place of floats for interface level detection and control. There are a variety of these devices available which operate with differing frequencies and have different operational philosophies. As with guided wave radar, these devices can be installed internal to a vessel or in an external bridle. Generally these devices, if properly configured based on manufacturer recommendations, work reasonably well. The one thing that an operator needs to remember with these devices is that they provide only one number: the total capacitance along the active length of the probe. Everything else is an interpretation of the capacitance number. In essence, a capacitance probe only tells you how much water and how much oil is in contact with the probe along its entire active length. It does NOT tell you where this water or oil is physically located along the active length of the probe. Thus a high oil concentration in the water continuous portion of an interface emulsion or a high water concentration in the oil continuous portion of an interface emulsion can lead to an incorrect interpretation and reporting of the interface level elevation.

Water Legs as Interface Level Control Devices: A passive method for oil/water interface level control is the use of a bucket & weir configuration in a vessel or a water leg on a process tank (FWKO, Wash Tank, Skim Tank). The configuration for a bucket and weir separator which is accepting low Gas Volume Fraction fluids (GVF <5%) is shown in Figure 8.13 and the configuration for an external water bucket with a water weir for interface level control is shown in Figure 8.14.

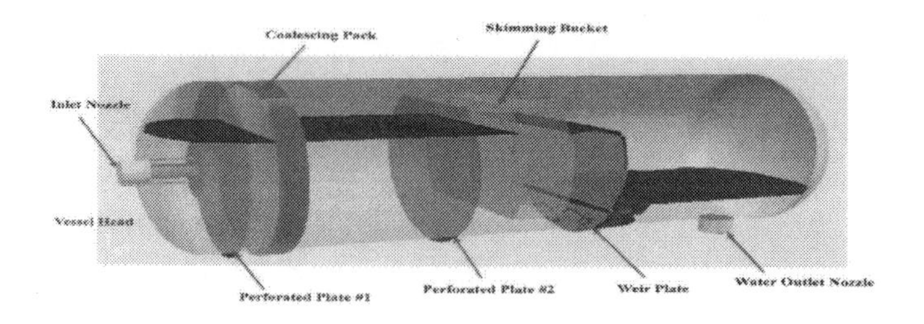

Figure 8.13 The Configuration for a Bucket & Weir Separator which is accepting low Gas Volume Fraction Fluids is shown

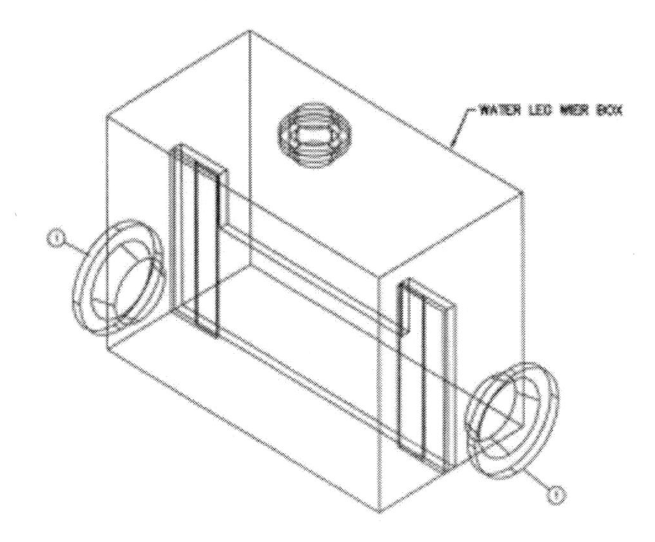

Figure 8.14 An illustrative configuration for the water weir in an external water bucket for a tank is shown. The oil pad in the tank is controlled by the difference between the elevations of the oil surface in the tank and the water surface in the water bucket. For clarity, the adjustable weir gate is omitted from the drawing

The depth of the oil pad in the vessel or tank with a bucket and weir configuration is determined by the following equation:

$$\text{Oil Pad Thickness} = \frac{(\text{Oil Level} - \text{Water Level})}{1(\rho_{oil}/\sigma_{wtr})} \qquad \text{Eqn (8.3)}$$

Where the oil level is typically (but not always) the height of the oil weir or skim bucket in the vessel or tank and the water level is (thought to be but in reality is not) the height of the water weir in a vessel or the height of the weir in the external water bucket of a tank.

The bucket & weir configuration or the use of a water leg for interface level control in a tank works well when the fluid flow rates are relatively constant and do not change rapidly. However, if the vessel or tank receives slug flow or widely varying flow rates, then this control strategy can break down due to varying crest heights over the water weir.

As an example, if the oil density ρoil = 0.80 at operating conditions and the water density ρwater = 1.00 at operating conditions, then the oil pad thickness for a configuration where the mechanical height difference between the oil and water weirs would be (Oil Weir ht. – Water Weir ht.) ÷ 0.2. if the mechanical height difference is 2.0", then the calculated oil pad thickness based solely on the mechanical difference in weir elevations would be 10". However, if the water crest height over the water weir at the design or nominal flow rate is 1.0", then the oil pad thickness declines to 5". If, due to slug effects other factors which control the influx of fluids to the vessel or tank results in the water crest height increasing even temporarily to 2.0", then the oil pad thickness in the vessel or tank drops to Zero and water would be skimmed into the oil bucket. Note: Expect significant water skimming into the oil bucket if the oil/water interface approaches within 3" to 6" of the weir height, depending on the quality of the interface.

The equations which govern the crest height for the flow of water over a weir are as follows:

For Notched Weirs:

$$Q = N\{4.28 \times C_d \times \tan(\theta/2) \times (h + k)^{5/2}\} \qquad \text{Eqn (8.4)}$$

(Kindsvater-Shen Equation)

Where,

Q = Flow Rate in ft^3/sec

C_d = Discharge Constant (0.58)

θ = V - Notch Angle in degrees

h = Head on the Weir in Feet (i.e., Ht. of water above the bottom of the notch)

k = Head Correction Factor in Feet (0.53)

N = Number of V-notches in the weir

For Flat Weirs:

$$Q = 1495\,(L - 0.2Hc)Hc1.5 \qquad \text{Eqn (8.5)}$$

Where

L = weir length if Feet

Hc = crest height of water in Feet

Q = flow rate in USGPM

8.3 Calculating Oil Droplet Retention Efficiency for a Horizontal Separator

For the vast majority of facilities, the condition of the fluid entering a separator or tank is not known and cannot be predicted a priori. The major contributors to this lack of predictability include:

- Actual flow conditions are typically not known
- The relevant physical chemical parameters of the incoming fluids are often not known in advance of the facility's design phase
- The control systems and piping configurations which influence the character of the fluids arriving at a platform or onshore facility are not determined until after the oil water separation train has been specified.

Although the Basis of Design oil and water flow rates, viscosities, and densities may be known for a specific facility, the distribution of oil droplets in water and water droplets in oil arriving at a 3-phase separator will not be known before actual operations commence. With these and perhaps other unknown factors in mind, calculating the oil content of the water entering the separator and predicting the oil content of water expected to be discharged from the separator is a more or less futile exercise.

What can be calculated for assumed ideal performance conditions, however, is the efficiency with which a distribution of droplet sizes is expected to be retained in the 3-phase separator. With this information, the selection of downstream technology options becomes a somewhat reasonable exercise. Also, a recommendation for the number of stages of water treatment which may be required becomes a facile exercise. It must be emphasized that although calculating the fraction of a given size of oil droplet which will be retained in a separator under ideal conditions is possible, how many oil droplets of a given size are physically present in the water entering the separator is still an unknown.

For the oil droplet capture efficiency calculation, it is necessary to be able to relate the water height in a horizontal vessel to the volume fill for the vessel. The equation (ignoring the head volumes) for this calculation is:

$$\text{Volume} = r2\cos\text{-}1[((r-h))/r] - (r-h)[\text{ABS}(2rh-h2)]1/2L \qquad \text{Eqn (8.6)}$$

Where:

r = ½ the diameter of the vessel (i.e., the radius)

h = the height of liquid in the vessel

L = the length of the vessel

ABS = Absolute Value

Using the equation above, a graph relating the volume fraction of liquid in a separator to the liquid level (distance from Bottom of Vessel) in the separator is shown in Figure 8.15.

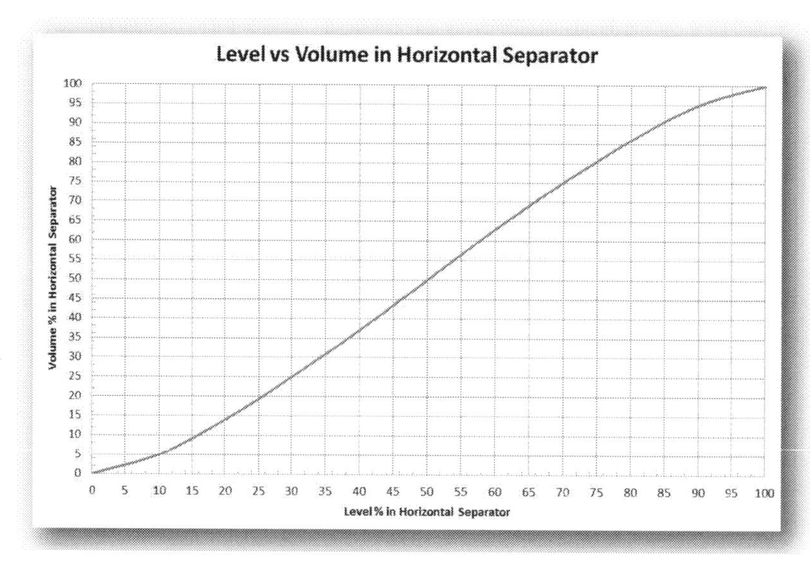

Figure 8.15 Liquid Level (% of Diameter) vs Liquid Volume (% of liquid packed volume) is shown for a horizontal vessel.

Using Figure 8.15, one can determine the fraction of the water in the separator at any given moment which is below any chosen level in the separator (expressed as, for example, inches from the Bottom of the Vessel, a.k.a. BOV). Figure 8.16 illustrates this for a 10 foot internal diameter horizontal separator which is operating at an oil/water interface level of 69 inch. For illustrative purposes, the following operating conditions are assumed:

Operating Temperature:	140oF
Water Viscosity at Temperature:	0.55 cP
Water Density at Temperature	1.025 gr/cc
Oil Density at Temperature	0.820 gr/cc
Theoretical Residence Time for Water:	11.9 minutes

- In this example, 100% of the water is below 69 inch which is 69/120 = 57.5% of the vessel diameter.
- Using Figure 8.16 or Equation (8.6), it is found that water occupies 60% of the vessel volume.
- The next step is to determine what fraction of the vessel volume is occupied, for example, by 90%, 75%, 50%, etc. of the water.
- As shown in Figure 8.16, 90% of the water will occupy 54% of the vessel volume (0.90 x 0.60 = 0.54). Also see Table 8.1.
- Returning to Figure 8.16, 54% of the vessel volume corresponds to a water level of 53% of the vessel's diameter – which is equivalent to a water level in the separator of 120 x 0.53 = 63.6 inch.
- The calculation can be repeated for as many levels as desired in order to complete the analysis. For illustrative purposes, water level volumes were calculated and are illustrated in Figure 8.16 for 75%, 50%, 25%, and 10% of the water in the vessel.

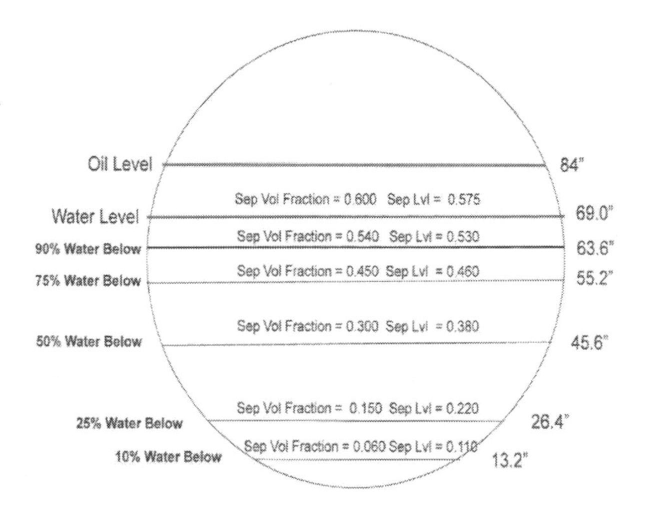

Figure 8.16 The levels in a 10' diameter separator for various fractions of water in the separator are illustrated for the case where the oil/water interface is set at 69"

To calculate the oil droplet separation efficiency

- The rise distance and the time available for all droplets of a minimum size to rise to the oil/water interface (where they are presumably captured) is first calculated.
- These quantities define the terminal rise velocity for a distribution of oil droplet sizes which will be retained in the vessel.
- Once the various terminal rise velocities are known, Stokes Law is used to determine what the smallest droplet size is that has sufficient time to rise to the interface from various levels in the separator. The calculation is illustrated in Table 8.2 using the operating conditions specified above.

Table 8.2. The calculation of oil droplet separation efficiencies in a 10' diameter horizontal vessel is illustrated

		Below	Below	Below	Below	Below	Below	Below
1	Vsl Level from BOV (in.)	69.0						
		↓						
2	Fraction of Vsl Diameter Occupied by 100% of the Water	58%						
		↓						
3	Portion of Vsl Volume Occupied by Specified Fraction of Water	60%	54%	45%	30%	15%	6%	0%
			↓	↓	↓	↓	↓	↓
4	Fraction of Vsl Diameter Occupied by Specified Fraction of Water (use Fig. 6-14)		53%	46%	38%	22%	11%	0%
			↓	↓	↓	↓	↓	↓
5	Vsl Level (in.) for Specified Fraction of Water		63.6	55.2	45.6	26.4	13.2	0
			↓	↓	↓	↓	↓	↓
6a	Oil Droplet Rise Distance (in.)		5.4	13.8	23.4	42.6	55.8	69
6b	Oil Droplet Rise Distance (cm)		13.7	35.1	59.4	108.2	141.7	175.3
			↓	↓	↓	↓	↓	↓
7	Gross Separation Efficiency (highly optimistic!)	85%	↓	↓	↓	↓	↓	↓
8	Actual Available Rise Time = Ideal Rise Time x Separation Eff.	10.1 minutes						
			↓	↓	↓	↓	↓	↓
9	Oil Droplet Rise Vel. For Separation (cm/min)		1.35	3.47	5.88	10.71	14.03	17.35
			↓	↓	↓	↓	↓	↓
10	Oil Droplet Diameter (microns) from Stokes Law calc.		35.7	57.2	74.5	100.5	115.1	127.9
11	Fraction of Oil Droplets that will be Separated = 100% - (Fraction of Wtr. Below the specified Lvl.), e.g., = 100 - 90		10%	25%	50%	75%	90%	100%

The results of this calculation indicate that 10% of the oil droplets with a diameter of 36 microns will be retained in the vessel, 25% of the oil droplets with a diameter of 57 microns will be retained in the vessel, etc. Using these calculated values, a capture efficiency curve can now be plotted for the vessel in a manner which is similar to such curves which will be illustrated in Chapter 10 (Hydrocyclones). The curve for this illustrative example is shown in Figure 8.17.

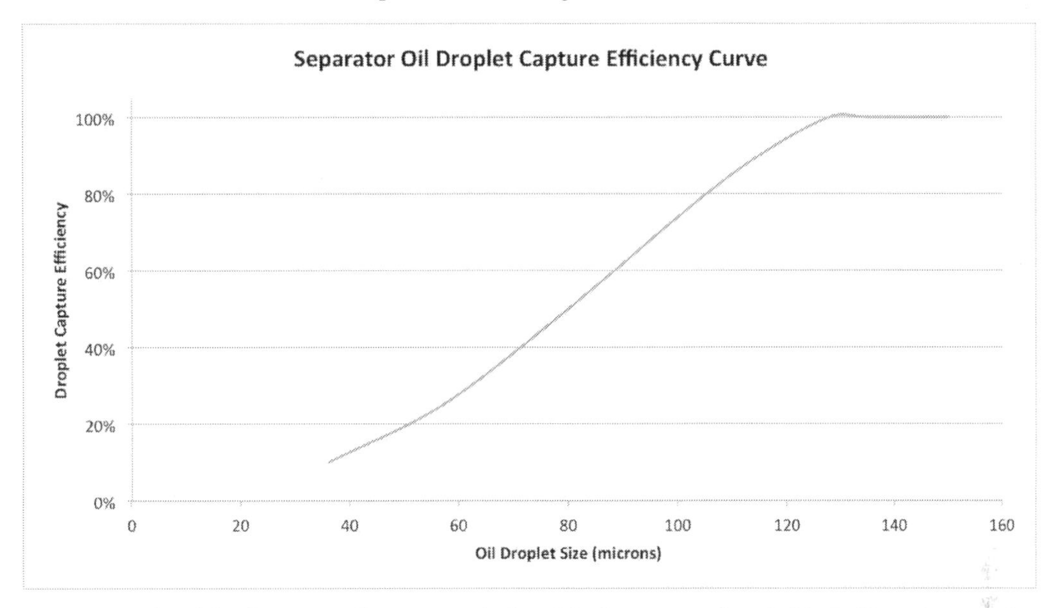

Figure 8.17. The oil droplet capture efficiency curve is plotted for the separator described in the illustrative example

The calculation thus indicates that the downstream water treatment equipment must be selected and designed to efficiently capture oil droplets < 128 microns and that, for example, only 50% of the oil droplets with a diameter of 75 microns and only 10% of the oil droplets with a diameter of 36 microns will be retained in the separator. Based on this information, the options for primary water treatment may include hydrocyclones or plate interceptors. However, if a skim tank is to be utilized, then oil droplet coalescence with the assistance of chemical treatment will almost certainly be required.

8.4 Process (FWKO, Skim or Wash) Tank Design Considerations

Onshore, large tanks are commonly utilized as free water knock-outs, wash tanks, or oily-water skim (clarification) tanks. Most are designed on the basis of 1 to 4 hrs of theoretical residence time for the fluid being processed – and many function with an actual residence time that is 10 to 25% of the theoretical residence time. As a general statement, the larger the tank, the greater is the challenge to actually utilize a large fraction (75 to 85%) of the tank's theoretical fluid residence time.

The first design decision required for a process tank is whether the tank should be designed for a vertical flow, a horizontal flow, or a hybrid circular flow pattern. Several designs exist in the literature for each option and several of the designs have been the subject of Computational Fluid Dynamic (CFD) simulation studies. Key lessons from the CFD simulations include the following:

- Fluids must be introduced in such a manner that flow is distributed over a significant fraction of the theoretical cross sectional area for flow in the tank, be it horizontal or vertical
- Fluids must be removed in such a manner that the outlet device configuration does not disturb the flow patterns in the tank and contribute to the formation of short circuiting fluid path lines

- Momentum effects and the desire of the fluid to exit the tank as quickly as the laws of physics permit will determine the actual flow path lines for the fluids – the concepts and wishes of the design engineer not withstanding

- Convection currents are likely to influence a tank's performance and these may not always be detected by a CFD simulation which typically assumes uniformity in temperatures and fluid densities

- For efficient oil droplet capture in a large tank, smaller oil droplet coalescence (diameter < 100 to 150 microns), usually with chemical assistance, is required

8.4.1 Designing Skim Tanks for Horizontal Flow

Two representative designs for skim tanks with horizontal flow are illustrated in Figure 8.18.

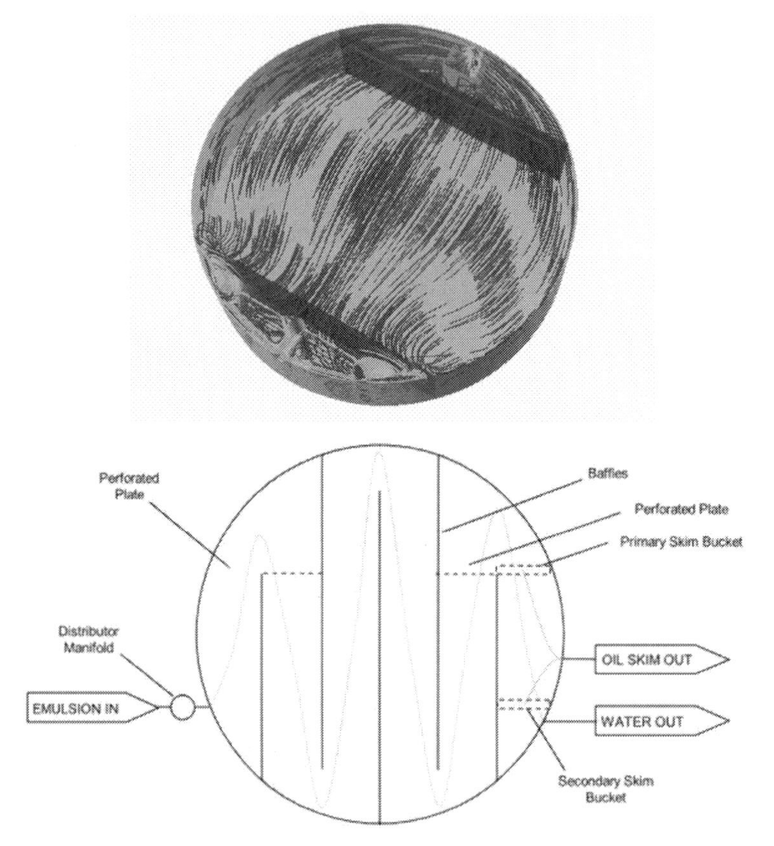

Figure 8.18 Two options are shown for the design of a skim tank with horizontal flow

On the left in Figure 8.18, two large perforated plate baffles are used to establish uniform flow across the tank. Note that with this design philosophy, the distribution of flow coming into the tank, and the distributed collection of clarified water for discharge from the tank are still required. On the right, solid wall baffles are used to restrict short circuiting of horizontally flowing water. The assistance of several perforated plate distributors is still required to eliminate short circuiting, as are the design of inlet and outlet configurations that do not adversely affect the development of spatially uniform flow paths in the tank. When conducting a CFD simulation of flow in the tank, it is important to monitor fluid flow paths at several levels in the tank in order to confirm that the inlet and/or the outlet devices do not disrupt the uniform distribution of fluids into and through the tank.

8.4.2 Designing Skim Tanks for Vertical Flow

A representative design for a skim tank with vertical flow is shown in Figure 8.19. With this design, incoming fluids are uniformly distributed to 1, 2, 4 or 8 cyclonic inlet devices, depending upon the size of the tank and design flow rate. This design is particularly suitable for tanks receiving 3-phase flow since primary gas/liquid separation takes place in the cyclonic inlets. Liquids exit the cyclonic units over a horizontal plate to insure that they are distributed over the horizontal cross sectional area of the tank.

Oil is skimmed from the tank via some number of oil buckets or an oil skim trough mounted on the tank wall. Water is removed from the tank via an array of water collectors which can be simple in design or, if the tank is in critical service, more complex. The illustrative water collection array shown in Figure 8.19 includes a header with a series of laterals. Water enters the laterals through holes in the bottom of the pipes. The size and distribution of holes for water entry requires that a detailed hydraulic model for the collection array be developed. For optimum efficiency, the water flux through all holes in the collection array should be within a few percent (e.g., within + 5 to 10%) of the median flux for all holes.

If the tank is not in critical service, then a simpler design for water collection can be used. An example would be the use of a number of twin-plate collection points designed such that water enters each collection point via a flow path having a horizontal orientation. Care must be taken with this design to insure that water flow velocities into the collection points is sufficiently low that vortexes do not develop. Typically the number of collection points for water will be the same as the number of cyclonic inlets utilized in the tank.

Figure 8.19 A representative design for a tank with vertical flow is shown. In this illustration, multiphase fluid enters the tank through a group of four cyclonic inlets

The diameter or cross sectional area of the tank designed for vertical flow is determined by the Stokes Law terminal rise velocity for the oil droplets to be removed in the tank. The design oil droplet removal diameter is simply the diameter of the oil droplet whose terminal rise velocity exceeds the average downward velocity of water at the design flow rate of the tank. The design retention time in the tank is typically 60 to 120 minutes with the median "actual" retention time in a well-designed tank being

70 to 80% of the design retention time. Design oil droplet removal diameters typically range from 100 to 300 microns with significant oil droplet coalescence in the tank being required for efficient water clarification. The injection of a water clarifier into fluids entering the tank may be required to facilitate oil droplet coalescence and improve water clarification.

If one calculates the D100 for a tank with horizontal flow or with vertical flow, one finds that the calculated ideal minimum droplet size for 100% removal is the same for both vertical and horizontal flow. However, whereas some smaller droplets will be removed from the upper portions of a tank with horizontal flow (see the calculation above for oil droplet removal in a horizontal separator vessel for an example). Droplets smaller than the D100 will only be removed in a vertical flow tank if oil droplet coalescence occurs. Field experience indicates that this generally does take place during the 60 to 120 minute retention time in the skim tank, probably due to oil droplet coalescence in the cyclonic inlets and the presence of a band of slowly rising oil droplets which collide and coalesce with smaller oil droplets in the feed water. In practice, there appears to be little discernible difference in the performance of tanks designed for vertical flow and those designed for horizontal flow – although there does not appear to be direct comparative data to quantify this statement.

The decision to select a design for a tank for horizontal or vertical flow will depend in part on the ratio of the tank diameter to the fluid height in the tank. If the tank diameter is > twice the operational fluid height in the tank, then a horizontal fluid flow configuration would tend to be preferred. However, if the tank diameter is < 2X the operational fluid height, then a vertical flow tank should be the configuration of choice. It should be also noted that the performance of a skim tank can be augmented by the implementation of a dissolved gas flotation assist. Dissolved gas flotation is discussed in some detail in Chapter 12 (Flotation). For skim tanks the source of dissolved gas flotation bubbles can be either the result of a pressure drop from an upstream pressure vessel, or by the installation of DGF pumps which recycle and saturate under pressure 15 to 25% of the water flowing through the tank. In either case, the pressure drop for the gas saturated water must take place close to the tank inlet – keeping in mind that DGF bubbles will begin to form within a few (4-5) seconds after pressure reduction.

References to Chapter 8

1. K. Arnold, Senior Technical Advisor, Worley Parsons, online discussion group of the Projects Facilities, and Construction discipline of the Society of Petroleum Engineers.

2. K. Arnold, M. Stewart, Surface Production Operations, Gulf Publishing (2008).

3. D.G.K. Murti, H.R. Al-Nuaimi, "Renovate produced-water-treating facilities to handle increased water cuts," SPE-22831, paper presented at the SPE ATCE, Dallas (1991).

4. R.H. Perry, D.W. Green, J.O. Maloney, Perry's Chemical Engineers' Handbook, 7th Ed., Mc-Graw-Hill, New York (1997).

5. R.W. Chin, "Oil and gas separators," Chapter 2 of Petroleum Engineering Handbook, Volume III, Ed.: K.E. Arnold, Society of Petroleum Engineers, Richardson, TX (2007).

6. S.G. Rochelle, M.T. Briscoe, "Predict and prevent air entrainment in draining tanks," Chem. Engineering, p. 37 (2010).

CHAPTER NINE

Interceptors CPI, TPI, PPI

Chapter 9 Table of Contents

9.1 Plate Interceptors: CPI, TPI & PPI – Practical Applications ... 37
9.1.1 Typical Configuration ... 38
9.1.2 Mechanism of Separation ... 40
9.1.3 Design Differences and Characteristics ... 43
9.1.4 Representative Performance Data ... 44
9.1.5 Operation and Maintenance ... 45
9.1.6 Troubleshooting Procedure ... 46
9.1.7 Application in Oil Spill Cleanup ... 46
9.1.8 Benefits and Drawbacks ... 47
9.2 Plate Interceptors – Theory, Design and Modeling ... 47
9.2.1 Cut Size ... 47
9.2.2 Fouling Mechanism and Model ... 51
9.2.3 Skim and Reject Fluid Separation ... 53
References to Chapter 9 ... 56

9.1 Plate Interceptors: CPI, TPI & PPI – Practical Applications

This chapter discusses several related technologies: Corrugated Plate Interceptor (CPI), Tilted Plate Interceptor (TPI), and the Parallel Plate Interceptor (PPI). These all operate on the same principle. Oily water and solids flow through a set of parallel plates. The oil droplets rise to the plate at the top of the channel. The solids sink to the plate at the bottom of the channel. The oil droplets continue to rise along the bottom of the plate until they reach the top of the plate pack. By that time, they have coalesced into larger droplets or into a continuous stream of oil. They are then captured and separated from the produced water and routed to a suitable oil stream in the process. The solids settle to a catch bucket and are typically removed manually. The distance that the oil has to travel in order to be captured is just the height of the channel. The closely spaced parallel plate arrangement provides this short separation distance. In contrast, the oil droplet rise distance in a gravity separator is defined by the height of the water layer in the vessel or tank (bottom of vessel to the oil/water interface). The plates also provide a coalescing medium for the oil droplets that is not present in a gravity separator or tank.

Dedicated Vessel Installation versus Internal Element: Plate packs can be installed in a dedicated vessel or they can be installed as internal devices in most three-phase separators, skim tanks, or API separators. In this chapter, the discussion will mostly focus on the installation of plate packs in a dedicated vessel. Most of the principles discussed, and the performance calculations can also be applied to plate pack installations in other equipment.

Needless to say, CPI units have a larger footprint than hydrocyclones and the oil droplet cut size is larger than it is for a hydrocyclone or flotation. Before the advent of the hydrocyclone, and before deep water oil and gas production, it was common to employ CPI units on platforms in the shallow water region of the Gulf of Mexico and in the North Sea [1]. CPI units are still used for onshore operations, both upstream and in the downstream refinery industry. They are also used on FPSOs in deep water locations where deck space and weight is not prohibitively expensive and where wave motion does not degrade their performance. As oil production shifted offshore to deeper water, operators abandoned CPIs due to size and weight and the fact that they only remove relatively large oil drops. CPIs were also phased out with the introduction of hydrocyclones which remove oil drops down to much smaller diameter.

As discussed below, plate separators are effective at removing oil from water when the oil drop size is relatively large (in the range of 40 to 70 microns and larger). They can handle relatively high oil concentrations (up to a few thousand mg/L) and moderate flow surges and oil concentration excursions do not disrupt the separation efficiency. They require only a low motive pressure to overcome internal hydraulic pressure loss.

Fouling and Cleaning: Due to the close plate spacing and the high surface area provided by the plate packs, CPIs can become fouled and become plugged by solids. Water which contains wax colloids, precipitated asphaltenes, or oil coated solids (e.g., some scale mineral precipitates) can be problematic for CPI performance due to plugging of interstitial plate space by retained solids. If this happens, the separation efficiency goes down. In severe cases, the plate pack may collapse due to the weight of material that has collected on the plates. Operability features include nozzles for cleaning the plate pack in-place, and an easy way to remove the plate pack. These should be specified by the customer and should always be provided unless the oil in the water has high API gravity (34 and above), and has essentially no solids. The ease of cleaning is an important operability feature that the customer should require.

Dissolved Contaminants: A CPI does not have the ability to separate dissolved contaminants. The CPI is a gravity separator which basically has three effluent streams – dispersed/coalesced oil droplets, treated water, and solids. Based on cut off droplet size and typical inlet droplet size distributions, the outlet oil concentration would be of the order of 1000 ppm. If oil droplets coalesce, then wet oil with a few percent water cut may be skimmed from the CPI vessel.

Floating Platforms: On floating platforms and FPSOs, sloshing compromises oil/water separation. The consequence is that the separator design must generally be larger to accommodate separation inefficiency and to provide for more distance between the control levels. There is not much experience with CPI or plate pack interceptors in general on floating platforms or vessels. Where they have been used is mostly on FPSOs which typically have more space and weight available than a TLP, spar or rig. To mitigate the effect of wave motion, at least one vendor offers a liquid-packed design. While this eliminates the liquid/gas interface, it will not prevent sloshing of the oil/water interface. Perhaps carefully designed baffles would help. The design of appropriate baffling is not straight forward, however, and will generally require the use of CFD modeling and simulation.. Also, placement of the unit near the centerline of motion and low toward the center of gravity of the FPSO also would help.

The plate pack design including cross sectional area, plate spacing, and residence time (plate pack length) is based on Stokes law which has been found to be reasonably accurate provided that the Reynolds number and Froude number are low enough to ensure laminar flow without vortices. Regular maintenance/cleaning is required for best performance.

Corrugated Plate Interceptors (CPI) are designed to remove small oil droplets or fine solids from produced water. For effective oil removal, the CPI is designed for downward flow through the coalescing plates which are installed at a 45o angle. For effective solids removal, the CPI is designed for upward flow through the coalescing plates which are installed at a 60o angle to minimize solids accumulation between the plates. Typical plate spacing is 1" for oil droplet removal and 1 to 2" for solids removal.

9.1.1 Typical Configuration

In this section, the basic design of a typical CPI is described. In order to describe the basic design of a plate interceptor unit, it is necessary to select one of the designs that are available on the market. There are several different vessel designs and configurations available. For our purposes here, we have chosen a fairly common design. Various manufacturers incorporate features that allow more effective and less time consuming cleaning of the plate pack. Those features are not added here but are, generally speaking, very important in the operation of such units. The design that is considered here is just a simple basic design. The following items correspond to the items in the schematic drawing in the Figure below:

1. The mixture of oil, water, and solids enters the side of the device.

2. The mixture enters an initial separation chamber which allows the heavier solids to drop immediately to the bottom of the solids collection chamber. The removal of these solids is discussed below.

3. The oily water then proceeds through a distributor plate. This is usually a perforated plate, or a plate with slots in it. The purpose of this plate is to arrest the forward momentum of the incoming liquid, and to distribute the liquid from the small diameter circular cross section of the inlet nozzle to the large rectangular cross section of the plate pack. There are variations in the details, but all designs have in common a plate with holes or slots with a limited of open area. It should be noted that if the open area of the perforated plate is

excessive, then the plate will not effectively distribute the flow over the full cross sectional area of the inclined plate pack.

4. Downstream of the distribution plate, the oily water enters the plate pack. Due to the pressure drop from the inlet (1) to the outlet (8), and the fact that the plate pack is sealed at the bottom, the fluid enters the plate pack in a downward direction. If the plate pack were not sealed at the bottom, the oily water would be able to bypass the plate pack altogether. Within the plate pack, oily water flows down and separated oil simultaneously flows upward to where it is captured (5). Based on field experience, all CPIs which target oil removal have a downward flow while CPIs which target solids removal have an upware flow. For CPI units, field experience trumps intuition. Inside the plate pack, the liquid flows through relatively narrow rectangular channels. The channel spacing between plates can be as low as 2.5 cm or as large as 10 cm. Oil droplet removal efficiency is improved with reduced spacing but plugging by solids deposition is more likely. Oil droplets migrate upward due to Stokes Law, and collide with the bottom side of an upper plate. Solids migrate downward, again due to Stokes Law, and settle on the top of the lower plate of the channel. Once oil and solids have risen to or settled upon their respective plates, they continue to migrate along the plate. The oil migrates upward and the solids migrate downward. This is shown schematically in Figure 9.1 below.

5. Eventually the oil reaches the top of the plate pack where it is then released from the plate pack. At this point, the oil has coalesced into fairly large drops which rise quickly.

6. The oil rises to the oil collection chamber. The level in the oil collection chamber is a function of the hydrostatic head of the oil (6) and water (7) chambers. There is a spillover weir where oil flows into a separate oil discharge chamber. This last chamber is not hydraulically connected to the oil (6) and water (7) chambers. It has a nozzle that allows withdrawal of oil from the unit (not shown). This is a common design for not only CPI units but also for oil/water separators.

7. Water chamber for hydrostatic control of the oil chamber.

8. The clean water outlet. This nozzle can be placed at several different locations along the side of the unit.

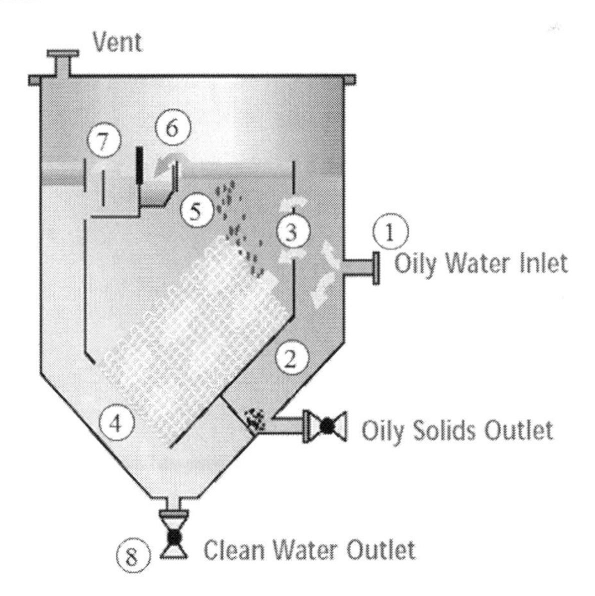

Figure 9.1 Schematic diagram of a corrugated plate pack unit. The numbers refer to items in the text description

The CPI shown in the figure above is more or less representative of most commonly used designs. There are several features that may distinguish one vendor's design from another. These features include the size, and angle of the plate pack, the number of plates and the distance between the plates. These variables are related to one another as discussed in the section on design. Vendors also distinguish themselves with their operability features which are related to keeping the plate pack clean, and to providing a means for removing any fouling material that might buildup. The figure above shows the configuration for a conventional CPI which is designed for down-flow and oil removal. At least one manufacturer has altered this design by installing the parallel plates in a horizontal cylindrical vessel which can be rotated to place the plates in a vertical orientation for jet washing at regular intervals.

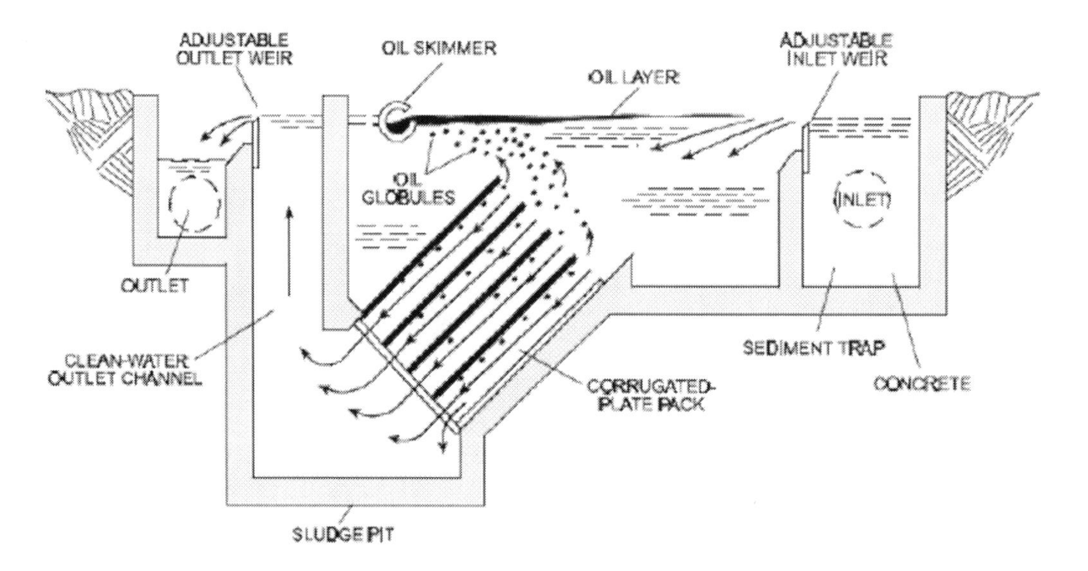

Figure 9.2 The installation of a Corrugated Plate Pack in an API Separator is illustrated [2]

A typical size for the plate packs is 8' long, 4' wide, and stacked 2' high. However, this can change from one manufacturer to another. The capacity of the plate packs is defined by the Reynolds number for flow between the plates with a preferred value of $N_{Re} = 1500 + 500$. The D50 for oil droplet removal by a plate pack is in the range of 20 to 50 microns depending upon the density difference between the oil and water and the temperature (viscosity) of the water.

9.1.2 Mechanism of Separation

Corrugated Plate Interceptor (CPI), Tilted Plate Interceptor (TPI), or Parallel Plate Interceptor (PPI) all have in common the use of a plate pack. The plate pack provides the following important functions:

1. The plate pack provides a narrow bounded geometry such that the oily water flow occurs at a low Reynolds number. This ensures laminar flow without eddies and without vortices. Within this quiescent state of flow, the oil droplets can migrate upward due to buoyancy according to stokes Law.

2. The narrow gap between plates provides a short distance for oil drops to travel (up) and solid particles to travel (down). This distance is referred to as the channel height (h) – which is different than the perpendicular distance from one plate to another. The channel height is

related to the plate separation distance (L) and the angle of inclination of the plate pack (θ). These variables are different depending on the vendor and on the application.

3. Within each channel, the oil rises until it hits the bottom surface of the upper plate. Once this occurs, the intention is for the oil droplet to coalesce against the oil film that is already there. This oil film is moving upward due to its buoyancy.

The distance that the oil droplets and solids particles have to travel is greatly reduced compared to a conventional separator. This is referred to as the Stokes Law distance, or distance to separation. In addition, the oil that gets captured has a much greater chance of being coalesced into larger droplets as a result of being captured on the plate surface. Thus, a CPI has significant effective surface area for oil droplet coalescence. The actual separation efficiency curve is discussed in detail below. The typical range of oil droplet separation is in the range of 30 to 50 micron droplets. Selection of up- or down-flow design depends on solids content. Most designs are such that the bulk water flows down and the oil droplets rise up. This creates a slight counter-current flow that the oil droplets and coalesced oil film must overcome. However, this counter-current flow is of significant benefit when it comes to separation of the bulk water from the bulk oil that has accumulated on the plates.

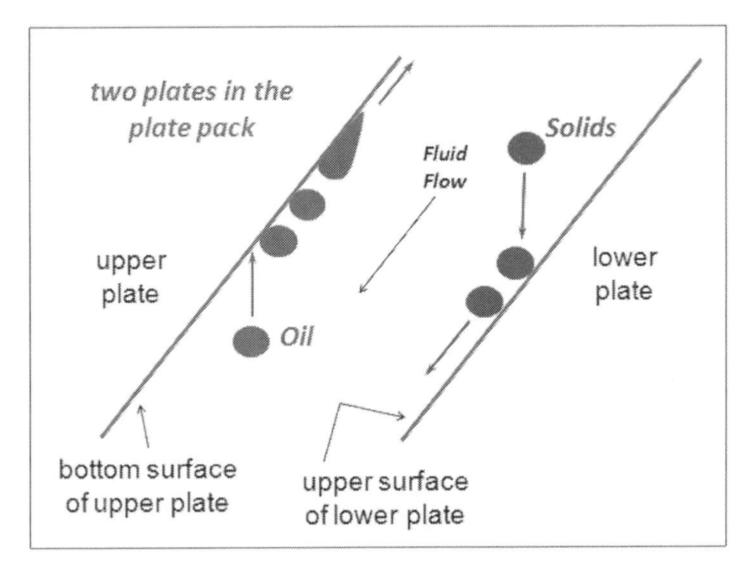

Figure 9.3 Schematic diagram of two plates within the plate pack and the motion of oil droplets and solid particles. The effectiveness of a corrugated plate or parallel plate interceptor is derived from the short settling distance for solids and short rise distance for oil droplets before encountering the surface of a plate

Figure 9.4 A picture of a corrugated plate pack. This figure is from [3]

Plate packs are designed for one orientation of flow or the other. Liquid flow can be parallel to the axis of corrugation or perpendicular to it. Shown in the figure below is a picture of a small section of corrugated plate. If water flows into the side of the plate pack that is shown, i.e. into the openings shown between the plates, this is referred to as flowing parallel or along the axis of corrugation. If the water flows in a zig zag flow perpendicular to the axis of corrugation then there is a greater tendency for solids to drop out and build up in the valleys of the corrugations. Those valleys face upward and will catch the solids and allow them to build up over time.

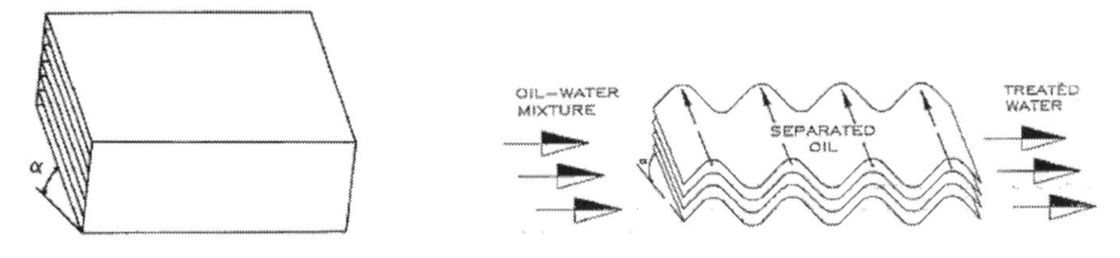

Figure 1. Basic shape of a plate separator.

Figure 2. Corrugated plates for a plate separator.

Figure 9.5 A schematic diagram of a corrugated plate pack also showing the flow orientation relative to the corrugations. In this case the flow is said to be perpendicular to the axis of corrugations. This figure is from [3]

The early designs of plate interceptors used parallel flat plates at an angle of 45 degree to the direction of water flow. These devices were commonly called parallel plate interceptors. Since then, various improvements to the basic parallel plate interceptor have been devised. These include corrugated plate interceptors and cross flow pack separators. Corrugated plates provide more surface area than parallel plates. Also, the corrugations provide a channel for the film of oil to flow upward. The channel allows the oil film to develop a somewhat thicker film with less flow resistance than with a parallel plate.

PPI and CPI are typically used where the feed oil concentration is in the range of 1,000 ppmv to a few thousand ppmv. Separation efficiency will depend on drop size, flow rate, and density difference. Typically 80 % of drops in the range of 40 micron are separated. Higher separation efficiencies are achieved with larger drops. Emulsions will decrease the effectiveness of the CPI. Typically, the design basis for a CPI is 100% removal of oil droplets of 30-60 microns or larger. The average effluent quality at the outlet would be in the region of 20 - 50 ppm residual oil in water depending on:

- the distribution of oil droplets in the inlet stream
- difference in specific gravity between water and oil
- temperature of the water stream
- turbulences
- housekeeping and maintenance.

Plate Pack Interceptors, or Plate Coalescers rely on gravity separation but the distance the oil drop must travel is greatly reduced and coalesce of the oil drops is enhanced by the presence of plates.

Drop Travel Distance for Separation: The distance that oil drops must travel in order to be separated from water is an important parameter. The shorter the distance, the greater the chance that drops will have sufficient residence time for separation. The use of parallel plates provides a means of significantly shortening the distance required for separation. When parallel plates are used, the distance is only the distance between the plates. Once a drop hits the plate, it is expected that it will coalesce into a continuous oil film and flow to the top of the vessel for separation.

Parallel Plates, Corrugated Plates & Cross-Flow Pack:. The early designs of plate interceptors used parallel flat plates at an angle of 45 degree to the direction of water flow. These devices were commonly called parallel plate interceptors. Since then, various improvements to the basic parallel plate interceptor have been devised. These include corrugated plate interceptors and cross flow pack separators. Corrugated plates provide more surface area than parallel plates. Also, the corrugations provide a channel for the film of oil to flow upward. The channel allows the oil film to develop a somewhat thicker film with less flow resistance than with a parallel plate.

PPI and CPI are typically used where the feed oil concentration is in the range of 1,000 ppmv to a few thousand ppmv. Separation efficiency will depend on drop size, flow rate, and density difference. Typically 80 % of drops in the range of 40 micron are separated. Higher separation efficiencies are achieved with larger drops.

Shallow Water GoM Implementation: The PPI and CPI were used somewhat successfully on shallow water platforms in the Gulf of Mexico. The devices tend to be somewhat bulky which is not a significant drawback for fixed structures onshore or in shallow water. As oil production ventured into deepwater, where floating facilities required compact light weight design, plate coalesce equipment was not used. The drawbacks (bulky, heavy, fouling tendency) outweighed the advantage of shorter travel distances for oil drops.

9.1.3 Design Differences and Characteristics

There are different designs which differ in some fundamental ways. One of the most fundamental differences in design is whether the oily water flows upwards through the plate pack or downward. There are pros and cons to each arrangement. Over the years, the design has tended to migrate to one configuration in particular. That is the downward flow through the plate pack. This will be described here.

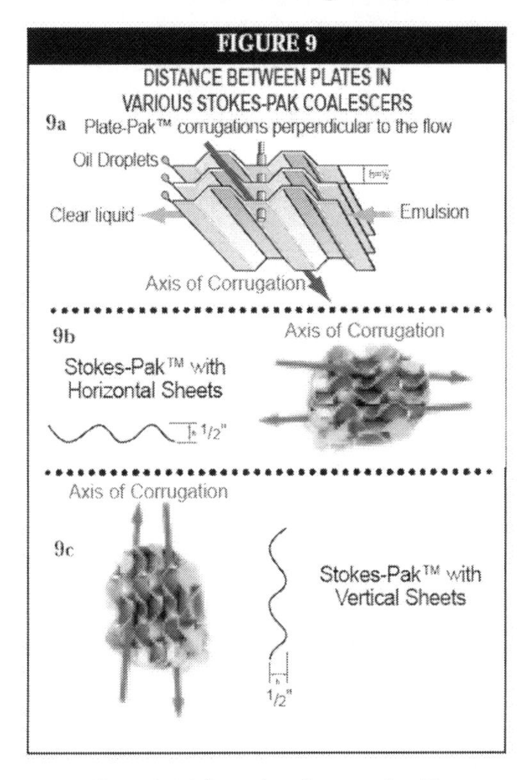

Figure 9.6 A few styles of corrugation [3]

Corrugated Plate Interceptor (CPI) or Tilted Plate Interceptors (TPI) both have a "plate pack" – i.e. plates stacked parallel to one another. The CPI uses a chevron style. The Stokes Law distance for separation is reduced to roughly ≤10 cm – the vertical distance (not the perpendicular distance) between the plates. The figures below illustrate some of the differences in the plate separator designs provided by various vendors.

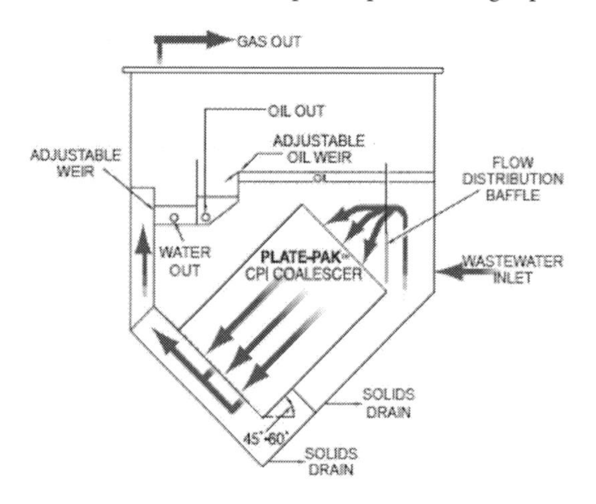

Figure 9.7 Schematic diagram of the flows within a CPI unit. This figure is from [3]

9.1.4 Representative Performance Data

Representative performance data has been generated and is shown in the figure below. The separation efficiency was measured for a unit with parallel plates and a unit with corrugated plates [4]. As shown, there is essentially no difference in the performance of these two plate configurations. It is noted that the design of these units were not typical of that for CPI. The channel height was longer and the length of the plate pack was shorter than typical application. Thus the maximum droplet diameter for 100 % separation is about 100 micron. This is larger than typical installations of CPI. The design of such units is discussed below.

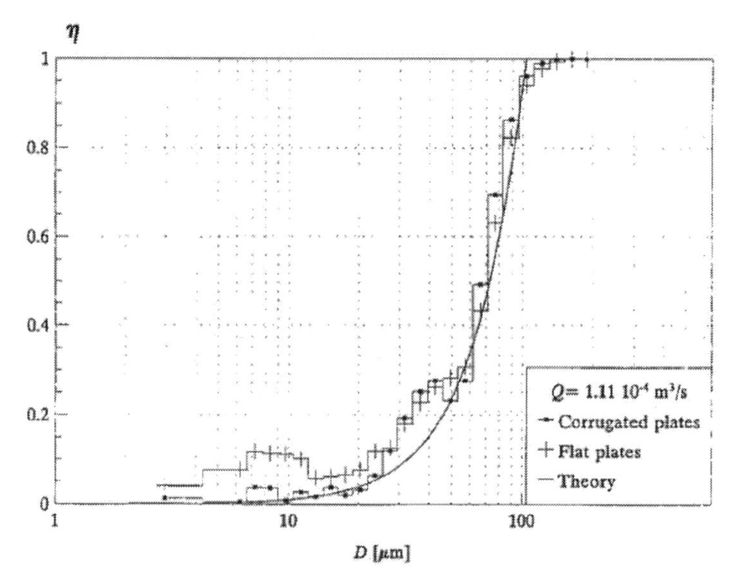

Figure 9.8 Separation efficiency as a function of oil droplet diameter (micron). Data for a corrugated plate interceptor, and for a parallel (flat) plate interceptor are shown. Also shown are calculation results using a design model. This model is discussed below. This figure is from [4]

There are two important data points. The first is the drop diameter at which 50 v% of the oil will be removed. In the Figure shown above this value if 70 microns. The second important data point is the drop diameter at which 100 % separation will occur. This value is 100 micron. As shown in the Figure corrugated plates and flat plates performed similarly.

The distance that the oil drop must travel in order to be captured is quite small compared to a primary separator or skim tank. However, CPI separation efficiency is still dependent on retention time and retention time is dependent on the available volume of the unit, which normally cannot be changed once installed. It is important therefore to ensure that the unit is not undersized to start with. Once the unit is installed, there is very little that can be done to enhance performance. In some designs a new plate pack can be installed. The other thing that can be done is to add flocculating chemical which increases OPEX cost and may contribute to plugging of the narrow passages between the plates.

9.1.5 Operation and Maintenance

The single most common problem with plate pack interceptors is fouling or buildup of oily solids on the plate pack.

Plugging & Fouling: All plate interceptor devices require laminar flow within the plate pack, and relatively closely spaced plates in order to provide an advantage over settling tanks. These requirements are a due to the separation mechanism or Stokes settling, as discussed above. As with all produced water treating equipment, oil droplets, oily solids, and particles covered with production chemicals have a tendency to stick to available surfaces such as those on the closely spaced plates. Sticky oil, containing resins or asphaltenes, and oily solids cause fouling and plugging problems within weeks. Once fouling occurs, then short-circuiting of fluids through the plate pack tends to occur as well which reduces the actual residence time for fluids and hence the separation efficiency.

Fouling in Heavy Oil Applications: Based on the performance data and design information given in this chapter, it would seem that CPIs would be an ideal technology for use in produced water recycling for steam flooding, and other heavy oil produced water applications. However, this is not the case. Heavy oils with high viscosity at moderate temperatures, has a tendency to foul CPIs and related devices. The mechanism of fouling is discussed in detail below.

Operability Features: Plate pack separators have relatively high surface area of material. This is inherent in the separation mechanism. Unless the oil is high API and solids-free, the plate pack is likely to become fouled with oil and oily solids. This requires cleaning. Some vessel designs allow for water or steam jet cleaning using permanently installed nozzles. This is an example of an operability feature that makes it easier to maintain and clean the vessel. In addition some vessel designs make it relatively easy to remove the entire plate pack for replacement or cleaning.

Fouling of the coalescing plate/media is normally an issue and the removal and cleaning of plate packs should be factored into the design. Solids buildup in CPI tanks also creates an excellent environment for bacterial activity which increases the risk of corrosion in the vessel. Bacteria growth also retains oil and can reduce the quality of the water discharged from the CPI. You may consider CPI vessel geometry with sloped bottoms or ensure that a properly positioned jetting system is incorporated into the design.

Collapse of the Plate Pack: Collapse of the plate pack can occur if the plate experiences significant fouling and the mechanical strength of the plate pack is not sufficient to accept that level of fouling. The mechanism of fouling is relatively well known from a chemical engineering standpoint. The two important variables are the contact angle of the oil against the plate material and the viscosity of the oil. If the oil contains solid particles it will have a relatively high viscosity. In that case, it will build up on the plate and will not flow. A simple pour point test can be used to determine if oil will build up on the plates or if it will flow and be carried out of the plate pack. This type of testing is simple and effective but is almost never carried out. A CPI should not be used when the oil to be removed is "sticky" (stickiness has been discussed in Chapter 3 - Emulsions).

9.1.6 Troubleshooting Procedure

As previously discussed, the main problem with CPI units is fouling of the plate pack. When this occurs flow becomes restricted to those plates and those sections of plates that are not fouled. Due to plugging off of those fouled sections, the remaining volume available for flow is smaller and hence the oily water will have a higher velocity through the plate pack. This condition decreases the separation efficiency of the plate pack.

Thus separation efficiency is the best diagnostic test to determine if the plate pack is fouled. As discussed in several other sections, it is helpful to carry out a benchmark test on a new unit or recently cleaned unit in order to establish the expected separation performance for the given crude oil and oily water, process conditions, etc.. When such benchmark tests are available for comparison, then subsequent performance data can be compared against original as well as theoretical performance. In Chapter 16 (Troubleshooting), this is referred to as the Technical Limit of the equipment.

9.1.7 Application in Oil Spill Cleanup

CPI for Oil Spill Cleanup: As part of oil spill contingency plans, some oil and gas companies have designed and built large barges with CPIs as one of the important oil/water separation units. In this application, space and weight are somewhat constrained. Also, there is quite a bit of other equipment such that it is not typically possible for the CPI to be placed near the centerline of the barge. Thus, wave motion can be a concern.

The objective of an oil spill cleanup unit is to extract large volumes of contaminated water rapidly and clean it to a reasonable specification of a couple hundred ppm oil in water. In this application, it is most important to remove most of the oil from the surface of the water before it breaks up into small oil droplets which are then very difficult to separate. For this application, it is likely that the oil will be found in large droplets and freely floating on the surface of the water. This would argue in favor of a CPI.

CPI for Oil Spill Cleanup in Cold Weather Conditions : For a low temperature application, it is important that the design of the CPI take into account the higher oil and higher water viscosity. Also, the crude oil may be weathered which would further increase both its viscosity and density. The increased viscosity of the crude oil will increase its tendency to foul the surfaces of the plate pack. As crude oil builds up, the channels will become plugged which will increase the water velocity in the remaining unplugged channels. This will reduce separation efficiency. Eventually the plate pack may collapse due to the weight of the crude oil plugging the channels. This mechanism of fouling is one of the operational problems with CPI/PPI.

Although vendors have developed CPI with cleaning ports and jetting, which work well in many applications, for high viscosity crude oils (such as that due to arctic temperature water), these features may not be adequate.

9.1.8 Benefits and Drawbacks

The benefits of using CPI and related technologies can be summarized as:

- better separation efficiency for dispersed contaminants than primary separators;
- easy to design and operate – mechanism of separation is well known and is robust (will work in a wide range of applications);
- CPI units require very low feed pressure.

The drawbacks can be summarized as:

- essentially no capability to remove dissolved contaminants;
- sensitivity to fouling for high viscosity or solids-containing crude oils. Such fouling can in some cases be overcome with good design and operability features;
- a CPI may have some sensitivity to slugging, sloshing and wave movement which can limit its performance on floating offshore facilities
- compared to hydrocyclones, CPIs and related technologies do not provide separation for smaller oil droplet diameters;

9.2 Plate Interceptors – Theory, Design and Modeling

9.2.1 Cut Size

The first design parameter to calculate for a CPI is the droplet diameter for which there will be 100% separation.. For smaller droplets, there will be a percentage of separation. Stokes Law along with the hydraulic parameters for a specific CPI unit can be used to make this calculation.

The channel height is an important variable. In order for droplets to be separated, they must rise to the bottom of the upper plate before the bulk water flow is discharged from the plate pack. Once they collide with the bottom of this plate, they must then flow along the bottom of the plate to the channel opening where they can then be separated. The final flow is an important part of ensuring that fouling of the plate pack does not occur. This is discussed below. For now, the rise height is considered as shown in the Figure below

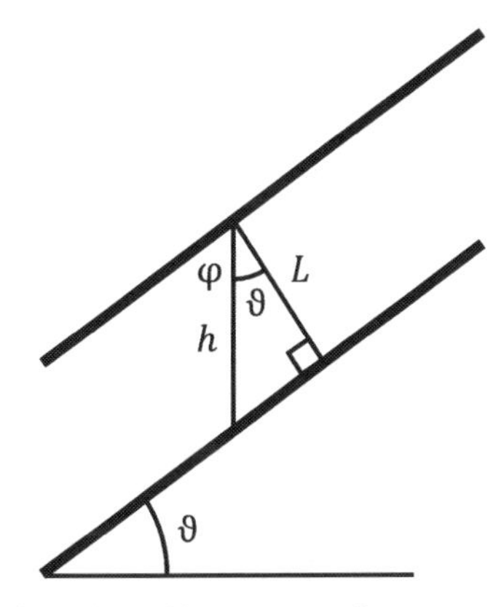

Figure 9.9 Geometry of the channel between two plates. L is referred to as the interval between the plates (or plate separation). h is referred to as the maximum rise height.

Maximum Rise Height: The word "maximum" is used in describing the rise height because the actual rise height required for separation for any given oil droplet depends on where it is initially in the channel. For example, it may enter the channel at the midpoint of the gap between the upper and lower plates. If this is the case, then the oil droplet need not rise the entire distance to hit the bottom of the upper plate. For any given droplet, the maximum rise height that it could experience is given by the vertical distance between plates.

$$h = L / \cos\theta \qquad \text{Eqn (9.1)}$$

where:

- L = plate separation distance (perpendicular, not vertical, distance between plates) (m)
- q = angle of plates (degrees)
- h = vertical distance between plates (maximum rise height) (m)

If, for example, the plates are inclined at an angle of 45 degrees, then cos(θ) is 0.707 (square root of 2) and h = L / 0.707 = 1.4 L. Thus, the vertical distance that the droplet must travel (*h*) is 40 % longer than the perpendicular distance between the plates (*L*). The equations discussed here are also valid for droplets in the channels of a disk-stacked centrifuge or a high rate clarifier. In the case of the centrifuge, Stokes Law is calculated using the force due to centripetal motion, rather than gravity. They are analogous as well to the trajectory of droplets in a conventional separator or skim tank where the droplet must rise a certain distance in order to be captured by the oil phase.

Flow Velocity: the velocity of the fluid in the channels is important for a couple of reasons. First, the flow must be slow enough that the oil droplets have time to rise to the upper plate. This will be calculated below with the use of Stokes Law. Second, the flow must be slow enough that it is laminar. Any turbulence will divert the oil droplets away from the shortest path to the upper plate. The flow velocity is calculated as the ratio of the volumetric flow rate divided by the collective or total cross-sectional area of the channels. Most plate pack dimensions are given in terms of the outside dimensions of the unit. The important dimensions as far as flow is concerned are the channel inside dimensions. For practical purposes these dimensions are similar.

The total cross-sectional area of the channels is given by the following formula:

$$A_c = N_c W_c L_c \qquad \text{Eqn (9.2)}$$

where:

- *Ac* = internal cross-sectional area of flow (m^2)
- *Nc* = number of channels
- *Wc* = internal width of a channel (m)
- *Lc* = plate separation distance (m)

This formula uses the plate separation distance since that is the distance that is perpendicular to the flow. As a numerical example, the number of channels is assumed to be 50. The internal width of the plate pack is approximately that of the outer width (W_C) which in this example is assumed to be 1.0 m. The plate separation distance (L_C) is 0.04 m. Thus, the cross-sectional area is: A_c = 50 x 1.0 m x 0.04 m = 2 m^2.

The linear velocity is given by the formula:

$$V_c = Q_W / A_c \qquad \text{Eqn (9.3)}$$

where:

Vc = linear velocity of flow (m/sec)

Q_W = volumetric flow rate of water (m^3/sec)

The volumetric flow rate is also known as the throughput. In this case throughput is assumed to be 130 m^3/hr (130 x 6.29 x 24 = 130 x 151 = 20,000 BWPD). The linear velocity is thus (130 m^3/hr) / (2 m^2) = 65 m/hr (0.018 m/sec). This is a relatively low velocity. The reasons for requiring this low velocity are discussed below in relation to the Reynolds Number and the Froude Number.

Residence Time in the Plate Pack: Now that the linear velocity is known, the residence time can be calculated.

$$t_R = H_c / V_c \qquad \text{Eqn (9.4)}$$

where:

Hc = length of the plate pack (m)

Vc = linear velocity of fluid in the plate pack (m/sec)

t_R = theoretical residence time (sec)

Continuing with the numerical example, the linear velocity was calculated as 0.018 m/sec. The length of the plate pack (H_C) has not yet been specified. In this example, it is assumed to be 1.4 m. For calculation of theoretical residence time, it is customary to use a slightly smaller effective length. This accounts for unstable and chaotic flow that occurs near the entrance and exit of the plate pack. If the effective length is 10 % less than the actual length, then the "effective length" is roughly 1.25 m. Using these values results in a theoretical residence time of t_R = 1.25 / 0.018 = 69 seconds (1.2 minutes). This value will be used to calculate the cut size.

Cut Size (d_{100}): There is a droplet diameter for which larger droplets will have 100 % separation, and smaller droplets will have less than 100 % separation. This droplet diameter is referred to as the cut size, or d_{100}. In gravity separators, CPI and centrifuges, the cut size is a convenient parameter to use in specifying equipment and in system design calculations [3]. The concept of cut size is not applicable for a hydrocyclone or flotation unit, for various reasons explained in the respective chapters.

The physical meaning of this droplet diameter is the following. This droplet diameter is large enough that no matter where a droplet enters the channel (in the up / down direction), it will rise to the bottom of the upper plate and be separated just immediately before the bulk water exits the plate pack. This diameter can be theoretically calculated. Stokes Law is used together with the value of the maximum rise height, and the residence time of bulk flow in the plate pack. The basic equation is the following. The rise velocity is given by Stokes Law which is discussed in Section 3.2.3 (Stokes Law Settling Rate):

$$u_{d_{100}} = \frac{g(\rho_w - \rho_d)d_{100}^2}{18\mu} \qquad \text{Eqn (9.5)}$$

where:

r_w = density of water phase (kg/m³)

r_o = density of oil phase (kg/m³)

d_{100} = cut size (m)

μ = viscosity of water phase (Pa sec = N sec / m²)

g = the gravitational constant (9.81 m/sec²)

In words, this equation says that there is a droplet diameter for which the CPI is capable of removing 100 % of the droplets. This diameter is the unknown variable in the equation. As an example calculation, consider a CPI which is designed to clarify water with a density of 1100 kg/m³ at 140 F. The viscosity of the water at this temperature is 0.50 cP (mPa sec). The oil droplets to be removed have a density of 800 kg/m³ at the specified temperature. Now that all of the parameters have been specified, the cut size (d_{100}) can be calculated

$$u_{d100} = \frac{(9.81\ \text{m/s}^2)(1100\ \text{kg/m}^3 - 800\ \text{kg/m}^3)(d_{100}\ \text{x}10^{-6}\text{m})^2}{18(0.5\text{x}10^{-3}\ \text{kg/(ms)})} = 0.163\ \text{x}\ 10^{-6}\, d_{100}^2$$

Eqn (9.6)

The rise velocity u_{d100} is related to the residence time and vertical channel height.

$$h = t_R u_{d_{100}} \qquad \text{Eqn (9.7)}$$

where:

u_{d100} = oil drop rise velocity for a droplet having 100 % separation efficiency (m/sec).

Substituting this equation into the equation for Stokes Law gives the following:

$$d_{100} = \sqrt{h/(t_R\ \text{x}\ 0.163\text{x}\ 10^{-6})} = \sqrt{0.057/(68\text{x}\ 0.163\text{x}\ 10^{-6})} = 29\ \text{micron}$$

Eqn (9.8)

In the case of an oil droplet that has a diameter d_{100} it can enter the channel at the bottom plate and have enough vertical velocity and residence time to rise from the bottom plate to the top plate before the bulk water exits the plate pack. In fact, it will arrive at the bottom of the top plate at precisely the end of the plate pack just before the bulk water is discharged. For other starting positions droplets of the same size will reach the upper separation plate earlier. All oil droplets with a diameter larger than d_{100} will be separated. This is one of the parameters used to characterize a CPI separator.

Reynolds Number: In order for the flow to be laminar the Reynolds number must be less than about 400. To calculate the Reynolds number, the Hydraulic Radius for flow between the plates must be determined. The Hydraulic Radius is the ratio of the cross-sectional area to the wetted perimeter:

Cross-sectional area between each pair of plates is:	1 m x 0.04 m = 0.04 m^2
Wetted Perimeter is:	(2 x 1 m) + (2 x 0.04 m) = 2.08 m
Hydraulic Radius (R_H):	0.04 m^2 / 2.08 m = 0.02 m

The Reynolds number can now be calculated from the Reynolds number equation:

$$N_{RE} = \frac{\rho V_C R_H}{\mu} = \frac{(1100\ \text{kg/m3})(0.0183\ \text{m/sec})(0.02\ \text{m})}{0.0005\ \text{Pa sec}} = 800 \qquad \text{Eqn (9.9)}$$

This value of the Reynolds number satisfies the requirement for laminar flow. The Froud Number must be calculated next.

Froud Number: In order for the flow to be stable, the Froud Number must be less than 1.

$$N_{FR} = \frac{V_C}{\sqrt{R_H g}} = \frac{(0.0183\ \text{m/sec})}{\sqrt{(0.02\ \text{m})(9.81\ \text{m/sec}^2)}} = 0.04 \qquad \text{Eqn (9.10)}$$

9.2.2 Fouling Mechanism and Model

A simple model is developed in this section that shows clearly the mechanism of fouling and the role of oil viscosity in fouling. The model provides a mechanistic understanding of how fouling occurs and it provides a guide for determining when fouling is likely to be a problem. In order to develop the model, a more precise definition is needed.

Definition of Fouling: For our purposes, fouling is defined as a build–up of oil on the underside of the upper plate, or a build-up of oily solids on the top of the bottom plate. The fluid mechanics of these two situations are similar. The case of oil on the upper plate will be developed in detail.

Mechanism of Fouling: From a high level view point, the mechanism of fouling is as follows. A certain amount of oil is separated from the produced water as the water traverses downward through the channel. That oil collects on the bottom of the upper plate and flows upward along the plate to the entrance of the plate pack where it disengages and continues to flow upward until it is captured in the oil bucket. The amount of oil that makes its way up and out of the plate pack must equal the amount of oil that is separated from the produced water in the channel. If the oil separated from the produced water in the channel is greater than the flow of oil moving upward, then there is a net accumulation of oil on the bottom of the plate. This net accumulation of oil is responsible for fouling.

The Mass of Oil Separated: The first quantity that must be calculated is the mass of oil that is separated from the produced water as it flows through the channel. The quantity of interest is the total mass collected per unit time. This can be calculated in various ways. In order to make the calculation simple, Stokes Law details will not be used. Instead, it will be assumed that an overall separation efficiency is known. The overall separation efficiency is of course dependent on the Stokes Law variables, and could be calculated if desired. But for our purposes here, this will not be necessary.

$$F_C = Q_C(C_{in} - C_{out}) = Q_C C_{in} \varepsilon \qquad \text{Eqn (9.11)}$$

where:

F_C	= volumetric flow rate of oil captured in the channel (m^3/sec)
Q_C	= bulk flow rate of produced water in the channel (m^3/sec)
C_{in}	= concentration of oil in the produced water entering the channel (kg/m^3)
e	= separation efficiency for oil in the channel ($([C_{in} - C_{out}]/C_{in})$)

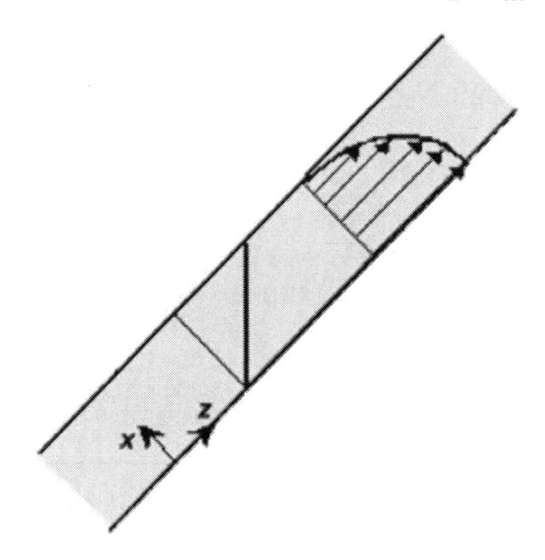

Figure 9.10

$$F_F = u_z \delta W_c = \frac{\rho_o g \delta^3 W_c}{6\mu_o} \cos\beta \qquad \text{Eqn (9.12)}$$

where:

F_F	= volumetric flow rate of oil in the film at the top of the channel (m^3/sec)
W_C	= width of the channel (m)
m_o	= viscosity of oil (Pa sec = kg / m sec)
b	= angle of the plate pack (degrees)
d	= oil film thickness (m)
g	= gravitational constant (m/sec^2)
u_z	= velocity of oil film along the surface of the channel (m/sec)
ρ_o	= density of the crude oil (kg/m^3)

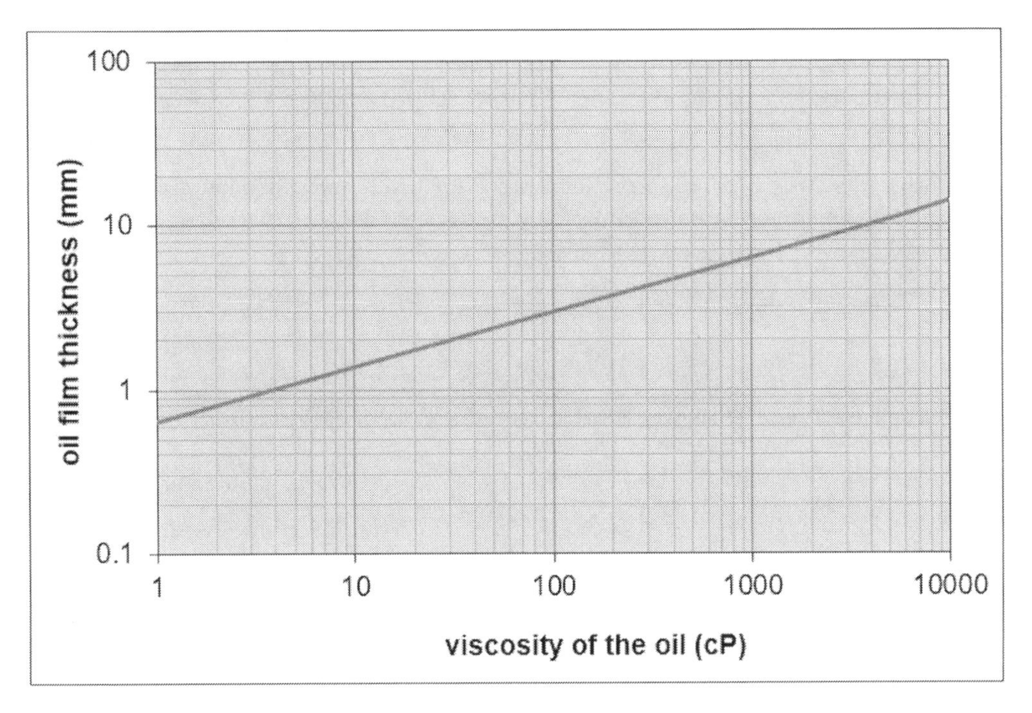

Figure 9.11 Oil film thickness along the upper plate as a function of the oil viscosity. As shown, higher viscosity oil will drain more slowly from the plate pack. This will result in a thicker oil film. When the oil viscosity is very high, oil film builds up and creates a fouling situation.

As shown in the figure above, the oil film thickness is a function of the oil viscosity. It is also a function of the design parameters of the plate pack such as the angle of inclination, the separation efficiency, and the volumetric flow rate of produced water. For high viscosity oils, the thickness of the oil film is relatively high. A thick oil film chokes the flow of produced water which then increases the velocity of the water. This higher velocity then pushes the film in the downward direction which causes further buildup of oil along the bottom of the upper plate. This is how fouling occurs.

9.2.3 Skim and Reject Fluid Separation

The sizing and design of a separator which can clarify skim fluids from skim tanks, flotation cells, etc. and reject fluids from a hydrocyclone or filter deserves special consideration. All too often these fluids are recycled directly or indirectly within a process with the resulting degradation of produced water quality. It is also not unusual for skim and reject fluids to be returned to an upstream separator where solids and chemically stabilized emulsions accumulate at the oil/water interface and periodically upset the entire oil/water separation process, resulting in the discharge of off-spec. wet oil as well as poor quality water which cannot be discharged overboard or, if injected, can seriously foul injection wells.

Typically, skim and reject fluids are 90+% water and can be clarified in a properly designed vessel. The relatively small volumes of wet oil (5 to 20% BS&W) from the reject vessel can be readily bled into the sales oil from a facility with minimal impact on oil quality. For example, if 500 mg/liter of 35o API oil are recovered from 100,000 BWPD, the volume of oil which would need to be mixed with sales quality oil is 59 bbls. If this oil has a BS&W of 10%, then the volume of water added to the oil is < 6 bbls. If this wet oily emulsion is uniformly mixed with, for example, 10,000 BOPD, then the impact on the BS&W of the sales oil is an increase of 0.06%.

The objective of the clarifier vessel is to separate reject and skim fluids into a wet oil fraction as described above and a clarified water fraction which can be recycled back to the first unit operation in a water treatment process. This water should not be recycled back to a 3-phase separator or other vessel/tank which is responsible for primary oil/water separation.

A typical design parameter for such a clarifier vessel is to provide about 10 minutes water residence time and 2+ hrs of oil residence time in a vertical vessel which is designed to receive a fluid consisting of gas-oil-water-solids. The vessel is best designed with

- A cone bottom for solids collection and disposal
- An on-line solids collection or solids removal device
- A suitable 3-phase inlet (e.g., a cyclonic or other suitable device)
- A cross sectional area which permits oil droplets of 175 + 25 microns to rise against the downward velocity of the water

As an example, one can calculate the size of a suitable vertical vessel for the following operating conditions:

Oil density at temperature	900 grams/liter
Water density at temperature	1015 grams/liter
Water viscosity at temperature	0.5 cP
Design value for oil droplet separation	175 microns
Volume of reject/skim water	5,000 BWPD
Volume & Quality of water treated	100,000 BWPD with 400 mg/liter of oil
Water retention time for clarification	10 minutes
Oil retention time for dehydration	4 hrs.

Using Stokes Law, the terminal rise velocity for oil droplets is calculated to be 23 cm/min = 0.75 ft/min. Thus the reject clarifier vessel must have a diameter such that the downward velocity of water in the vessel does not exceed 0.75 ft/min.

- 5000 BWPD = 19.5 ft^3/min
- 19.5 ft^3/min ÷ 0.75 ft/min = 25.9 ft^2 which is the required cross sectional area of the separator
- A 6 ft diameter vessel has a cross sectional area of 28.3 ft^2 and would thus be suitable for use as a reject fluids clarifier for the specified service

The oil pad height in the vessel is determined by the volume of oil in the reject fluid over a 4 hr. time span.

- 400 mg/ltr ÷ 0.900 gr./cm^3 = 444 PPMv of oil in the water being treated
- 444 PPMv x 100,000 BWPD = 44.4 BOPD or 7.4 bbls (41.6 ft^3) of oil in 4 hrs.
- With a cross sectional area of 28.3 ft2, the oil pad would need to be 41.6/28.3 = 1.5 ft to provide the desired 4 hrs of retention time for dehydration.

The water height in the vessel required to provide 10 minutes of retention time is similarly calculated:

- 5000 BWPD = 34.7 bbls/10 minutes or 195 ft3 of water in 10 minutes
- A water depth of 6.9 ft will provide the required residence time in the 6 ft diameter vertical vessel

If it is assumed that the liquids in the vessel will occupy 70% of the Seam-Seam distance, then the total Seam-Seam height of the clarifier vessel will be

- Liquid Ht. = (6.9 ft. for water + 1.5 ft. for oil) ÷ 0.7 = 12.0 ft.

References to Chapter 9

1. S.E. Rye, E. Marcussen, "A new method for removal of oil in produced water," SPE 26775, paper presented at the Offshore European Conference, Aberdeen (1993).

2. US Army Corps of Engineers, Afghanistan Engineer District, "AED Design Requirements: Oil / Water Separators," US Army Corps of Engineers (2010)

3. ACS Separations and Mass Transfer Products, ACS Industries, Houston, TX

4. W.M.G.T. van den Broek, R. Plat, M.J. van der Zande, "Comparison of plate separator, centrifuge and hydrocyclone," SPE 48870 (1998).

CHAPTER TEN

Hydrocyclones

Chapter 10 Table of Contents

10.0 Introduction 61

10.1 Deoiling Hydrocyclone – Practical Applications 61

10.1.2 Introduction 64

10.1.3 Mechanism of Separation – Performance Variables 67

10.1.4 Design Differences and Characteristics 71

10.1.5 The PDR: Pressure Differential Ratio: 74

10.1.6 Material Balance – Reject Oil Concentration 76

10.1.7 Process Configuration and Control 77

10.1.8 Representative Performance Data 82

10.1.9 The Effect of Gas Breakout with the Hydrocyclone 85

10.1.10 Operation and Maintenance 86

10.1.11 Troubleshooting Procedure for Hydrocyclones 90

10.1.12 Benefits and Drawbacks of Deoiling Hydrocyclones 96

10.2 Deoiling Hydrocyclone – Theory, Design and Modeling 97

10.2.1 Stokes Law in Rotational Acceleration 98

10.2.2 Inlet Fluid Flow 100

10.2.3 Tangential Velocity and Vortex Analysis 103

10.2.4 Shape of Separation Efficiency Curve – Trajectory Analysis 108

10.2.5 The Hydrocyclone Number & Performance Curve 110

References to Chapter 10 113

10.0 Introduction

There are several types of cyclonic devices used in the oil and gas industry for separation of contaminants from water. When discussing these technologies, it is important to select names that will allow the various devices to be discussed individually without causing confusion. As with so many other subjects in the oilfield, the terminology that is commonly used is confusing. In this book, three important technologies will be discussed. Each will have its own unique name [1]. They are:

1. Deoiling hydrocyclones, or simply hydrocyclones. These devices are designed to separate dispersed oil from water.

2. Liquid / liquid cyclone separators. These devices provide bulk separation of oil and water, usually in percentage concentrations of oil in water, as a compact substitute for a primary separator. In this book, this device will not be referred to as a hydrocyclone.

3. Desanding cyclone, sand cyclone, solid / liquid cyclone separator or simply desander, The primary function of this device is to remove solid particles that have higher density than water.

In order to avoid confusion, the term hydrocyclone will be used exclusively in this book to refer to a deoiling device for the removal of dispersed oil droplets from water. Within each of these three categories, there are several types, models, designs and brands. When necessary, the individual brands and designs will be referred to separately. This terminology and these distinctions in the technology are important. In some cases very hardened perceptions exist about the performance of one type of cyclonic device versus another. But the fact of the matter is that these three technologies are so different from each other that they cannot be compared one to the other. If, for example, a hydrocyclone has not worked well at a particular location there may be factual and verifiable reasons for this. This subject is discussed in several places in this book. However, it cannot be assumed therefore that a desanding cyclone will not work on the basis of poor hydrocyclone experience. They are two completely different technologies.

10.1 Deoiling Hydrocyclone – Practical Applications

Driving Force for Separation: In a hydrocyclone, the separation of oil droplets from water occurs as a result of large centrifugal forces which are generated by inducing a swirl motion in the fluid inside the device. The angular acceleration due to the swirl motion enhances the effect of the density difference between the dispersed oil droplets and water. The oil, being lighter than water migrates to the axial center of the device. As shown in the figure below, the oil is then directed to a reject port (shown as light overflow in the figure). The water exits through an effluent port (shown as dense phase underflow in the figure).

Pressure: There are three pressures that are important to understanding the fluid flow in the device: feed or inlet pressure, treated effluent or underflow pressure, and reject or overflow pressure. The feed or inlet pressure is provided either by an upstream separator (preferred) or by a suitably selected pump (acceptable if properly selected). The underflow and reject are controlled by external valves and the valve control systems. The feed pressure is always the highest of the three pressures. The high feed pressure pushes fluid through the device. The pressure of the fluid exiting the reject port is always the lowest pressure because it taps into the low pressure in the axial center of the cyclonic motion of the fluid in the hydrocyclone. The low pressure in the fluid core permits the reverse flow of oily water to the reject port, which will be discussed in the next paragraph. The pressure of the treated effluent or underflow is a function of the volumetric flow rate of liquid through the hydrocyclone and is always intermediate between the value of the feed and the reject pressures. Successful operation of a hydrocyclone and high oil/water separation efficiency depends on the proper selection and control of these three pressures, among other things, as discussed below.

Factors that Affect Oil Droplet Migration: The migration rate of oil droplets to the central core (or axis) is controlled by the same variables in Stokes Law (density, viscosity, droplet diameter), with one important exception. Stokes Law is based on gravity being the driving force for migration. In a hydrocyclone the centrifugal acceleration of the swirl motion provides the driving force. This force is many times, in some cases 1000 to 2000 times greater than the force of gravity.

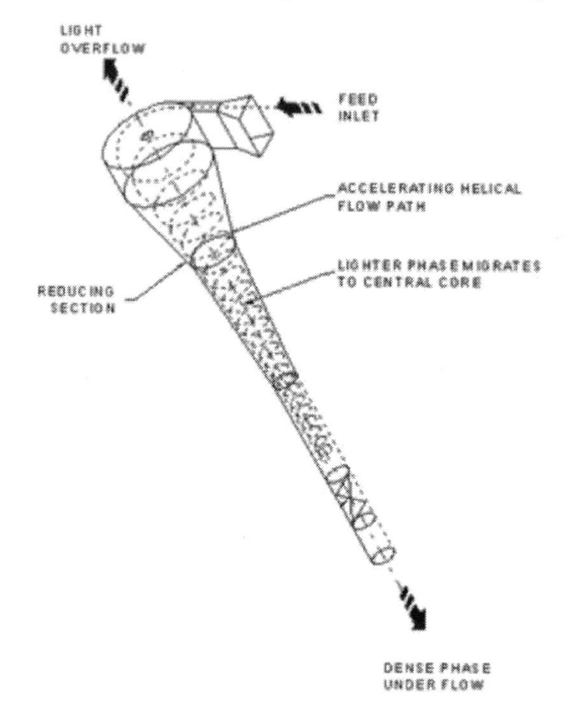

Figure 10.1 Schematic diagram of a single cyclone. Such a cyclone is typical of that used in packaged deoiling hydrocyclone units where several dozen such cyclones might be installed. This figure is shown here to give a basic idea of the geometry and fluid path of a typical deoiling hydrocyclone. Oily water enters at the top in a tangential inlet (feed inlet). Spiral flow is generated by the cylindrical geometry and the tangential inlet. The fluid spirals down the length of the device. Oil droplets migrate to the center axis due to a density difference between oil and water (lighter phase migrates to central core). The oil is drawn out of the device at the light overflow orifice (reject port). The clean water effluent exits at the dense phase underflow (water effluent port).

Historical Development: Hydrocyclones are regarded as 'standard technology' for most offshore operations, and have been used successfully onshore as well. They are compact, and therefore weigh less than other water treating equipment such as a Free Water Knockout or multistage flotation unit. The compact design is one of the benefits of using hydrocyclones. They are also used to some extent onshore to further clean the effluent discharged from large settling tanks or corrugated plate interceptors. The details of the design for onshore versus offshore units are usually quite different.

Desanding cyclones were developed by the minerals and mining industries long before (deoiling) hydrocyclones. The fundamental principles of a hydrocyclone are similar to those of a desanding cyclone. However, their geometries are different. This is due to the fact that sand is heavier than water, whereas oil is lighter. It is also due to the fact that there is a greater density difference between sand and water (usually about 1.3 to 1.6 gr/cm3) than there is between oil and water (usually about 0.05 to 0.3 gr/cm3). The centrifugal force of the swirl motion pushes the sand to the outer circular edge of the device. Thus, the contaminating sand moves in the forward direction and is discharged through an opening (apex) at the bottom of the cyclone. The bulk fluid (water effluent) reverses flow and is discharged through a relatively large discharge port. An important difference between sand/water separation and oil/water separation is the shear sensitivity of oil droplets. Sand particles are relatively insensitive to shear.

Initial attempts to develop an effective deoiling hydrocyclone were based on modifications to the design of solids separation cyclones. This strategy was unsuccessful. Although the fundamental principles are similar, the size, shape, pressure drop, and flow rate requirements of a deoiling hydrocyclone are significantly different from those of a solids separation cyclone. Efforts to modify the design of the solids separation cyclone such that it could be used for oil/water separation were derivative and did not go far enough. As history would show, a major departure from the desanding cyclone design was required.

In the late 1970's and early 1980's Derek Colman and Martin Thew [2], at Southampton University recognized that a deoiling hydrocyclone, with good separation efficiency, was possible [3]. Rather than use the design (size, shape, residence time, etc) of the desanding cyclone as a starting point, they instead used the fundamental principles of cyclone separation devices. The concepts of Hydrocyclone Number, Euler Number, Reynolds Number, Centrifugal Head, and particle trajectory had already been developed, measured and related to the shape and residence time of a cyclone separator. Some of these ideas were borrowed from basic fluid mechanics (Euler Number, Reynolds Number, and Centrifugal Head) and others were developed specifically for cyclone separators (Hydrocyclone Number, trajectory analysis). Those principles had been developed mostly for sand and other solids materials (coal, catalyst particles, etc). However, the principles were fundamental and could therefore be applied equally well to develop an oil/water separation device.

The fundamental concepts pointed to a design that is much longer, with a much smaller entrance port, a relatively large involute section, and two longer narrow tapered sections, and which requires a much higher pressure driving force and pressure drop than the desanding cyclone. This significantly different shape was required to provide enhanced force while minimizing droplet shear. Thus, the result of the Colman and Thew work was not only a practical hydrocyclone design, but a set of design rules that led to a range of hydrocyclone designs that could be tailored to different conditions of flow rate, separation efficiency and cost.

First Installations: The first installation of the hydrocyclone was in the North Sea in 1985 on the Piper platform. Shell installed its first hydrocyclone in 1987 on the North Cormorant platform. Within just a few years, the deoiling hydrocyclone became 'standard equipment' as a compact and efficient means of treating produced water offshore. In terms of technology uptake, the hydrocyclone gained very rapid acceptance, particularly in the North Sea. Companies like Shell and Chevron embraced hydrocyclones within three or four years of their initial development.

Hydrocyclones in Deepwater versus the North Sea: The acceptance of hydrocyclones in the deepwater region of the Gulf of Mexico took longer than in the North Sea. Even today, the hydrocyclone plays less of a role in deepwater Gulf of Mexico facilities than in the North Sea. This observation has been the subject of interesting discussions at various conferences and meetings.

Bothamley [4], and later Walsh and Georgie [5, 6] evaluated the possible technical and economic reasons that would explain why hydrocyclones have less success in the deepwater Gulf of Mexico. The analysis concludes with the idea that hydrocyclone implementation in the Gulf of Mexico had not been slow but rather the hydrocyclone has less applicability in deepwater where excessive shearing creates small oil droplets. In such systems, hydrocyclones provide less benefit than in North Sea-type systems. The detailed analysis is given in Chapter 20 (GoM versus North Sea).

Downhole Application: Hydrocyclones have also been tested for application downhole [7, 8]. The idea being to separate oil and water in the wellbore itself and discharge the oily water further downhole and pass the oil to the surface. In this application, the use of a hydrocyclone is less than ideal. One major problem for downhole application is that the device has a limited flow range over which it can successfully remove oil from water. As discussed below, several design features must be specified in order to match the feed, reject and treated effluent flow and pressure requirements. There are various ways to address this in a surface facilities situation. In a downhole installation, this would require replacing the device with a custom design every time there is a significant change in water cut. Over the life of a well, this could require several change-outs. Also, the cost of such equipment prohibits widespread application on each well in a field.

Oil Spill Cleanup: Liquid-Liquid hydrocyclones have been evaluated for use in oil spill cleanup [9]. In this application, the hydrocyclone would be installed on a barge. Oily water would be pumped from the sea surface and fed into the hydrocyclone. Hydrocyclones have the advantage that their performance is not sensitive to wave motion. The objective of this design is to handle a very large volume of water within a short time and discharge a treated water effluent to the sea. Another objective is to recover a fairly concentrated oily reject. The higher the oil concentration, the smaller the onboard storage tanks need to be. All things considered, hydrocyclones were found to be suitable for this application.

10.1.2 Introduction

Pressure Drop, Fluid Flow, Oil Droplet Migration: As shown in the figure above, the continuous water phase (product) flows down the length of the cyclone and out of the bottom. The pressure difference between the feed and the effluent discharge pushes the water in this direction. Due to the centrifugal force, the oil drops, which constitute the lighter phase, are pulled radially toward the central axis. The drops form a core of oily water. This core of oil flows in the opposite direction of the water effluent due to a pressure difference between the feed and the reject orifice, which is shown in the figure. This reject core is drawn out of the reject orifice at the top of the cyclone.

Overflow and Underflow: As shown in the figure, the reject orifice is also referred to as the overflow orifice. The water effluent port is referred to as the underflow. This terminology is a throwback to the early days of hydrocyclone development when typical installations were configured with the cyclone in the vertical direction. Today, many deoiling hydrocyclone systems are installed in a horizontal configuration but the historical terminology is still used. From a technical standpoint, there is no preference for either horizontal or vertical installation. The gravity force is relatively small compared to the tangential acceleration force. From an operational standpoint, it is typically more convenient to install and maintain a hydrocyclone unit in the horizontal configuration.

The figure below helps to establish the terminology that will be used through this chapter. It shows the tangential inlet (oil/water inlet); the treated water underflow; and the oil rich overflow. Also shown is the overflow orifice. This orifice is typically quite small (1 to 3 mm dia.) compared to the water-rich outlet. The size of the orifice is designed to provide the reject flow rate and to ensure that only fluid from the central core of the cyclone is discharged in this direction. The effect of pressure drops within the hydrocyclone on this flow rate is discussed in some detail below.

Various Sections: The largest diameter of the device is located at the inlet (involute) section. This section is relatively large in order to minimize inlet fluid turbulence and to allow the cyclonic flow pattern to be developed. Fluid acceleration (tangential velocity) occurs in the reducing, taper and parallel sections. The reducing and taper sections have a conical shape. The parallel section has a cylindrical shape.

There are several designs and manufacturers which differ in the diameter, angle of the tapered sections, and length of the various sections. Generally though, the device is tapered in order to maintain a roughly constant tangential fluid speed which maintains a roughly constant g-force as the fluids move through the device. The tapering overcomes fluid kinetic energy loss due to fluid friction.

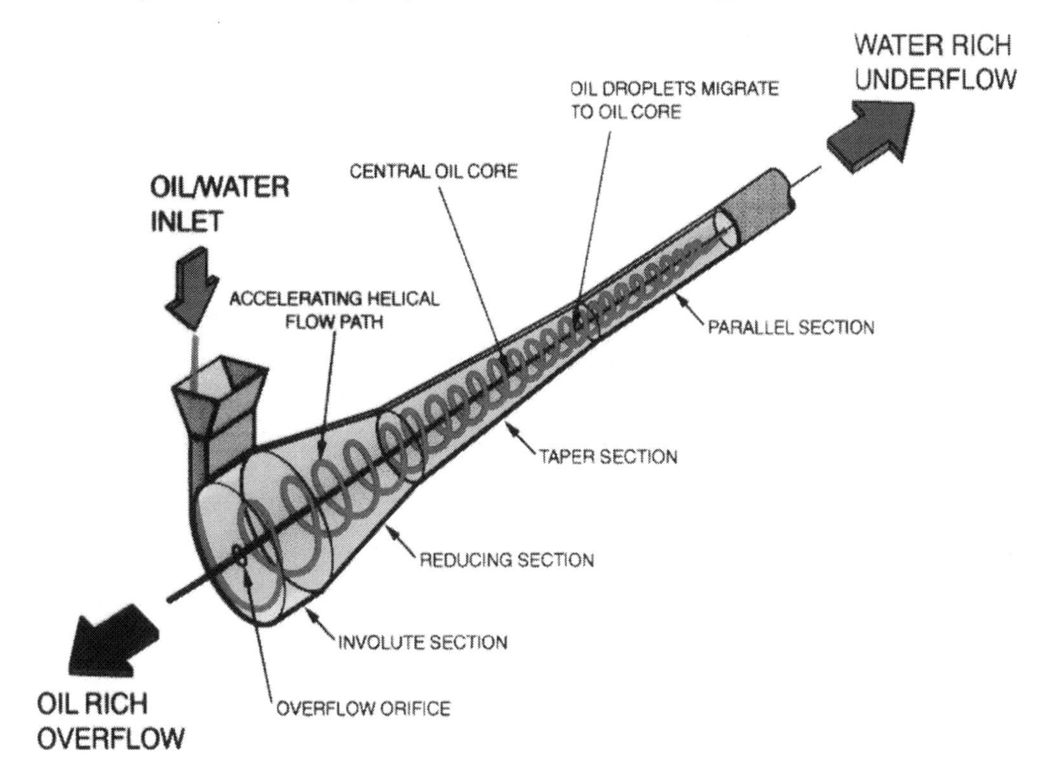

Figure 10.2 Schematic diagram of a single cyclone with labeled sections. In the text, reference is made to the different sections. This figure severs as the reference for the names of those sections.

Due to the intense swirling motion, several hundred to perhaps a couple thousand g-forces are applied to the oily water. Overall the residence time for water in the hydrocyclone is on the order of a few seconds. But given the intensity of the enhanced gravity forces, good separation efficiency can be achieved even in a short residence time.

Colman-Thew Design: The basic Colman-Thew design is given in the figure below [2, 10]. The key elements of the design are a small inlet nozzle diameter to establish a high linear velocity, a small cyclone diameter which results in high acceleration forces and minimizes the distance that the oil droplets must migrate to the central core, and a long body which maximizes the residence time, and which helps to maintain the oil core for the maximum length down the liner to maximize the probability of oil droplet capture. The Colman-Thew design is often referred to as a two-cone design. As shown in the figure, there is a steeply tapered conical section, connected to a shallow tapered section, which is then connected to a cylindrical tail section.

Various Designs: Most suppliers have a variety of liner geometries available. In general, the larger the D_H, the greater the capacity but the lower the centrifugal force generated within the hydrocyclone (a.k.a., the "liner"). Larger capacity means fewer liners are required to process a given flow rate and thus the capital cost is lower. But a lower centrifugal force means a lower separation efficiency.

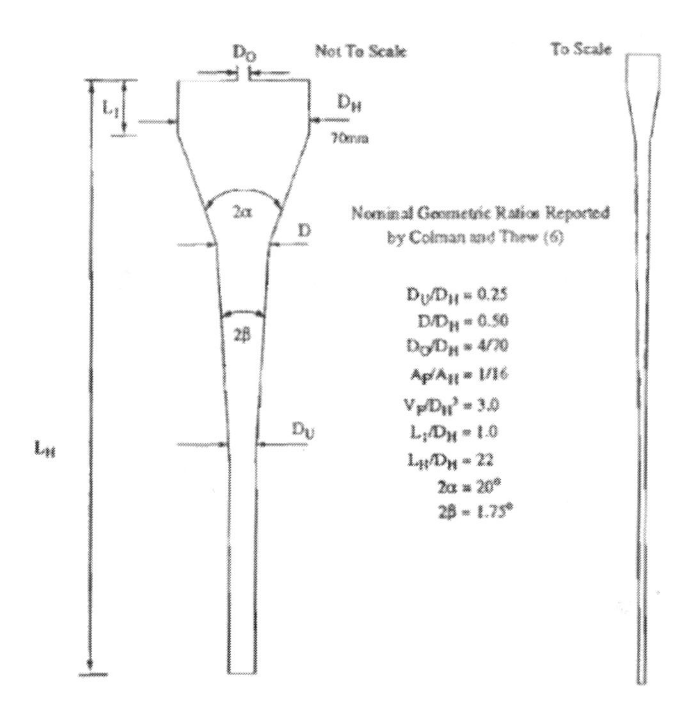

Figure 1. Schematic of the Commercial CT Hydrocyclone

Figure 10.3 Design parameters of the original Colman-Thew design.
The figure on the right hand side shows the dimension drawn to scale. Figure from [10].

Packaging of Liners: Most installations of hydrocyclones involve a pressure vessel that houses many liners. The most common packaged unit has a single feed stream, and a single product and reject stream. This is the common design and most people familiar with hydrocyclones have seen this design.

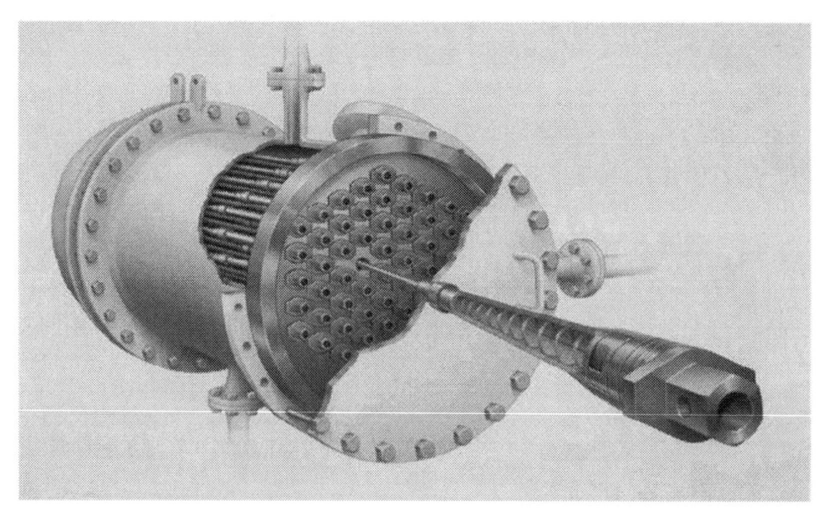

Figure 10.4 Packaging of many cyclone liners within a single pressure vessel. This cutaway diagram shows the reject header, and the feed section below the reject header. A single liner is shown in some detail including the o-rings which seal the liner into the reject header, and the rectangular feed port. Compact pressure vessel packaging of hydrocyclone liners requires much less space than individual liners for a relatively large number of liners. The main disadvantage is that the pressure vessel head must be opened in order to change the number of liners in operation – note the large number of bolts.

The pressure vessel packaged design is not necessarily the best design for a given application. Shown in the figure below is a system with stand-alone liners. Each liner has its own feed, reject, and product piping and valves to control the flow. The advantage of this design is that the number of liners in use can be adjusted easily by turning the individual valves. The main disadvantage of this system is the relatively large number of valves required and the space required.

Figure 10.5 Individual hydrocyclone liners (8 total). This system is manually operated and is used for batch cleanup of tank inventory. It allows easy variation of the number of liners but is not a compact design, has limited throughput, and requires a relatively large number of valves.

Packaging is an important aspect of hydrocyclone installation. Manufacturers have developed many varieties of packaging which have various desirable features. Some of these features are discussed further below.

10.1.3 Mechanism of Separation – Performance Variables

Separation of oil droplets from water in a hydrocyclone is based on the same fluid properties as with primary separators. As oil droplet diameter increases, separation efficiency increases. As oil density decreases, or as water density increases, the separation efficiency goes up. As temperature increases, the main effect is to reduce the viscosity of the water. This increases separation efficiency. All of these fluid variables (droplet diameter, density difference, water viscosity) are found in Stokes Law. However, as discussed in some detail below, Stokes Law is only one of the guiding principles required to understand oil separation in a hydrocyclone. The two other principles are the fluid mechanics of the swirling water and the trajectory analysis of the droplets suspended in the swirling water. These are rather complex subjects and are discussed in detail in the modeling section.

Separation Efficiency Curve: Shown in the figure below is an example of a separation efficiency curve for a typical deoiling hydrocyclone operating under design conditions (flow rate, pressure drop, PDR). There are two characteristics that are often used to compare one hydrocyclone to another. One is the d_{50} value. This is the droplet diameter at which roughly 50 % of the droplets are separated. For the particular separation efficiency curve shown in the figure, this value is just above 5 or 6 microns. The other is the d75 value. This is the droplet diameter for which 75 % of the droplets are separated. For the figure below, this is about 8 microns.

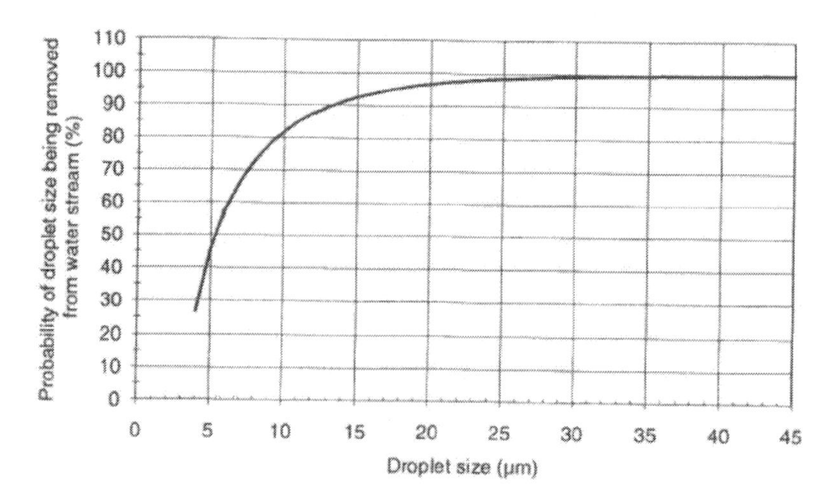

Figure 10.6 Separation Efficiency of a particular hydrocyclone design. The separation curve of a hydrocyclone is a unique curve for each manufacturers' design.

One of the interesting features of the figure above is that there is no discernible d_{100} or d_{min} value. In other words, there is no minimum droplet size for which 100% of the drops are separated. This is generally true of all hydrocyclones. Although they are considered primary separation equipment, and although Stokes Law influences the separation, the shape of the separation efficiency curve is significantly different from that of other primary separation equipment where a d_{min} or d_{100} is typically sharply defined as discussed in Chapters 8 and 9. The explanation for this has to do with the detailed fluid mechanics that occur within the device. This topic is discussed in detail below.

Separation, Capacity, Cost: In all hydrocyclones, there is a tradeoff between separation efficiency and capacity, which translates to cost. In other words, the improvement achieved by using a smaller diameter hydrocyclone does not come for free. By using a smaller diameter liner, more liners are required. More liners for a given water flow rate will cost more. This is one of the basic decisions that must be made when selecting a hydrocyclone.

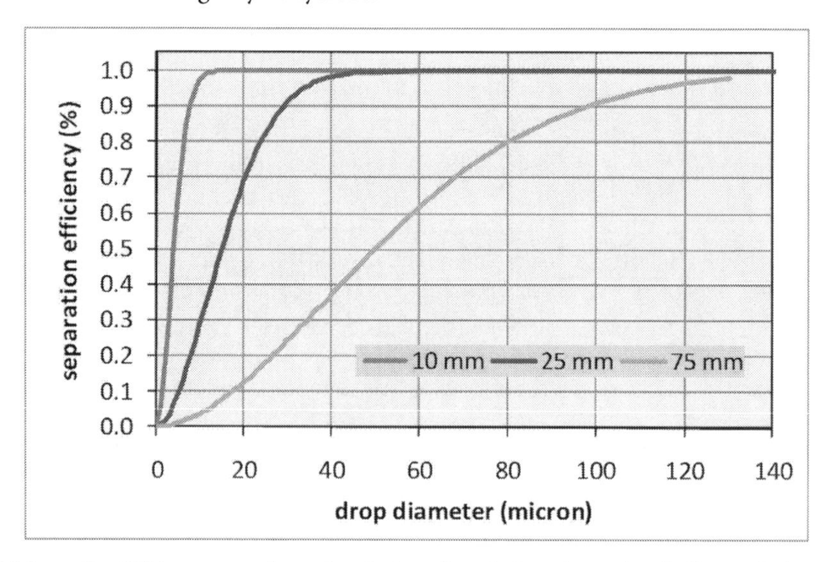

Figure 10.7 Separation Efficiency curves for various hydrocyclone designs. As previously discussed, these curves are indicative only. Actual curves depend on fluid characteristics, and on details of the design. See Section 10.2 (Deoiling Hydrocyclone - Theory, Design and Modeling) below for models that can predict Separation Efficiency performance.

Separation efficiency depends strongly on the diameter of the cyclone. The smaller the diameter, the higher the swirl velocity and the greater the driving force for separation. These curves were generated based on empirical correlations and refer approximately to the same density difference, and water viscosity. Differences in these variables will change the results [10].

Effect of Forward Flow Rate: As demonstrated in the next two figures, the effect of increased flow rate is ultimately to increase separation efficiency. This is only true however up to a point. As discussed below, beyond a certain flow rate, the separation efficiency decreases with any further increase in flow rate.

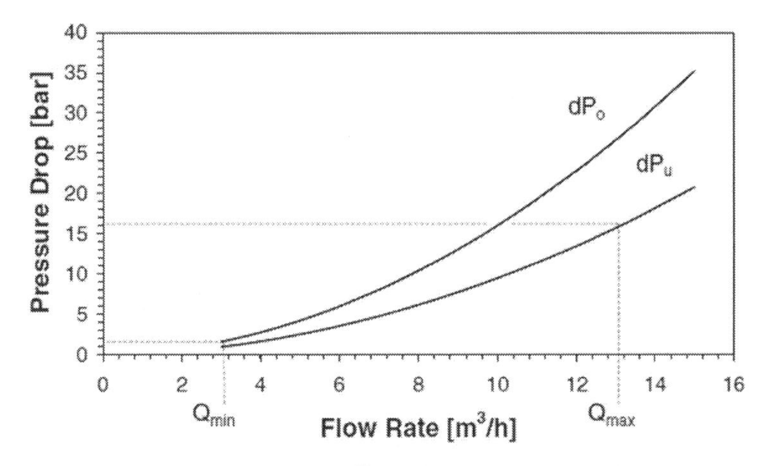

Figure 10.8 Typical differential pressures versus flow rate. dPu is the pressure difference between the feed and the underflow. dPo is the pressure difference between the feed and the overflow. At any given flow rate, dPo is greater than dPu. This is expected due to the fact that the reject port diameter is much smaller than the effluent discharge diameter.

Note that the reverse flow rate requires a higher pressure drop for a given flow rate than the forward flow. This is because the "diameter of the pipe" for forward flow is larger than that for the reverse flow. It is also a consequence of the fact that cyclonic flows generate a low pressure core.

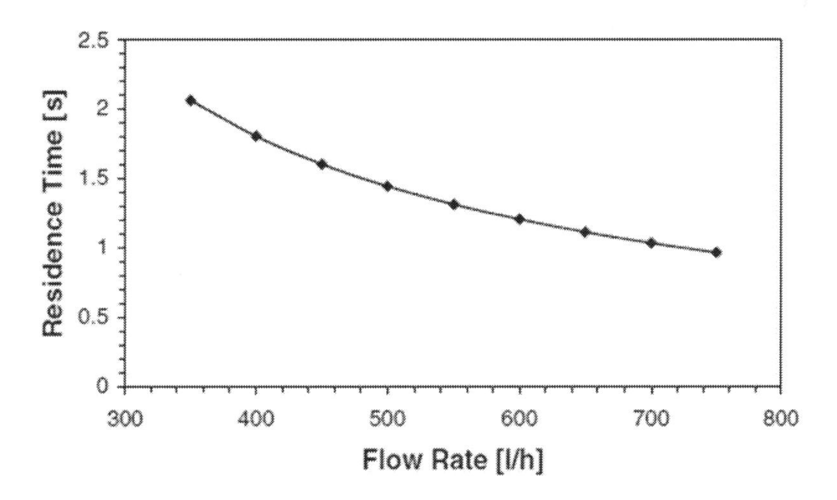

Figure 10.9 The residence time for a typical hydrocyclone is relatively short.
As with any flow through a pipe, the greater the flow rate, the shorter the residence time.

As shown in the figure above, the residence time for a typical hydrocyclone is relatively short. This results in a compact design. As expected, the residence time decreases as the flow rate through the device increases. This is what is found in separators as well. However, in a separator the reduction of residence time will typically lead to a reduction in separation of oil and water. This is not usually the case for a hydrocyclone.

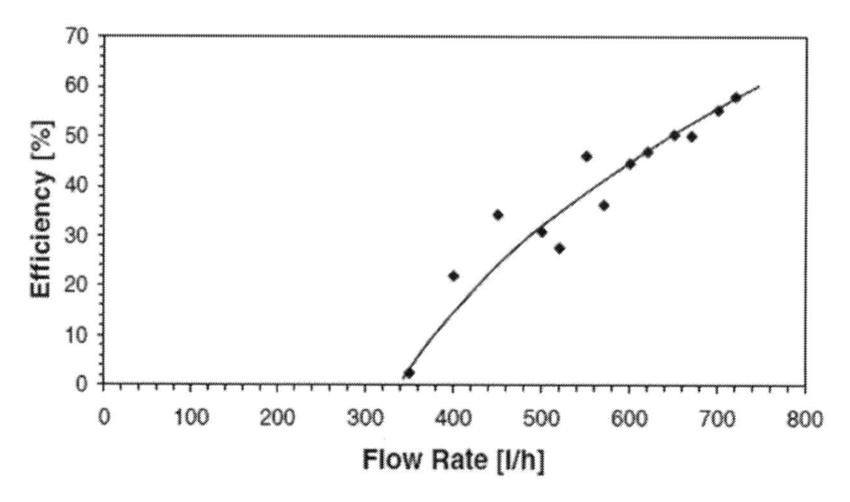

Figure 10.10 The separation efficiency as a function of the flow rate for a typical hydrocyclone. While the residence time decreases with increasing flow rate (previous figure), the separation efficiency increases. The explanation for this is that increasing flow rate enhances the intensity of separation forces. The separation efficiencies measured in this laboratory study are low compared to commercial units. Nevertheless, the data demonstrate the concept (efficiency increase with higher flow rate).

As shown in the figure above, the separation efficiency of a hydrocyclone typically increases as the flow rate increases. The explanation for this is that increasing flow rate increases the swirl intensity. This enhances the centrifugal force which is responsible for pulling the oil drops toward the axial center.

However, the curvature is negative. In other words, the rate of increase is not constant. The rate of increase in separation efficiency diminishes with increasing flow rate. This is due to a combination of effects. As flow rate increases, separation efficiency is promoted by increases swirl intensity and increased driving force. Conversely, as flow rate increases, separation efficiency is diminished by reduced residence time, and by an increase in shear intensity. This increase in shear intensity occurs at the fluid inlet, near the wall, and causes oil drops to shear into smaller drops thus generating some fraction of drops that are too small to be separated. These are gradual effects that result in the shape of the curve in the figure above. These gradual effects are not the cause of the dramatic drop in separation efficiency that occurs at yet higher flow rates.

Limited Operating Envelope – Meldrum Curve: Above a critical flow rate, which differs for each design, the separation efficiency drops off dramatically. This is shown in the figure below. The characteristic curve, shown in the figure, is sometimes referred to as the "Meldrum Curve" [24]. Meldrum was one of the first to publish high quality data on this effect. It has only been quantitatively understood in the past couple years [11]. Computational Fluid Dynamics has been used to demonstrate that shearing of oil droplets does not cause this effect. Above a critical Swirl Number, the fluid flow becomes unstable. When this happens, the oil core breaks up with much of it ending up in the effluent. Detailed understanding of this effect was delayed for many years due to the complexity of the fluid mechanics. Attempts to model the effect using Computational Fluid Dynamics (CFD) were hindered due to the fact that the most common turbulence closure models (e.g. Kolmogorov k-ε models) are not accurate at high Swirl Numbers. Another set of turbulence closure models (e.g. RANS) had to be used. These have only recently been developed to the point that they are practical [11].

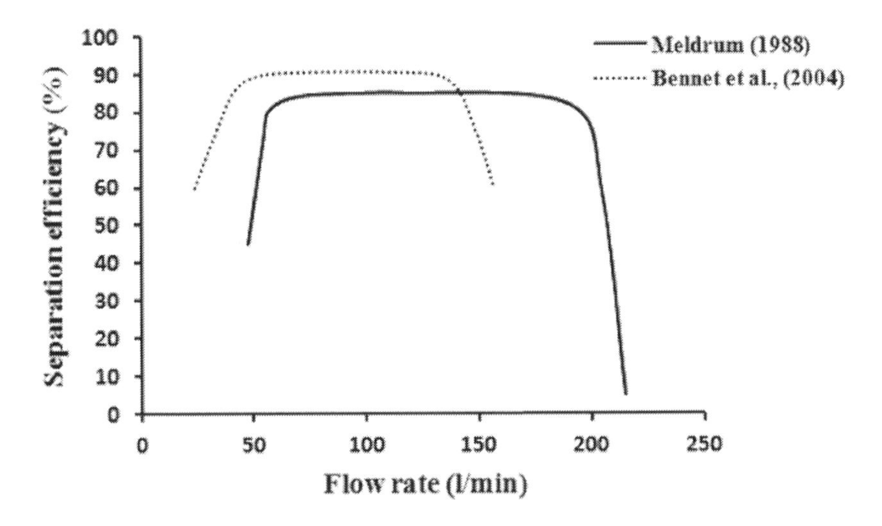

Figure 10.11 The separation efficiency as a function of the flow rate over a wide range of flow rates. The sharp decrease in separation is due to fluid turbulence which causes drops to break apart into much smaller drops which are not separated [3].

10.1.4 Design Differences and Characteristics

Perhaps the most illustrative way to describe the different hydrocyclone geometries that are now on the market is to start with the Colman-Thew geometry and discuss the major improvements that have been made. First, it is worthwhile to repeat what was said above about the original Colman-Thew model.

To a great extent, the Colman-Thew hydrocyclone design was practical and had good oil/water separation efficiency. It had a single round (or circular) inlet port. It had a short cylindrical section (involute), two conical sections, the first having a sharp taper, the second having a shallow taper, these were followed by a narrow cylindrical section. The oil reject port was a simple small-diameter hole without a vortex finder. These are the basic features of the design.

Liner Design: Each supplier has a variety of liner designs available. Among the many parameters that characterize the design are the involute diameter (also known as the major diameter), number, size, & design of inlet slots, reject orifice diameter, conical length and angle, and length of taper and parallel sections. Shown in the figure below are examples of several of the parameters required to specify a liner geometry.

The first commercial hydrocyclone involute section diameter was 60mm and had conical sections which were welded together. Higher performance, but lower capacity hydrocyclones with 35 to 40 mm involute section diameters were developed later and have become commonplace. The Table below gives parameters for some of the published designs.

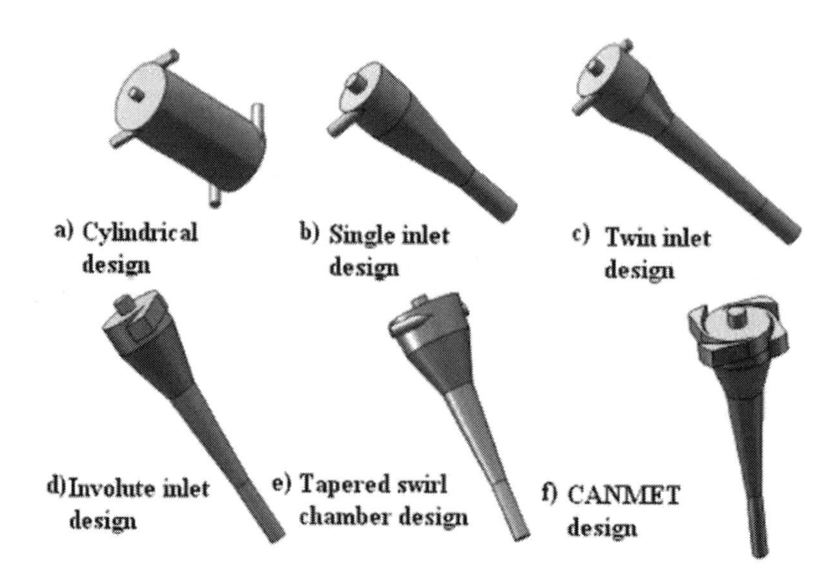

Figure 10.12 Various cyclone geometries that have evolved over the years.

Length of cyclone: Over time, empirical improvements to hydrocyclone design have been made. One of the important discoveries has to do with the length of the cylindrical section, or indeed the length of the overall device. As will be discussed, studies to optimize the length of the cylindrical section led to important improvements in the inlet nozzle and involute section. There are various theories in the industry regarding the function of the cylindrical section and the relationship between the length of that section and separation efficiency. One idea has it that the cylindrical section merely stabilizes the oil core so that the effect of fluctuations in flow and / or pressure are minimized. Other ideas are that the cylindrical section merely provides additional retention time for drops to migrate to the oil core. As discussed below, both of these ideas are correct, to varying degrees.

Initial work carried out at Southampton and elsewhere showed that increasing the length of the cylindrical section, beyond a certain length, had little effect on the separation efficiency. It was expected that the additional residence time afforded by a longer cylindrical core would allow smaller drops to migrate to the core and be captured in the oil core, and ultimately be rejected through the overflow reject port. Had this been the case, it would have increased the separation efficiency and reduced the drop size in the water effluent. But this did not happen. Extending the cylindrical section did not improve separation efficiency. When this was studied using clear plastic liners, it was revealed that the oil core disintegrated before the end of the cylindrical section. The oil core, having a reverse flow relative to the water flow, did not occur over the entire length of the device. Therefore extending the cylindrical section had no benefit on oil droplet separation per se. Without an oil core at the underflow end of the device meant that those oil drops that did migrate to the cylindrical center at the very end did not get captured in a reverse flow and were therefore not separated. Initial attempts to extend the oil core focused on reducing the reject pressure. This did have some limited benefit due to the increase reject flow rate but did not significantly impact the formation or length of the oily water core.

Upon closer study, it was discovered that the length of the oil core was closely related to energy dissipation due to turbulence. As easily imagined, fluid shear stress is high near the core and particularly at the oil core / water interface. After all, the oil and water are traveling in opposite directions. Efforts then focused on reducing turbulence in the inlet section. Actually, this was an obvious and intuitive goal from the very start. This continues to be the case today. To a certain extent though, the relationship between inlet turbulence and oil core length is sometimes overlooked or forgotten in current development efforts.

Fluid Inlet: The fluid enters the device tangentially at the cylindrical wall. This transforms the inlet forward momentum into the swirl velocity. The major improvement that came out of early work was in the design of the inlet slot(s) to the involute section of the hydrocyclone. Some manufacturers added a second or even a third inlet slot. The shape and design of the inlet is one of the areas where manufacturers claim competitive advantage. Thus, the inlet comprised of two or three symmetrically aligned inlet ports. The placement of the inlet slot in the axial direction was optimized in order to reduce shearing between the oil core and the inlet fluid. In the early designs, the shape of the inlet was circular. This was changed to rectangular. The mechanism by which this improves the design is difficult to describe in detail. Intuitively however, a rectangular shape reduces shearing at the edges of the injected fluid, so to speak. Finally, the cross sectional area of the inlet slot(s) was optimized. As described in Section 10.2 on Modeling and Theory, the size of the inlets establishes the inlet velocity of the fluid, which in turn impacts the swirl velocity. However, as discussed below, it must be recognized that the inlet velocity in the slot is significantly higher than the swirl velocity.

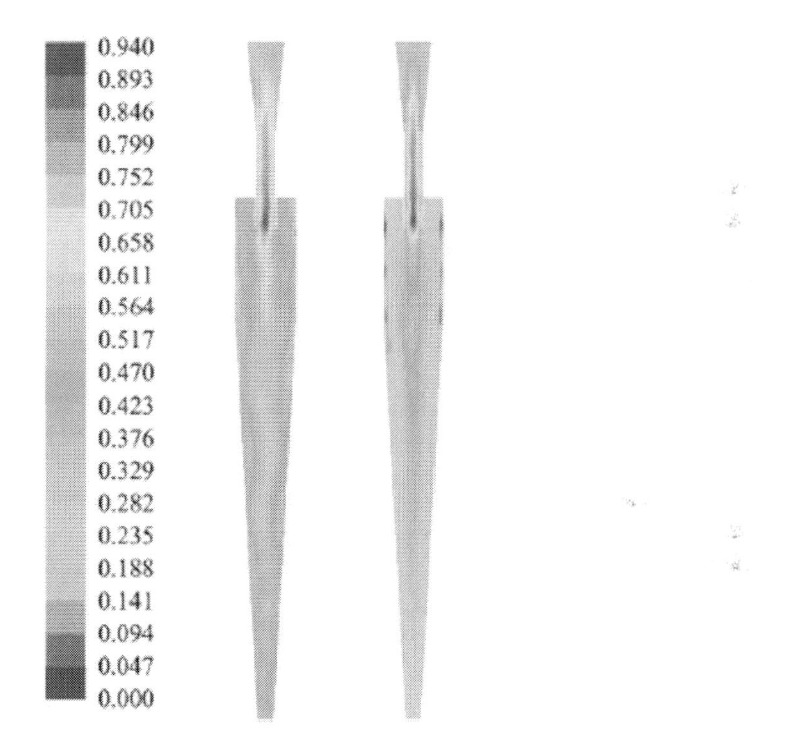

Figure 10.13 CFD result showing the axial velocity. The results on the left are for a cyclone with a single inlet. The right is for two inlets. Note the sinusoidal shape of the core in the cyclone on the left. This is due to the single inlet. The core on the right is much straighter. From [12]. These results are illustrative only since the application is water drops dispersed in diesel.

Table 10.1 Geometrical Dimensions of Models

D_{tap}	D_o	D_i	D_t	L_s	L_t	L	α°	β°	θ°	
0.5	≤ 0.07	0.175 (twin inlet)	0.25	1	10	~22.5	0	10	0.75	Colman and Thew (1983)
0.75	≤ 0.07	(Ds/12) x 0.5_s Single rectangular inlet	0.25	0.5	10	21	0	10	0.75	Colman and Thew (1983)
-	0.039	0.25	0.33	0-2	9	-	0	6	-	Young et al. (1994)
-	0.24	0.24	0.22	1	13.5	21	0	6	-	Wesson and Petty (1994), mini hydrocyclone D_s ≤ 10mm
0.75	-	Larger than the twin-inlet design	0.375	0.375	10	26-33	0	10	0.75	Thew (2000), single involute inlet
0.475	0.2 – 0.6	~0.15	~0.24	~0.36	~4.7	~13	<10	10	1	Belaidi and Thew (2003)

Involute Section: An inlet cylindrical section is required to reduce turbulence, reduce head loss, and to more gently introduce the swirling fluids into the cyclone. A variety of cyclone models have been introduced over the years. In general, the larger the involute diameter (D_H), the greater the capacity but the lower the centrifugal force.

Taper Sections: There are essentially two designs available for the tapered sections. The original Colman-Thew design has two conical tapers. The objective of the first taper is to accelerate the fluids to a high swirl velocity. Thus, the first taper is relatively short and has a relatively steep angle. The second taper is further down the cyclone and has a smaller conical angle. The objective of this taper is to maintain swirl intensity, in the presence of fluid friction and kinetic energy losses, by further accelerating the fluids.

Cylindrical (or Tailpipe) Section: There is quite a bit of discussion about the effect of a long tailpipe in a hydrocyclone. Most of the discussion has to do with the effect of the tailpipe to stabilize the flow of the oil core, or to help prevent the oil core flow from deteriorating with the flow rate fluctuates. These suggestions come from observations of see-through cyclones in operation with died fluids, and from analysis of CFD simulations. These suggestions are probably correct to some degree. However, the main function of the tailpipe section is to increase residence time and to provide separation. This is clearly demonstrated below in the section on modeling. It is shown there that the tailpipe section is responsible for a significant fraction of the separation efficiency for small drops.

10.1.5 The PDR: Pressure Differential Ratio:

The Pressure Differential Ratio is defined as the reverse pressure difference divided by the forward pressure difference. It is defined in the figure and text below. There are several reasons why this is an important variable.

As discussed previously, there are three pressures that are important to understanding the fluid flow in a hydrocyclone: feed, effluent (forward), and reject (reverse). The feed pressure is always the highest pressure. This pushes fluid through the hydrocyclone liner. The pressure of the fluid exiting the reject port is always the lowest pressure. This induces a reverse flow of oily water to the reject port. The pressure of the effluent is maintained at a value between the feed and the reject pressures. The PDR is the ratio of these pressure differences and it is one of the variables that is typically controlled in a hydrocyclone installation.

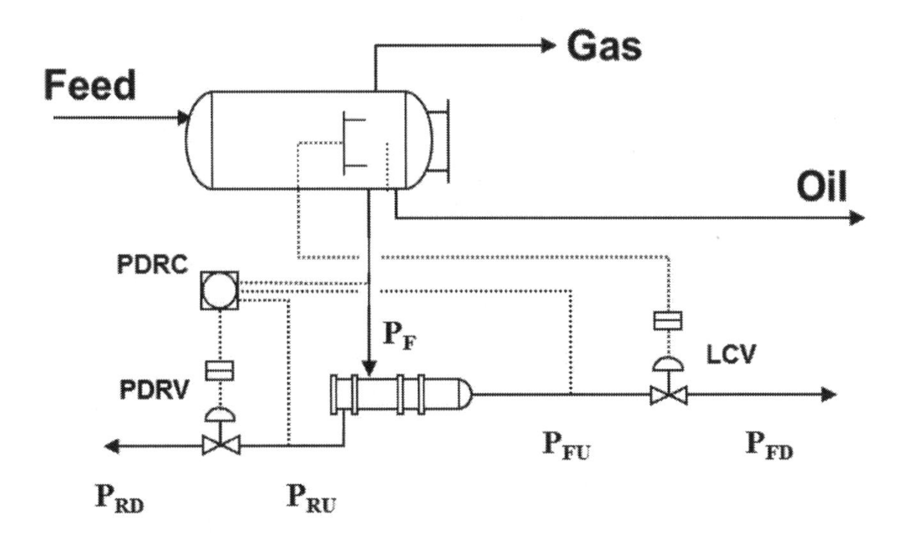

Figure 10.14 Process and Instrumentation Diagram for a typical hydrocyclone installation. A three phase separator is shown. Oily water discharge is routed to a hydrocyclone pressure vessel. The interface or level control valve (LCV) for the oil/water interface in the separator is shown downstream of the hydrocyclone, installed on the hydrocyclone effluent. A Pressure Differential Ratio Controller is shown, together with the relevant pressures. This figure is referred to extensively in the text.

Definition of Pressures:

PF	Feed pressure
PFU	Forward pressure upstream of LCV
PFD	Forward pressure downstream of LCV
PRU	Reverse pressure upstream of PDRV
PRD	Reverse pressure downstream of PDRV

PDR (Pressure Differential Ratio):

$$PDR = \frac{PF - PRU}{PF - PFU} \qquad \text{Eqn (10.1)}$$

PDR must be greater than 1 in order to force the oil in the core to go in the reverse direction of flow. For a given hydrocyclone geometry (reject port diameter, effluent port diameter, overall diameter and taper angle(s), length, etc) the PDR controls the amount of fluid that is rejected.

The figure below provides insight into the relationship between PDR, reject flow rate, and separation efficiency. The two diagrams shown in the figure characterize this relationship. The diagram on the right hand side is the relation between the reject ratio and the PDR. The reject ratio is the ratio of the reject flow rate to the feed flow rate. It is shown in the diagram as a percent. Typical values are in the range of a few percent. It has been mentioned already that the main purpose of controlling the PDR is to ensure adequate flow rate through the reject orifice. This figure shows the relationships for two particular hydrocyclone designs.

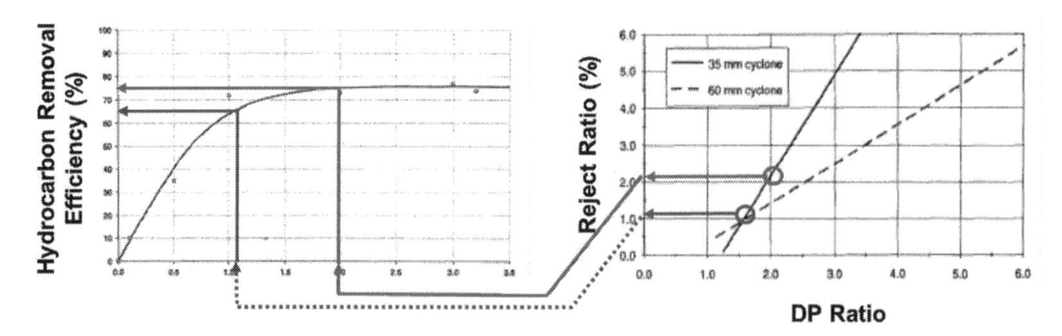

Figure 10.15 Two diagrams which together demonstrate the effect between DP Ratio (PDR) and Separation Efficiency. The two of these variables are tied together through the Reject Ratio. By setting the PDR, a certain Reject Ratio is established (right hand diagram). This Reject Ratio then determines, along with other variables, the Separation Efficiency (right hand diagram).

The diagram on the left hand side of the figure gives the relation between separation efficiency and the reject ratio. As with most water treatment equipment, the higher the reject rate, the better the separation efficiency. However, as might be expected, there is a region of diminishing returns where further increase in reject does not significantly increase separation. This might be expected on intuitive grounds and as shown in the diagram, this is indeed the case. Thus, there is no point to increase the reject flow rate beyond a few percent. In fact, it is detrimental to the performance of a hydrocyclone to raise the reject rate above 10 %.

Putting these two diagrams together results in an understanding of why PDR is typically set in the range of 1.6 to 2.0. Begin with the x-axis of the diagram on the right hand side with a PDR of about 1.6 for the 35 mm diameter hydrocyclone. This corresponds to a reject ratio of just about 1.2 percent which is read off of the y-axis. Now proceed to the x-axis of the diagram on the left hand side of the figure. This x-axis of the left hand diagram is the same as the y-axis of the right hand diagram. They are both reject ratio (in percent). A reject ratio of about 1.2 percent corresponds to a separation efficiency of about 65 %. Now consider a PDR of about 2, on the right hand diagram. The reject ratio is about 2.2 percent. Which corresponds to a separation efficiency of about 75 % on the left hand diagram. This example demonstrates the relation between PDR and separation efficiency. That relation is tied together by the effect that PDR has on reject ratio, which in turn has an effect on separation efficiency.

It must be stressed however that the diagrams used here and the values quoted are for only one particular model of hydrocyclone. These values differ, sometimes by a significant amount, for other hydrocyclone designs. As shown in the figure, note that the 35 mm hydrocyclone has higher reject ratio for a given DP Ratio. Although it is not indicated here, the 35 mm hydrocyclone has a higher ratio of reject orifice diameter to underflow diameter.

10.1.6 Material Balance – Reject Oil Concentration

Here the opportunity is taken to confirm some calculations on the hydrocyclone performance evaluation based specifically on Material Balance. Using the Material Balance equation in the hydrocyclone:

$$Q_F\, C_F = Q_P\, C_P + Q_R\, C_R$$

For an instance, assume one has

- $Q_F = 10\ m^3/hr$ with a feed OIW concentration (C_F) of 100 mg/L ;
- Reject ratio of 2% ($Q_R = 0.2\ m^3/hr$);
- Effluent OIW concentration (C_P) of 20 mg/L.

Using the volume balance equation $Q_F = Q_P + Q_R$, allows one to determine the effluent flow rate:

$$10\frac{m^3}{hr} = Q_P + 0.2\frac{m^3}{hr}$$

$$\boldsymbol{Q_P = 9.8\frac{m^3}{hr}}$$

To determine the OIW concentration at the reject, the Material Balance Equation can be used:

$$C_R = \frac{Q_F C_F - Q_P C_P}{Q_R}$$

$$C_R = \frac{10\frac{m^3}{hr}\left(100\frac{mg}{L}\right) - 9.8\frac{m^3}{hr}\left(20\frac{mg}{L}\right)}{0.2\frac{m^3}{hr}}$$

$$\boldsymbol{C_R = 4,020\ \frac{mg}{L}}$$

Thus the oil concentration in the reject flow is about 0.4%. Another way to consider this is that 80% or 80 mg/liter of the oil in the feed is concentrated into 2% of the water as a reject flow – a 50X concentration increase. Or expressed as a simple equation

- (80 mg/liter removed from the feed) x 50 = 4000 mg/liter in the reject.

10.1.7 Process Configuration and Control

By themselves, hydrocyclones are relatively simple. They involve no moving parts, do not require auxiliary utility streams (gas, electrical power), and do not generate a solid waste product. Nevertheless, proper installation of a hydrocyclone into an oil/water separation system is somewhat complicated. Considerable attention to details of pressure, flow rate, and process control are required. Those subjects are discussed in this section.

Correct feed flow rate & number of liners: The manufacturer will provide a liner curve like the one shown below. The curve gives the number of liners on the y-axis and the feed flow rate on the x-axis. This curve indicates the number of liners required for a given feed flow rate. As shown in the figure, as the number of liners in service increases, the spread or range in the allowable flow rate also goes up. This occurs as a result of the cumulative range of flow rates for each liner. The length of the horizontal bar for each number of liners covers the range of flow rates over which that number of liners is expected to provide acceptable oil removal performance.

There is a variety of designs that allow adjustment of the number of liners. The simple method is to open the head of the pressure vessel and manually insert or take out liners. Those liners that are removed must be replaced with a "blank." A blank is a device to fill the hole created by the liner that was removed. It ensures that that there is no flow between the feed section and the reject section. The number of liners installed, and not blanked-off, must be verified.

More sophisticated packaged units have internal compartments which have different numbers of liners. By compartmentalizing the overflow and underflow chambers of the vessel, the liners in each compartment can be placed into or taken out of service by opening or closing valves. The operator can adjust the number of compartments that are in operation without having to open the unit. Therefore he is able to adjust the number of liners to match the flow.

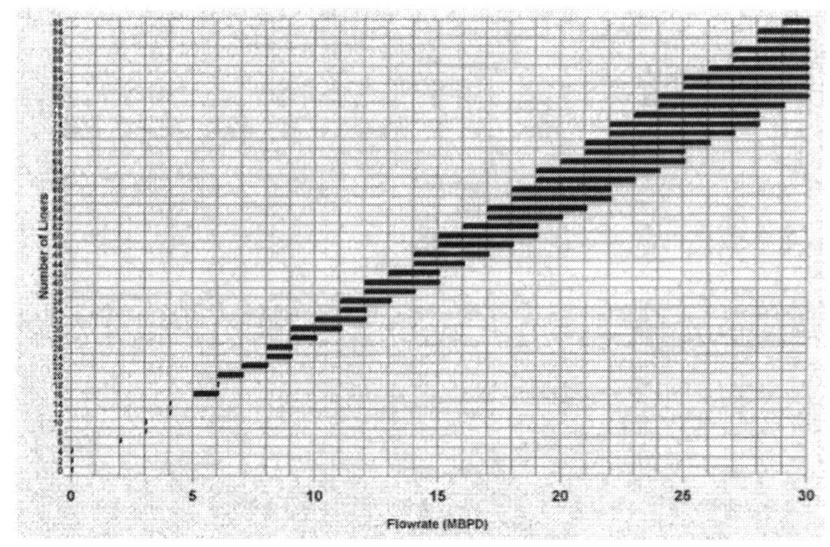

Figure 10.16 Typical capacity curve which relates the overall flow rate to the number of liners required for that flow rate. This curve is used to determine how many liners should be in operation for a given feed flow rate.

Two or more units are sometimes used in parallel in order to allow continuous adjustment of the number of liners. In a parallel installation of two units, the first unit is typically twice the capacity of the second unit. At low flow rates, the smaller unit will be used alone. The larger unit will be bypassed. At intermediate flow rates, the larger unit will be used alone, with the smaller unit in bypass. At the highest flow rates, both units will be used. This allows a reasonable variation of flow rates without requiring the opening of the pressure vessels to adjust the number of liners.

Where in a process stream should a hydrocyclone be installed? It is generally recommended that a hydrocyclone be installed on the oily water discharge of all three-phase separators. The hydrocyclone should be installed upstream of the Interface Level Control Valve for the separator. By placing the hydrocyclone upstream of the valve, enhanced oil/water separation can be achieved before the fluid experiences the shearing action of the valve.

As described below, hydrocyclones on GoM deepwater facilities are typically installed on the water discharge of the FWKO. Thus, the feed pressure to the hydrocyclone is the operating pressure of the FWKO, plus any hydrostatic pressure that may exist. Table 5 gives the type of hydrocyclone installed and the operating pressures for several GOM platforms.

Table 10.2 Hydrocyclone installation in Deepwater GoM

Platform	No. of units	Hydrocyclone Make	Liner Model	Upstream Pressure (psi)	Underflow Pressure (psi)	PDR
Bullwinkle	4	Baker Hughes	K	725	280	1.6
Auger	1	Kodiak	K	150	105	2.9
Mars	3	Kodiak	K	285	120	1.6
Ursa	3	Baker Hughes	K	384	190	1.9

Placing the hydrocyclone in this location has the following consequences. First, the available pressure is limited to the operating pressure of the FWKO vessel which is classified as low pressure (in the range of 250 psi MAWP). This can limit the PDR, which causes a limit in the percentage of fluid rejected. In general, a reject rate above 5 %, preferably 7 %, is required to achieve the necessary outlet water quality. Above 10 % is not practical from the standpoint that the volume of reject becomes too high for most deepwater process systems. There is also some field (anecdotal) evidence that withdrawing more than 10% of the feed from the reject orifice will disturb the internal hydraulic flow of the hydrocyclone liner and degrade its performance.

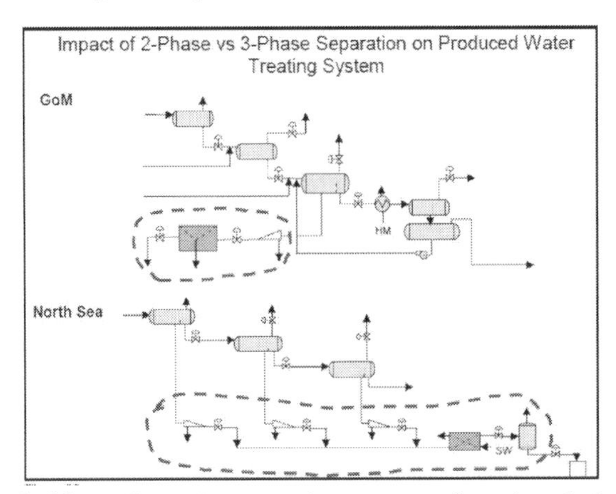

Figure 10.17 Typical offshore oil/water/gas separation process configurations. In deepwater GoM, high pressure and intermediate pressure inlet separators are two-phase. Oil and water are both discharged through a single nozzle hence there is no opportunity to operate a hydrocyclone. In the North Sea all stages of separation are three phase and are installed with hydrocyclones on the oily water discharge of the separator.

A degassing vessel of some kind must be incorporated downstream of a hydrocyclone unit to safely release gas arising from the pressure drop through the hydrocyclone. The degassing unit can be a flotation unit, provided that the gas venting system (including relief valve) are properly sized to handle gas breakout occurring from the pressure drop through the hydrocyclone. If a dedicated flash drum or degassing vessel is used, there can be a benefit from added oil / water separation via dissolved gas flotation. Thus, it is a best practice to install oil skim facilities in the flash drum.

How to determine the required pressures: The material below provides guidance to ensure that sufficient pressure is available for hydrocyclone operation. The pressures are defined below and are shown in the schematic figure 10.14 above. There are three important requirements:

Requirement 1:

Forward pressure drop for a given flow rate and number of liners must meet the manufacturers requirements in order to achieve good separation, where:

FPD = PF –PFU

Requirement 2:

The target PDR must be equal to or above the manufacturers requirement in order to achieve good separation, where:

PDR = (PF–PRU)/(PF –PFU)

Requirement 3:

Both the PFU and the PRU must be sufficiently high that fluids exiting the hydrocyclone can be pushed to their next destination. (Note: although this seems to be patently obvious, hydrocyclone units have been installed on more than one occasion with insufficient PRU to move the reject fluid to a receiving vessel.)

Where:

PF: Feed pressure to the hydrocyclone can be calculated as the separator operating pressure plus or minus any hydrostatic head between the liquid level in the separator and the feed to the hydrocyclone, minus any hydraulic losses between the top of the liquid and the hydrocyclone feed. The hydraulic losses are usually minor and related to the discharge nozzle and elbows or bends..

PFU: Forward pressure upstream of the level control valve.

PF: Feed pressure is given by the separator operating pressure and any hydrostatic head between the liquid level in the separator and the feed to the hydrocyclone

FPD = PF -PFU

Forward pressure drop. This is given by the flow rate, number of liners, and the liner flow curve (provided by the vendor).

PDR = (PF - PRU) / (PF - PFU)

Target PDR. This is typically set between 1.6 and 2.0

PRU Note that once the above three values are determined, there is only one variable left in the PDR equation (PRU). Thus, the PDR equation is used to calculate the required PRU.

There are two important pressures left to be determined:

PFD = PFU – DPLCV

Forward pressure downstream of the LCV (effluent stream). DPLCV is the pressure drop across the LCV. The full range of possible flow rates and pressure drops must be determined based upon the C_v curve for the valve. The water effluent stream must be routed to a location with a pressure less than or equal to the minimum PFD, as calculated above.

PRD = PRU – DPPDRV

Reverse pressure downstream of PDRV (reject stream). As shown in the figure, the PDRV is the PDR control valve. The PRD is the lowest pressure in the system. This stream must be routed to a location with a pressure less than or equal to the minimum PRD, as calculated above

The discussion above defines all of the relevant pressures, the relationships between one pressure and another, and the pressure requirements in a typical hydrocyclone system. These pressures start with the feed and end with the final reject pressure downstream of the PDR control valve. The most convenient sequence of calculations does not necessarily follow the order of presentation above.

To design a system, it is helpful to start with an understanding of the feed pressure. The feed pressure is the highest pressure in the hydrocyclone system. It is usually determined by gas compression requirements of the separator and is an external constraint for the hydrocyclone system. The next two pressures to consider are the two lowest pressures which are the reject stream, downstream of the control valve, and the effluent stream, downstream of its control valve. With these endpoints defined (feed and discharge), the available operating pressure range for the hydrocyclone is determined. The available pressure is then compared with the required pressure as specified by the manufacturer. If the hydrocyclone pressure requirement exceeds the available pressure range then the hydrocyclone cannot fit into the system unless a modification can be made.

Various modifications can be considered. These include adding a pump to boost the feed pressure, or finding alternative lower pressure destinations to which the reject and effluent streams can be discharged. The use of a booster pump is discussed below.

Process Control: Successful control of the hydrocyclone is critical to achieve good oil and water separation. In a typical installation, there are two controllers. The first controls the LCVC (Level Control Valve Controller). It controls the oil / water interface level in the separator by opening and closing the LCV (level control valve downstream of the hydrocyclone effluent discharge). The second controller is the PDRC (Pressure Differential Ratio Controller).. Control of the level or the interface level in the upstream vessel is the primary control variable in the hydrocyclone system. The LCV must always be allowed to throttle open or throttle closed in order to maintain the liquid or interface level at a desired set point or, more preferably, within a desired range. The PDRC then responds to control the PDRV so that the Pressure Differential Ratio (PDR) for the hydrocyclone is maintained.

Based on the above control requirements, the Pressure Differential Ratio Controller (PDRC) needs to receive input from three pressure sources. The PDRC's only function is to maintain the Pressure Differential Ratio (PDR) of the hydrocyclone by opening and closing the PDRV (Pressure Differential Ratio Valve).

An aside note is appropriate here. For a hydrocyclone system to function well, changes to the flow rate through the unit must be "relatively" slow. Since few, in any, upstream vessels receive flow at a constant rate, allowing the liquid or interface level in the upstream vessel to vary over a programmed range will help the system to absorb slugs while dampening changes to the LCV position over short periods of time. This in turn helps to maintain a more constant (but not absolutely) constant flow through the hydrocyclone itself.

The figure below shows time trends of the pressures and forward flow rate for a hydrocyclone installation where the PDR is controlled to a pre-set value. The Feed Flow Rate (Q) was manually increased by manually opening the LCV. As shown in the figure, by opening this valve, the effluent flow rate increased. The underflow pressure also dropped as a result of opening this valve. The control system responded by opening the overflow (reject) control valve. This decreased the overflow pressure, then maintaining the PDR at the set point. This resulted in more flow to the reject, but at constant PDR so the reject percent would have stayed constant.

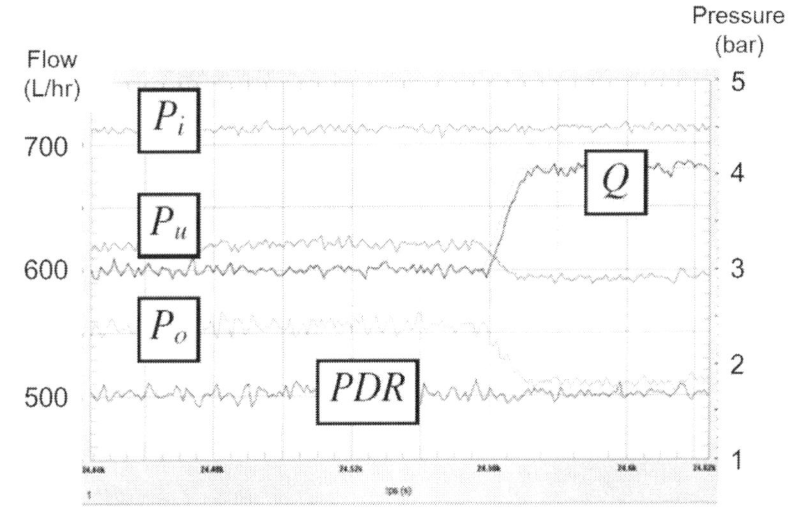

Figure 10.18 This figure shows the result of a change in the underflow pressure. The underflow control valve was manually opened. This is evidence by the drop in the underflow pressure (Pu). The caused the forward flow rate (Q) to increase. The control system responded by opening the overflow control vale (Po) in order to maintain a constant PDR. The feed pressure (Pi) was held constant [13].

Process control of hydrocyclones is one of the most misunderstood subjects related to hydrocyclones. The Pressure Differential Ratio (PDR) is an important, though often misunderstood variable. Often in discussions of hydrocyclone operation, the question arises, what is the optimum PDR? As will be discussed below, this is really the wrong question. The proper question is, what is the optimum reject ratio? The reject ratio has a significant impact on the performance of the device. The PDR has an impact only in so far that it influences the reject ratio. There are other influences that will also be discussed.

There is far too much emphasis placed on control of the PDR, and not nearly enough thought to provision and control of the forward pressure drop, and the reject flow rate.

It is the forward pressure drop that provides the driving force for separation. Forward pressure drop determines the inlet fluid velocity. That inlet velocity is converted to swirl motion by the geometry of the inlet and involute section. The swirl motion provides the centrifugal force which causes oil drops to migrate to the axial center of the device. As discussed previously, if drops reach the axial center before the bulk of the fluid exits the cyclone, then the oil will eventually be discharged out of the reject port. Altogether then, it is the forward flow that drives oil drops to the reject. Ensuring adequate forward pressure drop is a critical detail in process configuration.

While PDR and reject flow rate are related, the control of PDR by no means ensures a specified reject flow rate. PDR is literally a ratio of pressures. It is not a ratio of flow rates, which is really what is needed. Thus, it is recommended that in addition to control of PDR, the actual reject flow rate should be measured and the reject ratio should be calculated. However, measuring this flow rate may not be straightforward due to dissolved gas breakout which is always present in the reject stream and due to the presence of variable oil and fine solids content in the reject fluid. The use of a non-invasive flow measurement technology would be preferred but instantaneous rate measurements are likely to be misleading due to the presence of gas slugs.

Pumped hydrocyclone systems: The use of a booster pump has been shown to provide good results provided that the pump is carefully selected to minimize shear of oil droplets. In general, pumps generate shear which reduces the oil droplet size. However, low shear positive displacement or non-pulsing progressing cavity pumps, and moderate shear centrifugal pumps are available for this service. The selection of pumps for water treatment service is discussed in Section 3.5.3 (Shearing Through a Pump). Schubert [14], and Flanigan and co-workers [15, 16] show that centrifugal pump efficiency should be maximized in centrifugal pump selection in order to minimize oil droplet shearing. Flanigan further shows the advantages of using a positive displacement pump in hydrocyclone operation [15, 16]. This subject has been discussed in some detail in Section 3.5.3.

10.1.8 Representative Performance Data

SPE 48992 (Khatib, 1998) [17] gives pilot data, field trial data and the initial performance experience in operation for hydrocyclones in the deep water Gulf of Mexico. She gives performance data for the Krebs 2-inch, and Vortoil K-Liner.

Experience for hydrocyclones use in the deep water Gulf of Mexico for the last ten years or so is given in [18]. Experience with the K-Liner and other models is included. The paper also includes a discussion of process configuration (hydrocyclone placed downstream of FWKO water discharge with the FWKO LCV placed downstream of the hydrocyclone), and best practices for design and operation (quick opening davit – very important, degasser downstream, automated back flush, etc.).

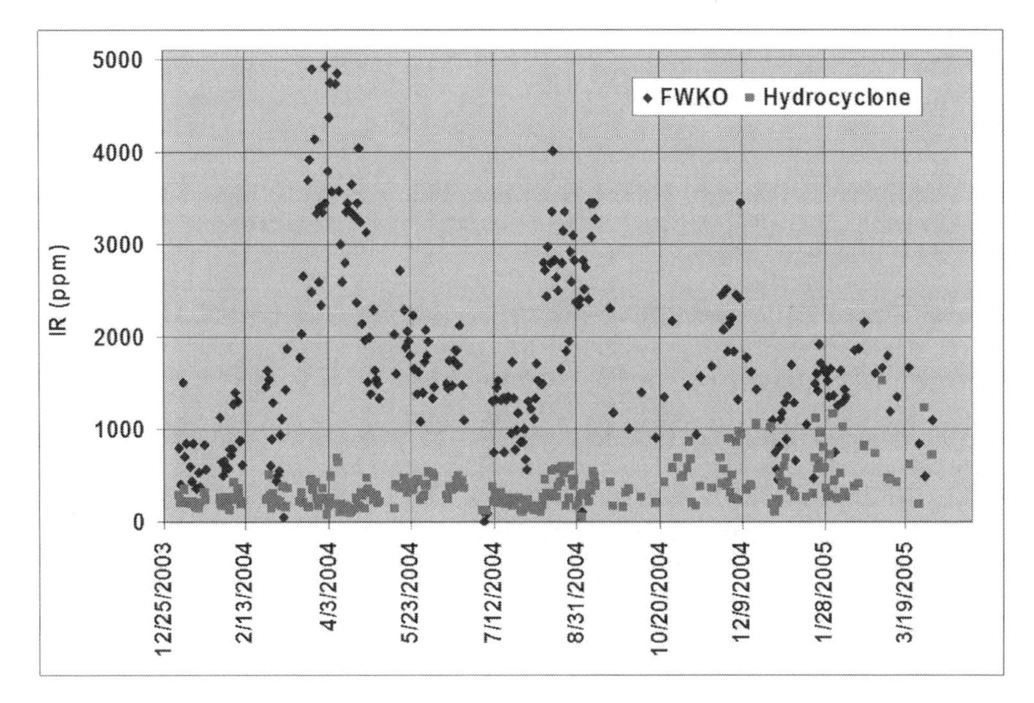

Figure 10.19 Field data for hydrocyclone performance on the Auger platform in 2004 [18]. Oil concentration as measured using an infrared device are shown for the FWKO discharge (diamonds, hydrocyclone feed) and the hydrocyclone discharge (squares). Note that oil concentration at the FWKO discharge (diamonds) attains very high values in some cases. The primary benefit of the hydrocyclone is to eliminate these very high values. This is a highly beneficial aspect of this hydrocyclone. The gradual deterioration of hydrocyclone performance (squares) that started in October 2004 was due to a reduced backwash frequency and lower flow rate without adjustment of the number of liners.

Pertinent observations based on the data in figure 10.19 include the following:

- Initial water handling experience in 2nd half 2003 was less than ideal
- Scaling & plugging of reject orifices prevented effective hydrocyclone utilization
- Starting early 2004 – performance began to improve
- Chemical & Chem Eng improvements in water treating were implemented:
    - identified scale formers & implemented a solids treating strategy
    - eliminated sludge-forming process chemicals
    - implemented improved emulsion breaker chemistry targeted at oil-wet solids

Despite the improvements, Hydrocyclone maintenance was still problematic due to a lack of operability features. Recommended features include

- automated backwash
- Quick-Open enclosure on the hydrocyclone vessel
- spare liners for quick change-out & offline cleaning

Roughly 70 to 90% separation efficiency can be expected from a properly sized, operated and maintained hydrocyclone system over a wide range of feed OiW concentrations. Note that under steady flow and pressure conditions, efficiency often improves with higher OiW feed. Most likely this occurs because the higher oil concentrations are contained in larger oil droplets which are readily removed by the hydrocyclone.

It should be noted that in most GoM deepwater installations the hydrocyclone is not intended to produce overboard discharge quality water. Instead, the hydrocyclone is intended to shave the peaks in oil content and to provide an overall pre-conditioning of water prior to a downstream flotation unit. This can been seen from the data shown in figure 10.19 for the Auger platform. Note the high concentrations coming into the hydrocyclone and the consistency with which the unit reduces the oil concentration. This type of service (peak shaving) is typically accomplished with a relatively larger reject orifice and a relatively high reject ratio. Final polishing of the oil content down to overboard discharge levels is then provided by the flotation unit.

Degassing Vessel Downstream of Hydrocyclones: Some experience with degassing vessels is given in Table 6. A degassing vessel can be incorporated downstream of hydrocyclones to safely release gas that breaks out of the oil / condensate due to the pressure drop through the hydrocyclone and level control valve. Such a vessel is designed with a pressure control system and a pressure safety valve. This control / safety system improves the process integrity of the flotation unit which typically operates at a somewhat lower pressure than the upstream separator. Experience has shown that the Degassing Vessel also functions as a flotation unit due to the effect of gas breakout. Such benefit must however be determined by design since the gas/water ratio of the gas that breaks out will depend on the operating pressure, the volatility of the hydrocarbon (condensate has a large effect). Also, the bubbles will be small since they evolve from dissolved gas. Small bubbles have limited upward velocity and have a tendency to get swept into the water discharge along with any oil that they have captured.

Table 10.3 Hydrocyclone performance experience from various Shell facilities [18].

Location	Hydrocyclone Inlet (ppm)	Hydrocyclone Outlet (ppm)	Degasser Outlet (ppm)	Hydrocyclone System
N. Sea	400	17		35 mm liners No degasser
N. Sea	600	25		60 mm liners No degasser
N. Sea	400	45	30	35 mm liners
US	520	39	20	K-Liner Pumped Feed
US	2310	160	6	35 mm liners Pumped Feed
US - Auger	1500	200	100	K-Liner FWKO Feed
Dubai	327	88	39	35 mm liners
N. Sea	102	30	17	G-Liner Production system

10.1.9 The Effect of Gas Breakout with the Hydrocyclone

There have been a number of studies of the effect of gas breakout [19 – 22]. There are two effects to consider. The first is a steady state effect. The second (discussed below) is the transient effect. Under steady operating conditions, gas breakout will typically occur to some extent due to the pressure drop imposed on the fluids as they traverse through the hydrocyclone. Live hydrocarbon fluids naturally evolve gas when subjected to a pressure drop. This is one of the main differences between field performance of a hydrocyclone and the performance of a hydrocyclone in a lab or test unit. In some laboratory testing, gas is deliberately added to the feed fluids to mimic the effect of gas breakout of live fluids [20]. But, for the most part, laboratory testing is carried out with dead fluids containing no gas.

When gas bubbles form within a hydrocyclone, they migrate to the oil core at a faster rate than that of the oil drops. In doing so, they have a tendency to collide with oil drops and carry them to the oil core. For limited gas volumes, gas breakout may actually improve oil/water separation by causing a gas flotation effect.

As the volume of gas increases, at some point the gas will form a continuous core at the axial center of the hydrocyclone. This gas core will exit the oil reject port. When this happens, the oil core forms around the outside of the gas core. This can have a detrimental effect since it may have a tendency to choke the oil reject port thus limiting the actual liquid reject flow rate. For example, if a hydrocyclone is designed to reject say 2 % of the volumetric liquid feed flow and gas comprises half of this volumetric flow, then the oil reject flow rate will be only 1 %, only 50 % of the design intention. This was studied and quantified by Khatib [17]. The results are given in the figure below. As shown, there can be a substantial detrimental effect of gas on the separation efficiency.

This choking effect must be considered in the overall design of the hydrocyclone (flow rates, pressure drop) and in particular in the design of the reject port diameter. The gas choking effect can be overcome to some extent by increasing the diameter of the reject port in order to accommodate the presence of the gas. Smyth and Thew [21] recommended a reject orifice diameter of 3 mm in the Colman-Thew hydrocyclone design in order to overcome the presence of gas, up to 40 Sv% of the feed flow, where Sv % is the percentage of gas under standard conditions as a percent of the total volumetric flow. A feed of 5 bar, with 40 Sv%, would be the same as 8 Av% (actual gas volume percent). But even when this is done, there is a second detrimental effect.

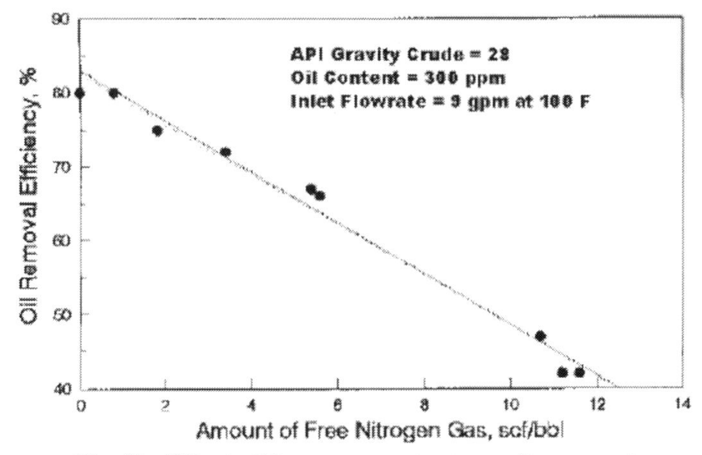

Figure 10.20 The effect of free gas content (scf/bbl) on oil removal [17]

The second effect to consider in relation to gas breakout is that it will cause flow instabilities. As discussed, when the gas volume is high, the gas will form a continuous stream in the core. Determining what is meant by "high gas volume" is discussed briefly below. Even if the liquid flow rate and the various controlling pressures are maintained fairly steady, the gas/liquid interface may be highly unstable. Maintaining a steady reverse flow oil core is critical to effective oil separation. The presence of gas, and the instability of the gas/liquid interface, significantly disrupts this flow. This will lead to instabilities in the radial location and flow rate of the core. In the typical case, in which flow and pressure have a tendency to fluctuate, the presence of gas will amplify these fluctuations and cause instability in the oil core which will have a significant negative impact on performance. This effect is common in the field. Only limited pioneering laboratory work has been conducted.

In carefully controlled laboratory experiments, Belaidi and Thew [20] found that separation efficiency for a conventional hydrocyclone design is a weak function of the gas volumetric flow rate. In that paper, Kai is the percentage of the air volume flow rate at hydrocyclone inlet pressure to the total flow rate at cyclone inlet pressure i.e air and liquid [23]. At 40 Sv% (gas volume under IUPAC standard conditions percentage of flow) there was a decrease in separation efficiency of only a few percentage points (from say 90 % to 87 % separation efficiency). While this is a relatively small effect, it does not simulate actual field conditions where pressure and flow rate are continuously fluctuating.

In general, field experience shows that when gas volumes reach roughly 20 Sv% (volume percent of inlet flow based on a correction to Standard conditions, the effect of gas is detrimental on oil/water separation efficiency [19]. For a hydrocyclone operating at a feed pressure of say 5 bar, the critical value of 20 Sv% is equivalent to 4 Av% (based on actual gas volume percentage of total volumetric flow).

Gas composition should also be considered. Carbon dioxide is about 20 times more soluble in water than is methan. Thus if the produced gas composition includes carbon dioxide, substantially more gas evolution can be expected. Similar considerations apply to gas which includes H_2S. The second consideration is temperature. As the water temperature increases, the solubility of all gases decreases. Thus gas evolution can be expected to be much less for hotter waters than for cooler waters.

10.1.10 Operation and Maintenance

Solids can deposit and accumulate on the internal surfaces of a cyclone. Fluid velocities tend to be relatively high inside of a cyclone but there is a wide range of solids that can deposit (FeS, mineral scale, asphaltenes, oil wet sand, wax, etc). Solids can accumulate in the cylindrical and conical sections, but especially in the reject port.

Asphaltene and paraffin stick to solids particles (mineral precipitates, sand, iron solids, etc) forming a conglomerate material. The best response is to use a solvent soak and flush with an asphaltene solvent to break the matrix and dislodge the material from the metal wall.

Scaling of reject nozzles and blockage by debris: The reject flow rate is a direct function of the diameter of the reject nozzle and the pressure drop across the nozzle. In other words, a larger reject nozzle diameter or a higher pressure drop will result in a higher reject flow rate. These are the two fundamental parameters that determine the reject flow rate.

The use of PDR to control the reject flow rate is only a convenience. The relation between PDR and reject flow rate was discussed already in Section 10.1.5 (The PDR: Pressure Differential Ratio) and in figures 10.14 and 10.15. The PDR is easy to measure and is easy to control. However, it is not a direct or fundamental measure of reject flow rate. If the reject orifice becomes partially blocked with scale, then the effective reject diameter is smaller and hence the reject flow rate will be smaller for a given pressure drop. The PDR is usually set to 1.6 or greater to ensure that the reject flow rate is sufficient to ensure good separation efficiency. However, if there is deposition in the orifice which restricts the reject orifice diameter, then a higher value of PDR would be required until the hydrocyclone can be chemically cleaned (clean in place), or until the hydrocyclone pressure vessel is opened and the liners and pressure vessel compartments can be manually cleaned. The only way to determine if a higher PDR is required is to carry out a test on-site. The procedure for this test is given below.

Figure 10.21 Photograph of a hydrocyclone unit with the head plate taken off showing an accumulation of sludge and tar. The individual cyclone liners (shown on the deck) were manually cleaned with aromatic solvent to move sludge from the reject ports.

Figure 10.21 is a photograph of a hydrocyclone unit with the head plate removed and an accumulation of sludge and tar. The following table gives a composition of the tar-like material that accumulated. To maintain performance, a hydrocyclone unit must be back-flushed on a frequent basis and may occasionally need to be cleaned with either steam or a chemical solvent.

Table 10.4

Solvent	Percentage of Weight Loss (%)	Spot Tests
Deionized Water Wash[1]	12.5	
Xylene Wash[2]	6.1	
Acetic Acid Wash[3]	0.1	Iron Carbonate – Negative
Hydrochloric Acid Wash[4]	42.3	Iron Sulfide – Positive
Acid Insolubles[5]	39.1	

[1] Includes substances soluable in water such as salts

[2] Includes substances soluble in xylene such as paraffin, oil and organics

[3] Includes substances soluble in dilute acetic acid such as carbonate scale

[4] Includes substances soluable in 15% HCl acid such as iron sulfide or iron oxide

[5] Includes substances insoluble in 15% HCl acid such as sulfate scale, sand and silicates

During a back-flush operation, produced water is diverted to the reject chamber of the hydrocyclone vessel so that the water flows backwards across the reject orifices in order to remove any non-adhering deposits. The appropriate frequency of back-flush differs from one installation to another. Typically, back-flush frequencies vary from one or more times per day to once every few or several days. It is best to determine the appropriate back-flush frequency by observation of hydrocyclone performance. It is recommended to always automate the back-flush operation such that operators do not have to turn the valves manually. This will ensure that the back-flush actual does take place regularly. The time for the backflush is short with manufacturer recommendations varying from a few seconds (10 – 15 seconds) to a few minutes (2 – 3 minutes).

Fouling causes the liners to build up an inner coating or scale. It also causes the reject orifice to become partially or completely plugged. The detection of fouling is based on the fluid flow equation for a liquid packed system:

$$\frac{\Delta P}{L} = \frac{\rho}{k}\frac{Q^2}{D^5} \qquad \text{Eqn (10.2)}$$

in which Q is the volumetric flow rate (gal/min, m^3/hr, bbl/day), D is the pipe internal diameter, and ΔP is the pressure drop per unit length. By rearranging this equation we obtain the following relationship for the cross-sectional area:

$$A \sim \left(Q^2 / \Delta P\right)^{2/5} \qquad \text{Eqn (10.3)}$$

in which A is the effective (actual) cross-sectional area available for flow in the presence of scaling or build-up of fouling material. By using the appropriate volumetric flow rate and pressure drop (right hand side of equation), this equation can be applied to detect any deterioration in flow.

The physical meaning behind this equation is based on the fact that deposition of scale and solids reduces the cross sectional area. As shown in the figure below, a plot of the relative areas shows several downward trends. Each of these downward trends follows immediately after a chemical wash. During the chemical wash, the hydrocyclone is injected with chemical solvent during a back-flush operation. After a few of these washes, the slope of the lines goes down. In other words, the fouling rate decreases after a few chemical washes. This is due to a gradual cleaning of the internal surface which reduces the rate of deposition. Thus, it is easier to keep the unit clean than it is to clean it after it has been fouled.

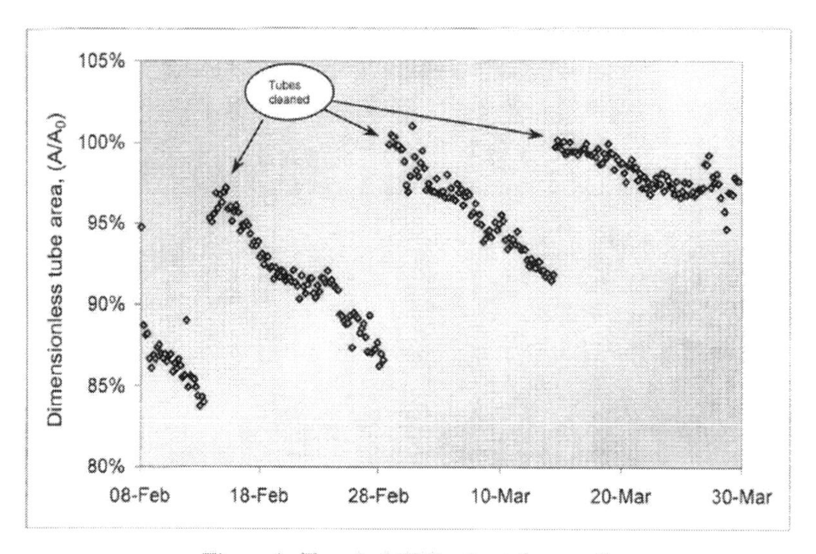

Figure 10.22 Dimensionless tube area as a function of time. This figure demonstrates the effect of fouling (downward sloping lines), and the effect of periodic chemical wash (periodic reestablishment of 100% flow area).

Quick Opening Davits: In order to change the number of liners in operation, the pressure vessel must be opened. In the conventional pressure vessel design this requires that the bolts on the pressure head be opened, and subsequently closed once the number of liners has been adjusted. The process is simplified if Quick-Opening closures are installed on the vessels containing the liners.

Figure 10.23 Skid-mounted compact pressure vessel packaging of hydrocyclone liners. Note the quick-opening davits instead of bolted heads. This design makes it much easier to open and change the number of liners in operation.

Operating Practices: As shown in figure 10.19, Auger hydrocyclone performance was very good from at least December 2003 through October 2004. During this time period, it was noted that operating procedures for the hydrocyclone were being followed and the operators and engineers both took ownership of the unit. This included daily backwashing, routine cleaning of the liners and adjustment of the number of active versus blank liners, and monitoring and adjustment of the available pressures in order to maintain a PDR above 1.6. Also, the chemical vendor had an effective treatment program in place to prevent scale deposition throughout the facility and in particular across the hydrocyclones. The trend in figure 5 shows that performance was good during this time. However, starting in roughly October 2004, maintenance procedures were relaxed for a number of reasons and hydrocyclone performance began to deteriorate. From this it was learned that the hydrocyclones on Auger had a significant benefit on water quality but that operability would need to be improved in order to ensure proper operation within an environment where there are numerous competing interests for operator attention.

Based on operations at a number of offshore and onshore locations, several Best Practices for Hydrocyclones can be listed. These include:

- Use hydrocyclone for removal of oil drops greater than 10 micron
- Install a flow meter on the feed or the underflow and provide monthly surveillance of flow volumes and adjust the number of liners to match the flow requirements. For this purpose, a quick open enclosure and readily available spare liners are required.
- Install a flow meter on the reject. Do not rely entirely on the PDR to establish adequate reject flow since PDR assumes that there is no fouling – often this is a poor assumption.
- Establish a minimum flow. This can be achieved by use of recycle, if necessary.

- Avoid flow fluctuations. This can be achieved by feeding the hydrocyclone with a low shear pump or allowing liquid or interface levels to vary as they absorb volumetric slugs of feed liquids.
- Minimize solids to hydrocyclones
- Minimize gas since it will interfere with oil migration to the core. This includes not only dissolved gas breakout, but free gas carry under from the upstream vessel
- Clean regularly
- Back flush regularly
- Achieve steady flow which ensures that the reject core is well developed
- Increase reject port size to prevent blocking with solids
- Any new installation should have the following operability features:
    - Automated backwash
    - Quick-open enclosure
    - Spare liners readily available for quick change-out and offline cleaning

10.1.11 Troubleshooting Procedure for Hydrocyclones

This procedure is intended to provide a performance assessment for hydrocyclones. The assessment can be carried out for the purpose of:

1. establishing a benchmark for future performance comparison,
2. troubleshooting to improve performance,
3. monitoring to sustain performance.

For each of these purposes, the sampling, laboratory analysis and process data gathering is essentially the same. The only difference is the number and frequency of samples and tests.

Several issues are involved in the assessment of hydrocyclone performance. Each of these issues must be handled properly or the assessment will be incomplete and have little value for sustaining performance or for future troubleshooting.

It must be emphasized that troubleshooting a poorly performing hydrocyclone is very difficult without the baseline data that this procedure provides. It may seem unnecessary to take such baseline data when the system is working well, but without such data, troubleshooting later will be difficult.

In the next section, each of the issues is discussed. Sufficient background is given for each of the issues to allow customization of the performance assessment, depending on the details of the installation. The final section gives a recommended procedure for performance assessment.

Issues involved in hydrocyclone performance: In order to assess the performance of a hydrocyclone, several issues must be addressed:

1. Are the oil / water properties suitable?
2. What is the efficiency of oil/water separation, as a function of the operating parameters, and does the efficiency match or exceed the expected performance?
3. Are the correct number of liners installed for the given feed flow rate?
4. Are the liners fouled or are reject orifices partially blocked?
5. Does the hydrocyclone have the optimal PDR and reject ratio?
6. Are the feed flow rates and pressures steady, and are the back-pressures (underflow and overflow) steady?

Each of these issues is discussed presently below.

Fluid Properties: For hydrocyclone performance, the most important fluid properties are:

a. Oil drop size,
b. temperature (water viscosity)
c. density difference between dispersed phase and water,
d. oil droplet coalescence characteristics

The simplest way to assess all three properties at the same time is to carry out a desk top settling test. Desk top settling is always recommended for water treating projects. To a trained eye, desktop settling will reveal water treating problems without much further analysis.

Take a sample and record the time. Set it on a desk or other horizontal surface such that it will not need to be moved for a day. Record how much (vol %) oil separates at the following time intervals: 1, 5, 10, 30, 60, 120 minutes, followed by 6 hours, 12 hours, 24 hours. Also record how much material sinks to the bottom of the jar. Record the presence of small black particles (iron oxide or asphaltenes). If possible, take a picture of the sample at the indicated time intervals.

In order for a hydrocyclone to be effective, the sample must be clear within 12 hours. In other words, there should be no difference between the sample at 12 hours, and that at 24 hours and both should be clear. The sample must be 50 % clear in 6 hours.

If this is not the case, then the sample should be submitted for detailed analysis. The density of the oil and water in the sample should be analyzed and the solids content of the oil assessed. The presence of production chemicals such as corrosion inhibitor should be assessed. The dosage and chemical selection of water clarifier upstream of the hydrocyclone should be assessed.

If drop size analysis is available, then this can be a useful to tool to supplement the desktop analysis. However, drop size analysis is highly dependent on the proper use of the equipment, and on the availability of proper sampling connections.

Efficiency, Operating Parameters & Expected Performance: The oil in water removal efficiency for a hydrocyclone was defined previously. Whenever efficiency is measured, the operating parameters must be measured as well. Without the operating parameters, the measurement of efficiency has no meaning. The operating parameters include:

1. All flow rates: feed, product, reject
2. All pressures: feed, underflow, reject
3. Reject oil-in-water concentration
4. Temperature
5. Density of fluids (oil and water)

With these measurements, two material balances can be made. The feed flow rate must equal the reject plus the effluent. The volume of oil into the hydrocyclone must equal the volume of oil out. The volume of oil is calculated as the product of the flow rate times the oil in water concentration. The equations for material balance were given previously in Section 10.1.6 (Material Balance – Reject Oil Concentration). If drop size analysis is available, then this can be used to help determine if the measured efficiency is to be expected on the basis of actual drop sizes.

Correct feed flow rate & number of liners: The manufacturer can provide a liner curve. The curve gives the number of liners on the y-axis and the feed flow rate on the x-axis. This curve indicates the number of liners required for a given feed flow rate. The number of liners installed, and not blanked-off, must be verified.

If possible, the feed flow rate should be varied and the oil/water separation efficiency of the hydrocyclone assessed. Variation of feed flow rate usually requires manual adjustment of one or more bypass valves. This will help establish a performance curve (efficiency vs feed flow rate) and if indeed the correct number of liners are installed.

Fouling: Fouling has already been discussed. It is one of the most common causes of hydrocyclone performance problems. It is also one of the most difficult to diagnose unless proper monitoring has been implemented. Said another way, the detection of fouling is the greatest justification for a robust monitoring program. With a proper monitoring program, fouling can be detected.

PDR & Reject Ratio: The PDR and the reject ratio have been discussed extensively above. In any hydrocyclone installation, effort should be made to adjust the PDR to an optimal and appropriate value, without sacrificing the forward pressure drop. The higher the PDR, the greater the separation efficiency, up to a point. Beyond a certain value, there are diminishing returns in any increase in PDR. Also to keep in mind, a higher reject flow rate means that the equipment receiving the reject flow will be required to handle greater volume of flow.

> In assessing hydrocyclone performance, the PDR should be adjusted and the separation efficiency of the unit should be assessed as a function of PDR. If the performance of the hydrocyclone increases significantly for a modest increase in PDR, then consideration should be given to further adjustment, finding the optimal value, and making a permanent increase in the PDR set point.
>
> If however, a hydrocyclone is not generating expected performance, and an increase in PDR does not improve performance, then it is possible that the reject ports are partially blocked. To determine if this is the case, the unit may need to be opened and reject orifices inspected. Since this is a time consuming operation, there are a number of other steps that can be considered first such as an overall material balance or a solvent soak.

Steady Flow Rates: Rapid fluctuations in any of the pressures or flow rates associated with a hydrocyclone will cause significant performance problems. Whenever the pressure or flow rate changes rapidly, the vortex of oil in the center of the hydrocyclone breaks down. This will cause oil to be discharged to the effluent.

To determine if this is a problem, the pressure and flow rate trends must be analyzed. Flow rate and pressure trends will always show some degree of fluctuation. However, sometimes it is difficult to know with certainty how much fluctuation is acceptable. For this reason, a field test is recommended. There are two pertinent field tests.

The first and simplest field test is to open the effluent (product) sample tap and watch the fluid stream carefully. If vortex breakdown is occurring, you should see an occasional jet of oil in the sample stream. If this occurs to any degree, then pressure or flow variation is too great and the control systems which influence the hydrocyclone performance must be adjusted to eliminate this effect. This may include a change to the control philosophy for the liquid and/or oil/water interface levels in upstream vessels.

The second field test involves running the hydrocyclone in manual over-ride and manually setting the pressures and flow rates to constant values for a short length of time. During this manual operation, samples are taken and the performance is assessed. If the performance is better during the manual operation than during normal operation, then fluctuations are too severe and the control system must be adjusted to eliminate this effect.

Single Cyclone Test: A single cycle test involves setting up temporary piping to allow operation of a single cyclone liner to run in parallel with the hydrocyclone pressure vessel. It can be one of the most effective techniques to troubleshoot the performance of a hydrocyclone system. The objective of the test is to determine the overall performance of a given liner size under the conditions of pressure, flow rate, drop size, etc.. The test should be run over a range of conditions of pressure and flow. The range should bracket the actual conditions. Sensitivity to these process variables should be studied.

Separation efficiency for the single liner test should be compared to that for the hydrocyclone system. If the single cyclone has better efficiency, then typically there is fouling in the hydrocyclone pressure vessel. Single cyclone tests are also used to determine if different cyclone liners would be beneficial.

Overall Procedure for Performance Assessment: Descriptions and explanations of the different issues and tests involved in assessing the performance of a hydrocyclone are given in the section above. The following is a summary of the issues and the tests.

Table 10.5

Issue	Test
Suitable Fluid Properties	Desktop settling Analytical Characterization Drop size analysis (if available)
Efficiency	OiW of feed & effluent
Flow rate & Liners	Performance curve (efficiency vs flow rate)
Fouling	Trend of $[Q^2/\Delta P]^{2/5}$
PDR & Reject Ratio	Material Balance vs manual PDR adjustment
Steady Flow	Monitor effluent stream for slugs of oil
Overall performance	Single Cyclone Test

The tests are presented below in the format of a field procedure. Note that some of the tests are combined together. The entire performance assessment can be carried out in roughly two days for each hydrocyclone. If the facility has multiple hydrocyclones, then some of the assessment work can be carried out in parallel. For example, four hydrocyclones can likely be assessed in four days rather than eight.

No procedure is given here for trend monitoring of $[Q^2/\Delta P]^{2/5}$ since this is done by either studying trend plots in the office or by obtaining trends from the DCS in the operations control room. Note that the statements "measure efficiency" and "measure oil in water concentration" in the procedure below are intended to mean, take samples for measurement of oil in water, or take readings from on-line instrument. In the case where on-line instrumentation is available, calibration of the instrument is required before each day of testing.

If manual sampling and laboratory analysis are used for oil in water measurement, then a minimum of three samples should be taken for each stream for each data point.

Procedure for Field-Assessment of Hydrocyclone Performance:

The material below provides a checklist of action items to carry out in troubleshooting a hydrocyclone unit. Explanation for the different steps is provided is the text above.

Desktop Settling, Samples for Subsequent Analysis, Effluent Monitoring:

1. Open sample ports for feed, underflow and overflow to purge sample lines
2. Monitor for fluctuations in appearance and for slugs of oil
3. If slugs of oil are found, then all operating parameters shall be recorded and efficiency determined (please get in touch with an expert – this is an important data point but sampling program must be customized)
4. Once data are taken, work with operators to eliminate surging and oil slugging
5. Take samples of feed, underflow (effluent, product), and overflow (reject)
6. Set aside two samples for each stream for subsequent analytical characterization
7. Take three additional samples for each stream and make observations of samples at the following time intervals:

   1, 5, 10, 30, 60, 120 minutes, followed by 6 hours, 12 hours, 24 hours

   Record appearance and the relative volume of oil

   Take digital pictures if possible

Drop Size, Efficiency & Material Balance:

1. Record flow rates, T, P of all streams around the vessel upstream of the hydrocyclone
2. Record flow rates, T, P of all streams around the hydrocyclone (feed, underflow, overflow)
3. Measure OiW at start of test for all streams (feed, underflow, overflow)
4. Measure drop size every few minutes for a period of 1 hour
5. Measure OiW at end of test for all streams (feed, underflow, overflow)
6. Obtain trend of flow rates, and pressures for time interval during tests
7. Review operating philosophy and trends to determine next appropriate time to repeat entire test. It is recommended to repeat the test at least once in a 24 hour period.

Performance Curve (efficiency vs flow rate):

1. Measure baseline efficiency for normal operating flow rate
2. Increase the feed flow rate 20 % from the baseline keeping the PDR constant, wait 10 minutes
3. Measure efficiency
4. Decrease flow rate 20 % from baseline keeping PDR constant, wait 10 minutes
5. Measure efficiency
6. Return flow rate to baseline, wait 10 minutes
7. Measure efficiency

Material Balance vs Manual PDR Adjustment:

1. Record flow rates, T, P of all streams around the hydrocyclone (feed, underflow, overflow)
2. Calculate and record Baseline PDR
3. Measure OiW for all streams (feed, underflow, overflow)
4. Place the control system in manual override with fixed flow rate (baseline) and PDR (baseline)
5. Adjust PDR to Baseline minus 0.2
6. Record flow rates, T, P of all streams around the hydrocyclone (feed, underflow, overflow)
7. Measure OiW for all streams (feed, underflow, overflow)
8. Adjust PDR to Baseline plus 0.2
9. Record flow rates, T, P of all streams around the hydrocyclone (feed, underflow, overflow)
10. Measure OiW for all streams (feed, underflow, overflow)
11. Return PDR to automatic control
12. Record flow rates, T, P of all streams around the hydrocyclone (feed, underflow, overflow)
13. Measure OiW for all streams (feed, underflow, overflow)

Procedure for Routine Monitoring: Routine monitoring shall include:

1. Surveillance of flow rates, pressures, and on-line oil in water via PI tags.
2. Surveillance of manual oil in water readings
3. Calculation of area (for fouling assessment)
4. Calculation of PDR
5. Calculation of efficiency

Frequency of Performance Assessment: The optimum frequency of performance assessment will vary depending on the results of routine monitoring. Initially however, performance should be assessed three times in order to get a representative understanding of the system. The spacing between these assessments is based on the nature of the system. The more dynamic the system, the more performance assessments that should be carried out and the more frequently. Some judgment is required.

Performance assessment shall be carried out at a minimum:

- Three times consecutively for a new or upgraded facility
- Every month for three months
- Every six months for two years
- Every year thereafter

If the initial assessment program indicates significant change from one assessment to another, then additional assessments shall be carried out until steady operation is achieved. While these frequencies may be excessively high, it must be kept in mind that the hydrocyclone is the workhorse of any oily water treating system.

10.1.12 Benefits and Drawbacks of Deoiling Hydrocyclones

The advantages and disadvantages of hydrocyclones are presented here.

Advantages of Hydrocyclones: Advantages of a hydrocyclone over other separation equipment is the generation of relatively high centrifugal forces without the maintenance requirement of rotating parts. Other advantages include:

- Small footprint and light weight compared to other technologies such as settling tanks and flotation systems.
- Hydrocyclone liners have no moving parts so maintenance requirements are simplified.
- Good turndown characteristics over the operating range of the installed liners.
- Vessels can have additional liners installed or removed to meet increasing or decreasing process flow rates.
- Vessels can be installed horizontally or vertically.
- Wide operating flexibility, suitable for both onshore, offshore and on FPSOs.
- Many vendors and competitive designs and cost situation.
- Easily retrofitted. Modular design allows capacity to be increased as required
- Can be designed for high pressure and should be installed upstream of control valves
- Insensitive to motion, ideal for floating vessels such as TLPs and FPSOs
- Newer high efficiency liners can often be retrofitted into older lower efficiency liner slots

Drawbacks of Hydrocyclones: The various problems and concerns that need to be addressed in selecting deoiling hydrocyclones for a produced water handling facility include the following:

- The sensitivity of the system efficiency to oil droplet size requires consideration. For example, hydrocyclones are generally unsuited to gas-condensate applications where the median oil droplet size tends to be small (<10 μm).

- Application and efficiency of the deoiling hydrocyclones in low pressure operation of less than 10 Bar (150 psi) can be sub-optimal , since there may be insufficient pressure available for optimal hydrocyclone performance

- Optimizing the operation of a hydrocyclone system with variation in flow rate may require regular adjustment in the number of active liners that should be operational in an individual hydrocyclone vessel.

- Deoiling chemical may be required to improve the droplet size distribution in cases where the deoiling efficiency is less than desired. Note: chemical treatment is generally successful only if it successfully induces oil droplet coalescence upstream of the hydrocyclone vessel

- The reject orifice can be prone to blockage in hydrocyclone liners, which reduces oil removal efficiency. Regular back flushing is required to reduce the potential for blockage.

- Solids such as sand or scale mineral debris from upstream piping can build up in the inlet zones of vessels where the velocities are low.

- A Steady flow is required to establish the reject core. If flow is variable, the reject core will breakdown and separation efficiency will deteriorate.

10.2 Deoiling Hydrocyclone – Theory, Design and Modeling

The discussion in this section is intended to provide a detailed understanding of how a hydrocyclone works. The material below is intended as background for fundamental study and design improvement. Hundreds of scientific papers have been written on the subject of hydrocyclones. The ultimate objective of most of the work has been to understand the head loss, oil droplet shearing, and ultimately improve the separation efficiency. It is not the intention here to provide a comprehensive review of the literature. Rather, the intention is to provide fundamental information that a water specialist needs to resolve design and operational issues for a hydrocyclone system. As we have done elsewhere in this book, the objectives of this section are summarized in terms of simple questions. These questions are:

1. What are the performance models available and how can they be implemented easily?

2. What is the basis for these performance models?

3. How can practical questions about performance, the effect of fouling, troubleshooting, etc. be answered using straightforward analyses?

4. Why, from a fundamental standpoint, do hydrocyclones perform as they do? Why, for example, does the separation efficiency curve look so different from that of other primary separators?

5. How significant is oil drop shear and drop-drop coalescence in typical hydrocyclones?

6. How can oil removal from high viscosity fluids from polymer flooding be modeled?

7. What is the likelihood that significant improvements in hydrocyclone design will be developed in the near future?

There are several subjects that need to be covered in order to answer these questions. While the questions above are given more or less in the order of importance, the presentation below is given in a different order, which starts with the fundamentals and builds on that knowledge. The material below starts with a discussion of the acceleration force in a rotating fluid. The next topic is the inlet nozzle. The design of this nozzle is important in determining the speed at which fluids enter the cyclone. That speed, together with the throat diameter then determines the tangential velocity, which in turn determines the angular acceleration which provides the driving force for oil droplet's or other particle's separation. Trajectory Analysis is presented since it provides an understanding of how a hydrocyclone actually separates oil and water. Shearing within a hydrocyclone is touched upon.

An important reason to discuss hydrocyclones in some detail is the likelihood that new hydrocyclone designs will be implemented in the next several years with improved separation performance. There is a long-standing problem in the design of hydrocyclones which has recently been understood. It relates to the Meldrum curve [1] which was discussed in Section 10.1.3 (Mechanism of Separation – Performance Variables). The Meldrum curve is an observation that the effective length of current hydrocyclone designs is limited by the internal fluid dynamics. The cause of this limitation is now understood [11]. It is a matter of time before the problem is solved and the separation efficiency of hydrocyclones are further improved.

10.2.1 Stokes Law in Rotational Acceleration

The acceleration due to gravity is small compared to the acceleration force which is developed due to rotational motion in a hydrocyclone. As most introductory textbooks on physics describe, the force due to rotational acceleration is given by:

$$a_R = \omega^2 r \qquad \text{Eqn (10.4)}$$

Where:

w	angular or rotational speed, radians per unit time (rad/sec)
r	distance of the particle from the center (m)
a_R	acceleration due to rotational motion (m/sec^2)

The physical picture behind this equation is most readily imagined as a ball attached by a string to a rotating stick. As the stick, string, and ball rotate, the string exerts a force on the ball which forces it to maintain a circular motion about the stick. The distance from the ball to the stick (length of the string) is r. The speed of rotation is ω. The units of ω are radians/second. There are 2π radians in one circumference of a circle. The acceleration due to rotational motion has the expected units of length / second^2. This is a good analogy for the fluid mechanics of a hydrocyclone since the ball can be thought of as an oil droplet. The rotational force acting on a droplet will provide the driving force for separation. However, this analogy should not be taken too seriously. It will be shown that while the rotational force does increase as a function of distance from the center (stick), the rotational velocity as a function of distance from the center is a complicated function and does not follow a straight line as suggested by the stick and strong analogy.

In the above equation, the rate of rotation is characterized by the frequency with which the ball makes a revolution around the stick. For the purpose of calculating force, it is more useful to consider the speed with which the ball is moving. This is referred to as the tangential speed. The rate of rotation is related to the tangential speed as follows:

$$V = \omega r \qquad \text{Eqn (10.5)}$$

Where:

V tangential speed (m/sec)

The quantity V is the speed of an object as it rotates about a central point. One of the important aspects of V is that it increases as r increases, even if the angular rotation rate (ω) is constant. Substituting the above into the original equation for angular acceleration, the following relation is obtained:

$$a_R = V^2 / r \qquad \text{Eqn (10.6)}$$

Where:

a_R acceleration due to rotational motion (m/sec^2)

Using Newton's Second Law, $F = ma$, the rotational force is calculated from the equation above:

$$F_R = ma_R = mV^2 / r \qquad \text{Eqn (10.7)}$$

Where:

F_R force due to rotational motion (N = kg m/sec^2)

As shown, the rotational force is proportional to the rotational acceleration given by V^2/r.

This term will play an important role in the mechanics of the hydrocyclone. It is this rotational force that acts on the oil and water and provides the driving force for separation. It is important to understand the extent to which the rotational force varies as a function of the distance from the rotational center. Returning to the physical model of a ball attached to a rotating string, now consider several balls on the same string. All of the balls have the same rotations per second but each ball differs in the distance between the ball and the stick. As shown in the equations above, it is apparent that the rotational force is greatest for those balls furthest away from the stick. This is more difficult to appreciate by examining equation (10.7) alone where it appears that the acceleration varies with $1/r$. It is necessary to appreciate that V is proportional to r and thus increases as r increases. Thus, the dominant effect is that the rotational force increases as r increases, for constant ω.

In the case of oil droplets surrounded by produced water, the rotational force the relative mass is given by the volume of the oil droplet times the density difference of the oil and water. The final expression for the body force is given by:

$$F_R = ma_R = \left(\frac{V^2}{r}\right)\left(\frac{\pi}{6}d^3\right)(\rho_w - \rho_o) \qquad \text{Eqn (10.8)}$$

<u>Stokes Law in a Rotational Acceleration Field</u>: Stokes Law is based on a balance of forces. When an oil droplet reaches its so-called terminal velocity, the body force due to centrifugal acceleration is balanced by the surface drag force. On the basis of the above discussion, the gravitational acceleration is replaced with the rotational acceleration. The following equation, which applies to fluid drops in a hydrocyclone, is a modified form of Stokes Law:

$$u = \frac{(\rho_w - \rho_d)d^2}{18\mu}\left(\frac{V^2}{r}\right) \qquad \text{Eqn (10.9)}$$

Where:

u = oil drop rise velocity

ρ_w = density of water phase

ρ_o = density of oil phase

d = diameter of oil drop

μ = viscosity of water phase

In this equation, the gravitational acceleration (g) is replaced by the rotational acceleration (V^2/r). It will be convenient later to rewrite the equation in a form that uses the Rotational Gravity Number which is defined as:

$$N_{RG} = \frac{V^2}{rg} \qquad \text{Eqn (10.10)}$$

Where:

N_{RG} = The Rotational Gravity Number (dimensionless)

g = acceleration due to gravity (9.81 m/sec^2)

The physical meaning of this dimensionless number is that it gives the number of 'g' forces generated by the rotational motion of the hydrocyclone. Stokes Law is rewritten in terms of the Rotational Gravity Number as follows:

$$u = \frac{g(\rho_w - \rho_d)d^2}{18\mu}\left(\frac{V^2}{gr}\right) = \frac{g(\rho_w - \rho_d)d^2}{18\mu}N_{RG} \qquad \text{Eqn (10.11)}$$

This last equation is a very simple way of writing Stokes Law for a hydrocyclone. The effect of rotational motion is entirely provided as a multiplicative factor times the original Stokes Law. The Rotational Gravity Number is calculated below after a discussion of how rotational motion is generated in a hydrocyclone.

10.2.2 Inlet Fluid Flow

In this section, the inlet design of a hydrocyclone is discussed. The primary importance of this topic is that the inlet fluid velocity controls the "swirl velocity" in the hydrocyclone which, as was discussed above, provides the driving force for separation. In the case of primary separator vessels and tanks, the way that the fluids are introduced into the separator has an enormous impact of the separation performance of the device. It is important that the design of the inlet to a vessel or tank arrest the momentum of the incoming fluid and guide the fluid gently to the settling zone of the separator.

In the case of a hydrocyclone, it is important to introduce the fluid with a high velocity, to gently convert the forward velocity to a rotational or "swirl velocity", and to minimize the interaction between the incoming fluid and exiting reject fluid so that these streams do not disrupt and contaminate each other. As discussed in the Section on hydrocyclone design, different cyclone diameters are used to achieve different separation efficiencies and different flow handling capacities. When a customer selects a particular cyclone model, these are the main characteristics that he is selecting. But there are other details that differentiate one design from another. The inlet fluid nozzle design is onc of those important design features. The inlet fluid nozzle diameter or the slot cross sectional area determines the inlet fluid velocity. The inlet fluid velocity and the involute diameter determines the swirl intensity, and therefore determines the separation efficiency. Thus, nozzle design is as important as involute design and the two go together. Nozzle design also has a significant impact on the mixing of inlet and reject fluids mixing and on the turbulence in the inlet of the cyclone. Further, the design of the inlet section must promote a stable core from the entrance of the cyclone all the way along the axial length of the cyclone.

Inlet Fluid Speed and Swirl Velocity: The nozzle fluid speed is an important quantity because it is the source of momentum to drive the fluid inside the hydrocyclone in a swirl motion. It is the swirl motion and resulting tangential velocity that generates the large centrifugal force that replaces gravity in the separation of oil and water. For the purpose of the present discussion, the inlet of the hydrocyclone is defined as the inlet nozzle or entrance slot, and the cylindrical head, as shown in the figure below. The inlet nozzle of the hydrocyclone is placed such that fluids enter tangentially into the throat (involute) of the cyclone. It is located at the top (when looking at a vertical cyclone) of the hydrocyclone liner. The equivalent diameter the or cross sectional area of the inlet nozzle (D_i), together with the volumetric flow rate (Q_i), determines the speed at which fluid enters. The relation is given in the equation below.

$$V_i = \frac{Q_i}{(\pi / 4) D_i^2} = \frac{4Q_i}{\pi D_i^2} \qquad \text{Eqn (10.12)}$$

Where:

Q_i	volumetric flow rate of the hydrocyclone inlet (m^3/sec)
D_i	equivalent diameter of the inlet nozzle (m)
V_i	speed of fluid in the inlet nozzle entering the hydrocyclone (m/sec)

The equation above is a relatively simple relation. It indicates that the speed of the fluid in the inlet nozzle equals the volumetric flow rate divided by the cross sectional area of the inlet nozzle.

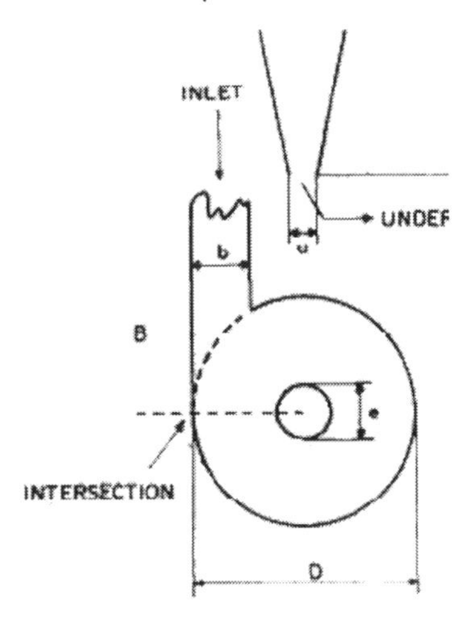

Figure 10.24 Bird's-eye view of a hydrocyclone.
The tangential inlet nozzle is shown together with a cross-section of the involute and the reject port.

One of the obvious assumptions behind the above equation, as shown in the figure, is that there is only one inlet nozzle. When there are two inlet nozzles the equation is modified as follows.

$$V_i = \frac{2Q_i}{\pi D_i^2} \qquad \text{Eqn (10.13)}$$

These equations establish the speed with which fluid enters the hydrocyclone. The speed with which the fluid enters the hydrocyclone has a direct impact on the swirl velocity and hence the separation power of the device. However, the inlet fluid speed does not equal the swirl velocity of the fluid inside the hydrocyclone. The inlet fluid speed influences the swirl velocity. A higher inlet fluid speed results in a greater swirl velocity. Wolbert et al [25] estimate that the swirl velocity is about half of the inlet fluid speed but this is only a rough estimate. But there is no simple formula for relating the two.

Inlet Reynolds number: The inlet Reynolds Number for a hydrocyclone is also a useful quantity. It is defined as:

$$N_{RHi} = \frac{\rho V_i D_i}{\mu} \qquad \text{Eqn (10.14)}$$

Where:

ρ	density of fluid (kg/m^3)
V_i	linear velocity of fluid (m/sec)
D_i	diameter of the inlet nozzle (m)
μ	viscosity of fluid (Pa sec = kg/m sec)

The subscript *RHi* refers to Reynolds Number for hydrocyclone at the inlet conditions. The N_{RHi} is dimensionless and it is straightforward to calculate. Once a value is known, it can be used to compare one set of conditions versus another, and to compare one hydrocyclone design to that of another. It is worthwhile to calculate both the inlet velocity and the inlet Reynolds number for a typical set of conditions. An example of this calculation is given here.

$Q_i = 0.5$ L/sec $= 0.0005$ m^3/sec

$D_i = 0.01$ m

$V_i = 4Q_i / \pi D_i^2 = 6.4$ m/sec

$\mu = 1\text{x}10^{-3}$ kg/m sec

$N_{RHi} = \rho V_i D_i / \mu = 64{,}000$

This calculation has some interesting features. First, the values chosen are well within the typical range found in a commercial hydrocyclone design. Yet, both the inlet velocity and the inlet Reynolds Number are large. In typical pipelines and flow lines, the fluid velocity is designed and generally operated at less than a few meters per second, primarily to prevent erosion. The high inlet velocity in hydrocyclones explains why materials of construction for the inlet and involute are typically tungsten carbide or another hard, erosion-resistant metallurgy. It is also worthwhile pointing out that the inlet Reynolds Number is well within the turbulent region (which starts roughly in the range of about 2,000 for flow in a cylinder).

One of the natural questions is to what extent do the high velocities in a hydrocyclone cause shearing of the oil drops. Computational Fluid Dynamics have been used to answer this question [26]. Results are given in the figure below. The location of the greatest shear in at the entrance to the liner (axial distance = 0 m), and near the wall (radial distance = 0.02 m). This is where the fluid enters the liner through the nozzle. As shown, the maximum shear rate is about 700 m^2/sec^3. As discussed in Section 3.4.2 (Turbulence Energy Dissipation Rate), this is a very small value. The analysis suggests that shear within a hydrocyclone is relatively minor.

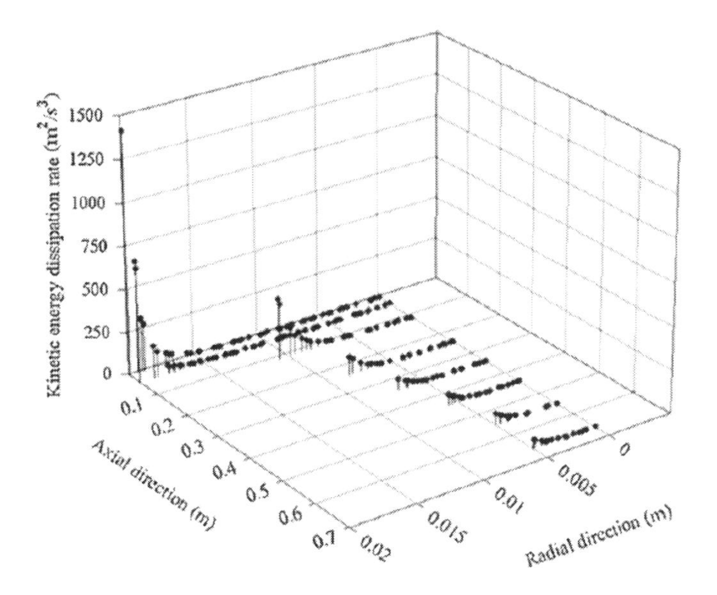

Figure 10.25 Kinetic energy dissipation rate versus the radial position for different axial distance from the top wall of standard hydrocyclone

The next issue to discuss is how much centrifugal force is generated by the tangiential velocity. Oil droplet separation from water, of course does not actually begin until the fluid starts spinning. Thus, the important quantity is the rotational speed of the fluid inside the cyclone. This is a complex problem in fluid mechanics. CFD has been used to develop a model for the flow field in a hydrocyclone. Before discussing those results, it is helpful to discuss flow field development from an intuitive perspective using physically motivated, although approximate analyses.

As discussed already, the fluid velocities inside of a hydrocyclone are significant. The highest velocities are typically at the outer edge, near the outer radius, and in the axial direction they are highest near the feed. Depending on the feed flow rate, and the geometry of the hydrocyclone, the fluid velocity can be as high as 10 m/sec. In pipe flow, this would be a significant fluid velocity.

10.2.3 Tangential Velocity and Vortex Analysis

As mentioned already in Section 10.2.1 (Stokes Law in Rotational Acceleration), the analogy of particles rotating on a string cannot be used directly to analyze the fluid motion in a hydrocyclone. The reason for this is that the particles on a string will all rotate at the same rotational velocity. In a hydrocyclone the rotational motion is provided by fluid motion. As such rotational velocity is a complicated function of the geometry of the device, the viscosity of the fluid, and the bulk flow rate.

There are two ideal models for the rotational velocity profile. Both can be described by the equation below by setting n equal to + or -1. The free vortex ($n = +1$) is physically similar to a swirl that is generated upon draining a sink, or the swirl motion of a tornado. As shown in the figure below, the tangential speed at the outer edge is low, and it is high at the center. The tangential speed decreases as the radial distance increases from the center to the outer edge. The effect of a surrounding fluid is negligible. The equation suggests that at the center, the velocity is infinite but this is never the case in practice due to fluid viscosity.

At the opposite extreme is the case of the forced vortex (n = -1). The physical picture is that of a solid disc that is rotating, such as a vinyl record on a record player. As shown in the figure below, the tangential speed varies linearly from the center to the outer edge.

Ideal Vortex: The most basic analysis of the tangential motion of fluid in a hydrocyclone starts with the vortex equation:

$$u_\theta = c_1 / r^n \qquad \text{Eqn (10.15)}$$

Where:

u_θ	tangential speed at any radial location (r) and at any axial location (z)(m/sec)
c_1	unspecified constant (dimensionless)
r	radial location measured from the axial center (m)

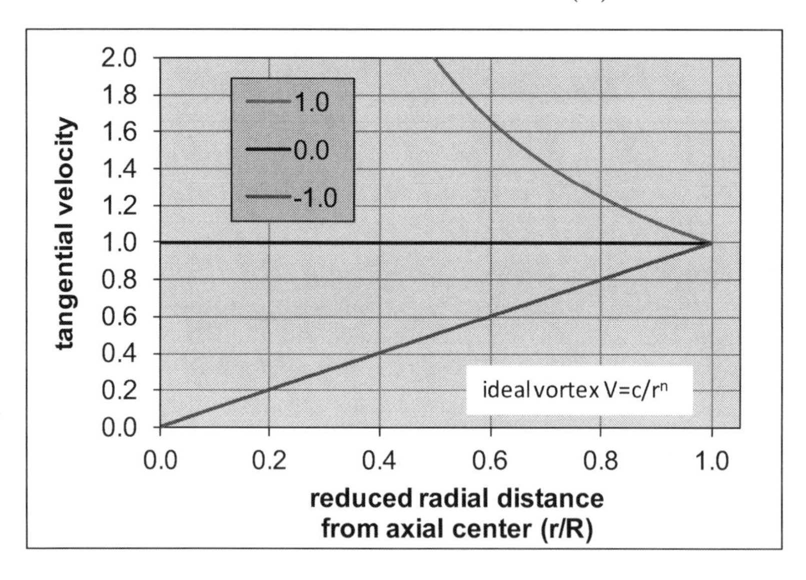

Figure 10.26 Tangential speed profile for various values of n in the vortex equation.

The characteristics of these two ideal vortices are summarized below. The vortex equation has been applied in the literature to analyze the tangential flow in a hydrocyclone [25]. By varying the value of n, the vortex equation can be manipulated. In the literature, n is usually adjusted to optimize the agreement between experimental and predicted separation efficiency. However, there is a major drawback to using this approach. While the vortex equation is somewhat flexible, it is not flexible enough. It does not allow a maximum or minimum in the velocity as a function of distance.

As shown in the figure below, the tangential velocity behaves as a forced vortex near the center where the speed is relatively low, and therefore the Reynolds Number is relatively low, and the effect of viscosity is relatively high. As one moves away from the center of the vortex, the tangential speed increases with respect to the radial distance. This the tangential speed behaves as a free vortex near the involute wall where the fluid speed is high, the Reynolds Number is high, and the effect of momentum far out-weighs the effect of viscosity. At the involute wall, the no-slip boundary condition still applies, but due to fluid turbulence, the effect of viscosity is compressed into a narrow boundary layer. This has the consequence that the vortex equation does not accurately describe the fluid motion across the entire cross-section. There is no single value of n that allows for a transition from forced vortex to free vortex. This is illustrated in figure 10.27 below.

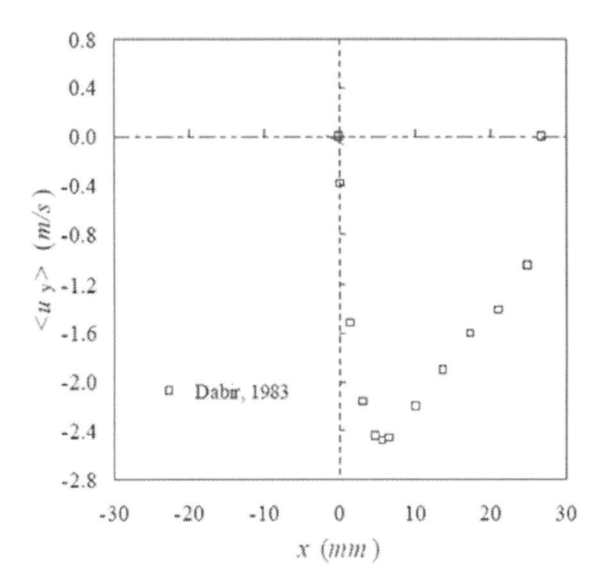

Figure 10.27 Calculated and measured tangential speed profiles for a Colman-Thew hydrocyclone. Note that near the center (x=0), the speed behaves as a forced vortex, and that progressing out in the radial direction the speed becomes more like a free vortex. Finally, near the wall, there is a thin boundary layer which ends with the no-slip condition at the wall. The shape of this tangential velocity profile gives rise to an s-shaped separation efficiency curve.

Combined-Vortex Model for Tangential Speed: In reality, the actual velocity profile is a complicated problem in fluid mechanics. The presence of turbulence, together with the high vorticity is a challenge. There are two approaches that have been developed in the literature to calculate more realistic velocity profiles. One approach is to use a more realistic though still relatively simple model for the tangential speed profile. At the other extreme, sophisticated turbulence modeling is carried out using Computational Fluid Dynamics.

An approximate analytical model for the tangential speed profile is given here based on the work of Reitma [27]. This analytical model is then used to predict the separation efficiency. In doing so, a physically meaningful model is developed. The model captures the fundamental physics and is not overly complicated. From this approach it is hoped the reader will obtain a more detailed understanding of how a hydrocyclone works.

The derivation starts with the Navier-Stokes and continuity equations in cylindrical coordinates. The main variable of interest is the tangential velocity u_θ. The equation below describes how the tangential velocity varies with the radial position across the cross-section of a hydrocyclone:

$$\rho u_r \left[\frac{\partial u_\theta}{\partial r} + \frac{u_\theta}{r} \right] = \mu \left[\frac{1}{r} \frac{\partial}{\partial r} \left(r \frac{\partial u_\theta}{\partial r} \right) - \frac{u_\theta}{r^2} \right] \qquad \text{Eqn (10.16)}$$

Before presenting the solution to the differential equation, it is worthwhile to point out some features. First, if viscosity is relatively small in comparison to the velocities, then the right hand side of the equation is set equal to zero. This leaves just the left hand side. The solution to this equation is the free vortex discussed above. Thus, viscosity cannot be ignored. It is worthwhile to note that the Euler equation cannot satisfy the inner boundary condition (no-slip). If fluid viscosity is high, relative to particle velocities, then the left hand side is set equal to zero. The solution to this equation is the forced vortex which was discussed.

There are two boundary conditions for this equation. The first boundary condition occurs at the outer edge of the vortex. This corresponds to the inside wall of the cylindrical involute section of the hydrocyclone. At the wall, the velocity is zero. This is referred to as the no-slip condition. The no-slip condition is difficult to solve mathematically. Instead, the outer boundary is idealized such that it is assumed to have a rotating velocity equal to V_0. This is not physically correct. However, both the experimental data and CFD simulation results show that the tangential velocity near the wall is not strongly impacted by the presence of the wall and the no-slip condition. The tangential velocity profile is more strongly influenced by the fluid motion away from the wall. The effect of the wall is only seen at very close distances to the wall, i.e. in the transition boundary layer.

Solving the above equation, taking into account the boundary conditions, gives the following equation for the tangential velocity as a function of radial position:

$$u_\theta = \frac{1-(r\lambda+1)e^{-r\lambda}}{r(1-(1+\lambda)e^{-\lambda})} \qquad \text{Eqn (10.17)}$$

where:

$\lambda = \mu / (\rho\, u_r)$.

The advantage of this equation is that it has the correct shape of tangential velocity profile. It has a single fitting parameter (λ) which can be used to adjust the shape of the curve to match experimental data. Since it is a relatively simple equation it can be used in trajectory analysis to calculate separation efficiency. In the figure below, it is shown that this has the correct behavior as discussed previously in relation to experimental data and CFD results.

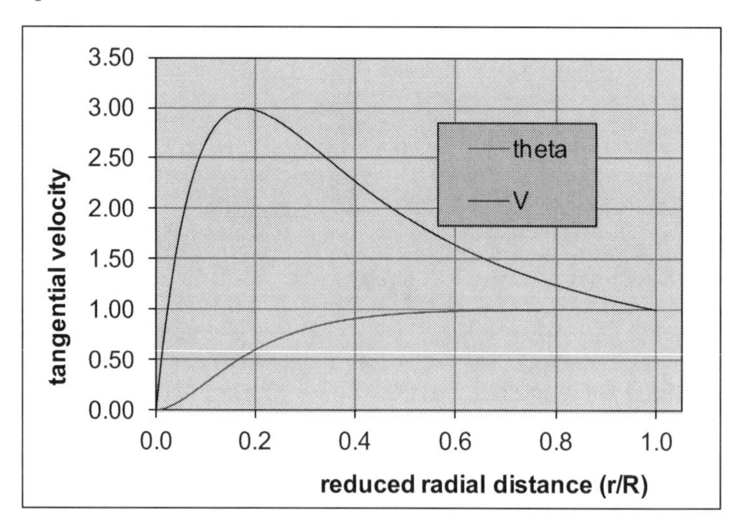

Figure 10.28 Tangential velocity (m/sec) profile as a function of radial distance from the axial core

The figure above shows that the equation has the correct shape of the tangential speed as a function of the radial distance. As already mentioned, the equation above is only approximate in that it does not model the boundary layer at the wall and it does not therefore show that the tangential velocity is zero at the wall, as would be expected by the no-slip boundary condition.

Nevertheless, as will be demonstrated, the analytical form of this velocity profile is extremely useful in understanding the mechanism of a hydrocyclone. In fact, much can be learned by applying the tangential profile to a simple cylindrical cyclone.

Rotational Gravity Number: Now that a tangential velocity profile has been derived, the driving force for separation can be calculated. The following calculation is intended to be indicative only. In other words, the values used will not be relevant to any particular cyclone design. The shape of the curve will be roughly indicative of the inlet region of a small diameter cyclone operating at high pressure drop and hence high fluid velocity. The intention is to demonstrate how to calculate the Rotational Gravity Number using the equations previously derived.

$$N_{RG} = \frac{u_\theta^2}{rg} = \left(\frac{\tilde{u}_\theta^2}{\tilde{r}}\right)\left(\frac{V_R^2}{Rg}\right) \qquad \text{Eqn (10.18)}$$

Where:

N_{RG} = The Rotational Gravity Number (dimensionless) defined above.

g = 9.81 m/sec

R = 0.04 m

V_R = 3 m/sec

N_{RHi} = $\rho V_i D_i / \mu$ = 64,000

$\tilde{u}_\theta = u_\theta / V_R$= dimensionless tangential velocity

$\tilde{r} = r / R$= dimensionless radial distance

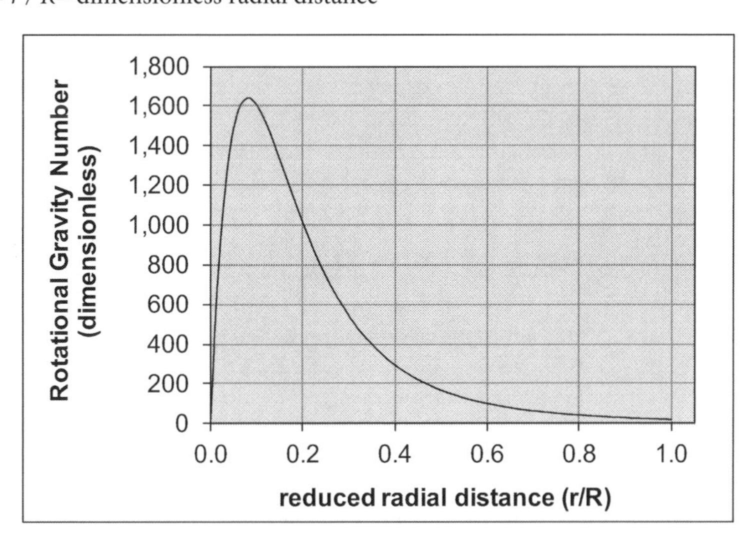

Figure 10.29 Rotational Gravity Number as a function of reduced radial distance.
The Rotational Gravity Number is the ratio of the driving force for separation to the gravity driving force.
Thus, NRG is the number of "g-forces" applied by a hydrocyclone.

As shown in the figure above, the 'g-forces' in a hydrocyclone can be well over one thousand. They are highest near the center, and trail off toward the wall.

10.2.4 Shape of Separation Efficiency Curve – Trajectory Analysis

Shape of the Separation Efficiency Curve: In the section on primary separators, the separation efficiency curve was derived by simple physical arguments. The physical picture was that of a drop moving upward by Stokes Law. By considering different drop sizes, together with the geometry of the vessel, a separation efficiency curve was derived. Although it was not mentioned in that section, the overall procedure used is referred to as trajectory analysis. In the case of a gravity separator, an intuitive approach, based on physical arguments, was sufficient to derive the separation efficiency curve.

The analysis was simplified by using the plug flow model. Deviations from plug flow were then taken into account by a hydraulic efficiency factor. Another reason why the analysis was simple is the fact that the driving force for separation does not vary as a function of the position of the drop. In other words, gravity is a constant throughout a primary separator. One consequence of this is the ability to calculate a minimum drop size for separation. This is a very useful engineering concept which is used frequently in designing a separator, and in designing a water treating process.

In the case of a hydrocyclone, a more complicated analysis is required. In a hydrocyclone, the tangential velocity is not a constant, in any direction. The tangential velocity varies with radial distance. This was demonstrated in the previous section. It also varies as the fluid moves down the axis of the cyclone whose diameter is getting progressively smaller. This is caused by the competing effects of geometry (taper), which is used to increase tangential velocity, and the loss of rotational kinetic energy due to fluid friction. The tapering shape of a cyclone is intended to compensate for the loss in rotational kinetic energy but some variation still occurs. In other words, the driving force for separation is not a constant along the path that the drop travels either radially or axially. As will be demonstrated, this variation of driving force has a profound effect on the separation efficiency curve.

Trajectory Analysis: The objective of Trajectory Analysis is to derive a relationship between drop size and separation efficiency. It is used to relate the separation efficiency to the fluid characteristics (drop diameter, water viscosity, and density difference) and the flow characteristics within the hydrocyclone. A specialized trajectory analysis is required for the hydrocyclone since the driving force varies over the trajectory of the oil drops. In the material below, the consequence of a constant driving force is considered, within the framework of trajectory analysis, and it is shown that a constant driving force is indeed responsible for the parabolic shape of the separation efficiency curve for gravity separators.

Trajectory analysis, as the name implies, is a way of predicting the trajectory of a particle due to the forces acting on it. Mathematically, force is used to calculate acceleration. Acceleration is integrated with respect to time to determine velocity. Velocity is integrated with respect to time to determine the path traveled by the particle. The velocity can be a vector in three dimensional space and likewise, the path determined in this manner can traverse through three dimensional space. If the quantity of interest is the distance in a particular direction, then the component of velocity in that direction is integrated rather than the total velocity which includes components in the radial and axial directions.

In the case of a hydrocyclone, trajectory analysis is used to relate fluid characteristics to the separation efficiency curve. The analysis given in this section is general in the sense that it applies for all fluid properties and all geometries of hydrocyclone. The only assumption is that there are drops of a light phase which are suspended in a heavier continuous phase. The derivation assumes that the fluid is subjected to a varying driving force for separation along its trajectory. Once this general result is obtained, simplifications, and geometries are considered to add insight to the general result.

The analysis given below is based on work by Rietema from the Koninklijke laboratory of Shell Internationale Research Maatschappij (SIRM) [27 - 30], Colman-Thew at Southampton University [31], and by Wolbert et al. [25]. The fact that these references span nearly 35 years gives a clue to the complexity of the internal workings of a hydrocyclone. The later reference is particularly useful in understanding the framework of trajectory analysis, but the vortex equation was used which resulted in a separation efficiency curve that appears like that of a primary separator with a maximum drop size cut off.

The fluid flow in a hydrocyclone has five components:

V: tangential velocity of the fluid

W_w: axial flow of the oily water toward the underflow (water effluent) nozzle

W_o: axial flow of the oil reject core toward the overflow (oil reject) nozzle

U: radial (tangential) flow of fluid

u_p: radial flow of drops and particles due to density difference

The general equation for the trajectory of an oil droplet is:

$$\frac{dr}{dz} = \frac{dr / dt}{dz / dt} = \frac{U + u_p}{W} \qquad \text{Eqn (10.19)}$$

Where:

r	radial position (m)
z	axial position (m)
t	time (sec)
U	radial velocity of the fluid (m/sec)
W	axial velocity of the fluid (m/sec)
u_q	tangential velocity (m/sec)
u_p	(inward) radial velocity of oil drop (m/sec)

The term, dr/dz is the radial motion of oil droplets as the fluid proceeds along the axial direction of the cyclone. This is the important quantity in designing a hydrocyclone. The term, $dr/dt = U + u_p$ is the velocity of a particle in the radial direction. There are two drivers for this radial motion. The first is the conical shape of the cyclone – this determines U. As the diameter of the cyclone decreases in the axial direction, the fluid moves inward in the radial direction. The other driver for radial movement is the relative motion of oil drops relative to the bulk motion of fluid – this determines u_p. The term u_p is calculated from the modified version of Stokes Law applicable to a hydrocyclone. The other main term in the equation is related to the axial motion of the fluid. The fluid spirals down the z direction. The motion is given by: $dz/dt = W$.

The next step in the trajectory analysis is to integrate the trajectory equation along the path that the drop takes through the cyclone. To do this, the r-component of the trajectory equation is integrated with respect to time. Physically, this integral equals the radial distance traveled by the drop from the moment it enters the involute section, to the moment it reaches the oil core. The time interval is given by the time it takes to traverse the entire length of the cyclone. The particular droplet of interest has a diameter which is large enough to allow the drop to just reach the oil core at the end of the cyclone.

In a sense, the oil drop diameter is the real unknown in this equation. In practice, the calculations are carried out numerically, not analytically.

The final step in the trajectory analysis is to actually calculate a separation efficiency curve. The separation efficiency for all drops having diameter *di* can be calculated in terms of the ratio of inlet nozzle areas, as follows.

$$E(d_i) = \frac{R_i^2 - R_c^2}{R_d^2 - R_c^2} \quad \text{Eqn (10.20)}$$

Where

$E(d)$ = separation efficiency for droplets of diameter d

R_i = inlet radius for the drop of diameter di (m)

R_c = radius of the oil core (m)

R_d = radius of the involute section (m)

Three of the four terms on the right hand side are defined and can be calculated in terms of the geometry of the hydrocyclone. The oil core is some small fraction of the radius of the device. The radius of the involute section is typically known for a hydrocyclone. The fourth term depends on *Ri* which must be calculated from the integral equation above. This calculation can be carried out for all droplet diameters. Once this is done, the separation efficiency curve is defined.

An example of a separation efficiency curve is given in figure 10.6 above. Note that it has the characteristic s-shape that is typically found for hydrocyclones. Thus trajectory analysis, based on a modified form of Stokes Law, does give the correct shape of the separation efficiency curve. The quantitative accuracy of the analysis has been evaluated in the literature. The fundamental methodology of the calculation has been confirmed.

10.2.5 The Hydrocyclone Number & Performance Curve

One of the by-products of trajectory analysis is the recognition of a dimensionless number which has become known as the Hydrocyclone Number:

$$N_{Hy} = \frac{Q_f \Delta\rho (d_{75})^2}{\mu D^3} \quad \text{Eqn (10.21)}$$

N_{Hy} = Hydrocyclone Number (dimensionless)

Q = volumetric flow rate (m^3 / sec)

Dr = density difference between water and oil (kg / m^3)

d_{75} = drop diameter for which 75% of drops are separated (m)

m = viscosity of the water (Pa sec => kg / m sec)

D = involve diameter of the cyclone (m)

The parameter d_{75} is key to this relationship. It specifies a particular drop diameter. More precisely, it is defined as the drop diameter for which 75% of the oil volume will be separated. The other 25% of oil volume is discharged in the effluent (product) water. There was no particular reason why the value of 75 percent was chosen. Another value, of say 50 percent, could equally well have been chosen. In that case, the value of the Hydrocyclone Number would be different. Mathematically any convenient value in the range of about 20 to 80 percent could have been chosen. It is a feature of hydrocyclones however that the 100 percent separation value cannot be chosen. The s-shape of the separation curve precludes the use of a d100 value. The figure below is a plot of the calculated Reynolds Number and Hydrocyclone Number.

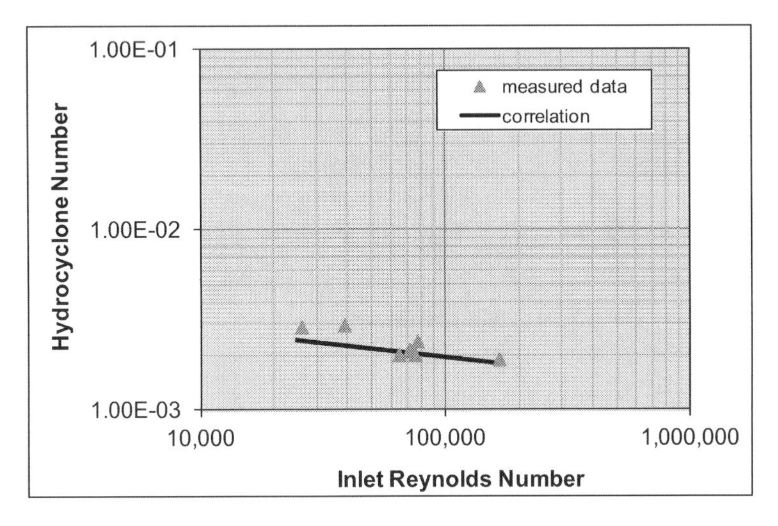

Figure 10.30 The Hydrocyclone Number as a function of the Inlet Reynolds Number. As shown, the Hydrocyclone Number is a weak function of the Reynolds Number. This suggests that the Hydrocyclone Number is indeed mostly dependent on the geometry of the device, rather than the flow conditions.

When SI units are used, no unit conversion factors are required. An example of the calculation is given by the following. In this calculation, essentially all of the parameters have been measured. The oil drop diameter that is used is the d75 value. That is the drop diameter for which 75 % of the oil volume is separated in the cyclone.

$$N_{HY} = \frac{(990\,\text{kg/m}^3 - 820\,\text{kg/m}^3)(14.5\text{x}10^{-6}\,\text{m})^2(1\text{x}10^{-3}\,\text{m}^3/\text{s})}{(0.55\text{x}10^{-3}\,\text{kg/(ms)})(30\text{x}10^{-3}\,\text{m})^3} \qquad \text{Eqn (10.22)}$$

$$N_{HY} = 2.41x10^{-3}$$

This value of the Hydrocyclone Number represents the Colman-Thew design. It is a constant for that particular design, and all liners that are geometrically similar. Defining geometric similarity is complicated but a clue can be obtained from figure 10.3. where dimensions are given in dimensionless units. Since the Hydrocyclone Number is a function of geometry, it provides a useful guide in understanding performance over a range of flow rates, density difference, and water viscosity. Within certain limits, if the flow rate (Q) increases, then one of the other variable must decrease in order for the Hydrocyclone Number to remain constant. If all other variables are constant, then the droplet diameter will decrease. This is expected on physical grounds as discussed already in relation to Figure 10.10, where it is seen that the separation efficiency increases as the flow rate increases. Likewise, if the water viscosity increases, as for a decrease in water temperature, the droplet diameter will need to increase, again on physical grounds this is expected. The higher the water viscosity, the slower droplets travel.

The equations presented above provide a powerful, if somewhat complicated model for estimating the performance of a hydrocyclone. The Hydrocyclone Number, together with the Euler and Reynolds Numbers, provide an essentially complete model that relates flow rate, separation performance, and water viscosity to the geometrical shape of a cyclone. Although geometrical shape does not explicitly enter into the formulas, it influences the numerical relation between the three dimensionless numbers.

In order to provide a better description of this theoretical framework, it is worthwhile to apply the formulas in some actual calculations. This will help to elucidate how this theoretical framework actually works. As shown in the figure, the Hydrocyclone Number has a relatively weak dependence on the Reynolds Number, as expected.

The particular equations used in this discussion are based on measurements carried out at the Orkney Water Treating Center. The Hydrocyclone Number is calculated for a given value of d75. Separation efficiency is then calculated from the following empirical equation:

$$S(d) = 1 - \exp[c_1 (\tilde{d}_{75} - c_2)^{c_3}] \quad \text{Eqn (10.23)}$$

$S(d)$	=	separation efficiency curve, as a function of drop diameter (fraction)
$\tilde{d}_{75}$	=	d/d_{75} = reduced drop diameter (dimensionless)
c_1, c_2, c_3 =		-2.05, 0.39, 0.75 respectively (dimensionless)

The value of d_{75} and hence S(d) is calculated from the equation above. A plot of S(d) versus d is given in figure 10.6 above. The performance (separation efficiency) given in the figure is characteristic of a commercial hydrocyclone liner such as the Krebs K-liner.

References to Chapter 10

1. J.C. Ditria, M.E. Hoyack, "The Separation of Solids and Liquids with Hydrocyclone-Based Technology for Water Treatment and Crude Processing," SPE-28815 (1994).

2. D.A. Colman, M.T. Thew, "Correlation of separation results from light dispersion hydrocyclones", Chem. Eng. Res. Des., Vol. 61, pp. 233-240. (1983).

3. N. Kharoua, L. Khezzar, Z. Nemouchi, "Hydrocyclones for de-oiling applications – A review," Pet. Sci. Tech., v. 28, p. 738 – 755 (2010).

4. M. Bothamley," Offshore Processing Options for Oil Platforms," SPE 90325 (2004).

5. J.M. Walsh, W.J. Georgie, "Produced water treating systems – comparison between North Sea and deepwater Gulf of Mexico," SPE 159713 (2012).

6. J.M. Walsh, "Produced water treatment systems: comparison of North Sea and deepwater Gulf of Mexico," SPE 159713, paper published in Oil & Gas Facilities (2015).

7. P.H.J. Verbeek, R.G Smeenk, D. Jacobs, "Downhole separator produces less water and more oil," SPE-50617, paper presented at the 1998 SPE European Petroleum Conference, The Hague (1998).

8. J.A. Veil, "Feasibility evaluation of downhole oil/water separator (DOWS) technology," U.S. Department of Energy Report, Contract No.: W-31-109-Eng-38 (1999).

9. D.A. Hadfield, S. Riibe, "Hydrocyclones in large-scale marine oil spill cleanup," OTC-6504, paper presented at the 23rd Annual Offshore Technology Conference, Houston, TX (1991).

10. S.K. Ali, C.A. Petty, Deoiling hydrocyclones," Produced Water Society meeting, Houston (1994).

11. A. Motin, "Theoretical and numerical study of swirling flow separation devices for oil-water mixtures," PhD dissertation, Michigan State University (2015).

12. S. Schutz, G. Gorbach, M. Piesche, "Modeling fluid behavior and droplet interactions during liquid-liquid separation in hydrocyclones," Chem. Eng. Sci., v. 64, p. 3935 (2009).

13. T. Husveg et al. (U. Stavanger), "Performance of a Deoiling Hydrocyclone during Variable Flow Rates," Minerals Eng., v. 20, p. 368-379 (2006).

14. M.F. Schubert, "Advancements in liquid hydrocyclone separation systems," OTC-6869, paper presented at Offshore Technology Conference, Houston, TX(1992).

15. D.A. Flanigan, J.E. Stolhand, M.E. Scribner, E. Shimoda, "Droplet size analysis: A new tool for improving oilfield separations," SPE-18204, presented at the Annual Technical Conference and Exhibition, Houston (1988).

16. D.A. Flanigan, J.E. Stolhand, E. Shimoda, F. Skilbeck, "Use of low-shear pumps and hydrocyclones for improved performance in the cleanup of low-pressure water," SPE – 19743, SPE Production Engineering, August (1992).

17. Z.I. Khatib, "Handling, treatment and disposal of produced water in the offshore oil industry," SPE 48992 (1998).

18. J.M. Walsh, T.C. Frankiewicz, "Treating produced water on deepwater platforms: Developing effective practices based on lessons learned," SPE-134505, paper presented at the SPE ATCE, Florence (2010).

19. University of Calgary Produced Water Treatment Design Manual. Courtesy of Munib Ahmed (2010).

20. A. Belaidi, M.T. Thew, "The effect of oil and gas content on the controllability and separation in a de-oiling hydrocyclone," Trans. Inst. Chem. Eng., v. 81, p. 305 – 314 (2003).

21. I.C. Smyth, M.T. Thew, "A study of the effect of dissolved gas on the operation of liquid-liquid hydrocyclones." In: Hydrocyclones '96, Eds: D. Claxton, L. Svarosky, M. Thew, London: Mechanical Engineering Publications Limited, p. 49 – 61 (1996).

22. Q. Luo, J.R. Xu, "The effect of the air core on the flow field within hydrocyclones," Fluid Mechanics and its Applications, v. 12, p. 51 – 62 (1992).

23. Personal communication with H. Belaidi (2014).

24. H. Meldrum, "Hydrocyclones: a solution to produced-water treatment," SPE-16642, paper presented at the Offshore Technology Conference (OTC), November (1988).

25. D. Wolbert, B.-F. Ma, Y. Aurelle, J. Seureau, "Efficiency estimation of liquid-liquid hydrocyclones using trajectory analysis," AIChE J., v. 41, p. 1395 (1995).

26. S. Noroozi, S.H. Hashemabadi, "CFD analysis of inlet chamber body profile effects on de-oiling hydrocyclone efficiency," Chem. Eng. Res. Des., v. 89m p. 968 (2011).

27. K. Rietema, "Performance and design of hydrocyclones – IV. Design of hydrocyclones," Chem. Eng. Sci., v 15, p. 320 (1961).

28. K. Rietema, "Performance and design of hydrocyclones – III. Separating power of the hydrocyclone," Chem. Eng. Sci., v. 15, p. 310 (1961).

29. K. Rietema, "Performance and design of hydrocyclones – I. General considerations," Chem. Eng. Sci., v 15, p. 298 (1961).

30. K. Rietema, "Performance and design of hydrocyclones – II. Pressure drop in the hydrocyclone," Chem. Eng. Sci., v 15, p. 303 (1961).

31. D.A. Colman, M.T. Thew, "Correlation of separation results from light dispersion hydrocyclones," Chem. Eng. Res. Des., v. 61, p. 233 (1983).

CHAPTER ELEVEN

Centrifuges

Chapter 11 Table of Contents

11.0 Introduction 119

11.1 Mechanism of Separation: 121

11.2 Representative Performance Data: 123

11.3 Operation and Maintenance: 125

11.4 Summary and Conclusions: 127

References to Chapter 11 128

11.0 Introduction

As discussed by Khatib [1], disc-stack centrifuges have been widely used to treat industrial waste water for over 100 years. They are used in the dairy, brewery, pharmaceutical, and other food and beverage industries. Many of these applications involve the resolution of emulsions having high loading of suspended solids, and in some cases emulsions containing high molecular weight polymers composed of various sugar and starch functional groups, complexes of these polymers with multivalent cations, and the presence within these mixtures of oil-wet solids and small droplets of oil and grease. Suffice to say that centrifuges have been used successfully on some of the most difficult industrial water emulsions. In those industries just mentioned, centrifuges are the go-to technology for breaking difficult emulsions. Many of those emulsions have similarities to flow back from conventional and unconventional (shale) hydraulic fracturing operations in the oil and gas industry. Thus, it would seem likely that the oil and gas industry would make use of centrifuges. Unfortunately however, applications of centrifuges for produced water treatment have not been as widespread as might be expected.

Disc-stack centrifuges have been used in the oil and gas industry [2, 3] to:

- dehydrate crude oil;
- separate finely dispersed condensate droplets from produced water;
- dehydrate bitumen in tar sand production;
- process slop oil and oily drains at the fluid handling facilities.

Full-scale centrifuges have been installed at several gas producing platforms on the Netherlands continental shelf, on a North Sea platform (Tyra West and Tyra East), and at several locations in the Gulf of Mexico. The success of these installations can be summarized as follows. In almost all cases, the separation capability of these devices is impressive. As expected, oil droplets of just a few microns can be removed. However, in stranded locations (such as deep water), the reliability, required maintenance, and required presence of specialists to start up and troubleshoot the devices have been too much for successful operation. As discussed below, all of the large disc stack units installed on Shell facilities in the deep water Gulf of Mexico were dismantled after relatively short time in use. One location, nearer to shore, and employing significantly smaller units, has been run in intermittent service for many years with favorable experience. Other successfully operated units are also discussed below. The main factor that determines success is the ability to provide specialist staff for operation and maintenance of the units. In near-shore locations this is feasible. In most offshore locations, where it is expensive to provide adequate staff, these units have not been successful.

As shown in the Figure below, the mixture of oil, water, and solids enters the top of the device in the inner most tubing. It flows down the center of the device and into a channel that is open to the bowl (disc stack). The mixture is distributed across all of the discs. The fluid does not enter the disc stack at the radial center since that position is occupied by oily reject. Instead, the fluid enters at a radial position that is at some intermediate radial location.

Once the fluid enters the disc stack, the light contaminant (oil) moves to the radial center which the heavy components (water and solids) move to the outer periphery. Since the discs are spinning at a high rate, the fluid gets flung out toward the outer radial edge of the discs. Due to the high centrifugal force, the solids and water are pulled to the outer edge and the oil is pushed toward the inner edge.

Another way of saying is as follows. The heavier contaminants are pulled toward the underside of the upper disc (periphery). The oil is pushed to the upper surface of the bottom disc (center of rotation). The gap between any two discs is very small. The combination of high centrifugal force, and short distance required for effective separation, results in very small particles being separated effectively.

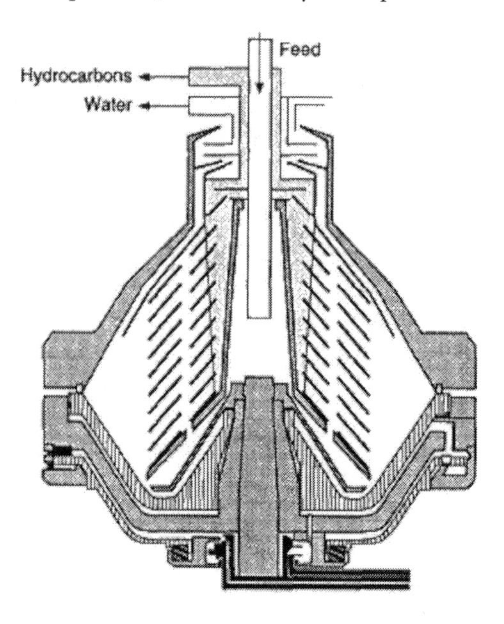

Figure 11.1

The figure 11.2 below shows the outlet at the top of the device where oil and water are shown leaving the bowl from the inside and outside respectively. The high separation efficiency of a disc stack centrifuge is due to two features. The first is of course the high centrifugal force derived from the high speed rotation. The second feature is that the phases, and contaminants have only a short distance to travel in order to be separated. The larger the number of discs, the shorter the distance of travel. Also, maintaining a short gap between discs helps to suppress secondary flow within the gap. The distance between each disc is typically less than about 1mm.

In the case where the design provides a pressurized liquid discharge, both the treated water and the oil leave the bowl, in separate chambers, under pressure. This eliminates the need for downstream pumps. In most such designs, the kinetic energy of the discharged liquid is used to build pressure.

In those design that have an automatic solids discharge, the bowl has a double conical shape. At its widest part it has a number of rectangular solids discharge ports. A sliding bottom is used which is moved a few millimeters up and down to control the flow of the solids. When the sliding bottom is in the upper position, a closed space is formed. In this mode the device collects solids. When it is in its lower position, the discharge ports open and solids are ejected by centrifugal force into the surrounding collecting cover.

Decanting Centrifuge: Decanting centrifuges are used primarily to thicken and dewater sludge. They are used in many industrial applications such as:

1. Oil /solids separation (oil well drilling, refining, de-watering)
2. Industrial and biological wastewater treatment
3. Food processing (olive oil, wine, fruit juice)
4. Chemical slurry

In addition, they are applicable to produced water treatment. Removal of heavy metals often involves an intense chemical treatment that results in a sludge that must be dewatered and disposed of. A decanter-centrifuge can be used to produce a cake with a minimum water content. In this application, where the water content of the sludge is minimized, some contaminant particles are left in the water phase.

They are also very effective as a tool in handling well flow back. During a flow back, temporary rental equipment is often used to treat the flow back fluids and keep it from contaminating the main produced water treatment system. In this application, the decanting centrifuge can be manned continuously and adjusted to achieve the optimal separation. For this purpose, there may or may not be sludge involved. But even when there is no sludge, the centrifuge, if properly adjusted, can separate a brine from an emulsion with far less weight and space then a settling tank or weir box. In this application, where the objective is to separate an emulsion from a water phase, the water has a tendency to remain somewhat contaminated and the emulsion has a tendency to be somewhat water-wet. Nevertheless, the compact size and weight of the device provides a benefit over gravity settling. The water can be cleaned by traditional methods. The emulsion will likely require chemical treatment, heat, and settling time. Having a much smaller volume to treat as an emulsions is an enormous benefit.

The basic type of centrifuge used for sludge thickening is the solid-bowl centrifuge. The solid-bowl centrifuge consists of a long bowl, normally mounted horizontally and tapered at one end. Sludge or emulsion is introduced into the unit continuously, and the solids concentrate on the periphery. An internal helical scroll, spinning at a slightly different speed, moves the accumulated sludge toward the tapered end where additional solids concentration occur and the thickened sludge is discharged. Under normal conditions, thickening can be accomplished by centrifugal thickening without polymer addition, but very regularly polymers are added to improve the system performance. The performance of a centrifuge is often quantified by the concentration achieved in the thickened solids and the TSS (Total Suspended Solids) in the concentrate. The principal operational variables include the following:

1. Characteristics of the feed sludge,
2. Rotational speed,
3. Hydraulic loading rate,
4. Depth of the liquid pool in the bowl,
5. Differential speed of the screw conveyor, and
6. The need for polymers to improve the performance.

Because the interrelationships of these variables will be different for each type of sludge, bench-scale or pilot-plant tests are recommended.

11.1 Mechanism of Separation:

The principles of separation are similar to those already discussed for the hydrocyclone in Section 10.2.1 (Stokes Law in Rotational Acceleration). Some of these principles are repeated here. For swirling flow, Stokes Law is given by:

$$u = \frac{(\rho_w - \rho_d)d^2}{18\mu}\left(\frac{V^2}{r}\right) \qquad \text{Eqn (11.1)}$$

Where:

u = oil droplet rise velocity (m/s)

V = rotational speed (m/s)

ρ_w = density of water phase (kg/m^3)

ρ_o = density of oil phase (kg/m^3)

d = diameter of oil drop (m)

μ = viscosity of water phase (Pa-sec)

Comparison of this equation with the usual form of Stokes Law that is commonly used for gravity separation shows that the gravitational acceleration (g) is replaced by the rotational acceleration (V^2/r). It will be convenient later to rewrite the equation in a form that uses the Rotational Gravity Number which is defined as:

$$N_{RG} = \frac{V^2}{rg} = \frac{\omega^2 r}{g} \qquad \text{Eqn (11.2)}$$

Where:

N_{RG} = The Rotational Gravity Number (dimensionless)

ω = angular or rotational speed, radians per second (rad/sec)

The physical meaning of this dimensionless number is that it gives the number of 'g' forces generated by the rotational motion of the centrifuge. The equation shows the importance of the radius (or diameter) of the centrifuge. An example calculation will help to clarify the forces and settling rates involved.

The angular velocity of a droplet located at say 0.2 m (r = 0.2 m) from the center axis, with a rotational speed of 5,000 revolutions/minute ($\omega = 2\,\pi\,5{,}000\,/\,60$ radians/second = 524 rad/sec) will give a value of the Rotational Gravity Number of:

$$N_{RG} = \frac{V^2}{rg} = \frac{(524\ \text{rad/sec})^2\ 0.2\text{m}}{9.81\ \text{m/sec}^2} = 5590 \qquad \text{Eqn (11.3)}$$

The driving force for separation is therefore 5,590 times stronger than gravity. As discussed in Section 10.2 (Deoiling Hydrocyclones – Theory, Design and Modeling), this enhanced separation force results in good separation efficiency for small particles and droplets and it can be used to provide small compact equipment given the short residence time required under such high driving force for separation. However, there are also drawbacks which concern reliability, proper operation, and the fact that a centrifuge is a complex rotating machine. This is a significant difference compared to a deoiling hydrocyclone.

The Table below was adapted from Faucher and co-workers who provide a practical comparison of the important parameters in comparing disc stack centrifuges to other separation technology [4].

They point out that the driving force is only relevant to the distance that the oil droplet must travel in order to be separated. The disc stack centrifuge combines high driving force together with short distance to achieve high separation efficiency for small oil droplets.

Equipment	Driving Force (g-factor)	Settling Distance (inch)	Droplet Size Removed (Dv50)
Gravity separator	1	> 40	> 40
Hydrocyclones	1000 to 2000	> 602	> 12
Disc stack centrifuges	6000	> 602	> 2

11.2 Representative Performance Data:

The table and figure below confirm what has been said above. The separation performance of a disc stack centrifuge is usually outstanding, when they have not broken down. Very small droplets can be sepatated.

Table 11.1 Slop Oil treatment from wet oil tank centrifuge in Deoiling mode at 81°F crude API gravity = 38°

Flowrate BPD	Inlet Dispersed Oil Concentrate ppm IR	Outlet Dispersed Oil Concentrate ppm IR
1056	370	19
1207	137	2
1207	116	2
1207	113	3
1207	101	31
1207	2815	30
755	100,000	132
755	90	18
755	120	26
755	106	17
755	101	22
755	90	3
755	97	3
755	106	4

Data in the above table from Khatib [1]. While there is considerable fluctuation, the centrifuge performance is good. Separation efficiency is above 90 % for the most part.

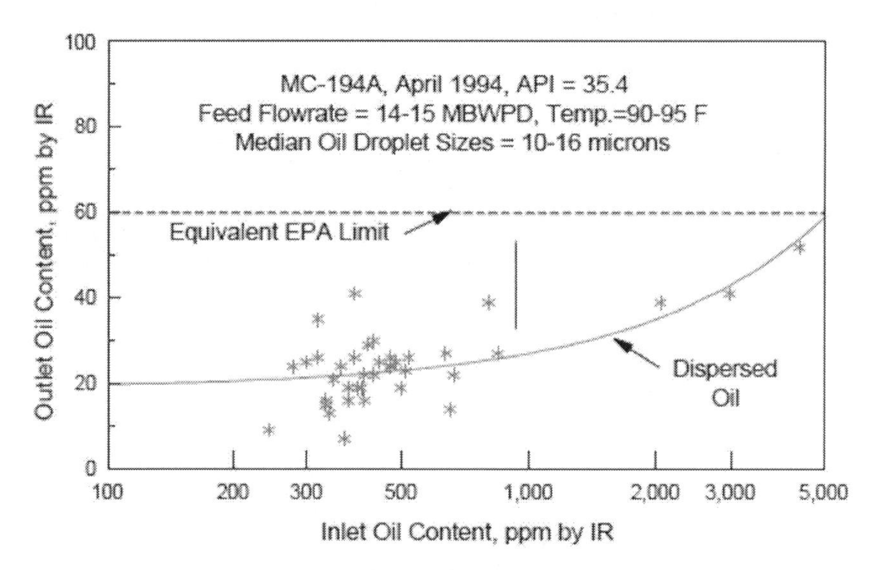

Figure 11.3 Evaluation of the LEO Centrifuge

Data in the above figure from Khatib [3]. These data show that the LEO centrifuge has outstanding separation performance over a wide range of oil-in-water concentration. The following figure is from van der Zande and co-workers [2].

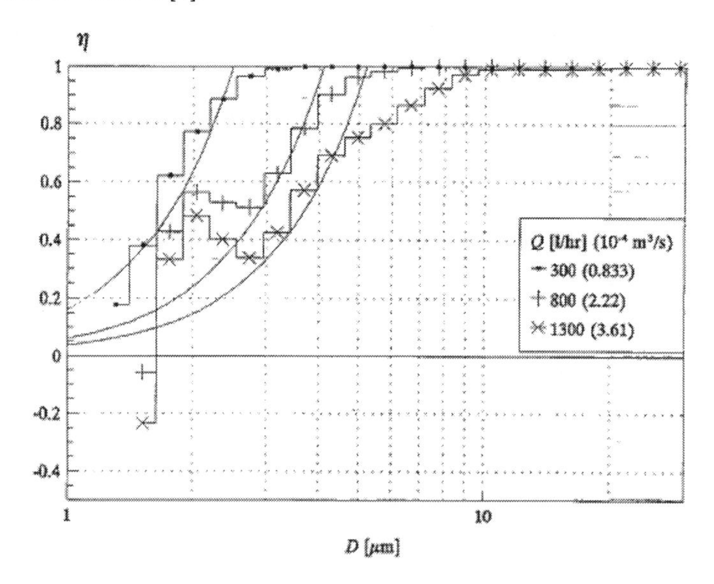

Figure 11.4 Example of experimentally determined removal efficiencies for a centrifuge with vertical separator plates

Van der Zande [2] and co-workers provide the following analysis:

The equations as presented for the plate separator are also valid for the centrifuge, with the understanding that the acceleration of gravity, g, has to be replaced by the centrifugal acceleration, @zr. This yields for the critical oil-droplet diameter:

18Hpvx

D,' =Ap(i)2rL

Also in this case the channel height, H, has to be corrected

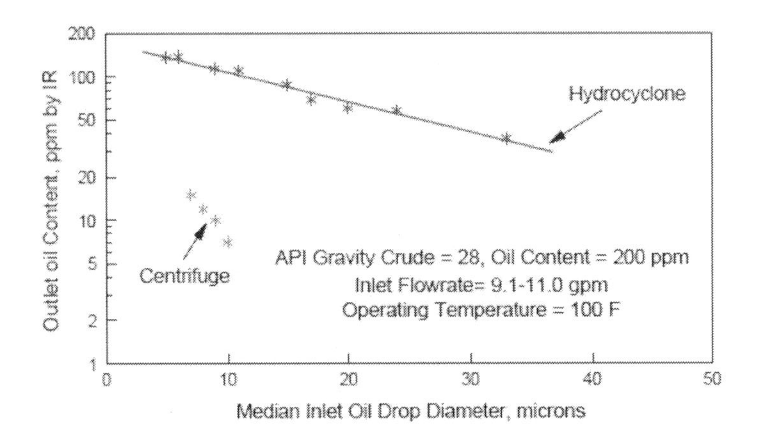

Figure 11.5 Performance comparison of hydrocyclones and centrifuges

The data in the figure above is from Khatib [3].

11.3 Operation and Maintenance:

Petrobras: Petrobras has applied disc stack de-oiling centrifuges with some success [5]. The inlet OiW concentration was between 20 to 200, with an average of 80 ppmv. The centrifuge discharge was 15 ppmv, with essentially no fluctuations. No information was given as to the amount of water being treated, nor the capacity of the units. There was no mention of the mechanical reliability.

Mobil The Netherlands: An early shallow water offshore test was carried out by Mobil in response to water treatment problems in the P/6-A gas field [6]. A high concentration of aromatics in the gas resulted in a high concentration of dissolved aromatics in the water. This increased the hydrocarbon content of the water even when the concentration of dispersed hydrocarbon was on target. In order to avoid expensive technology to remove the dissolved hydrocarbon, various technologies were tested to reduce the dispersed hydrocarbon to as low a value as possible. Corrosion inhibitor further complicated the situation by reducing the oil droplet size due to shear, and by preventing the coalescence of the oil droplets (see Chapter 4 for a discussion of corrosion inhibitor effect on produced water). The Dv50 value of the oil droplets was 6 micron. This is indeed too small for a hydrocyclone to be effective. The testing confirmed this. A disc stack centrifuge was found to be effective in achieving the target total oil-in-water concentration.

Maersk Tyra Installations: As discussed by Rye and Marcussen [7], Maersk Oil and Gas installed disc stack centrifuges on the Tyre East and Tyra West platforms in the Norwegian sector of the North Sea in the early 1990's. These platforms are part of the Danish Underground Consortium (DUC) which also includes the Dan and Gorm fields. The reservoirs are composed of chalk and production is mostly gas. The oil API gravity of production from DUC is 34. Acid number is low (0.3 mg KOH/g) and the asphaltene content is low as well (0.16 wt %). Centrifuge technology was identified as a viable candidate following oil drop size measurements on the platforms. Choking of the Tyra wells from 2,200 psig down to 1,200 psig was causing small oil drops to form. The authors do not report the measured drop diameters but they do comment that drop diameters were small enough to cause separation decline for typical hydrocyclone liners.

Prior to installation extensive testing was carried out. Research and pilot testing was carried out at the Orkney Water Test Center. Feed droplet size was systematically varied and separation efficiency measured in the pilot tests. After this laboratory testing, an extended field trial of one year was carried out Permanent installation of centrifuge skids was then carried out on the Tyra East and Tyra West platforms in late 1992 and early 1993. Each of the two platforms had centrifuge skids installed with capacity of 12,000 BWPD. Each skid was comprised of 2 x 6,000 BWPD centrifuges. The units were installed downstream of a produced water flash vessel. The reject oily water was routed to the slops tank. The treated effluent water was discharged overboard.

Liquid Capacity	2 x 6,000 BWPD
Oil in Water in	max. 1,000 mg/l
Oil in water out	max. 40 mg/l
Pressure rating	80 psig
Motors	2 x 45 kW
Skid weight	12 tons
Dimensions	3.2 x 3.5 x 3 m
Feed pumps capacity	3 x 6,000 BPD
Motors	3 x 11 kW
Skid weight	4.5 tons
Dimensions	1.7 x 3.8 x 2 m

After about six months operation, the initial experience was documented by Rye and Marcussen [7]. The equipment was described as complicated. Effluent treated water averaged between 10 to 20 mg/L oil concentration. This was the anticipated treatment performance. The maintenance routine included:

- complete shutdown 4 times / year
- complete dissassemble, clean, inspection and replacement of worn parts
- typical shutdown and maintenance time: 12 hours x 2 crew

The main observation from these shutdowns is an extensive buildup of chalk on the discs. As a result of this excessive cleaning requirement, a Clean-in-Place skid was purchased and installed. This allows the unit to be flushed with citric acid without dismanteling it. Further, only one centrifuge need be CIP cleaned at a time with the other centrifuge still in operation.

Shell Deep Water Experience: The overall experience with disc stack centrifuges on Shell deep water facilities has been poor. This experience is discussed in some detail in Chapter 19 (Applications – Deepwater Best Practices) Section 19.1.5 (Disc-Stack Centrifuges). The experience is summarized in the table below. In general, the units did a good job of removing oil and solids from water when they were in operation. However a combination of problems caused the uptime to be very low.

Table 11.2 Summary of Operating Experience with Disc Stack Centrifuges

Location	Water Treating Effectiveness	Mechanical Reliability	Status
MP-252	Good	Difficult to maintain	Working
Auger	Not Tested	Failed immediately	Removed
Mars	Good	Failed	Removed
NaKika	Good	Failed	Removed
Holstein	Good	Failed	Removed

The centrifuge on Mars suffered some of the same problems as the unit on Auger plus a set of control system problems, fouling of the disc stack by sticky solids. The rotating equipment problems together with the control problems were sufficient difficult that maintenance staff spent an inordinate amount of time working on the unit. A buildup of solids in the disc stack would cause an imbalance fault which required that the unit be disassembled for cleaning. This only increased the maintenance requirements to the point that the platform could not keep the centrifuge running with the available staff.

- Probably be treated with induced gas flotation if an appropriate flotation polymer is utilized
- Insufficient on-board personnel were available for regular maintenance on the centrifuges
- Large units (higher forces) failed in actual use
- Smaller units (1300 BWPD) clean water successfully, but require significant maintenance and operator training

11.4 Summary and Conclusions:

Disc-stack centrifuges have been widely used to treat industrial waste water for over 100 years. They are used in the dairy, brewery, pharmaceutical, and other food and beverage industries. Suffice to say that these industrial emulsions difficult to treat and represent some of the most difficult emulsions in industry. Thus, it would seem likely that the oil and gas industry would make use of centrifuges. Unfortunately however, applications of centrifuges for produced water treatment have not been as widespread as might be expected. In almost all cases, the separation capability of these devices is impressive. In many cases, oil can be removed down to a micron or a few microns.

However, in stranded locations (such as deep water), the reliability, required maintenance, and required presence of specialists to start up and troubleshoot the devices have been too much for successful operation. As discussed in some detail in Chapter 19 (Applications - Deepwater Best Practices) Section 19.1.5 (Disc-Stack Centrifuges), all of the large disc stack units installed on Shell facilities in the deep water Gulf of Mexico were dismantled after relatively short use. One location, nearer to shore, and employing significantly smaller units, has been run in intermittent service for many years with favorable experience. Other successfully operated units are also discussed below.

References to Chapter 11

1. Z.I. Khatib, M.S. Faucher, E.L. Sellman, "Field evaluation of disc-stack centrifuges for separating oil / water emulsions on offshore platforms," SPE 30674 (1995).

2. W.M.G.T. van den Broek, R. Plat, M.J. van der Zande, "Comparison of plate separator, centrifuge and hydrocyclone," SPE 48870 (1998)

3. Z.I. Khatib, "Handling, treatment and disposal of produced water in the offshore oil industry," SPE 48992 (1998)

4. A. Finborud, M. Faucher, E. Sellman, "New method for improving oil droplet growth for separation enhancement," SPE – 56643, paper presented at the SPE ATCE, Houston, TX (1999).

5. G. Cavalcanti, "Water Treatment for Brown Fields," 3rd International Seminar of Oilfield Water Management, Rio de Janeiro (7-9 June 2010).

6. F.J. op ten Noort, J.P Etten, R.S. Donders, "Reduction of residual oil content in produced water at offshore gas production platform P/6A," SPE 20882 paper presented at the Europe 90 conference, The Hague, Netherlands October (1990).

7. S.E. Rye, E. Marcussen, "A new method for removal of oil in produced water," SPE 26775 (1993).

CHAPTER TWELVE

Flotation

Chapter 12 Table of Contents

12.0 Introduction ... 133

12.1 Flotation Practical Applications ... 133

- 12.1.1 Introduction – Broad Overview ... 133
- 12.1.2 Mechanism of Separation & Performance Variables ... 135
- 12.1.3 Chemical Application ... 138
- 12.1.4 Selection of Flotation Gas ... 142
- 12.1.5 Material Balance – Reject Flow Rate and Concentration ... 143
- 12.1.6 Process Configuration, Rejects Handling, Control ... 145
- 12.1.7 Operation and Maintenance ... 147
- 12.1.8 Troubleshooting Procedure & Bench Top Flotation Testing ... 148
- 12.1.9 Benefits and Drawbacks of Flotation ... 154

12.2 Flotation Equipment ... 155

- 12.2.1 Design Considerations ... 155
- 12.2.2 Bubble Coalescence ... 159
- 12.2.3 Classification of Designs (MIGF, HIGF, VIGF, VDGF, etc) ... 161
- 12.2.4 Horizontal Multistage Mechanical IGF (MIGF) ... 162
- 12.2.5 Horizontal Multistage Hydraulic IGF (HIGF) ... 170
- 12.2.6 Vertical IGF (VIGF) ... 172
- 12.2.7 Dissolved Gas Flotation (DGF) ... 177
- 12.2.8 Gas Flotation Tank ... 182
- 12.2.9 Sparger Systems ... 182

12.3 Flotation Theory, Design and Modeling ... 183

- 12.3.1 Modeling of Flotation Performance ... 184
- 12.3.2 Collision Frequency (F_{coll}) ... 184
- 12.3.3 Bubble Rise Velocity (V12) ... 186
- 12.3.4 The Bubble / Water Interface ... 187
- 12.3.5 Capture Mechanisms and Spreading Coefficient ... 188
- 12.3.6 Capture Efficiency (E_{capt}) ... 190
- 12.3.7 Modeling the Batch Flotation Process ... 193
- 12.3.8 Correlating Performance of Continuous Flotation Cells ... 196

12.3.9 Sweep Factor 196
12.3.10 Flux Factor 199

References to Chapter 12 202

12.0 Introduction

In this chapter, a detailed understanding of flotation is provided. The subject is approached from several perspectives – practical applications, equipment performance, design and modeling. The questions that are answered in this chapter include:

- how flotation works;
- how to evaluate equipment performance;
- how to improve the performance of a flotation unit;
- hot to improve the performance of a system that includes a flotation unit;
- how to troubleshoot;
- how to design equipment.

The chapter provides a wide range of details, from the basics through the most advanced.

Gas flotation has been used in the mineral processing industry for over a century [1, 2]. During the past 50 or so years, flotation has been applied to the processing of recycled paper to remove ink, toner particles and other waste contaminants from paper fibers [3]. Its use in the upstream oil and gas industry dates to roughly the 1960's and today is widespread. It is used extensively offshore as the final stage of water treatment for overboard discharge. It is used onshore as a critical pre-treatment stage in recycling produced water for steam flood. It is used in some parts of the industry to remove oil, solids and algae from impoundment water upstream of a desalination unit. It is used for solids and polymer removal in recycling water for hydraulic fracturing.

The term 'Gas Flotation' is sometimes used as if it is one technology. It is not. There is a wide range of flotation technologies available today. Some have greater than 90 % separation efficiency, but are large and heavy; others have 50% or less separation efficiency and are compact and relatively inexpensive. It would not be sensible to select a small and inefficient flotation unit for an application that requires high separation efficiency. However, the authors have seen this many times. This statement does not apply to all companies. For example, at least one company that operates in the deepwater Gulf of Mexico has a consistent history of selecting large multistage flotation units in deepwater despite their large size and high cost. All things considered, this is the correct selection for the particular operating conditions encountered by that company in that location. The units perform well and consistently help the operator to meet the water quality requirements for overboard discharge. The objective of this chapter is to provide guidance in selecting the most appropriate flotation technology for a given application.

12.1 Flotation Practical Applications

12.1.1 Introduction – Broad Overview

There are many different designs of flotation available on the market [4]. Some flotation equipment delivers large bubbles and a large volume of gas, while other equipment produces small bubbles and a small volume of gas. While it would be most advantageous to have a machine that delivers small bubbles and a lot of gas, unfortunately there is no machine on the market which does this. The reason for this is partly related to mechanical configuration and to the cost of generating gas bubbles. It turns out that the cost to generate gas bubbles is roughly inversely proportional to the gas bubble size. Large bubbles are inexpensive to generate. Small bubbles are more expensive. Thus, there is a tradeoff between gas volume and bubble size. Capture efficiency increases as the bubble size decreases because larger bubbles move too fast to capture the slow moving oil drops. But this effect is less pronounced when flocculating chemical is added.

The following table gives the range of variables for commercially available flotation equipment on the market. No particular machine (model of flotation) has all minimum or maximum values. This slide illustrates the differences that can be found in the marketplace from one flotation technology to another.

Table 12.1 Range of values for important variables in flotation

Parameter	Minimum	Maximum
Gas/Water Ratio	0.12	8.5
Residence time (minutes)	0.5	4
Height/Width ratio (m high/m diameter)	1	4
Bubble size (microns)	30	1,200
Reject percentage	0.5	7

In the E&P industry, there are several factors that drive the selection of a particular flotation technology. These include:

- total cost (including the indirect costs of weight, space and chemical injection)
- schedule
- operability
- after-sales service
- separation efficiency

Among these factors, separation efficiency is often the most difficult to obtain reliable and trustworthy information about. In this chapter data and models are presented to help predict performance of new installations and to troubleshoot existing installations. Model calculations will be combined with field data to explain why some flotation designs provide greater separation efficiency than others.

Preferred Process Line-up: Flotation is a secondary process. It relies on contact between gas bubbles and the contaminant particles. As discussed previously, in this book, primary separation involves migration of oil and solids as a function of the density difference of these contaminants with water. Stokes Law usually applies to primary separation. Since density difference of the contaminants is not the main factor in separation efficiency for flotation, it is not classified as primary separation.

If a hydrocyclone unit is present in the system, flotation is located downstream of the hydrocyclone unit. It may be located upstream of various other primary, secondary and tertiary processes such as centrifuge, filtration, water softening, etc.. The designations of primary and secondary do not indicate the location of the technology in the process configuration.

A process flow diagram for a deep water Gulf of Mexico platform is shown in the figure below. Ultimately, the reject must be recycled in order to capture the oil contained in the oily water recycle. Here, flotation reject is routed to a Slop Tank for chemical treatment and settling time. This reduces recycled chemical, and provides additional oil/water separation before recycle into the rest of the system.

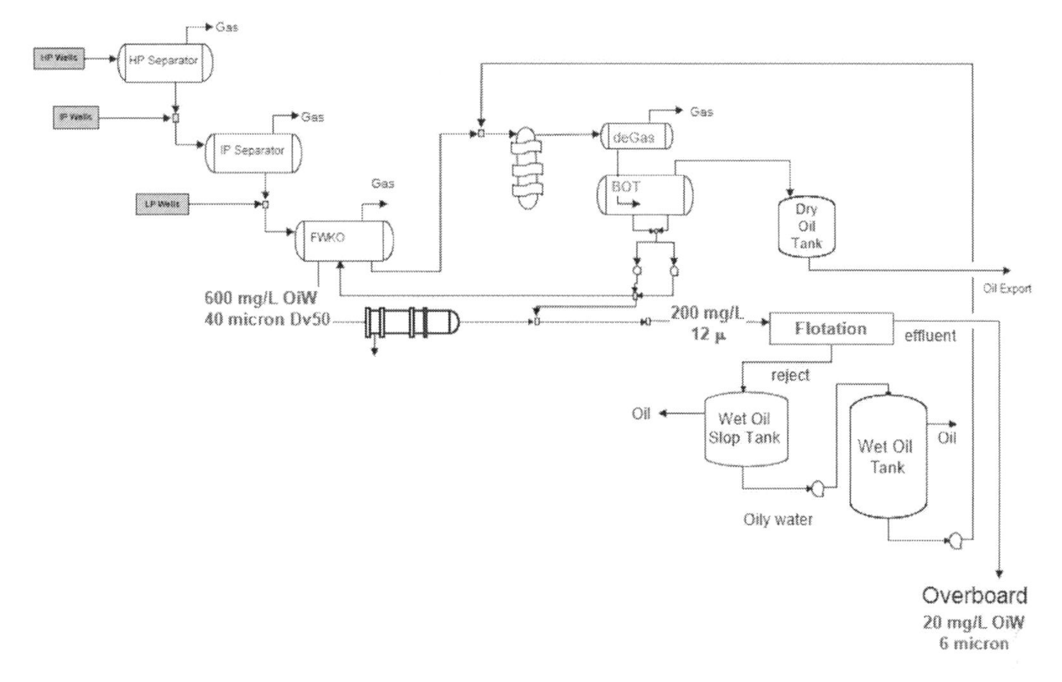

Figure 12.1 Offshore deepwater GoM example process lineup for flotation. In this example, flotation is downstream of the hydrocyclone and just upstream of the point of overboard discharge. Representative data from a particular platform are shown. A wide range of oil-in-water concentration and oil droplet size are encountered from one installation to the next.

12.1.2 Mechanism of Separation & Performance Variables

To begin the discussion of different flotation technologies, there must already be a common understanding of how flotation works. There is a slight difference in the way that oil is separated versus the way solids particles are separated. Oil separation is discussed first. Bubbles rise and crash into oil drops. Some of the oil drops are captured by the bubbles. The bubbles continue to rise to the water surface where they join other bubbles to make an oily foam. The foam continuously collapses allowing pools of oil to form. The oil is swept off, scraped off, weired off, pushed off or allowed to fall off the surface of the water into a separate trough.

Relative to the oil drops, the gas bubbles rise rapidly. In the figure below, the collision path is shown for two bubbles. As shown, each bubble may have several oil drops in its collision path. The number of oil drops that are in this collision path divided by the rise time defined the collision frequency. The collision frequency depends on the concentration of oil drops, the concentration of gas bubbles, and on the projected areas of the oil drops and the gas bubbles.

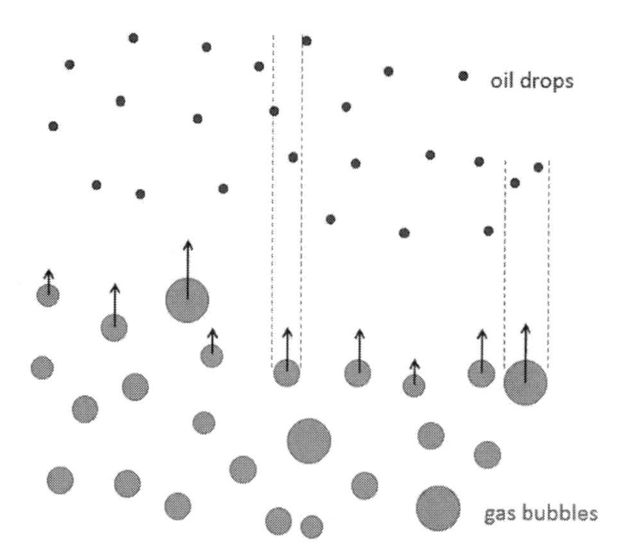

Figure 12.2 Schematic illustration of gas bubbles rising (blue dots) and oil droplets. The collision path of two of the bubbles is shown

In addition to collision frequency, capture efficiency is important. When an oil drop and a bubble collide, the oil drop doesn't necessarily get captured by the bubble. The oil drop can slide off the surface of the bubble, or be carried around the bubble by the hydrodynamics of the flow. Whether or not an oil drop gets captured depends on the complex fluid dynamics around the rising bubble, and on the surface chemistry between the bubble and drop.

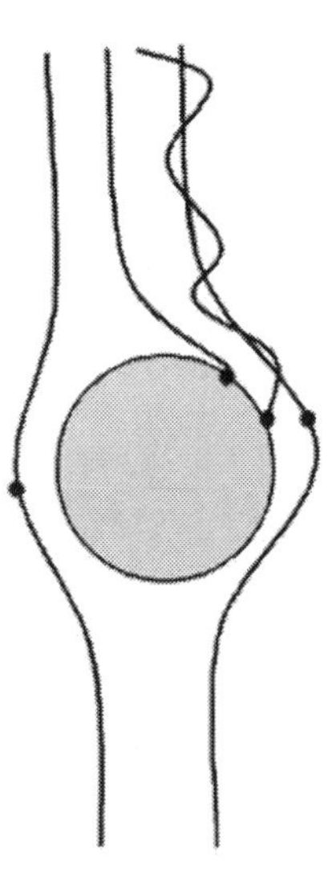

Figure 12.3 Schematic illustration of a single flotation bubble is shown with four oil droplets and their trajectories. Three of the oil droplets follow simple trajectories. One of those trajectories is close enough to the bubble that collision occurs. A fourth oil droplet follows a path influenced by fluid turbulence which results in this case in a collision with the bubble. Turbulent and non-turbulent trajectories would not typically be found together in a particular flow

Oil Separation: Bubbles with their captured oil drops rise to the water surface where an oily froth is formed. There are several mechanisms in the design of flotation devices for removing the oil layer from the top of the water. The important point for flotation is that once the oil has been deposited on the water surface, there is a very high probability that it will be separated from the produced water.

Solids Separation: The capture mechanism for solid particles is similar to that for oil drops in the sense that gas bubbles rise through the produced water and collide with the solids particles. Those particles that get captured by the bubbles are then carried to the water surface. Unlike oil, once a solid particle is deposited on the water surface, there is a chance that it will sink from the surface given the fact that most solids particles are denser than the produced water. This is not the case with all solids particles since some have oil attached and may be neutrally buoyant or even lighter than water. Also, solids particles can agglomerate at the water surface into a loosely agglomerated solid mass. When these solid agglomerates sink there is a finite probability that they will be discharged with the treated effluent. This is one of the common mechanisms for the so-called "rose bud sheens" that can sometimes be seen in the treated water discharge from an offshore platform. Rose buds are small, round and isolated sheens. They are usually due to sinking of oily solids from the water surface inside the flotation unit into the treated water effluent.

Important Variables in Flotation: Flotation is a contact separation process. It relies on the contact of gas bubbles with oil drops and solid particles. As discussed in detail below, the important variables are:

- gas / water ratio (volumetric ratio);
- bubble size (diameter);
- staging (single, dual, multi-stage);
- chemical application (surface chemistry);
- contaminant size and surface chemistry.

Gas/Water Ratio and Bubble Size: In recent years there has been an emphasis on the generation of small bubbles, in the range of 30 to 100 microns. Small bubbles have high capture efficiency. But they are somewhat costly to generate. Being costly, some flotation designs that generate small bubbles only do so with a relatively small volume of gas. When that is the case, it is possible that performance (oil/water separation) will suffer even though small bubbles with high capture efficiency are used. Thus, small bubbles do not necessarily lead to better performance.

A quick analysis of the total surface area of the bubbles can show the relative importance of bubble size versus gas volume. The greater the bubble surface area, the greater the likelihood that oil droplets will be captured. The following formula gives the surface area of the bubbles per unit volume of feed water.

$$S_g / V_w = \frac{C_1}{d_b} \frac{Q_g}{Q_w} \qquad \text{Eqn (12.1)}$$

where:

S_g	=	gas surface area (m^2)
V_w	=	water volume (m^3)
d_b	=	average bubble diameter (m)
Q_g	=	gas flow rate (m3/sec)
Q_w	=	water flow rate (m3/sec)
C_1	=	dimensionless constant (roughly equal to 6)

This parameter, Sg/Vw is the surface area of gas per unit volume of water. It is directly related to the surface area of bubbles that is available for capturing contaminant particles or droplets per unit volume of water. There are essentially two ways to increase this parameter. One is to decrease the bubble diameter. The other is to increase the gas / water ratio. Both of these changes would have the same percentage or relative effect. If the bubble diameter is halved, the parameter would go up by a factor of two. If the gas / water ratio is doubled then the parameter would go up by a factor of two. Thus, both bubble diameter and gas / water ratio are important, not just bubble diameter. As will be discussed in detail below, other parameters are also required to predict separation efficiency. The capture efficiency has a complex relationship with the bubble diameter which must also be taken into account.

Staging: The use of multiple stages in a flotation unit is a very effective way to increase separation efficiency. If for example each stage had 50 % efficiency, then each would leave 50 % of the incoming oil in the product water sent to the next stage. For four chambers in series this equates to an ideal efficiency of 0.5 x 0.5 x 0.5 x 0.5 or 94 % overall efficiency.

12.1.3 Chemical Application

Chemical application for any type of flotation is critical. Leech and co-workers at Shell and Envirotech Inc. [5] evaluated several performance variables and determined that flocculation chemical addition is the most important performance variable. Performance can be enhanced significantly. Enough time must be provided for flocculation to occur. Also, no shearing can be tolerated between the point of chemical addition and the flotation unit. In other words, chemical injection must occur downstream of all valves and pumps. Sometimes this will limit the retention time for the chemical. Shear will destroy any chemical effect.

On the Bullwinkle platform (deepwater US), the mode of chemical injection was critical in chemical dispersion and dramatically affected oil removal efficiency. Initially the chemical was being injected in its "neat" form (in liquid state as provided to the platform). Performance of the flotation unit was poor. After modification to a pre-diluted injection mode by a 20:1 dilution with fresh water, oil removal greatly increased while overall chemical injection concentration was greatly reduced. This occurred because the chemical in use tended to encapsulate into small "blobs" when injected neat into the salt water, while pre-dilution with fresh water achieved immensely greater chemical dispersion in the fluids to be treated. Better mixing resulted in greater overall chemical performance and much improved performance of the flotation unit.

The role of a flocculating agent is shown in the figure below. As shown, in the ideal case there are two steps in the flocculation mechanism. In the first step, a molecule or network of molecules attaches to one oil drop. In the second step the flocculating agent molecules on one drop attach to those on another drop. The final result is referred to as a floc.

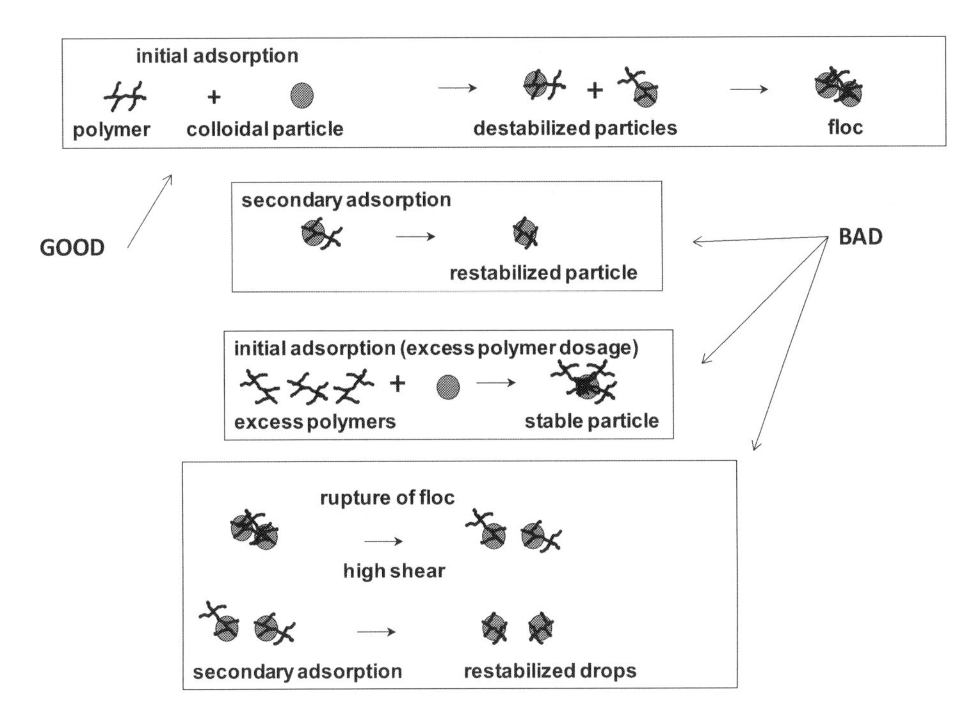

Figure 12.4 Mechanisms for the interaction of flocculant and oil droplets. The top reaction path is the most favorable for flocculation. The other reactions paths lead to droplet coating without the desired flocculation of other droplets

Flocculation is used extensively in industrial and municipal water treatment to remove solid particles. In a typical application, the water with flocculating agent already added will be allowed to settle in a large clarifier. If the reaction is effective, it is possible to see large (10 to 20 cm or larger) whitish cloud-like objects floating in the water. These are the flocs of polymer molecules attached to solids and each other in large networks. It is apparent that these flocs are delicate objects. Thus, once the reaction has been initiated, the mixture should not be subjected to shear. This is discussed further below.

The following data was taken during a chemical field trial for the selection of flocculating agents upstream of a flotation unit. The Jorin Visual Particle Analyzer was used to measure drop size. Drop size was measured upstream of the flotation unit and downstream of flocculant application. At the same time, the separation efficiency of the flotation was evaluated. The data provides a correlation between flocculation, drops size, and flotation performance.

The flotation unit used for the test was a Unicel which is described below. During the test, the flow rate and hence residence time was held constant. The residence time in the reactor tube was three minutes. Without chemical, the separation efficiency is 23 %. With proper chemical selection and dosage, the separation efficiency increases to > 80 %. Note the presence of flocs of oil drops in the range of 64 to 84 micron. Feed oil drop size Dv50 = 25 micron. Effluent oil drop size Dv50 = 14 micron. (compare with Mars Wemco Effluent Dv50 = 6 micron). The importance of flocculating agent is well known, at least qualitatively. The data demonstrate in a quantitative way the importance of flocculating agents in achieving high separation performance in a flotation unit.

Three different chemicals were used at various concentrations. The Jorin drop size data are shown in the Figure below. Altogether there are 8 different combinations of chemical and concentration.

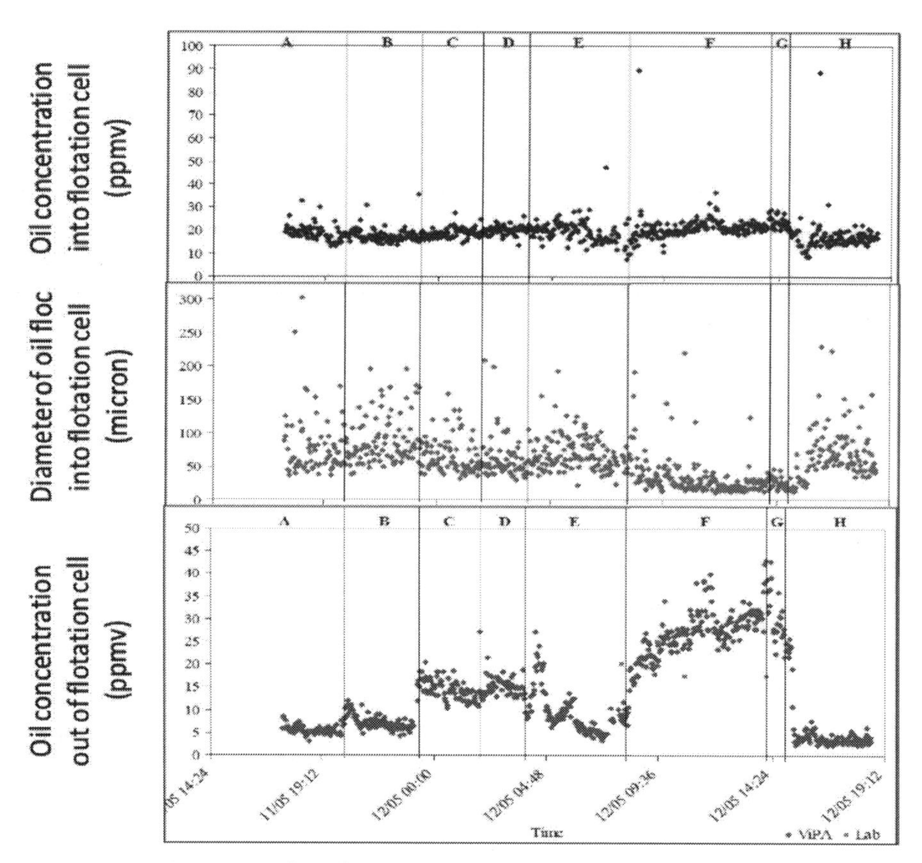

Figure 12.5 Online oil concentration into and out of a flotation unit. Also shown in oil droplet diameter data downstream of flocculant addition and upstream of the flotation unit.

Table of chemical selection and concentration.

Test Segment	Chemical	Concentration (ppmv)
A	1	16
B	2	12.8
C	2	9.6
D	2	8.0
E	1	16
F	3	6.5
G	3	9.6
H	1	16

The data shown in the figure and Table above can be analyzed as follows. The separation efficiency is plotted as a function of the drop diameter. This is done without regard for which chemical or chemical concentration is used. Of course the data could also be used to determine which flocculating agent and concentration is most effective at flocculation, and therefore at flotation. However, the figure below simply shows the relationship between flocculation and flotation performance. The larger the number of droplets in a flocculated mass, the greater the flotation effectiveness.

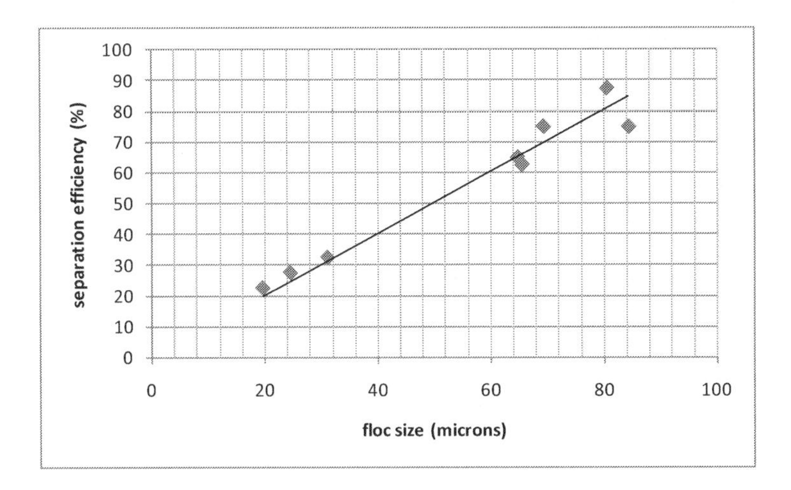

Figure 12.6 Online oil concentration into and out of a flotation unit.
Also shown in oil droplet diameter data downstream of flocculant addition and upstream of the flotation unit

Frother Chemicals: Strickland [45] carried out studies of chemical additives to determine their effect. These chemicals were short chain alcohols, not flocculating agents. He found that above 3 % TDS (30,000 mg/L) none of the conventional frother chemical had any effect. However, below this value of TDS, the short chain alcohols improved separation efficiency. The effective chemicals reduced the bubble size.

Upstream Process Chemicals: As shown in the figure below, the selection of demulsifier for oil dehydration can dramatically affect the performance of the flotation unit.

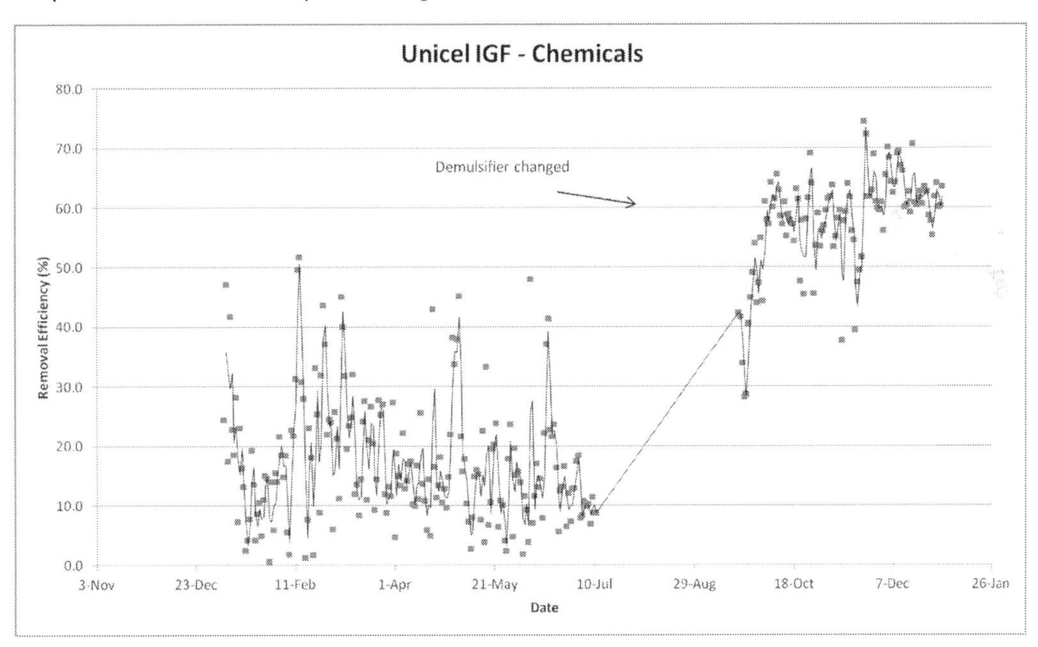

Figure 12.7 The oil / water separation efficiency of a Unicel IGF over the time period when the upstream oil dehydration demulsifier was changed out. As shown, the second demulsifier has much less detrimental effect on the flotation efficiency

12.1.4 Selection of Flotation Gas

In most upstream oil and gas applications, natural gas is used for flotation. Various gasses can be selected for flotation including hydrocarbon gas, nitrogen, air, or carbon dioxide. Although air can be used, it is not common to do so in upstream oil and gas applications due to the hazards of fire and explosions when hydrocarbons and oxygen are mixed, and the fact that most produced water streams would form significant quantities of solids upon expose to air. The oxygen in air will cause corrosion, and it will oxidize dissolved ferrous iron to the ferric form which is essentially insoluble. Solids precipitation will then be a problem. The use of air in a flotation process is only viable if the produced water has already been contaminated with air. This is almost always the case in refinery applications where waste water will usually have been treated in an aerobic digestion pond. Thus, dissolved air and induced air flotation is common in refineries but not in upstream oil and gas facilities.

Natural gas is the preferred choice for flotation gas because it has a natural affinity for oil drops and oil coated solids. This effect is discussed below in Section 12.3.6 (Capture Mechanisms and Spreading Coefficient). Natural gas also does not precipitate solids from the produced water. When natural gas is used, it is most economical to have an internal recycle process such that only a small fraction of gas is consumed by the flotation process, and the gas volume required for flotation is internally recycled. In an upstream oil and gas facility there are many different sources for such gas. The main considerations in selecting a source of gas are available / required pressure for the flotation unit, gas volume, and richness of the gas. The later consideration has to do with the BTEX content and whether or not contact with rich gas will result in aromatic hydrocarbons getting dissolved into the produced water.

Dissolved Contaminants: Flotation systems are not designed to remove dissolved contaminants. In some cases, flotation units do remove volatile dissolved organics. In those cases, the gas used for flotation has a lower concentration of the volatile organic compounds than the equilibrium concentration of these compounds (such as the BTEX compounds). In this case, volatile organics vaporize from the produced water into the flotation gas and if the gas is not recycled, these volatile compounds can be stripped from the produced water. If the flotation gas is recycled, these volatile components build up and no net stripping occurs. Conversely, the opposite can also be true. That is, it is possible that the flotation gas contains a higher concentration than the equilibrium and organics can dissolve from the gas into the produced water.

Gas Selection - Effect on the Bubble Diameter: The bubble diameter that is generated will theoretically be different depending on the gas that is used. If an eductor is used to generate the bubbles, then the average bubble diameter will depend on the turbulence intensity in the eductor and on the water/gas interfacial tension and density of the gas. As discussed by Rawlins [6], there is a slight difference in the interfacial tension, depending on the gas that is used. The difference is small but it appears that the use of air will give a slightly higher water/gas interfacial tension. According to turbulence theory, this will result in slightly larger bubbles for air. The salinity of the water and the temperature has a greater effect than the selection of gas.

For most bubble generation devices, bubble diameter does depend on temperature. This is due to the fact that the water/gas interfacial tension decreases with an increase in temperature. Practically speaking this means that pilot testing should be carried out at system temperature. For onsite work this is not an issue. Any pilot unit that uses a slipstream directly will have close enough temperature. Any testing where a batch sample is taken, then the tests need to be run soon in order to have the same temperature as the system. The water / gas interfacial tension increases with salinity. This results in a slightly larger bubble diameter, for typical bubble generation devices, all other factors being equal.

The Effect of Pressure: Most flotation units are operated at relatively low pressure. Generally speaking, pressure does not have a great effect on most liquid/liquid separation processes. A moderate increase in pressure does not have a significant effect on interfacial tension, wax appearance temperature, solubility of mineral solids and migration of interfacially active components to the interface. Large increases in pressure can have an effect on thermodynamic equilibrium, particularly where solids formation occurs. But in general the effect is limited. There is some effect of the operating pressure on the stability of asphaltenes. But this effect usually occurs at pressure near the fluid bubble point. By the time the fluids reach the flotation system, that effect is negligible. All things considered, liquid / liquid separation is not strongly dependent on the absolute pressure.

Gas pressure does have an effect on bubble diameter. Since gas is compressible, the higher the pressure, the smaller the gas bubble. This does have an effect on flotation. Also, the size of gas bubbles that nucleate in a dissolved gas flotation unit is smaller at lower pressures. Both of these effects are discussed quantitatively in Section 12.3 (Flotation Theory, Design and Modeling).

Gas Density: Theoretically, a heavier gas will rise more slowly than a lighter gas. The rise velocity of a gas bubble depends on its gravimetric density, as well as other factors, as given by Stokes Law. The gravimetric density is measured in kg/m^3 and so depends on molecular weight and the molar density. Molar density depends on pressure, through the ideal gas law. The gravimetric density difference, which controls the rate of rise, all other things being equal, is calculated in the following table. As shown, the gravimetric density difference between commonly used gasses has negligible difference from one gas to another (1,098.8 versus 1,099.3).

Table 12.2

Gas	molecular weight (gr/mol)	molar density (mol/m3)	gravimetric density (kg/m3)	density difference (kg/m3)
nitrogen	28	44.6	1.25	1,098.8
methane	16	44.6	0.71	1,099.3
natural gas	19	44.6	0.85	1,099.1

12.1.5 Material Balance – Reject Flow Rate and Concentration

As with most water treatment equipment, use of a flotation unit results in a treated effluent stream and a reject stream. In many activities related to troubleshooting and design, it is important to have a rough idea of the concentration of oil in the reject stream and the flow rate of the reject, relative to the inlet feed flow rate. Most equipment suppliers will specify the recommended reject rate. It is usually expressed as a percentage of the feed flow rate. In some cases, it is specified as a percentage of the design flow rate. This is the case typically for units that employ a spillover weir for reject. Once a flotation unit is commissioned and put into service, the separation efficiency should be evaluated as a function of the reject rate. Once this relationship is established, then a material balance of the entire system should be performed to determine the optimum reject rate for the individual unit (flotation) unit and for the system as a whole. It is counterproductive to run a water treatment unit at high reject rate if the rest of the oil and water system cannot handle the high reject flow. The starting point for system optimization is to develop a good understanding of the performance of the flotation unit as a function of the reject rate. One of the challenges in determining a material balance for a flotation unit is to measure the flow rates. Since flotation streams are usually liquid packed, it is possible to obtain accurate results from a clamp-on flow meter.

The following illustrative mass balance calculations are carried out. A Wemco model 76 is chosen for these calculations. This type of flotation unit is discussed in Section 12.2.4 (Horizontal Multistage Mechanical IGF). This particular model has a design inlet flow rate of 25,725 BWPD. This equates to 2.8 m3/minute. The reject rate is recommend to be between 5 to 10 % of design feed flow rate. The calculation is carried out for an actual operating flow rate of 20,000 BWPD. Also, it is assumed that the oily feed water contains 200 mg/L of oil, and no solids. For the purpose of this illustration, the reject rate is chosen to be 7 %. Typical Wemco performance in this application would be around 95 % separation efficiency.

Assumptions:

Q_F = 2.2 m^3/minute

R = 7 % of design feed flow rate

E = 95 %

Using the Material Balance equation for a generic flotation unit:

$$Q_F = Q_E + Q_R \qquad \text{Eqn (12.2)}$$

where:

Q_F = feed flow rate (m3/min)

Q_E = treated effluent or product flow rate (m3/min)

Q_R = reject flow rate (m3/min)

E = separation efficiency (%)

R = reject percentage $R = Q_R \text{ x } 100 / Q_F$ (%)

writing the material balance in terms of the concentration of oil in water:

$$Q_F C_F = Q_E C_E + Q_R C_R \qquad \text{Eqn (12.3)}$$

where:

C_F = feed concentration of oil-in-water (mg/L)

C_P = product concentration of oil-in-water (mg/L)

C_R = reject concentration of oil-in-water (mg/L)

Using this, and the other values given, the reject flow rate is:

$$Q_R = 0.07 \text{ x } 2.8 \, m^3/min = 0.2 \, m^3/min \qquad \text{Eqn (12.4)}$$

The values of the other variables are:

Q_R = 0.2 m^3/minute

Q_E = 2.0 m^3/minute

In addition, the separation efficiency is given by:

$$E = (C_F - C_E) \text{ x } 100 / C_F \qquad \text{Eqn (12.5)}$$

Using this equation allows calculation of the treated effluent (product) oil-in-water concentration:

C_F = 200 mg/L

C_E = 10 mg/L

This allows the calculation of the reject oil concentration:

$$C_R = (Q_F C_F - Q_E C_E)/Q_R = (440\text{ g/min} - 20\text{ g/min})/(0.2\text{ m}^3/\text{min}) = 2{,}100\text{ g/m}^3 = 2{,}100\text{ mg/L}$$

Eqn (12.6)

the material balance can be verified:

$Q_F\, C_F \quad = 440$ gr/min

$Q_R\, C_R \quad = 420$ gr/min

$Q_E\, C_E \quad = 20$ gr/min

These calculations illustrate a couple of points about flotation and about doing calculations for produced water in general. First, the oil concentration in the reject is usually relatively low, compared to typical hydrocyclone reject. Thus, the location for recycle must be chosen carefully. It would not make sense to route this reject directly to an oil stream since there is so much water associated with this stream. Also, this reject stream will be laden with water treating flocculant.

As discussed previously, when doing calculations in produced water it is almost always easier to convert to metric units immediately. This is because so many of the quantities of interest are metric based, such as concentration (mg/L), drop diameter (micron), viscosity (cP), interfacial tension (dyne/cm), etc.. While these units may seem unfamiliar at first, they do allow very quick calculation that can be done without spreadsheet or calculator.

12.1.6 Process Configuration, Rejects Handling, Control

As with most water treatment equipment, integration of a flotation unit into a facilities or process is critical to successful performance. The most important aspect of integration is handling the reject. From the standpoint of the flotation unit, the reject is just the reject. But from the standpoint of the overall system, the reject is a recycle stream. As such, it represents an endless list of possible problems in terms of compatibility, contamination, flow rate fluctuations, and so on. In some systems, the units and overall system functions very well but one of more reject streams are problematic and cause the entire system to struggle in many ways. As discussed in the section above, the concentration and flow rate of the reject stream should be known with reasonable accuracy (+/- 20 %). Also, it is critically important to the selection of slops handling to have an understanding of the concentration and chemical nature of the contaminants in the reject stream.

Objective of Reject Handling: The primary objective of reject handling is to further condition the oily water to make it suitable for recycling into the main process so that the oil can be exported with the primary oil stream. Ultimately, this is what must happen. The oil must be captured by or routed into the main oil stream since there is no other outlet for this oil. The same objective can be said of the oily solids contained in the reject. Also, there will typically be some concentration of residual water treatment chemical mixed with the other contaminants. The type of chemical and its chemistry may have an influence on the optimum destination for routing the oily reject from the flotation unit. This too may be required to exit the facility in the main oil export.

Options for Reject Handling: The oil that accumulates at the top of the water, is typically in the form of a froth, and often has a mixture of solids, oil, water treatment chemicals, and production and flow assurance chemicals from upstream. It is typically collected with other "slop oil" so that it can be subjected to further processing before attempting to route it to an oil stream. Additional processing may include settling, chemical solvent treatment, heating, or other process.

The oily water that is rejected from most flotation units is relatively coalesced such that there is a free oil layer (recall that free oil is oil that separates from a sample within one minute of standing). This free oil should be mixed with the main oil process stream in order to capture it. The water associated with this oil is usually quite contaminated with oil droplets. Depending on the concentration, flow rate, and contaminant chemistries it may be prudent to carry out an additional processing step prior to introducing this stream into the main process. The objectives of this treatment step might include the objectives mentioned above, as well as the transfer of water treatment chemical into the oil phase, or inactivation of this chemical.

There are several options for handling this reject stream. Depending on the objectives, the following equipment could be used:

- Slops tank
- Centrifuge
- Back-washable filtration
- Media filtration

A slops tank can be used with or without chemical treatment to coalesce the oil further into a more concentrated phase, and allow some additional concentration. This can be done by heating, promoting oil drop coalescence, settling the oil and water to make the oil more concentrated and the water somewhat cleaner, and by processing the oily water with some water treatment technology. The primary objective is to promote oil drop coalescence and to transfer the oil into the primary oil stream for export from the facility. The same objective can be said of the oily solids mixed with the oil. Also, there will typically be some concentration of residual water treatment chemical mixed with the other contaminants. The type of chemical and its chemistry may have an influence on the optimum destination for routing the oily reject from the flotation unit.

Slops or Reject Processing Tank: Reject handling can be greatly facilitated by the use of a properly designed slops tank. Emphasis is on the phrase "properly designed." In some process designs the slops tank is given short consideration – after all, it is only a slops tank. However, the performance of the entire system may depend strongly on the proper design and operation of a slops tank and associated equipment. Slops tank design involves residence time of each phase (oil and water) which determines size and shape of the vessel. Vessel internals will likely be important. Associated equipment might include chemical injection. Given that the Slops Tank is located on the reject side of the flotation unit, it is usually designed for low pressure. This implies the need to use one or two transfer pumps to discharge fluids from the tank to the destination. If the flotation reject stream is intended to be transferred to some point in the upstream side of the process, pressure will be required. For this purpose a low to moderately low shear centrifugal pump should be used with variable speed drive and not valves that would likely cause shearing.

It is likely that the rejects from a flotation unit will contain oily solids. Flotation can be quite effective at removing oily solids from the water. The term oily solids is intended to refer to solids that have at least some of their surface wet by oil components such as resins, asphaltenes, waxes etc.. These solids will remain oil wet despite most attempts to use dispersants. Further, some of these solids will be neutrally buoyant thus making their separation from water in the recycle system somewhat difficult. Thus, it is important to ensure that these solids are routed into the main oil stream so that they do not become recycled and build up in the system. It is important to note that in many produced water systems, the flotation unit is the most effective separation process for oily solids and handling the flotation reject is the most important aspect of removing solids from the system.

Any chemical treatment applied to the oil that is collected in the slops tank should aim to retain the solids in the oil phase. The best outlet for oily solids from a facility are in the oil itself. For this reason, one of the preferred chemical treatments in a Slops Tank is an aromatic solvent such as HAN (heavy aromatic naphtha). This will reduce the viscosity and stickiness of the oil phase without causing the solids to drop out.

If the solids fed to the flotation unit are water-wet, and remain so through the flotation unit, they may stay in the water and be discharged with the treated effluent. Provided that these solids remain relatively free of oil, they will not contaminate the overboard discharge, nor will they cause a sheen, and will not contribute to the Total Oil and Grease of the treated effluent discharge water.

12.1.7 Operation and Maintenance

Sand Accumulation: Accumulation of sand in the bottom of the flotation unit is a typical problem. Only a couple of models are designed to easily remove sand. As discussed elsewhere, the most benign consequence of sand accumulation is the volume that it occupies. This volume reduces the residence time of oily water in the unit. In some cases, this is not a significant effect because the sand level is below the effective reaction zone. In other cases, the presence of sand changes the fluid flow and reduces the height of the reaction zone. However, as mentioned, this is the least detrimental effect of sand.

Sand Scouring: A related problem with sand is the scouring that occurs. Most flotation units are constructed of carbon steel with a baked on epoxy coating. The coating is usually hard and abrasion resistant. However, all such statements are relative. With time, and with a steady stream of sand particles, the epoxy coating can become worn away. Then this happens, the carbon steel becomes exposed and corrosion is initiated. Corrosion produces dissolved iron which can combine with hydrogen sulfide

Bacteria: One of the more detrimental effects is that it provides an almost ideal environment for biological contamination. The pore spaces of the sand provide a shelter for microbes both in terms of the fluid flow but also in terms of the dosing of biocide. Once a biological colony has built up in an accumulation of sand, the sand becomes cemented together due to the biopolymers in the biofilm. Most biocides do not penetrate such a biomass. Once this happens, it becomes very difficult to remove the sand accumulation without a shutdown and mechanical cleaning. The other detrimental consequence of such sand/biomass is the generation of hydrogen sulfide from SRB, and corrosive acid from the APB.

It is prudent to periodically test for bacteria upstream and downstream of the flotation unit. MPN and / or ATP methods are useful for this purpose. There should be no increase in bacteria count, regardless of method used to test. If there is an increase, it is indicative of a potentially serious condition. Keep in mind that such tests are based on planktonic (free floating) bacteria. If free floating bacteria increase from feed to discharge it is usually indicative of a significant build up of sessile bacteria in the unit.

Scale due to CO_2 Breakout: Another common problem with flotation units is the buildup of calcium and magnesium carbonate in the discharge line and control valve. The flotation unit is usually located at the tail end of the water treatment system. It is also typically operated at relatively low pressure. The carbon dioxide partial pressure is typically at the lowest value of the entire system. Thus, the produced water pH is at its highest value. If the produced water contains a significant concentration of carbonate and Type II Earth ions (Ca, Mg, etc), then carbonate scale can buildup.

12.1.8 Troubleshooting Procedure & Bench Top Flotation Testing

This procedure is intended to provide a performance assessment for a flotation unit. The assessment can be carried out for the purpose of:

1. establishing a benchmark for future performance comparison,
2. troubleshooting to improve performance,
3. monitoring to sustain performance.

For each of these purposes, the sampling, laboratory analysis and process data gathering is essentially the same. The only difference is the number and frequency of samples and tests. Establishing a benchmark generally requires a more complete assessment of performance. Monitoring over time requires a large number of sampling but on a daily basis monitoring requires the least amount of data.

Performance Variables: As discussed above, the main performance variables for flotation are: bubble size, gas / water ratio, relative chemistry between the oil droplets, solids, and bubble surface, number of stages, and the characteristics of the incoming produced water. These units typically provide a consistent gas/water ratio, without adjustment, and allow a visual determination of the extent to which the gas bubbles are able to capture oil drops. Most bench top units do not provide a quantitative assessment of separation efficiency. That is much more difficult because it requires control of the gas/water ratio and the bubble size.

Baseline Data: It must be emphasized that troubleshooting a poorly performing flotation unit is difficult without baseline data. It may seem unnecessary to take baseline data when the system is working well, but without such data, troubleshooting later will be difficult. With good baseline data, troubleshooting is greatly facilitated by simply evaluating "what is different now compared to when the unit was working properly."

Issues involved in flotation performance: In order to assess the performance of a flotation unit, several issues must be addressed. Each of these issues must be handled properly or the assessment will be incomplete and have little value for sustaining performance or for troubleshooting. The troubleshooting procedure is intended to address the following issues:

1. Is the chemistry of the system suitable for the generation of bubbles of the right size and surface chemistry?
2. Is the volume of bubble generation adequate / as expected?
3. What is the efficiency of oil/water separation, as a function of flow rate, reject percentage?
4. Is the correct reject percentage being used?
5. Is there fouling of the level indicator or control valve?
6. Are the feed flow rates, pressures, and oil-in-water concentrations steady?

Each of these issues is discussed presently below.

Observation Ports: There is a very great difference between a flotation unit that has view ports and one that does not. The status of the foam height, the foam uniformity, weir height relative to foam height, and the amount of oil being floated to the surface of each cell, the gas/water ratio, and the chemistry of the gas/oil interaction can be quickly qualitatively assessed in a unit that has view ports. Figures 12.8 and 12.9 below show the difference between bubbles that have good interaction with the oil and bubbles that do not. Both of the pictures show adequate bubbles. If a flotation unit has inadequate bubbles, it is usually obvious from simple visual observation. Figure 12.8 is an example of a situation where there is a good volume of bubbles but the bubbles are not capturing oil. This can be contrasted with Figure 12.9 where there is a thick mat of oily bubbles on the water surface. This indicates that the bubbles are capturing the oil. It is helpful to train the operators to recognize what the bubbles should look like.

Foam versus Froth: It is somewhat pedantic to make a distinction between foam and froth. They are both basically gas bubbles interspaced in a liquid matrix. However, there are some company cultures where a distinction is made and quite a big deal is made out of the presence of foam versus froth in the flotation unit. For that reason, the following distinction is offered. A foam consists of relatively large bubbles that tend to sit on top of the oily patches on the surface of the water. A froth is composed of a high concentration of relatively small bubbles that are well dispersed in the oil patches and which give the patches of oil a creamy texture. A good cappuccino or flat white will have a smooth creamy froth layer of steamed milk on the surface of the coffee.

From the water treating perspective there is little difference in the separation efficiency for a froth versus a foam. In most cases, the presence of large bubbles only indicates coalescence of bubbles once they have arrived at the surface. However, in some cases, the presence of large bubbles in the oily patches on the water surface is indicative of poor control of bubble size, or excessive coalescence of bubbles in the path to the surface. This can have a significant effect on separation efficiency. If the flotation unit allows observation or sampling of the bubbles just below the surface, then this would be a prudent thing to check.

In any case, the oil content of the surface is critical. When the chemistry is correct, a thick oily foam will be seen. When the chemistry is not correct, a bubbly layer without oil will be seen. This later situation indicates that oil droplets are not sticking the gas bubbles. The figure below illustrates the situation. Production chemicals such as corrosion inhibitor, antifoam, hydrate inhibitor, scale inhibitor, and other production chemicals can result in adverse surface tensions which can cause this problem. Incorrect selection of coagulant or flocculating agent can also contribute to the situation. The best way to address this situation is to select a coagulant and / or a flocculating agent using a bench top flotation unit. This is a small scale unit that generates gas bubbles and provides a relative indication of the affinity between gas bubbles and oil droplets. Using samples taken far upstream in order to avoid production chemicals as much as possible, the procedure is to first reproduce the problem and then adjust the water chemistry to eliminate it.

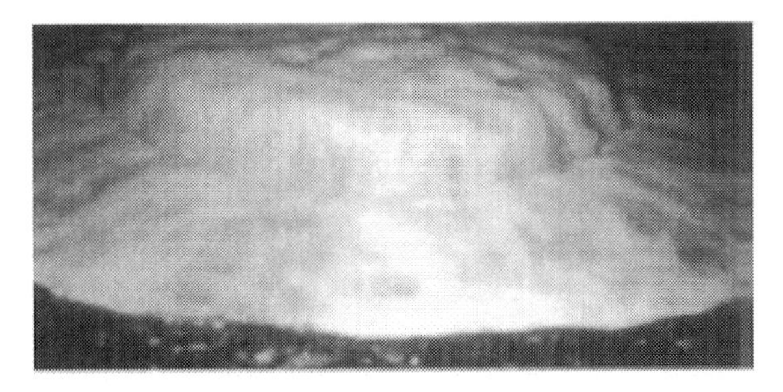

Figure 12.8 Flotation bubbles at the top of a single stage compact flotation unit. The bubbles float to the top and then spread out radially. There is very little oil apparent in the bubbles which indicates poor capture efficiency. As can be seen, there is a layer of oil surrounding the circular plume of bubbles which accumulated over a long period of time and does not indicate good capture efficiency.

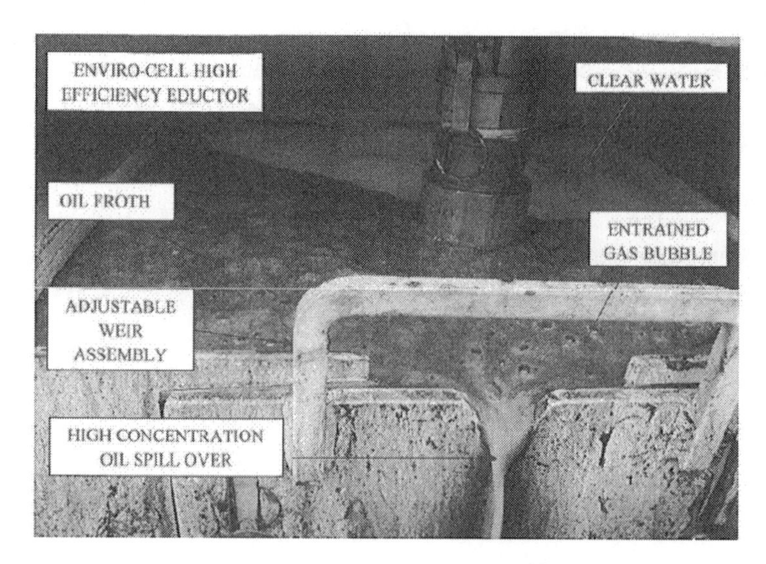

Figure 12.9 Shows the top of a stage of a multistage flotation unit. There is a high concentration of oil in the bubble froth. This is typical of a flotation unit that is operating with good capture efficiency.

One of the tests that operators sometimes run is to "burp" the flotation unit to test whether the gas lines are unobstructed. One of the causes of poor flotation performance is scale buildup and plugging of the gas lines. Burping a flotation unit involves quickly opening the gas discharge line from the unit. This lowers the pressure in the gas discharge which increases the gas pressure drop across the unit. The operator looks for a rapid evolution of gas as a result of this action. If the response is delayed or not apparent, then there is probably an obstruction in the gas lines feeding the unit or in the transmission lines within the unit.

Free Oil Droplet Size: Oil droplet size in the feed stream is obviously another important parameter. If the oil droplets are too small, then separation efficiency will be low. The determination of drop size can be carried out with desk top settling, or with one of several instruments that are available for this purpose (e.g. Jorin, Advanced Sensors, Canty). Desk top settling is always recommended for troubleshooting water treating projects. An entire system can be evaluated within an hour or so. By taking pictures of the results, a catalog of system performance can be built up over time. This will help later investigations to determine what has changed. To a trained eye, desktop settling will reveal water treating problems without much further analysis.

The procedure for Desk Top Settling is given elsewhere. A brief summary is given here. Take a shear-free sample and record the time. Set it on a desk or other horizontal surface such that it will not need to be moved for a day. Record how much (vol %) oil separates at the following time intervals: 1, 5, 10, 30, 60, 120 minutes, followed by 6 hours, 12 hours, 24 hours. Also record how much material sinks to the bottom of the jar. Record the presence of small black particles (iron oxide or asphaltenes). If possible, take a picture of the sample at the indicated time intervals. In order for a flotation unit to be effective, the sample must be clear within 12 hours. In other words, there should be no difference between the sample at 12 hours, and that at 24 hours and both should be clear. If the sample is not clear within this timeframe then the contaminants may be too small for flotation to be effective.

If this is not the case, then the sample should be submitted for detailed analysis. The density of the oil and water in the sample should be analyzed and the solids content of the oil assessed. The presence of production chemicals such as corrosion inhibitor should be assessed. The dosage and chemical selection of water clarifier upstream of the flotation unit should be assessed.

If drop size analysis is available, then this can be a useful to tool to supplement the desktop analysis. However, drop size analysis is highly dependent on the proper use of the equipment, and on the availability of proper sampling connections.

Relevance of Stokes Law: Flotation is not a gravity separation process. In a typical gravity separation process, Stokes Law governs the rise velocity of oil droplets which, together with the geometry of the separator, determines the separation efficiency. In a properly designed flotation unit, oil droplets still rise, but the velocity of the oil droplets rising to the surface is much smaller than that of the gas bubbles. In many quantitative models of flotation performance, the oil drops are assumed to be stationary with little loss of accuracy due to the fact that the bubbles rise much faster than the oil droplets. In fact, gas flotation units based on small bubbles such as that generated in dissolved gas flotation can in some cases perform poorly because the bubbles are so small that they do not rise fast enough and can be swept to the treated effluent. While small bubbles may have excellent oil droplet attachment, that is not all that is required for effective flotation.

In flotation, the Stokes Law variables (drop diameter, density of oil and water, and viscosity of water) are still important, but a number of other factors must be considered as well. Although the upward motion of the gas bubbles and oil drops is governed by Stokes Law, the rate at which oil drops rise to the surface is in most flotation designs not important because it is so slow in comparison to the rate of rise of gas bubbles. In flotation, the number and size of gas bubbles is the most important variable affecting performance.

Efficiency, Operating Parameters & Expected Performance: The oil in water removal efficiency for a flotation unit is defined as follows:

$$E = \frac{C_{in} - C_{out}}{C_{in}} x100 \qquad \text{Eqn (12.7)}$$

in which C_{in} is the oil in water concentration in the feed to the flotation unit, C_{out} is the oil in water concentration in the underflow (effluent product). To determine the efficiency all that is needed is sampling of the feed and product.

Whenever efficiency is measured, the operating parameters should be measured as well. The performance data is only representative of a single snapshot in time. The data have greater meaning if the operating parameters can be measured and recorded for future comparison. Not all of the operating parameters can be easily measured. The operating parameters include:

1. All flow rates: feed, product, reject
2. Operating pressure
3. Reject oil-in-water concentration
4. Temperature
5. Density of fluids (oil and water)

With these measurements, a material balance can be made. The calculation of material balance is discussed in Section 15.1 (Material and Flow Balance). If drop size analysis is available, then this can be used to help determine if the measured efficiency is to be expected on the basis of drop size.

Correct Feed Flow Rate: The manufacturer should provide a recommended flow rate. If possible, the feed flow rate should be varied and the oil/water separation efficiency of the flotation unit assessed.

This is rarely possible of course since the unit is typically critical to the water treatment system. Variation of feed flow rate usually requires manual adjustment of one or more bypass valves. This will help establish a performance curve (efficiency vs feed flow rate).

Fouling: Fouling can occur and cause operating problems in those units that employ packing or internal coalescing elements. In those units, fouling is difficult to detect. The most practical way to assess fouling is to make sure that a visual inspection is carried out as part of the facilities shutdown schedule. For many units, this would not require a vessel entry since most flotation internals can be seen from the outside of the man way without entering the vessel.

Steady Flow Rates: Rapid fluctuations in pressures or flow rates associated with a flotation unit will cause low separation efficiency. To determine if this is a problem, the pressure and flow rate can be trended. Flow rate and pressure trends will always show some degree of fluctuation. However, sometimes it is difficult to know with certainty how much fluctuation is acceptable. For this reason, a field test is recommended. There are two pertinent field tests.

The first and simplest field test is to open the effluent (product) sample tap and watch the fluid stream carefully. If vortex breakdown is occurring, an occasional jet of oil will be seen in the sample stream. If this occurs to any degree, then pressure or flow variation is too great and the control system must be adjusted to eliminate this effect.

Bench Top Flotation Test: In general, there are two objectives of carrying out flotation evaluations with small test units. The first is to study the chemistry of the interaction between water, oil drops, and gas bubbles. As mentioned, it is important that the surfaces of these three fluid phases allow attachment of oil droplets to the gas bubbles. Measurement of interfacial tension could, in principle, be done. However, measurement of interfacial tension requires instrumentation that is difficult to use in most field environments. It is not the most direct way to ensure that the chemistry is correct. The best way is to use a bench top test unit and carry out comparative studies with and without upstream production chemicals, and to test various vendor products such as foaming/frothing agents, surfactants, coagulants, etc.. For this purpose, the amount of gas and the bubble diameters do not need to be measured or carefully controlled. The simplest test unit requires a batch (grab) water sample and is usually run in a lab. The lab can be either onsite or offsite. An onsite lab has the advantage that the sample does not need to be transported and the time for aging and degassing is reduced. This type of unit is usually run in order to test the chemistry of the flotation process. It is not suitable for making a quantitative assessment of separation efficiency. However, it can be used to assess the stickiness of the bubbles to the contaminant, as well as to assess the effect of production chemicals, and flocculating agents.

The second objective is to make a semi-quantitative estimate of the separation efficiency. In order to do this, the test unit is connected to the process and a slipstream of water is run through it. The gas/water ratio must be measured and controlled. Also, the bubble size must be measured and controlled. This is a far more difficult objective than the first objective described above.

These two objectives require different bench units. However, any device that can be used in a semi-quantitative manner can also be used to evaluate the chemistry. In fact, chemistry evaluation should be the first phase of any onsite study. Descriptions of these bench units can be found in [7].

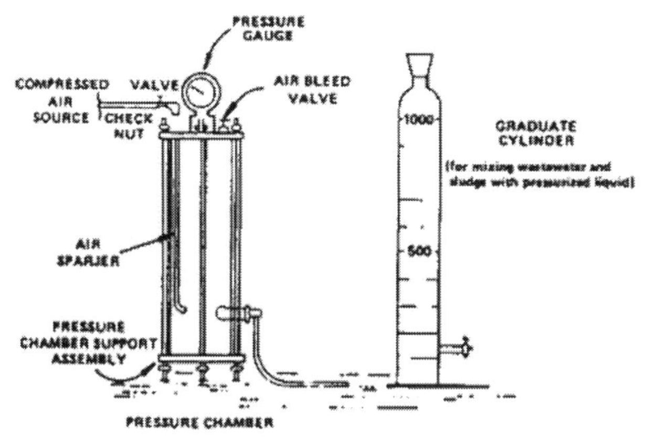

Fig. 1.7. DAF laboratory unit.

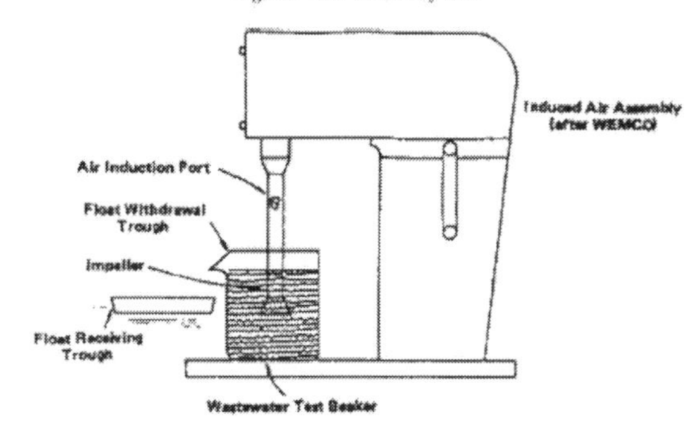

Fig. 1.8. Bench-scale IAF unit.

WEMCO (an equipment manufacturer) utilized bench-scale units of both IAF and nozzle air flotation systems. Descriptions of these test units and their use in laboratory studies are found in the masters' theses of Steiner (51), Ching (63), and Nipper (64). The IAF model is shown in Fig. 1.8.

Figure 12.10 Bench Flotation Simulation Units used in the lab or in the field [7]

Overall Procedure for Performance Assessment: Descriptions and explanations of the different issues and tests involved in assessing the performance of a flotation unit are given in the section above. The following is a summary of the issues and the tests.

Table 12.3

Issue	Test
Suitable Fluid Properties	Desktop settling Analytical characterization Drop size analysis (if available)
Efficiency	OiW of feed and effluent
Flow rate	Performance curve (efficiency vs flow rate)
Fouling	Visual inspection upon shutdown
Insufficient Reject	Material Balance
Non-Steady Flow (excursions)	Monitor (visually inspect) effluent stream for slugs of oil
Overall performance	Compare against benchmark

12.1.9 Benefits and Drawbacks of Flotation

The advantages and disadvantages of flotation units are presented here.

Advantages of Flotation units: Advantages of a flotation unit over other separation equipment is the generation of relatively high centrifugal forces without the maintenance requirement of rotating parts. Other advantages include:

- Small to moderate footprint, depending on the type of flotation unit.
- Flotation unit liners have no moving parts so maintenance requirements are reduced.
- Good turndown characteristics.
- Vessels can have additional liners installed to meet increasing process flow rates.
- Flotation units are insensitive to wave motion, which makes them suitable for FPSO applications.
- Vessels can be installed horizontally or vertically.
- Good retrofit capability.
- Wide operating flexibility, suitable for both onshore, offshore and on FPSOs.
- Many vendors and competitive designs and cost situation.
- Easily retrofitted. Modular design allows capacity to be increased as required
- Can be designed for high pressure and should be installed upstream of control valves
- Insensitive to motion, ideal for floating vessels such as TLPs and FPSOs
- New high efficiency liners can be retrofitted into older lower efficiency liner slots

Drawbacks of Flotation units: The various problems and concerns that need to be addressed in selecting deoiling flotation units for a produced water handling facility include the following:

- The sensitivity of the system efficiency to oil droplet size requires consideration. For example, flotation units are generally unsuited to gas-condensate applications where the median oil droplet size tends to be small (<10 μm).
- Application and efficiency of the deoiling flotation units in low pressure operation of less than 10 Bar (150 psi) can be poor, since there is not enough pressure to utilize the optimum benefit of the design.
- Optimizing the operation of the unit with variation in flow rate will require regular adjustment in the number of liners that should be used in the individual flotation unit vessels.
- Deoiling chemical may be required to improve the droplet size distribution in cases where the deoiling efficiency is less than desired.
- The reject orifice can be prone to blockage in flotation unit liners, which impacts on oil removal efficiency. Regular back flushing is required to reduce the potential for blockage.
- Solids such as sand can build up in the inlet zones of vessels where the velocities are low. This can ultimately reduce vessel capacity as sand builds up.
- Steady flow is required to establish the reject core. If flow is variable, the reject core will breakdown and separation efficiency will deteriorate dramatically.

12.2 Flotation Equipment

In this section, specific equipment is discussed and performance data are given. The section starts with a discussion of the challenges that designers face in terms of how and where to feed the contaminated water, and how and where to discharge the treated effluent water and the oily reject. All of these streams must be handled so that they do not contaminate each other. Other design challenges include how to suppress secondary flow within the flotation unit. Secondary flow will often disrupt the main separation mechanism.

Brand Names: As with other sections of this book, brand names, and vendor names are not typically identified for obvious reasons. However, laboratory and field performance data are presented. In most cases, an effort is made to present data for a type or class of equipment without identifying the equipment by name. When the data only apply to one specific brand or model, then the brand and model are identified.

12.2.1 Design Considerations

Before discussing specific designs, it is worthwhile to discuss the challenges that a designer is faced with. There are many factors that contribute to high separation efficiency. The scientific principles are discussed below in Section 12.3 (Flotation Theory, Design and Modeling). Those principles form the basis for the design of the equipment. In this section, the practical aspects of the design and operation of flotation equipment are discussed.

Separation efficiency is enhanced by maximizing the number of interactions between the contaminants (oil droplets and solid particles) and the gas bubbles, and the enhancement of the oil droplet / bubble interface so that the two fluids stick together. Design variables include:

1. Large contaminant oil droplets and oily solids particles
2. Uniform bubble distribution
3. High gas / water ratio
4. Small gas bubble

These four critical performance parameters are elaborated. Maximize the size of the contaminants to be removed (by flocculation or by minimizing shear). Ensure a uniform distribution of oily water and gas bubbles throughout the flotation cell and eliminate secondary motion of gas and water. Maximize the gas/water ratio which will ensure good bubble surface area for contact between the contaminants and the rising gas bubbles. Minimize the size of the gas bubbles while ensuring that the bubbles are large enough to migrate to the oil/water interface. Ensure that the bubbles are not swept out of the compartment or vessel.

In addition to these performance parameters, there are a few important practical considerations in the mechanical design. First, the oily water must be introduced in a way that does not disrupt the flotation process. Second, the oil and clean water must be discharged in a way that does not allow contamination with the feed and that does not disrupt the process. Third, the overall weight and space must be as low as possible, while achieving a high separation performance.

Gas/Water Ratio and Bubble Size: Regarding separation performance, the two most important design parameters are the gas/water volumetric ratio and the gas bubble size. Together they determine the number of bubbles, the rise velocity of the bubbles and ultimately they have an enormous impact on separation efficiency. From a design standpoint, there are essentially four ways to generate bubbles:

1. mechanically induced flotation
2. Hydraulically induced flotation (eductors)
3. Gas injection into a pump (microbubble)
4. Dissolved gas flotation

Each of these technologies requires different equipment, each has benefits and drawbacks, and each has different costs associated with them. From a performance standpoint, each of these technologies delivers different gas/water ratios and different bubble sizes. It turns out that there is no single technology that is unequivocally better than all others. This is part of the reason why so many different technologies have survived for so many years. A revealing analysis is obtained by simply plotting the gas/water ratio versus the bubble size for the major flotation technologies on the market. This has been done in the figure below.

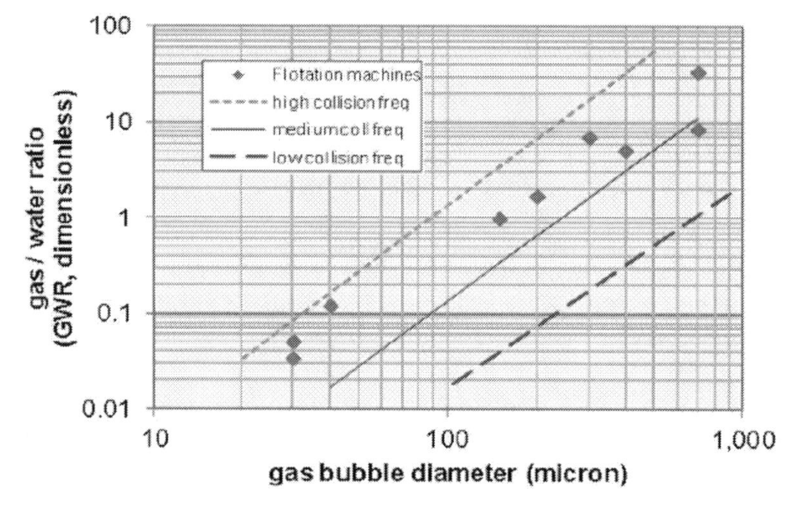

Figure 12.11 Gas/ water ratio (volumetric at standard conditions of the gas) versus gas bubble diameter (under standard conditions) for most of the major manufacturers flotation models. Note the relatively tight band of data points. Calculated oil droplet / bubble collision frequencies are also shown. The tight band of data points falls within a narrow range of collision frequencies

As shown in the figure, there is a tradeoff in the industry between bubble size and gas/water ratio. In general, small bubbles are preferred since they have high capture efficiency, as explained below. However, small bubbles are generally more difficult to generate than large bubbles. Thus, machines that generate small bubbles tend to generate a lower gas/water ratio. However, all things considered, the collision frequency between oil droplets and gas bubbles, also shown on the figure, is roughly constant across many different models of flotation units.

There are several ways to maximize the number of collisions with oil drops. One way is to generate a huge volume of gas with large bubble size. This is the approach embodied in the mechanical flotation design, such as the Wemco. Another way is to generate a modest amount of gas with moderately small bubbles. This is the approach embodied in the induced gas flotation units that use an eductor. The use of eductors requires one or more pumps but the pumps can be relatively inexpensive generic centrifugal pumps. A third way to maximize performance is to generate a small amount of gas with very small bubbles. This is the approach used in dissolved gas flotation. While DGF typically produces smaller bubbles than IGF, it relies on multiphase pumps to do so which are somewhat expensive.

Distribution of Oil Droplets and Gas Bubbles: The distribution of gas bubbles and water droplets in a flotation cell is critical. Poor distribution can cause severe channeling of bubbles through the water with limited bubble/contaminant contact. When bubbles are poorly distributed, they can rise in a column that sweeps adjacent water along with them. This plume will typically have a high velocity and will cause the formation of still zones where gas becomes trapped for a relatively long period of time. Water in these still zones typically swirls in eddy-like motion. CFD analysis has been used in the design of some eductor and bubble distributor systems. For example, the VersaFlo gas flotation device was designed with extensive use of CFD. This device is discussed below.

Water Flow Pattern as a Function of Gas Bubble Size: Although small bubbles are more effective for oil drop capture, vessel hydraulics provides a practical limit to the size of gas bubbles that can be used effectively for flotation. If the bubble diameter is too small the bubble will be discharged with the treated effluent. In some compact column flotation systems, a column of water flows downward while a column of gas bubbles flows upward. In such a system, the gas bubbles must be sufficiently large to overcome the downward flow of water, otherwise the gas bubbles will become entrained.

For compact flotation units, one of the design parameters is the nominal downward velocity for water in the unit. To overcome a 2 ft/min nominal downward water velocity, an oil droplet would need to have a diameter $\geq$300 microns, a gas bubble would need a diameter $\geq$120 microns, and an oil droplet associated with a gas bubble would need an effective hydraulic diameter in excess of 155 microns. Thus the lower limit for gas bubble diameter in the practical application of this type of flotation unit is in the range of 120 to 150 microns.

Swirl Motion and Secondary Flow: Several flotation designs are based on a vertical cylindrical vessel, and have tangential inlet nozzles that induce a swirl motion in bulk fluid inside the vessel. As the fluid enters the cylindrical vessel through one or more tangential nozzles, it is forced to spin around the inside of the cylinder. As discussed extensively in the chapter on hydrocyclones, swirl motion is readily induced by the use of tangential nozzles. However, unlike the case for a hydrocyclone, the swirl motion in a flotation unit typically does not have sufficient tangential velocity to cause appreciable migration of oil droplets to the center of the vessel. The swirl velocity in a flotation unit is too slow and the diameter of vessel (migration distance) is too large.

Estimating the tangential velocity of this swirl motion is difficult because it is only a small fraction of the inlet nozzle velocity. CFD has shown that the fluid velocity of the inlet jet slows down considerably due to friction with the fluid already in the vessel. There are no simple formulas for calculating this effect.

Swirl motion does however have an important benefit on the overall hydrodynamics of the fluid in the vessel. In any flotation process, the rising of gas bubbles drag a considerable volume of water upward. In almost all of these vertical cylindrical systems, the water discharge is near the bottom. Water must flow downward in some way in order to maintain mass balance within the system. Thus, some water gets pulled up by the bubbles, and it must then circulate down prior to discharge. To help visualization, this is the same mechanism that can be seen in a glass of beer immediately after it has been poured. This is particularly true of a glass of Guinness Stout poured into a Guinness glass. In that case, the carbon dioxide bubbles rise throughout the beer, dragging some beer upward in the glass. By mass balance some of the beer must then circulate downward. This is seen as bubbles falling near the inside surface of the glass.

This circulatory flow can be quite disruptive if not properly channeled or controlled in some way. This is the benefit of swirl motion. Fan et al. [8] carried out measurements of fluid velocity within a cylindrical flotation column. Lehr et al. [9] carried out Computational Fluid Dynamics simulations. Both groups confirmed that without swirl motion, the column of rising water has a corkscrew pattern with significant lost volume due to a chaotic meandering downward flow outside of the upward moving corkscrew flow. When swirl motion is present, the column of rising water is much more uniform and the falling water occupies a thinner region against the wall of the vessel. This leaves a greater cross-sectional area for flotation. The swirl motion ensures that a well-controlled flow pattern forms that allows the bulk water and bubbles to rise and pushes the falling water into a narrow gap near the wall in the space below the water inlet. The falling water is confined to a narrow region next to the wall.

Coalescing elements: Typical coalescing elements are made of high surface to volume ratio packing. By utilizing an open structure, random packing along contiguous surfaces, droplets have a opportunity to encounter a coalescing surface, but clogging by solids is minimized.

Skimming: Skimming is the process of collecting the oil from the water/gas interface and directing it to the reject piping. There are many ways to do this. The designs differ in the volume of oily water that is captured and rejected, and in the mechanical details and complexity. Conventional horizontal mechanical IGF, such as the Wemco, employ electrically driven skimmer paddles located at the edge of the water / gas surface, next to the oily water reject trough. The skimmer paddles rotate so as to pull the oily froth from the surface into the through.

Equipment used for the collection, transfer, or processing of skimmed fluids from the atmospheric conventional IGF machine should be designed to handle start-up, upset conditions, chemical optimization, and wave motion, in order to avoid flooded oil compartments and off spec water quality.

The skim flow rate for a spillover weir depends on the height of fluid above the weir and does not depend on the flow rate of liquid through the vessel. Therefore, at low flow rates, the skim volume can be higher as a percent of the forward flow than at higher flow rates.

Some newer IGF machine designs utilize a horizontal cylindrical tank which skims hydraulically using internal diverter plates and "V" notched overflow weirs, eliminating the rotating skimmer paddles. The cylindrical IGF machine can be equipped with either a conventional pneumatically driven level controller or with computer logic level control. Pneumatic level control in a cylindrical tank design is essentially the same as with the conventional atmospheric designed IGF units and must skim at a fixed level. With computer control logic and timers, it is possible to control the level at either a fixed point (constant skim) or at two fixed points (intermittent skim) on the control span.

The advantages of the conventional pneumatic control scheme in the cylindrical tank are:

1. Less CAPEX because of simpler controls,
2. Assurance of oil removal on a constant basis,
3. Smaller HP motor to drive the rotors of the aeration devices.

The advantages of the computer logic level control scheme are:

1. Potentially less skim volume because of more precise control,
2. Skim volume can be somewhat tied to flow rate rather than operating level,
3. Control logic allows the operator to either continuously skim or intermittently skim,

Because timed level control intervals are often employed for intermittent skim control, it is possible during times of low forward flow for the skim mode settings to cycle out and not remove oil - because low flow left the level below the overflow point. If an excessive oil pad accumulates on the surface of the liquid, it can impede froth formation and damage the oil removal efficiency of the IGF machine. Care must be used in setting the control logic to assure that the oil is removed during each skim cycle.

12.2.2 Bubble Coalescence

Coalescence of gas bubbles is a significant issue in design and operation of flotation units. As discussed, generating small bubbles is more costly than generating larger bubbles. Therefore in the design of a unit, it would be cost prohibitive to generate small bubbles only to have them coalesce into larger bubbles before they are used to remove contaminant particles. In the design of flotation units, there is usually an attempt made to introduce the gas bubbles into the water as close as possible to the contact (reaction) cell. Several designs introduce the gas bubbles directly in the contact cell itself. The MIGF (Horizontal Multistage Mechanical Induced Gas Flotation units generate the bubbles in the contact cell. In the eductor type units (HIGF – Hydraulically Induced Multicell Induced Gas Flotation) the eductors are sometimes submerged into the produced water. This introduces the gas into the water directly in the contact cell. Early designs of flotation units such as the Unicel did not do this. This is thought to be one reason for the relatively poor performance of such units.

Drop/Drop versus Bubble/Bubble Coalescence: The subject of liquid droplet coalescence was discussed in some detail in Section 4.5 (Coalescence of Oil Droplets). In that section it was pointed out that coalescence models are rather complex in general. The only way to make them useful for engineers is to select rather specific circumstances and simplify the equations for the specific situation. This restricts the range of application to only those circumstances for which the models were derived. Nevertheless, this allows simple and useful calculations to be made.

The coalescence model derived for oil droplets does not give accurate results for the case of bubbles suspended in produced water. The difference being that the interface between a gas bubble and water is fully mobile. In other words, the gas viscosity is so low that the water at the interface can move tangentially along the interface as the water film drains between the two bubbles. This is not the case for two oil droplets approaching each other in produced water where liquid viscosities are relatively high and mobility of the interface is therefore restricted.

Experimental Data for Bubble Coalescence: For the case of bubble coalescence no model will be given here. Instead the experimental data of Prince and Blanch will be presented [10]. These data were collected and reported for a specific set of experimental conditions which are discussed below. The values given will be used below in an illustrative calculation of the importance of bubble coalescence.

Prince and Blanch studied the rates of bubble coalescence and break-up in turbulent gas-liquid dispersions [10]. As they discuss, coalescence of two bubbles in turbulent flow occurs in three steps. First, the bubbles collide. This traps a small amount of liquid between the bubbles. This liquid then drains until the liquid film separating the bubbles reaches a critical and very small thickness. The third step is rupture of the thin liquid film which results in coalescence.

Measured coalescence rate data are given in the figure below. Bubble diameters ranged from 2,500 to 6,500 micron. This is considerably larger than the flotation bubble sizes that have been used in commercial flotation devices. Nevertheless, these results are useful to give an order-of-magnitude estimate for the coalescence rate of smaller bubbles. The figure below gives the coalescence rate as a function of salt concentration. As salt is added, the interfacial tension increases. This was discussed in section 4.4.2 (Surface and Interfacial Tension). Higher interfacial tension leads to longer film drainage times and thus, coalescence rate goes down as a function of salt content.

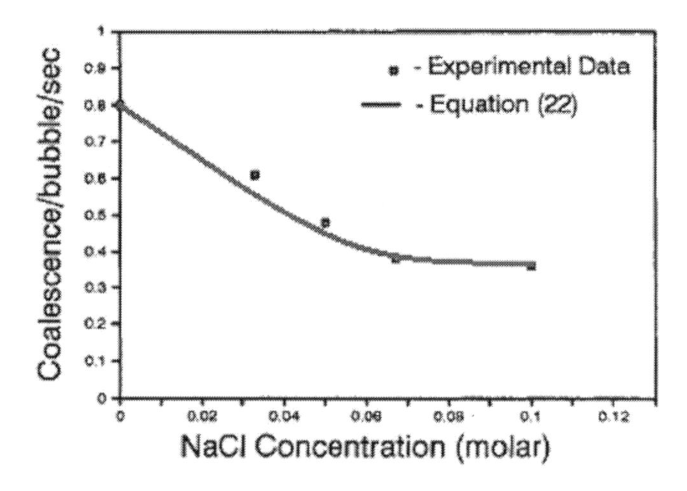

Figure 12.12 Gas bubble coalescence rate as a function of salinity from [10]

Note that the range of molar concentrations of salt is rather low compared to most oilfield brine. A 0.1 molar concentration of NaCl corresponds to 5,800 mg/L TDS.

The next set of results is for coalescence rate as a function of surfactant concentration, as shown in the figure below. Very little surfactant is required to make a dramatic drop in coalescence rate. The mechanism for this drop is worthwhile to discuss. First, surfactant will reduce the interfacial tension. Given the fact that an increase in interfacial tension accompanies the addition of salt, which led to a decrease in coalescence rate, something further must be happening here. In the case of surfactant, the interfacial tension does indeed decrease but the presence of surfactant also creates a steric barrier to coalescence. But a more important effect also occurs. When the bubbles collide, a thin film of water is trapped between the two bubbles. This film must drain and then collapse for coalescence to occur. As the film drains, it sweeps surfactant molecules away from the region between the bubbles and into the water / bubble interface just outside of the film. This sets up an interfacial tension gradient such that the interfacial tension is higher between the bubbles and lower just outside the film region. This interfacial tension gradient is analogous to a pressure gradient which resists the expansion of the interface between bubbles. As such, the drainage of the film is slower than if the surfactant was not present. Since film drainage is slower when surfactant is present, the coalescence rate is lower.

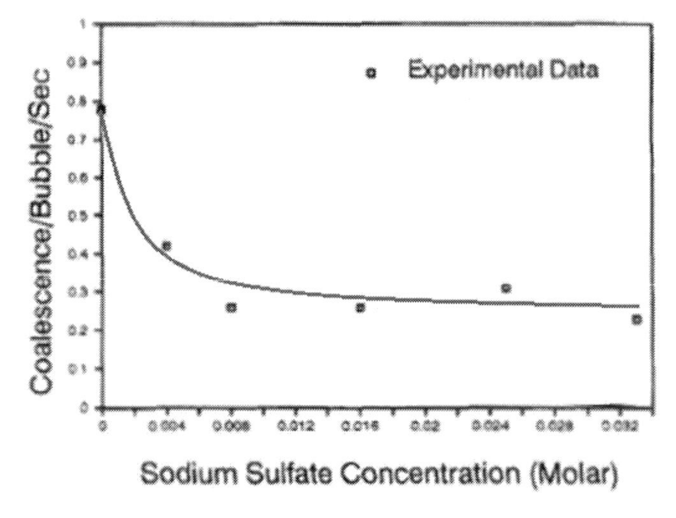

Figure 12.13 Gas bubble coalescence rate as a function of surfactant from [10]

Illustrative Calculation of Bubble Coalescence: The case that will be shown is that of moderate salinity (0.1 molarity), and some small amount of surfactant, equivalent to 0.02 M sodium sulfate. From the figures above, the coalescence rate can be estimated as 0.3 coalescence events per bubble per second. For the sake of illustration, the residence time will be 30 seconds. The initial average bubble diameter is 200 micron. Considering just one coalescence event, conservation of mass requires that the volume of the bubble formed after coalescence is equal to the sum of the volumes of the two bubbles that disappeared due to coalescence. This results in the following relation between bubble diameters: $d_2 = 2^{1/3} d_1$. Considering several of these events, the following general formula is derived for the bubble diameter after n successive coalescence events:

$$d_2 = 2^{n/3} d_1 \qquad \text{Eqn (12.8)}$$

where:

d2	= diameter of bubble created by coalescence (micron)
d1	= diameter of each of the two bubbles that collided and coalesced (micron)
n	= number of coalescence events

As an example calculation, assume that 9 successive coalescence events have occurred in 30 seconds (0.3 events per second x 30 seconds = 9 events). Given an initial bubble diameter of 200 micron, this gives a new bubble diameter of 1,600 micron. It must be noted that this calculation assumes that successive coalescence events occur, each time with the bubbles generated in the previous event. This is not what happens in reality. Nevertheless, a rough approximation of the rate of bubble size increase in a flotation is obtained. The net result of this analysis is that bubble coalescence is a significant process within the practical timeframe of introducing bubbles into a flotation unit. Thus, those designs that generate bubbles within the fluid are justified on the basis that bubble coalescence is minimized prior to contact with contaminant particles.

12.2.3 Classification of Designs (MIGF, HIGF, VIGF, VDGF, etc)

In this section a brief description of the different types of flotation unit on the market is given. For some purposes, only a rough classification and a rough breakdown of performance is required. This section is intended to fulfill this need. Also, it provides an overview of the main differences between one type of flotation and another.

Classification of Flotation Devices: There are different ways of classifying gas flotation systems. The conventional way is in terms of the bubble generation process that is used. In this classification scheme, there are four types of gas flotation in use in the upstream oil and gas industry:

1. Horizontal Multistage Mechanically Induced Gas Flotation
2. Horizontal Multistage Hydraulically Induced Gas Flotation
3. Vertical Induced Gas Flotation
4. Dissolved Gas Flotation

This classification is somewhat dated. There are new developments that do not easily fit into this scheme. They are discussed on their own below. Historically, mechanical IGF, hydraulic IGF, and DGF have been the main classifications of flotation systems. Other means of generating bubbles have been developed which do not fit into this categorization scheme. Hybrid approaches that generate both dissolved and very small dispersed bubbles have been successful and do not fit within this classification scheme. Also, gas sparging units, which are seldom used in the upstream industry, are outside of this classification.

This classification makes certain assumptions which will be explained in detail below. One assumption is that horizontal dissolved gas flotation is not suitable for upstream applications due to the high rates of water flow and the slow rise velocity of dissolved gas bubbles. The bubbles do not have time to reach the water surface and have a tendency to get swept into the effluent discharge. It is also assumed that compact flotation units are cylindrical vertical units. This geometry tends to make the greatest use of the weight and space required. The geometry of box-style horizontal units tends to waste space and weight to some extent. Almost all horizontal units are multi-stage. As mentioned all of these assumptions are discussed below.

12.2.4 Horizontal Multistage Mechanical IGF (MIGF)

The Wemco Depurator System was pioneered in the minerals mining industry for the beneficiation of ore. This original Wemco was a single cell unit. Over 15,000 such units were in use by 1971 in the minerals mining industry. The multi-cell Wemco Depurator was introduced to the oil and gas industry by the Tretolite Division of the Petrolite Corporation. The first installation was a 150 GPM (5,142 BWPD) unit for Gulf Oil Company in Bakersfield, California in 1969. These early units utilized a submerged rotor, disperser and stand pipe combination design.

Basic Design: A typical horizontal IGF with mechanically induced gas bubbles, consists of a rectangular tank with six chambers in series [11]. The first chamber, or cell, is essentially an inlet section that provides momentum reduction, and a quiet zone for flocs of oil drops to form based on injection of chemical flocculent upstream of the unit. This inlet (or quiet) cell is followed by four "active" cells where gas bubbles are generated and oil is floated to the top and rejected. Water flows from one cell to another. Each cell is partitioned from the adjacent cells by a plate that spans the height of the cell. The plate typically has a rectangular or oval section cut out of it at the bottom to allow water to flow from one cell to the next. In some designs, the water must travers a series of overflow and underflow weirs to move from one cell to another.

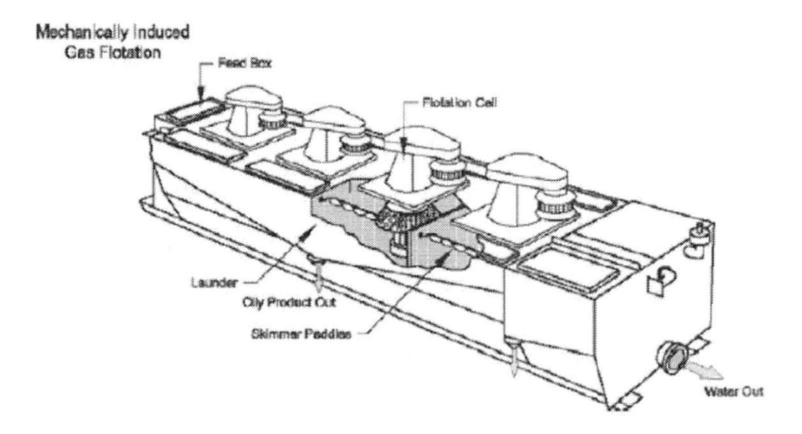

Figure 12.14 Schematic diagram of horizontal multistage induced gas flotation unit (Wemco).

Figure 12.15 Picture of a Wemco. Note the view glass which is used by operators to determine the state of gas bubbles and foam. The view glass has a manual operated windshield wiper (see handle) for cleaning the underside of the view glass. The four motor housings are clearly visible.

As shown in the figure, a motor is mounted above the atmospheric vessel. The rotor spins which forces water through the disperser at the bottom of the standpipe [11]. This creates a vacuum in the standpipe. The vacuum creates a vortex and forces gas to be sucked into the spinning water in the standpipe. The gas is sucked down the height of the standpipe to the rotor. Together the gas and water are swept out of the bottom of the standpipe by the rotor through the disperser. The disperser is perforated. The motion of water and gas through the perforated disperser shears the gas into small bubbles. The rotor, when energized, imparts a pumping action to fluids in the immediate vicinity. As these fluids are pumped away from the rotor through the disperser, a vacuum is formed in the stand pipe above the rotor, resulting in a gas transfer from the vapor space of the tank into the top of the rotor.

Since the rotor is open to water on the bottom, the fluids pumped from the bottom of the rotor are replaced by fluids from the bottom of the tank. Fluids and gas are mixed at the approximate center of the rotor length, and contact is made between the gas bubbles and oil droplets.

The rate of water flow through the entire unit is dictated by the discharge (level) control valve. The level indicator is located in the final cell, where there is no flotation gas. Most horizontal units have a slight decreasing hydrostatic head (level) to ensure that there is motive force for water to flow from the inlet to the outlet.

Effluent water is discharged from the bottom of the final cell. The water pathway into this cell from the upstream cell does not involve an open area at the bottom of the partition plate, as with the upstream cells. Such an open area would provide a direct path for water to flow from the last active cell to the water discharge, thus short circuiting the volume of the last cell. Therefore, there are instead a number of plates without cut-out sections. Instead these plates span the entire distance from the bottom to the top, but only part of the distance from one side to another. The plates are arranged in a staggered configuration thus providing a serpentine, or maze-like path for the water.

As mentioned, water discharge is at the bottom of the last cell. A vortex breaker is required at the water discharge. Due to the geometry of this cell, a vortex is very readily set up unless there is an effective vortex breaker.

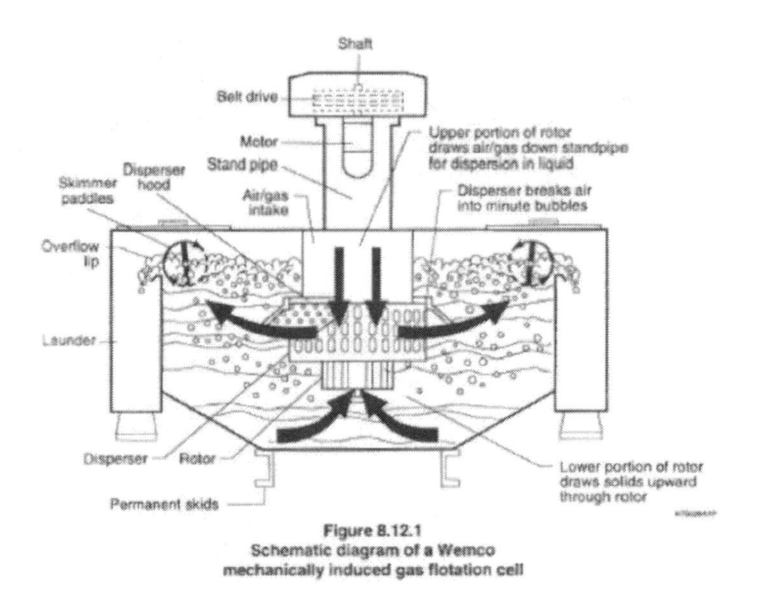

Figure 12.16 Flow pattern in a Wemco flotation cell

Because the unit uses rotating paddles to form a vortex to suck the gas into the water, it is referred to as a mechanically induced flotation unit. A schematic diagram of a unit is given in the figure above.

Klimpel Model for Multi-Cell (Multi-Stage) Efficiency: While each chamber has only moderate separation efficiency, the series arrangement provides an overall high separation efficiency. If for example each chamber has 50 % efficiency, then four chambers would equate to an ideal efficiency of 0.5 x 0.5 x 0.5 x 0.5 or 94 % overall efficiency. Of course, this calculation is based on ideal efficiencies. Actual efficiencies are somewhat different. For example, the actual efficiency of the first cell is typically higher than 50 %. This cell has the highest oil concentration, which means that there are many more oil drops available for capture. Typically the subsequent cells have lower efficiencies than the previous, due to the diminishing concentration of oil in each cell.

Actual efficiencies are presented below. It is not unusual for horizontal IGF to achieve greater than 90 % efficiency, when operated properly. Oil removal efficiency for various multi-cell IGF machines is a matter of historical record. These data may be used to develop a removal efficiency equation for a particular IGF machine based on the Klimpel kinetic equation.

Klimpel presented a mathematical relation that captures the series effect [12 – 14]. The formula for oil removal efficiency is expressed as follows:

$$E = 100\left(1 - \left[1 \div \{1 + Kt\}^{N}\right]\right) \qquad \text{Eqn (12.9)}$$

Where:

E = Efficiency of oil removal (%),

K = Coefficient (developed from historical data),

t = Liquid residence time in each cell (minutes),

N = Number of flotation compartments in the machine.

For the conventional DEPURATOR® design:

K = 1.2 (an average value)*

T = 1.0 minutes

N = 4

The calculation exercise becomes:

(E = [100] X [1 - {1 ÷ (1 + [1.2 x 1.0])^4} = 95.73% removal efficiency.)

The typical residence time per cell is one minute. The K value is dependent on the specific process employed. The K constant is a combination of two coefficients as applied to the specific technology in use. The first coefficient relates to total contaminant removal at infinite process exposure time, and the second coefficient relates to the speed at which the contaminant removal increases throughout the process.

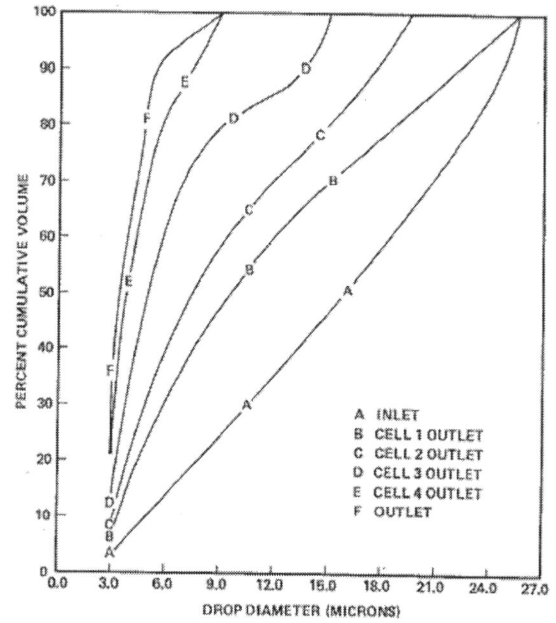

Figure 12.17 Measured oil droplet cumulative volume versus oil droplet diameter for each of the 4 active stages of a multistage horizontal flotation unit [5]. Note that the outlet, in this case, contains oil droplets that have diameters that are predominantly within the 3 to 6 micron range.

The Effect of Inlet OiW Concentration: As shown in the figure below, the separation efficiency decreases as the inlet OiW concentration drops below a certain value. The explanation for this is that the upstream process removed the larger oil droplets, leaving behind only the small droplets, which are more difficult to remove.

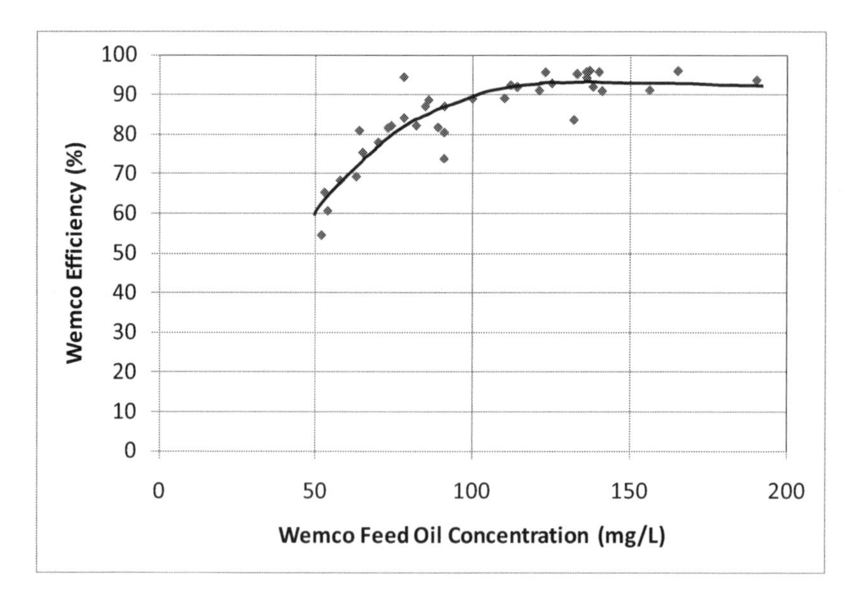

Figure 12.18 Impact of inlet OiW concentration on the separation efficiency of a Wemco Horizontal MIGF. See text for explanation.

Reject Rate for a Horizontal IGF: The reject flow rate in a horizontal IGF is relatively independent of the water feed rate. This is due to the mechanism of skimming and level control. Skimming is accomplished using a spillover weir. The height of fluid relative to the height of the spillover, and the relative location and speed of the rotating paddles, are the two main variables that control the flow rate of the rejects. Neither of these factors depend on the feed water rate. Instead, they depend on the design flow rate. So in effect, the flow rate of rejects is relatively independent of feed rate. This makes turn-down a bit of a problem since the reject percentage of flow increases as flow rate decreases. Actually, this is a bit of an overstatement and there is a bit more to the story.

Fluctuations in feed rate will still cause fluctuations in the reject flow rate. This is due to the fact that a sudden increase in flow rate, for example, will cause the level to rise. The level indicating device will then signal the LCV to open which will bring the level back down. Before the LCV has had time to respond, the higher level will cause more fluid to spillover and the reject rate will temporarily increase.

Conversely, if there is a sudden decrease in the feed flow rate, then the level will go down. Before the LCV has a chance to respond, the reject rate will decrease, and may in fact temporarily stop. This is a bad situation since it means that oil will accumulate at the surface. When this happens there is a greater chance for oil to get carried through to the downstream compartments and perhaps discharged with the effluent water.

To prevent this from happening, the level control set point, and the level control system (controller and valve) are usually adjusted to maintain a minimum reject flow rate for most fluctuations. It is this prudent control strategy that leads to excessive reject rate for horizontal IGF that are oversized for the application.

This means that an oversized unit, operating at a feed rate that is below its rated design capacity, will reject too much fluid. Although this is a somewhat common problem in the operation of horizontal IGF, a satisfactory solution to the problem has not been found. One installation in Alberta which faced this problem resorted to removing the rotating paddles. This seemed to improve the situation. However, Alberta is obviously a land-based operation. Whether or not this approach would work offshore remains to be seen.

Wave motion which causes roll of floating vessels has an effect on the reject rate from a spillover weir arrangement. Roll motion causes the level above the spillover weir to move up and down. This leads to fluctuations in the reject rate. Vertical flotation units have less of this effect. But also, baffles been included in the later designs of horizontal units which have reduced the effects of wave motion on the performance of horizontal IGF units. However, as will be shown presently, wave motion can still have a significant effect on fluctuations of reject rate.

Reject Flow Control: An automated feedback control system was installed on a Wemco located on a deep water GoM Tension Leg Platform. This Wemco was located on the top deck near the outer edge of the platform. In that location, it was particularly susceptible to roll motion. The control system is shown schematically in the figure below. The feedback system was very effective in eliminating the variability and allowing the Wemco to operate with a steady reject flow rate. A flow meter (FI) sends the flow rate to a PID controller which compares the measured flow rate to a set point. The controller then sends a signal to the level controller for the effluent discharge which adjusts the level control set point.

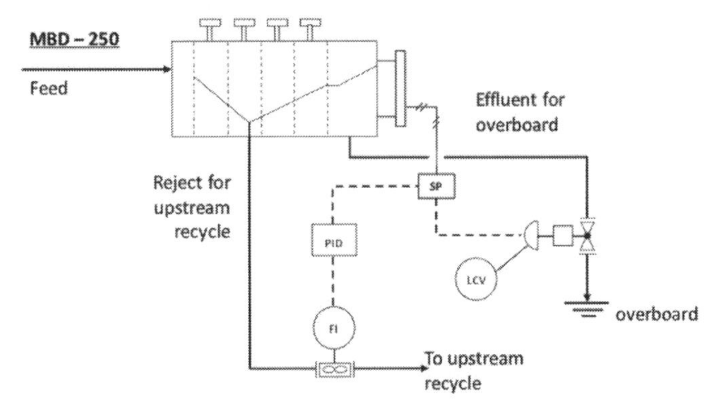

Figure 12.19 Wemco reject flow rate control system.

A field trial of the control system was carried out in 2010. As shown in the Figure below, when the control system was turned on, the reject flow rate stabilized. When it was turned off, the reject flow rate had significant fluctuations. Thus demonstrating the effectiveness of the flow control system.

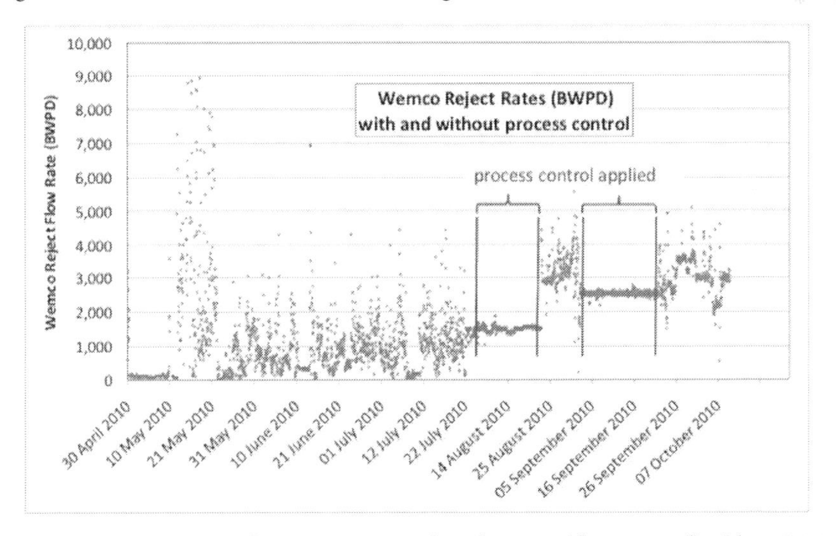

Figure 12.20 Wemco reject flow rate versus time for a deepwater (floating vessel) offshore GoM installation. A test of a new process control system was carried out. Process control was applied, then turned off, then applied again, and turned off.

Wemco – Individual Models: There are seven models of Wemco which vary in water treating capacity. The cell dimensions vary from one model to another. Typically a constant theoretical water retention time of one minute per active cell was applied. However, the models differ from each other in some important ways. The model numbers and flow rates are given in the table below.

Table 12.4 Individual Wemco models

Model No.	GPM	BPD	m³/minute
36	50	1,720	0.19
44	150	5,150	0.57
56	300	10,300	1.14
66	450	15,450	1.70
76	750	25,725	2.84
84	1,125	38,585	4.26
84X	1,450	50,000	5.49
120	2,250	77,175	8.52
120X	3,000	103,000	11.36
144X	5,000	171,400	18.93

Pressure Vessel: In order to reduce fugitive emissions, pressure vessels have been designed and built. One design is shown in the figure below.

Figure 12.21 Pressure vessel for multistage horizontal mechanical flotation. This flotation unit has two active cells, as shown.

Deepwater GoM Experience with Horizontal Mechanical IGF: Much of the design and operating experience regarding horizontal IGFU for the deepwater was transferred directly from decades of use in the shallow water GoM. C.A. Leech [5] describes the design capacities and operating parameters for horizontal IGFU at Eugene Island 158C, Eugene Island 259C, Eugene Island 331A, and Black Bayou. Flotation units at all of these facilities are similar in many respects to flotation units deployed on Shell deepwater facilities [15, 16]. While the use of Corrugated Plate Interceptors and three phase primary separation is more prevalent at the shallow water locations, and not deepwater, the use of horizontal IGFU was carried over directly from shallow water to deepwater. Also, the design capacities used in shallow water match those used in the deepwater.

While the importance of horizontal Induced Gas Flotation Units (IGFU) on deepwater Gulf of Mexico is taken for granted, such equipment is not necessarily part of the essential water treating equipment in other parts of the world. This is discussed in more detail in Chapter 20 (Applications – GoM versus North Sea).

Table 12.5 Summary of oil removal efficiency experience with Horizontal Rotor-Type IGF. Note that additional detail is given below in separate sections

Location	Water Rate (BWPD)	Oil Removal Efficiency (%)
Deepwater GoM	10,000	92
Deepwater GoM	15,000	93
Alberta	43,000	90
Bakersfield	10,000	95

Representative Performance Data: As a guide, the following data was obtained from a model 76 Wemco Depurator with a design capacity of 25,725 BWPD, operating at roughly 20,000 BWPD. The oil removal efficiency of this unit was 90 %. The feed contained roughly 200 ppmv oil in water. The discharge averaged roughly 20 ppmv.

Wemco Depurator Gas/Water parameters:

Gas/water ratio in each chamber:	8.5 std m3 gas/m3 water
Surface area of each chamber:	1.47 m2
Water rate:	2.84 m3/minute
Gas rate:	24.1 std m3/minute
Gas flux:	16.4 m3 gas/m2 surface area each chamber
Gas bubble diameter:	100 to 500 microns
Gas bubble number density:	1x1011 to 1x1013 bubbles/m3 water
Gas bubble surface area:	1,000 to 20,000 m2/ m3 water

The important parameters to note about Wemco Depurator design are the Gas/Water Ratio and bubble size. Both are relatively large compared to other flotation unit designs. The Wemco generates gas bubbles using a reliable method that generates a large volume of bubbles. The bubbles, it turns out are relatively large. The unit achieves high separation efficiency by utilizing a multi-stage design, with a large volume of gas bubbles. Other technologies such as DGF provide smaller bubbles but may not provide enough bubbles to be effective.

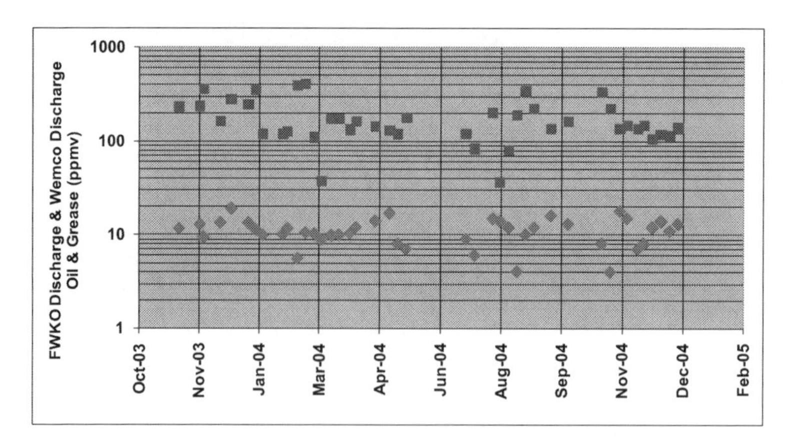

Figure 12.22 Oil in water concentration upstream (red squares) and downstream (blue diamonds) of a Wemco. This Wemco was installed on the Mars deep water offshore Tension Leg Platform. This level of performance was typical.

12.2.5 Horizontal Multistage Hydraulic IGF (HIGF)

The operation of a hydraulically induced gas flotation cell is similar to the mechanically induced gas flotation cell. However, instead of using a mechanically driven impeller to generate bubbles, a recirculated stream of treated effluent water is mixed with gas and the gas/water mixture is injected into the flotation unit. The capacity of the unit must be designed to handle the feed (untreated influent) plus the recirculated water.

The figure below gives a side view of a hydraulically induced flotation unit. This unit has 6 chambers. The middle four chambers have gas addition and are referred to as the active cells. Water enters the unit from the left side of the figure. Each chamber is separated by two steel plates which are referred to as weirs. The design shown in the figure uses an under / over weir arrangement whereby water flows through a notch at the bottom of one weir, then it flows up between the two weirs, then over the second weir into the next chamber. The final chamber is the discharge box (shown on the right hand side). Level is controlled in the discharge box. Since all of the chambers are in hydraulic communication, the level in the discharge box controls the level in all of the chambers. The chambers with gas have higher level (as shown).

The routing of gas is a bit more complicated than that of the water. In the design below, there is a common gas space throughout the vessel. Gas is pulled from this gas space into a device called an eductor. The eductor is mounted in a pipe that carries water. Each chamber has a water pipe with an eductor. The gas / water mixture enters each active cell at the left hand side of the cell. The water used in the eductor is recirculated from the clean water effluent or discharge. A pump is used to provide the motive force for this water recirculation.

An eductor is a device that uses venturi action to inject gas into a stream of water. A venturi is essentially a nozzle with a restricted diameter. The restricted diameter accelerates the water which creates a low pressure region allowing the gas to be sucked in. The amount of gas that is injected to the water depends on the venturi diameter, the gas pressure and the water flow rate. The water used in the eductor comes from a recirculation pump.

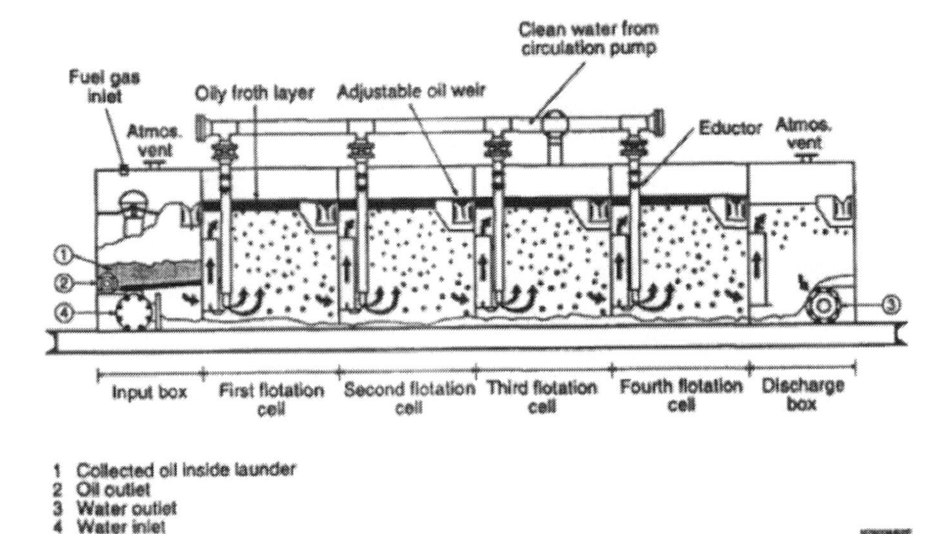

Figure 12.23 Horizontal Multistage Hydraulic Flotation Unit

Eductors: Eductors utilize the kinetic energy of one liquid to cause the flow of another liquid or gas, and operate on the basic principles of flow dynamics. This involves taking a high pressure motive stream (water in the case of IGF units) and accelerating it through a tapered nozzle to increase the velocity of the fluid that is put through the nozzle. There are three connections common to all venturi eductors.

1. **Eductor Motive Connection:** This connection is where the power for the eductor is generated, by increasing the velocity of the motive fluid.

2. **Suction Connection:** This connection of the eductor is where the pumping action of the eductor takes place. The motive fluid (water) passes through the suction chamber, entraining the suction fluid (gas) as it passes.

3. **Discharge Connection:** As the motive fluid entrains the suction fluid, part of the kinetic energy of the motive fluid is imparted to the suction fluid. This allows the resulting mixture to discharge at an intermediate pressure. The combined fluid leaves via the outlet.

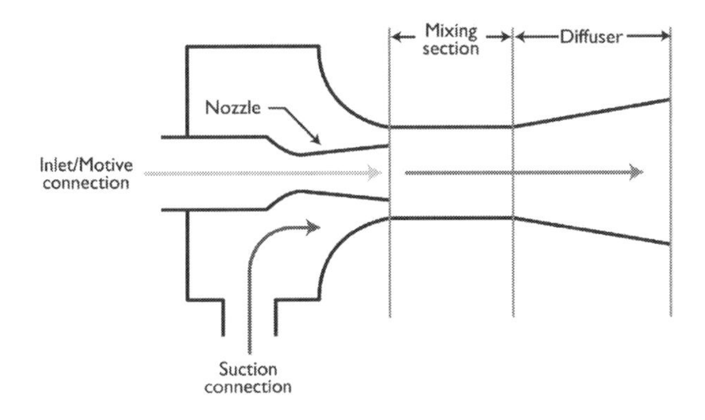

Figure 12.24 Eductor

12.2.6 Vertical IGF (VIGF)

Vertical Column flotation units were originally designed to address concerns over the effect of wave motion on deep water floating structures, and to reduce footprint and as a lower cost alternative to the large multistage horizontal units discussed above. Vertical flotation units were adapted for use in the North Sea and on FPSO applications where atmospheric tanks and horizontal vessels are typically not accepted. Column flotation units have performed well however single vessel applications can be prone to upsets offering no redundancy or protection against mechanical failure. When column flotation units are required to meet system design or space limitations, the use of two or more vessels in series should be considered. Currently, the main advantage offered by vertical IGF is the low cost, and reduced space requirement and weight.

Types of Vertical IGF: There are three basic types of vertical IGF. They are:

- Unicel (reactor tube)
- Radial Eductor
- Tangential Inlet / Solids Sump

While these designations are not particularly handy, they suffice as a substitute for using brand names and identifying the vendors by name. This allows a candid discussion of the strengths and weaknesses of the designs.

Unicel (Reactor Tube) Design: The classic eductor type vertical flotation system is the Unicel vertical IGF. With the use of an eductor, either a horizontal or vertical design is possible. Some designs rely on inlet pressure to drive the eductor (> 45 psig) such that no recycle or inlet pump is required. High volume / low feed pressure system do require a recycle pump to charge the eductor. Water with dispersed gas bubbles enter at bottom of the riser pipe. There is a coalescing element at the top of the pipe. An oily froth is spilled over the weir into the skim through, which is emptied periodically by open a discharge valve.

As with the other flotation devices discussed, the mechanical configuration and the geometry of gas and water contact is important. Feed water and gas enters the bottom inside of a cylindrical tube, which is referred to as a reactor tube. The water and gas mixture travel upward with the gas obviously traveling faster than the oily water thus resulting in collisions between the gas and oil droplets. Once the water and gas mixture arrive at the top of the riser tube, the intention is for the gas bubbles to disengage from the water and carry the oil over the spillover trough. The water then begins a downward decent, guided by the outer cylinder of the vessel, toward the product discharge which is at the bottom of the vessel.

Gas is introduced into the feed water stream using an eductor. One of the earlier Unical designs passed all the water through the eductor and into the flotation machine. The design was later changed to re-circulate clean water reportedly because oil droplets in the raw water were sheared into smaller droplets, making the process less efficient at removing oil. Some designs rely on inlet pressure to drive the eductor (> 45 psig) such that no recycle or inlet pump is required. High volume / low feed pressure system do require a recycle pump to charge the eductor. Water with dispersed gas bubbles enter at bottom of the riser pipe. There is a coalescing element at the top of the pipe. An oily froth is spilled over the weir into the skim through, which is emptied periodically by open a discharge valve.

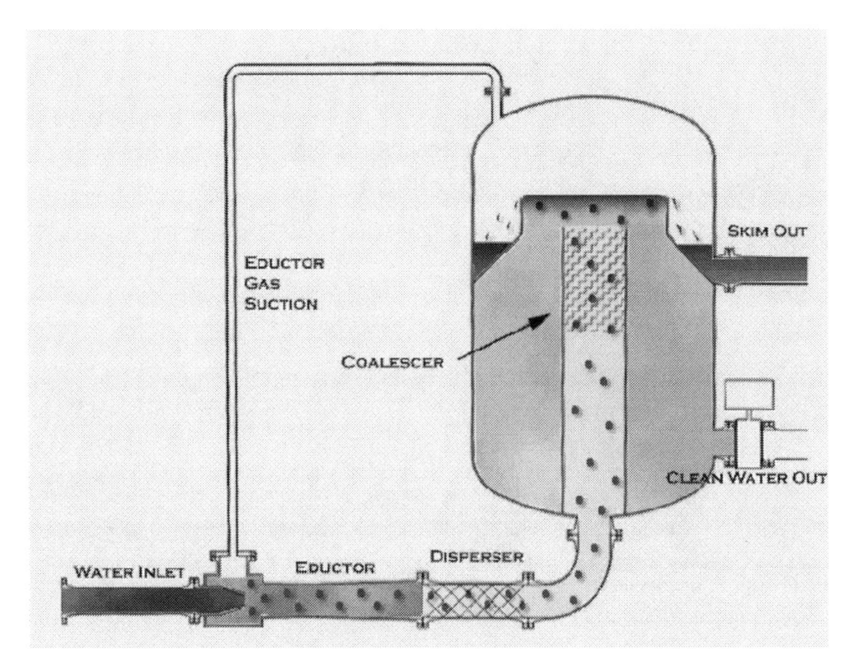

Figure 12.25 Unicel design of Vertical IGF.

Most of the gas / oily water contact occurs in the reactor tube. For any given model of Unicel, the residence time of oily water in the reactor tube can be calculated from a knowledge of the reactor tube diameter and height, the water volume and the actual gas flow rate. The gas actual flow rate is calculated from a knowledge of the gas standard volume and the pressure at the bottom of the riser tube. The pressure at the bottom of the riser tube is the sum of the water hydrostatic pressure and the vessel operating pressure. Various approximations in this calculation can be made by recognizing the fact that the gas density is essentially zero compared to that of the water, and the fact that there is a relatively low gas volume (10 to 20 %). One message from this calculation is that the gas/water reaction time is small compared to the overall water residence time. The difference is due to the relatively small reactor tube diameter compared to the diameter of the vessel. Thus, the reactor tube prevents gas entrainment in the clean water discharge, which is a required design objective for compact flotation. However, it is achieved at a relatively high cost in weight and space.

Coalescing Elements: The vertical configuration lends itself to the use of vessel internal elements that promote uniform distribution of gas, good gas bubble / oil drop contact, and coalescence of oil drops. What has resulted are sophisticated internal elements. While these units are still induced gas systems, they are significantly different from the classic horizontal or vertical IGF units.

Some vertical flotation units rely not only on gas bubble flotation but also coalescing achieved by vessel internal elements. The vessel internals come in many shapes and configurations but all are designed to provide surface area for oil drops to collide and coalesce. The internal devices are referred to as coalescing elements. These units can only be provided in vertical cylindrical design in order to accommodate the flow patterns necessary for coalescence. While coalescing elements do indeed improve flotation system performance, it is difficult to evaluate the effectiveness of such elements and selection of such systems is best made on the basis of field trial.

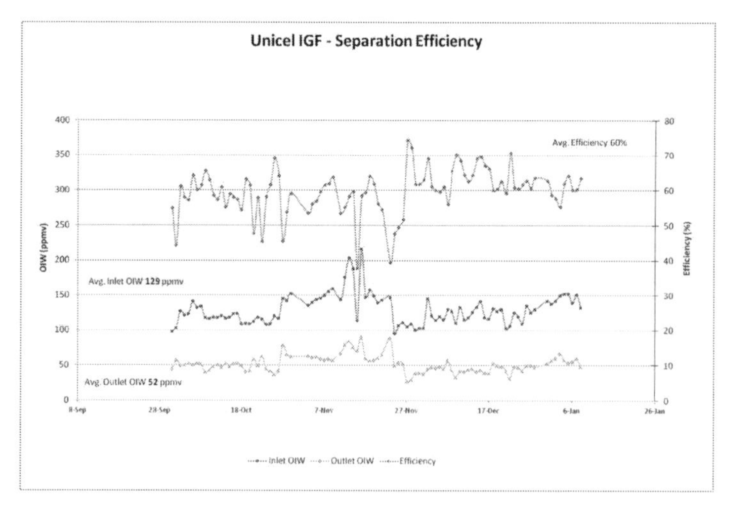

Figure 12.26 Separation efficiency for Vertical Unicel IGF

In one application, the original performance of the Unicel was poor. Significant improvement (from ~20% to ~ 60%) in the separation efficiency was achieved by changing the demulsifier in the FWKO upstream of the Unicel. The original demulsifier prevented the gas flotation bubbles from adhering to the oil droplets. With proper selection of the demulsifier, applied upstream of the Free Water Knockout, the performance of the flotation unit greatly improved.

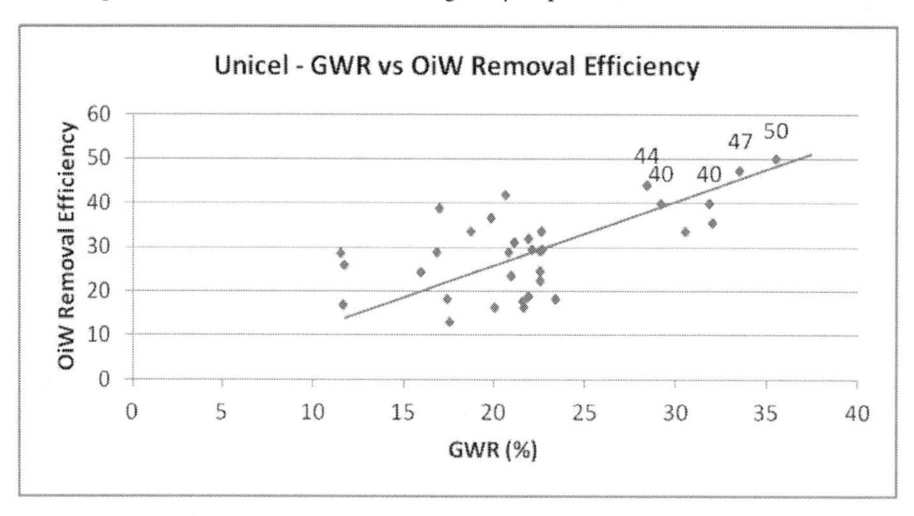

Figure 12.27 Impact of gas/water ratio on the flotation separation efficiency. As the Gas/water ratio goes up, the separation efficiency increases as well.

As shown in the figure above, higher gas pressure was applied which increased the gas/water ratio. With more gas, there was a proportional increase in separation efficiency.

Radial Eductor Design: In the early vertical IGF designs, using a conventional eductor system, gas distribution was poor. Most of the gas formed a plume of fast rising bubbles that drew water into a high velocity rising fluid channel near the center (centerline) of the vessel. Only a small portion of the water in the vessel was exposed to the gas plume. As discussed previously, gas bubbles rise at a fast rate. The drag force around the bubbles transmits upward momentum to the water. Not only did the original gas distribution lack uniform distribution, but it also set up recirculation zones that further increased the tendency toward short-circuiting.

To an extent, this was not unexpected. What was needed was a means to distribute the gas bubbles uniformly across the cross-section of the vessel. Using the computational fluid dynamics tool Fluent, and clear plexiglass models, a number of different inlet gas distribution systems were studied [17, 18]. What developed was a distribution system where the eductor was submerged into the water, with a radially-directed discharge. This increased the fraction of water that experienced gas sweep, and reduced short-circuiting by reducing recirculation zones.

Using a submerged gas eductor provides gas bubble generation immediately adjacent to the gas/water reaction zone. Thus, the bubbles do not significantly coalesce prior to water contact. Such coalescence would increase their size and decrease their number. Both of which have a negative impact on the oil treating efficiency. Also, larger bubbles have a higher rise velocity which increases the induced channeling. Typical residence times are on the order of 4 minutes.

The Radial Eductor design also utilizes a cyclonic inlet for the oily water [18]. At the top of the vessel there is a layer of floating oil. Gas slugs sometimes come into the vessel with the incoming produced water. The inlet cyclonic device protects this layer from being disrupted and re-entrained. As typical of inlet cyclonic devices, it breaks the momentum of the inlet fluids and gently guides the incoming stream into the bulk fluid already in the vessel. Also, as discussed elsewhere, the induced swirl suppresses unwanted secondary flow which can give rise to maldistribution of gas bubbles and oily water. Like most other vertical IGF units, the Radial Eductor unit can be fitted with a coalescing element at the top.

EPCON CFU Tangential / Swirl Inlet Design: The EPCON CFU technology is unique enough that it warrants identification by brand name. It has been developed in close cooperation with Norsk Hydro (now Statoil ASA), and has been specifically designed for application in the North Sea produced water market. It is cost-effective, compact and light weight.

The water enters the vessel horizontally in a tangential direction along the outer wall of the vessel. This inlet geometry creates a cyclonic effect inside the vessels. The inlet pipe has a reduced diameter section to increase the velocity of the water entering the vessels. Inside the vessel a vane is located to guide the water slightly downward to prevent disruption of the oil/water/gas interface.

At the oil/water/gas interface, the oil surface is forced towards the centre of the vessel, by the swirl motion. The gas and oil are routed through the reject line routed to the low pressure trap. This liquid/ gas interface will be held at a constant level by the suspended oil/gas pipe at the top of the vessel. No level controller is required. This reject flow rate is set at a constant value, covering the entire flow range of the system. The oil content in the reject flow typically ranges from 0.5 – 10%, depending on the inlet Oil in Water (OiW) concentration.

The treated water exits at the bottom of the vessel. A vortex-breaker is used. Back-pressure is required to obtain a required pressure drop across the reject valves. Typically the operating pressure for the CFU unit ranges from 0.5 bar (7 psi) and upwards, depending on the system requirements. Obviously this will be higher when adding more stages to the system.

Gas is injected into the oily feed water as shown. The water / gas mixture enters the unit near the top of the vessel. This tangential inlet induces a spiral flow. Since water exits the bottom of the tank, the overall water motion is in a downward spiral. The motion of bubbles is of course in an upward spiral. The moderate spiral flow has low angular velocity, and therefore it provides low centrifugal force such that essentially no oil / water segregation due to centrifugal force is achieved by this motion. However, the spiral flow provides two other functions.

The primary function of the spiral motion is to maintain an even distribution of bubbles and to suppress secondary flow due to potential bubble aggregation and jetting. This is an important issue in flotation unit design. The secondary design intent of the spiral motion is to provide an upward path for oil coated bubbles. To appreciate this idea, it is necessary to understand the function of the cylindrical pipe that is mounted parallel to the centerline of the vessel (shown in the figure). The pipe acts as a draft tube for oil coated bubbles, inducing an upward flow in the centre of the vessel and helping oil coated bubbles to rise to the top of the vessel. It also provides a barrier so that the oily bubbles are directed to the top of the vessel without encountering the cross flow of the tangential inlet. In this case, the mild spiral motion is sufficient to provide some migration of gas bubbles toward the centre and up through the draft tube. Thus, the overall flow of the oil coated gas bubbles follows the red arrows shown in the figure. The draft tube also acts as a coalescing plate for oil coated gas bubbles that migrate to its outer surface. In this case, an oil film forms which rises to the top.

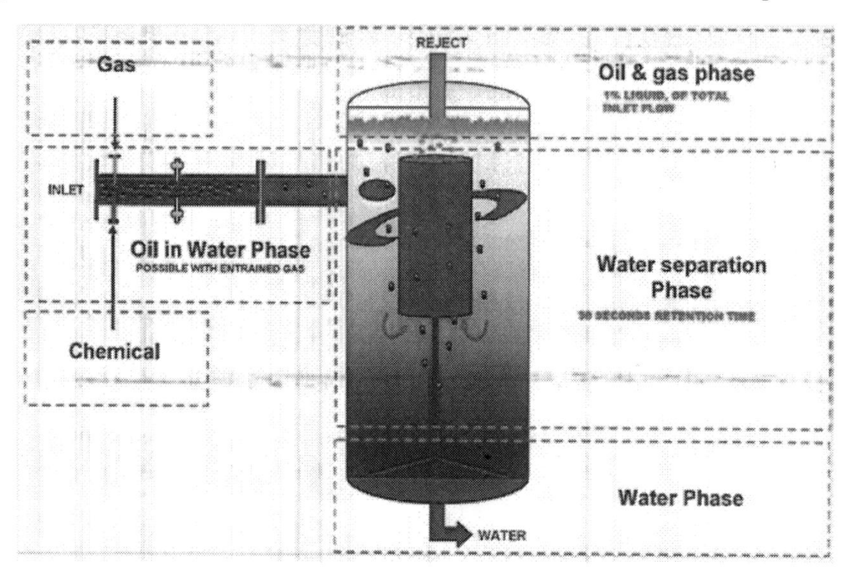

Figure 12.28 Schematic diagram of the Epcon CFU and internals.

The technology is designed to be modular system, which allows the end-user to add additional modules as the volumes of produced water increase. The system does not require auxiliary power and is extremely low on its maintenance intervals, not to say that it does not require any attention when in operation. This provides that the technology is the lowest available system in operational expenditure (OPEX) and is competitive in capital expenditure (CAPEX) to other systems available in the market. It has received several technology awards and has been announced to be the best available technology by the environmental authorities in Norway.

Residual Gas Breakout versus Gas Injection: Like all flotation units, the separation efficiency depends on the gas/water ratio. In some applications there is enough residual dissolved gas in the produced water that no additional gas is required. In that case, gas break-out alone is used for flotation. If this amount of residual gas is not sufficient to ensure optimum flotation efficiency, additional hydrocarbon gas can be injected into the feed water entering the flotation unit. A special designed gas/liquid mixer is located between the injection spool and the vessel inlet flange, which ensures a homogenous distribution of the injected gas in the water prior to entering the vessel. Jahnsen and Vik [19] provide technical details. The amount of gas is depending on the process conditions of the produced water and is typically up to 0.1 Am^3 per m^3 of water (actual meter cubed of gas per meter cubed of water or 10 v%) entering each vessel. A differential controller prevents backflow of the produced water into the gas injection line and will shut down entirely if the low level alarm on one of the vessels is activated.

Retention time of 30 seconds is typically used. The turndown ratio for the EPCON CFU technology is 30% and in some cases even as much as 20%. This means that the same oil removal efficiency can be achieved with only 30-20% of the design flow rates, which provides a stable operating window for the system.

The liquid/gas interface in each vessel is maintained at a constant level by the reject oil/gas pipe in the upper part of the vessel. A gas pocket in the top of the vessel will at all times be present, and will level out automatically as reject residue will escape through the reject pipe. This pressure balance is controlled by the pressure control valve at the water outlet in combination with the reject valve, which is set at a fixed value.

The operating pressure for one CFU vessel is minimum 0.5 bar (7 psi) and upwards. With the use of gas injection, the total pressure drop through the CFU unit is approximately 0.8 bar (12 psi) at the maximum flow rate the unit is designed for.

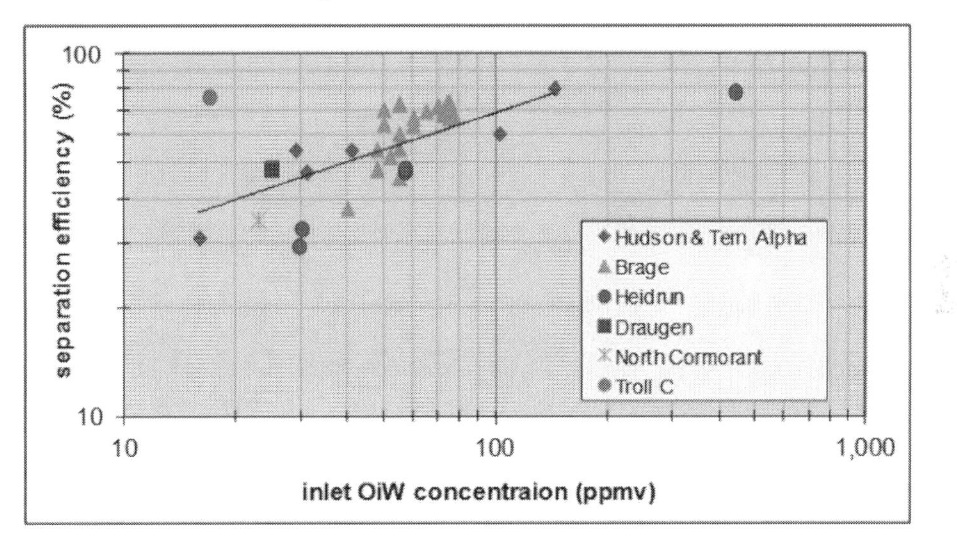

Figure 12.29 Separation efficiency of the Epcon CFU system [19].

The technology is a modular system and therefore it has no restrictions in terms of handling large volumes of water with a treatment result for OiW below 10 ppm. The technology also removes a significant amount of harmful soluble oil pollutants such as PAH's, alcylated phenols and BTEX components. When additional flotation gas is used in a controlled manner up to 80% reduction of PAH's and 88% reduction in BTEX can be achieved.

BTEX Stripping: The EPCON CFU has been evaluated for the removal of BTEX [20]. In this application a source of low BTEX gas was available for use in the flotation unit. The gas was not recycled. Instead, fresh gas was continuously fed to the unit. The concentration of BTEX in the influent water was 4 mg/L. The concentration in the treated effluent water was 0.5 mg/L. This equates to a separation efficiency of about 88 %.

12.2.7 Dissolved Gas Flotation (DGF)

In a dissolved gas flotation unit, gas is first dissolved into a portion of the water at relatively high pressure; the pressure is then released which causes gas to flash-out of solution forming small bubbles. A portion of the effluent water is usually used for generating and injecting the dissolved gas bubbles.

There are various mechanical devices that are used for generating the bubbles. One such approach is shown in the schematic Figure below.

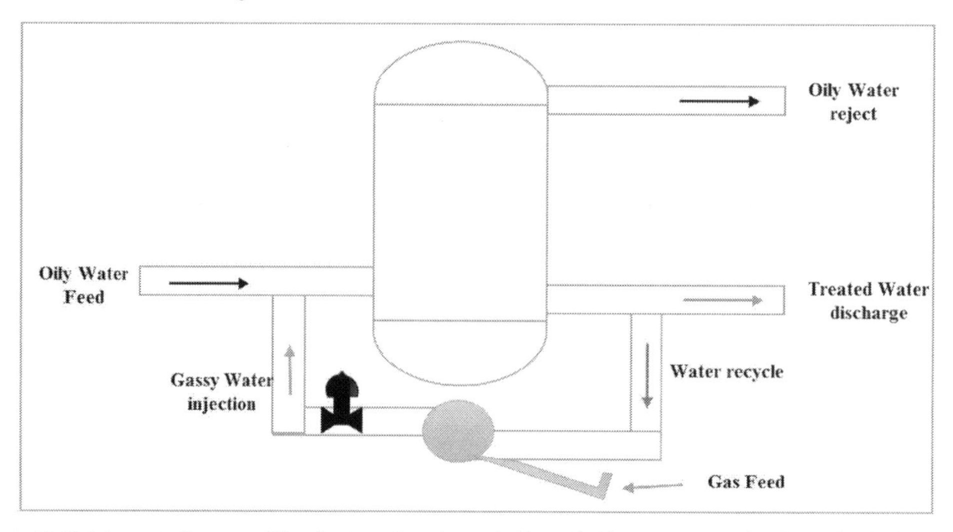

Figure 12.30 Schematic diagram of dissolved gas flotation unit. Treated effluent water is taken from the discharge line and fed into the dissolved gas flotation pump, together with gas. gas dissolves in the water. The pressurized gas then flows through a valve which breaks out the gas. The gas forms small bubbles. The gasified water is injected into the feed line. Variations of this configuration are provided by different equipment vendors

In the figure, a slipstream of treated effluent water is routed to a centrifugal pump. The percentage of this flow varies depending on the manufacturer. A range of 20 to 25 percent is common. Gas is injected into the suction of the pump. The gas is dissolved in the water due to the high pressure of the pump discharge. The gasified water then passes through a valve which drops the pressure and forces the gas to nucleate and form bubbles. This slipstream of water containing small gas bubbles is then injected into the feed stream of oily water which then flows into the vessel where the flotation process occurs. In some designs, the slipstream of water containing the small gas bubbles is injected into the vessel rather than the feed stream. The diameter of the bubbles is in the range of 20 to 60 micron, which is much smaller than typical of induced gas devices which are approximately five to ten times larger.

Reaction Chamber: In some designs there is a reaction chamber downstream of the valve. This reaction chamber allows undissolved gas to break out leaving only the small dissolved gas bubbles to be used for flotation. This allows less cost to be used in the design of the pump.

Shearing of Recycled Water: Unless there is an alternate source of clean water, which is almost never the case, the water used for generating the bubbles is the treated effluent water. This water, though cleaner than the feed water, still contains 10 to 20 mg / L of oil. That oil will be sheared to small drop size in the bubble generating pump. Once sheared, it is unlikely that these oil drops will be removed due to their small size.

Gas/Water Ratio: Using dissolved gas has a limitation in the amount of gas that can be injected. This is due to two factors. The first is a thermodynamic limit of gas solubility which depends on the temperature of the water and the pressure of the gas. As the temperature of the water goes up, the gas volume that dissolves goes down. This can often be a limitation in upstream applications where the fluids tend to be hotter than in downstream / refinery applications. The solubility of natural gas as a function of temperature and pressure is given in the figure below.

$$\Delta P = P_1 - P_2 \qquad \text{Eqn (12.10)}$$

where:

DP	= pressure differential (bar)
P_1	= pressure downstream of pump (pump discharge) (bar)
P_2	= pressure downstream of the valve (bar)

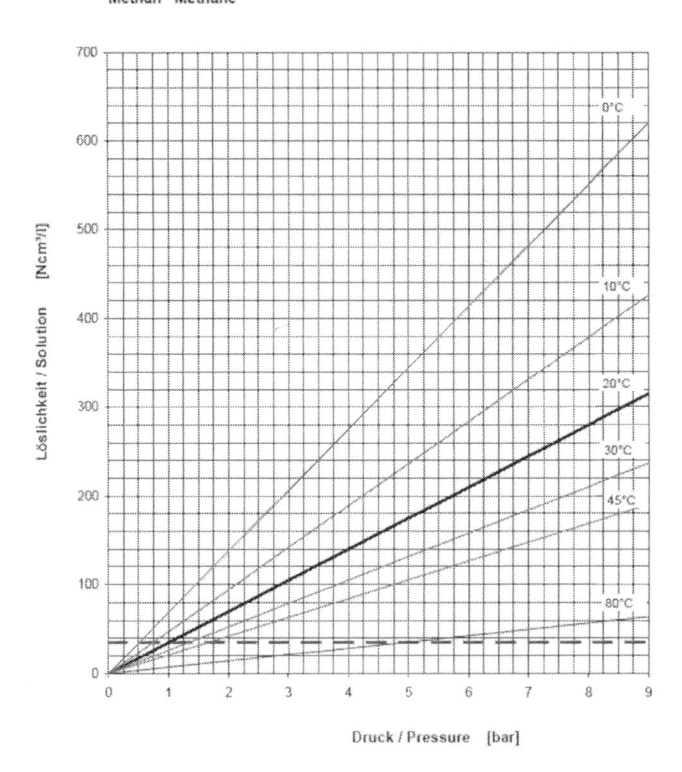

Figure 12.31 Gas solubility curves as a function of temperature and pressure.

The second factor is the practical limit of how much treated water can be recycled in order to inject the gas into the feed water. While the concentration of gas that can be injected is fixed by thermodynamics, a greater volume of gas can be achieved by using a greater side stream flow. Thus, a larger volume of gas bubbles can be injected into the water. The downside of this is that the larger the pump, the higher the added cost. This is the same case in hydraulically induced gas flotation. The volumetric flow rate of gas (under standard conditions) that is ultimately introduced to the water is a design feature which varies from one manufacturer to another but is generally in the range of 4 to 9 % of the feed water volumetric flow rate. This is considerably smaller than the gas volume induced by a Wemco Depurator where the total gas can be as high as 800 v%. Using dissolved gas increases the capture efficiency of the oil droplet / gas bubble interactions. But since there is a smaller gas/water ratio there are fewer bubbles. Thus, there is a tradeoff between higher capture efficiency and fewer bubbles. Whether or not a particular dissolved gas unit provides greater performance than a particular IGF unit depends on the details of design.

Slow Rise Velocity: One of the potential problems of DGF technology is that the gas bubbles do not rise fast enough. The unit depicted in the Figure is a vertical unit. The strengths and weaknesses of vertical units are discussed above. However, where dissolved gas is concerned, there is an additional problem with horizontal units. In upstream applications, as opposed to downstream / refinery applications, most water treatment equipment is designed and operated for higher flow rates. When a horizontal flotation unit used in an upstream application, the horizontal flow of water is relatively high. In fact, it is so high that dissolved gas bubbles do not have enough residence time to reach the water surface. There is a strong tendency for the bubbles to get carried to the treated water discharge without depositing the captured oil to the water surface. Also, in upstream applications, the horizontal flow rate through a flotation unit can be higher than in a refinery. In this case, the bubbles have a tendency to get swept through the system rather than reaching the top of the water phase.

There is one last issue to discuss regarding the generation of dissolved gas bubbles. As can be guessed from the schematic diagram in the figure above, the amount of gas that is injected into the pump is a function of water and gas feed pressures and the pump suction pressure. These variables are not controlled at all by phase equilibrium thermodynamics. Thus, in some cases, more than the dissolved gas volume is injected, and in some cases less than this volume is injected. In the cases where less than the thermodynamic limit is injected, there is a lost efficiency due to low gas / water ratio. In the case where greater than the thermodynamic limit is injected, the additional gas remains as gas bubbles through the pump and through the valve. This is not necessarily a problem but it does add a level of design inconsistency from one location to another.

Most dissolved gas flotation units are designed for roughly 45 psig (3 bar) differential pressure. This is the pressure difference between the pump discharge and the valve discharge. This pressure differential controls the amount of gas that breaks out as a result of the pump and valve combination. This is the same pressure that can be used in the figure on the x-axis. As can be seen from the figure, the solubility curves in the range of pressures of interest are all straight lines. Thus, there are two ways to calculate the amount of gas. Each gives identical results. The first way, is to use the pressure just mentioned and read off the value on the y-axis. The second way is to read the gas solubility (y-axis) for P1 (x-axis), and read the gas solubility for P2 and subtract these two solubility. The same value will be obtained.

The pressure downstream of the valve is essentially the pressure of the flotation vessel, with appropriate hydrostatic head difference. Thus, the actual pressure of the discharge of the A question that often is asked is: what is the effect of raising the operating pressure of the unit. As just demonstrated, as long as the pressure differential remains the same, the same amount of gas will be generated. However, at the higher pressure the bubble diameters will be somewhat smaller.

Effect of Pressure on Bubble Size: The following question often arises: what is the effect of higher pressure operation on the generation of gas bubbles. The following figure addresses this question to some extent. The size of the initial bubble generated is plotted as a function of the operating pressure. The smaller the initial bubble, the greater the number of bubbles and the smaller the average diameter.

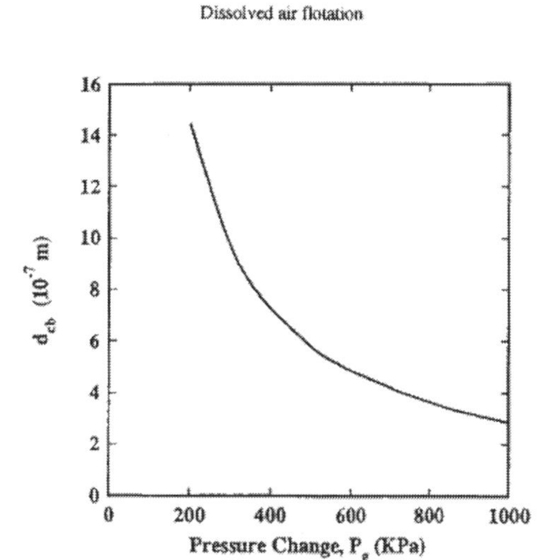

Figure 12.32 Dissolved gas bubble diameter as a function of the pressure change. From [21].

Enviro-Cell: DGF uses a single specialized pump to mix the air/gas mixture providing the bubbles. Specialized systems such as the Enviro-Cell™ IGF use a standard centrifugal pump in conjunction with an advanced eductor design to provide the gas bubbles. Additionally the Enviro-Cell uses multiple eductors in each cell to create a smaller bubble and greater gasification in each cell. Systems such as Enviro-Cell are considered hybrid designs between IGF and DGF. They hold promise for combining the best characteristics of both technologies but are relatively untested in the industry.

IGF vs DGF Experience In early applications of DGF (early 1990's), DGF units suffered mechanical problems. After having broken down several times, most operators refused to repair the units due to competing demands for maintenance personnel. Most operators of DGF units installed in that time period found the units to be difficult to operate correctly. In contrast, typical IGF units of that time period were relatively easy to operate and maintain. Thus, the initial experience with IGF versus DGF favored the IGF due to its relatively better reliability. However, over the years, the reliability of DGF has improved in part due to widespread application in downstream water treatment, in refineries and chemical plants. Today, there are several brands of DGF that compete well with IGF for reliability.

A Britoil report [44] contains the following statement: "It is interesting to note that in the North Sea, DGF systems are not generally employed. The most common designs are based on the induced gas concept. The small bubble size generated by DGF remains a problem as bubble rise rates are low." This is the same problem discussed above where it is noted that small bubbles have slow rise rate which can result in the bubbles getting swept into the treated effluent stream while also discharging any oil that has attached. The report goes on to say that there are design details that can partially overcome this problem. By installing vertical baffles, the DGF bubbles will be directed upward. Also, by orienting the injection nozzles upward, the nozzle discharge velocity will assist upward flow of the bubbles. Applying these concepts, Britoil was able to convert an existing hydraulic IGF to DGF. The increased capture efficiency of the small bubbles was an advantage since the feed water had a mean bubble diameter of 10 micron. Prior to the conversion the existing IGF had a separation efficiency in the range of 60 %. After the conversion to DGF with baffles and vertical nozzles, the separation efficiency went up to 90 %. A similar design concept was used in the CETCO CrudeSep DGF where inclined plates, similar to those of a CPI are used to reduce the distance required for DGF bubbles to travel. In certain applications, primarily where small oil droplets are encountered, the CrudeSep has been successful. Thus, the slow rise velocity of DGF bubbles can be overcome by clever design.

12.2.8 Gas Flotation Tank

One of the more economical and clever designs that has been implemented in recent years involves the addition of gas flotation to a skim tank.

Table 12.6

Date	Skim Tank Inlet (ppm)	Skim Tank Outlet (ppm)	Water to Re-injection (ppm)	Skim Tank Removal Efficiency (%)	Overall System Efficiency (%)
21/09/2005 11:00	240	20	15	92	94
22/09/2005 14:00	100	2	7	98	93
23/09/2005 02:00	64	2	2	97	97
26/09/2005 0:00	149	17	17	89	89
27/09/2005 0:00	411	37	21	91	95
28/09/2005 0:00	321	19	10	94	97
29/09/2005 0:00	420	25	18	94	96
30/09/2005 0:00	109	23	3	79	97
03/10/2005 0:00	130	18	9	86	93

12.2.9 Sparger Systems

When comparing sparged flotation technology to other flotation technologies, the sparged flotation suffers from two main problems. The first is poor distribution of gas bubbles. The second is the tendency of the sparger unit to scale-up. The size of the gas bubbles depends on the pressure drop across the sparging device vs. volume of gas that is being forced through the sparging device. This will be a generalized statement and probably impossible to put empirical size value on.

Sparged Gas Systems Sparged gas flotation is included in the category of induced gas. The modification of a degassing vessel or a separation tank by adding a sparge gas system is relatively common. Such systems can provide an improvement in oil separation efficiency. However, sparge elements are prone to plugging and fouling from solids and scale deposition. Also, the performance is not necessarily reliable since it is difficult to predict and control the gas bubble size and the distribution of gas bubbles. The finer the holes in the sparging element the more prone the element is to fouling and plugging. Excessive gas sparging can create strong poorly distributed flow patterns that lead to excessive channeling.

Wemco did laboratory studies to evaluate sparging technology versus single cell IGF eductor technology. The studies utilized clear tank containment designs that facilitated ease in observing bubble size and bubble motion from the various sparged and eductor induced gas transfer mechanisms. In general, bubbles from sparged inserts rose in a vertical path to the surface of the vessel. Primarily because there is no remaining energy to cause the bubbles to move side to side and they take the path of least resistance to the surface while eductor induced gas bubbles moved around randomly in the tank, why? Because the bubbles are small? Because of the converse of statement above – there is an additional energy source from the eductor or rotor that causes bubble random movement somewhat in relation to the energy input of the eductor motive pump. I thought that an eductor system does not have a pump? Eductor system terminology is reference to a venture eductor, which requires a motive fluid from a pump to induce the gas vs. a submerged rotor system that induces gas into the vacuum above the spinning rotor Sparged bubbles had no side-to-side motion, which is critical to effective IGF oil removal efficiency. Increased gas volume introduced through the sparger insert did not achieve better bubble "sweep", but resulted in larger bubbles and "rolling" of the liquid surface, neither of which is conducive to enhanced oil removal. Increased energy input from the eductor motive pump resulted in significant bubble migration and motion within the tank without rolling of the surface.

Because there is no dynamic mixing of the oily stream with the sparged gas curtain, large numbers of long, closely spaced sparging inserts are designed into the vessel in an effort to (1) realize full bubble "sweep" in the tank and (2) compensate for the bubble size and distribution inadequacies. The sparging inserts have small pores that are designed to deliver small gas bubbles, but unfortunately these small pores plug with solids and scale deposit on a regular basis. When the sparging inserts become plugged and ineffective in gas delivery, they must be removed from the vessel, cleaned, and re-inserted in order to restore maximum oil removal. Because of this maintenance requirement, which can occur as often as twice per month, it becomes necessary to install sparging inserts with 100% stand-by capacity. The process of removing sparging inserts is not a tidy exercise and is usually done by personnel wearing "rain suits" to shield their clothes and bodies from oily water spray that exits around the packing gland at each insert nozzle. While eductor maintenance is also required for IGF technology, the requirement is not as often.

A surprising bubble size phenomenon was observed during these tests in that small bubbles were seen to rapidly grow into larger bubbles within a few inches of the point of introduction. The final rising bubble size seemed to be impacted more by temperature and water salinity than by the size of the original bubble entering the water. Fresh water had significantly larger bubbles than saline water and hot water had larger bubbles than cold water – except in saline water where the temperature had less of an impact than if fresh water. This leads one to conclude that the size of pores in the sparging insert ultimately has limited effect on the flotation gas bubble size.

12.3 Flotation Theory, Design and Modeling

Separation of oil and solids using flotation has been extensively studied and reported in the mineral and mining technical literature. Much of this information relates to the removal of solid particles either to clean the water or to recover valuable minerals. With few exceptions, this enormous body of literature can be applied directly to the treatment of produced water for both oil and solids removal. As in previous chapters, the terms "contaminant" and "particles" will be used interchangeably with oil droplets and solids particles. Whenever an actual distinction is required, then oil droplets, oily particles, or solids particles will be used.

The models for flotation are presented first for batch processes, and second for continuous processes. In practical application, batch processes are not used. However, batch processes are often used in the field and in the lab to troubleshoot and to verify certain aspects of design such as the effect of bubble size and gas/water ratio. Modeling of batch processes is also relevant because the concepts are most easily understood and have direct relevance to the continuous processes used in actual application. After the models of bench flotation are given, then the models for continuous flotation processes are given.

There are two main objectives in presenting this material. The first is to provide an understanding of the processes that cause separation of oil and water by flotation. While there are several formulas that allow calculation of various quantities, the main objective of this material is to demonstrate the relative importance of various factors such as gas / water ratio, bubble size, particle size, the chemistry of the bubble / water interface, as well as the importance of flocculating agent.

12.3.1 Modeling of Flotation Performance

In this section, models of flotation performance are discussed. Flotation is a contact separation process. Contact separation processes can be modeled by considering the collision frequency, and capture efficiency [23]:

- collision frequency   Fcoll
- capture efficiency   Ecapt

These factors are defined in such a way that together they give the overall separation efficiency for a given particle size, gas/water ratio, and bubble size. The mathematical product of these factors results in a model for the separation of oil from water:

$$E_{sep} = F_{coll} E_{capt} \qquad \text{Eqn (12.11)}$$

In the material that follows, each of these factors is discussed and mathematical formulas are given to calculate them. These formulas are intended to provide order-of-magnitude estimates of these factors, and to provide a relative comparison of the importance of gas / water ratio, oil droplet size, gas bubble size, and attachment of gas to oil droplet depending on surface chemistry. The actual calculation of separation efficiency is carried out using a combination of these mathematical factors and empirical data from the laboratory and the field. It is not possible to derive a detailed mathematical model that can accurately predict separation efficiency. The processes are too complex and would require an impractical amount of information about the oil droplets, gas bubbles, and water. Instead, a practical approach is based on guidance from the mathematical theory together with measurements of actual flotation devices.

Collision Frequency (F_{coll}) The calculation of collision frequency is rather straightforward and is aided by the well established kinetic molecular theory of gases. It is a measure of how many contaminant particles are in the path of the bubbles. It does not account for whether or not the bubble and contaminant actually do collide. Collision frequency depends on the concentration of bubbles and contaminant particles, their relative size, and the rate of ascent.

Capture Efficiency (E_{capt}) When an oil drop and a bubble collide, the oil drop doesn't necessarily get captured by the bubble. The oil drop can slide off the surface of the bubble, or be carried around the bubble by the hydrodynamics of the flow. Whether or not an oil drop gets captured depends on the complex fluid dynamics around the rising bubble, and on the surface chemistry between the bubble, droplet and surrounding water. Collision efficiency is a measure of whether or not the contaminant actually does collide with the bubble and stick. It is based on a detailed accounting for fluid mechanics and surface chemistry very close to the bubble and contaminant. It is intended to provide an accurate and detailed correction to the collision frequency discussed above.

Some detailed modeling [22] of the flotation process includes other efficiencies such as the Attachment Efficiency, and the Stability Efficiency. These factors are attempts to include sub-processes in order to provide more accurate models. The physical effects of these processes will be discussed but not the detailed modeling.

12.3.2 Collision Frequency (F_{coll})

The first factor taken into account in modeling is typically an accounting of the number of bubbles and the probability that collisions will occur between the bubbles and contaminant particles. Physically it is the fraction of contaminant particles in the path swept by the bubbles. It is not focused on actual collisions, only of the number of contaminant particles in the path of the bubbles [45]. For this purpose, a collision is when the trajectory of the bubble surface intersects the particle surface. It does not account for an actual capture of the particle by the bubble, only that there is an overlap of trajectories. All of the other complex details of capture are considered in the Capture Efficiency which is discussed below.

In a typical flotation unit, collision frequencies are very large while capture efficiencies are very small. All flotation systems overcome poor capture efficiency to greater and lesser extent by adding chemical and by having a lot of bubbles which increases the collision frequency. In general, when large bubbles are used (e.g. 700 micron) capture efficiency is roughly 1 in 10,000. When small bubbles are used (30 micron) capture efficiency is roughly 1 in 100. Gas bubbles rise fast (large density difference with water) and the small oil drops tend to follow the currents which carry them around the bubbles. The oil drops follow the "streamlines" around the gas bubbles. The larger the bubble, the faster it rises and the fewer oil drops it will collect.

If a contaminant particle is in the path of a bubble, then it is said to be in the collision path. All particles that are in the collision path are included in the calculation of collision frequency even if the detailed hydrodynamics would carry the particle around the bubble without collision. Thus, from the standpoint of a detailed analysis, it is readily acknowledged that the calculation of collision frequency is not precisely equal to the actually frequency of collisions. It essentially ignores important details about fluid mechanics. It over-predicts the actual collision frequency and must be corrected by a much more complex term, the Collision Efficiency, which does take into account the detailed fluid dynamics.

The frequency of collisions between oil droplets and bubbles (or between solid particles and bubbles) can be calculated using the same model as already presented in Section 4.5.2 (Collision Frequency). That model was presented for the case of collisions between oil droplets. However the model is fundamental and can be applied to flotation provided that the appropriate properties for gas bubbles are used. The formula for the collision frequency is give by [23]:

$$F_{coll} = \frac{\pi}{4} n_o n_b \left(d_o + d_b\right)^2 V_{12} \qquad \text{Eqn (12.12)}$$

where:

n_g	number density of gas bubbles (number of bubbles per unit volume)
n_o	number density of oil drops (number of drops per unit volume)
d_b	diameter of gas bubbles
d_o	diameter of oil drops
V_{12}	relative velocity of bubbles and drops (Stokes Law)

The collision frequency depends on the concentration of oil drops, the concentration of gas bubbles, and their relative size. The number density of oil drops and bubbles (n_1, n_2) is multiplied by their cross-section area in order to calculate a total cross-sectional area for collisions. The relative velocity of oil drops and bubbles is important with respect to the residence time. If the bubbles are too small to travel the required distance for oil/water separation they will get swept into the treated effluent discharge and will carry oil with them. The above equation can be rewritten to emphasize the roll of the gas / water ratio:

$$F_{coll} = k_1 n_o (Q_g / Q_w) V_{12} / d_b \qquad \text{Eqn (12.13)}$$

where:

Q_g volumetric flow rate of gas

Q_w volumetric flow rate of water

k_1 proportionality constant

This equation shows the direct relationship between oil/bubble collisions and the gas/water ratio. A high gas/water ratio is required for high collision frequency. As discussed by Strickland [45], for a given gas rate the concentration of bubbles increases as the bubble diameter decreases. Note that the bubble diameter appears in the denominator of the second equation.

But it is not the entire picture. As the bubble diameter decreases the probability of each bubble to attach to an oil droplet goes up. This is not accounted for in the equation above. The two equations above provide insight into the physical mechanism behind the first step in the flotation process, collisions. Neither equation can predict the frequency of oil droplet capture since most collisions do not result in capture. This is due to the hydrodynamic interactions between colliding drops, the stability of the liquid film between the drops, and the presence, in some cases, of chemically and physically adsorbed materials on the surface of the drops which can impede capture. The capture efficiency is discussed below in Section 12.3.6 (Capture Efficiency). Ramirez and Davis [23] describe the model in detail.

12.3.3 Bubble Rise Velocity (V12)

In order to determine the collision rate of bubbles with oil droplets and solid particles, values for the bubble rise velocity are required. It helps therefore to have actual values to work with. Shown in the Figure below are values of bubble rise velocity over a wide range of bubble sizes.

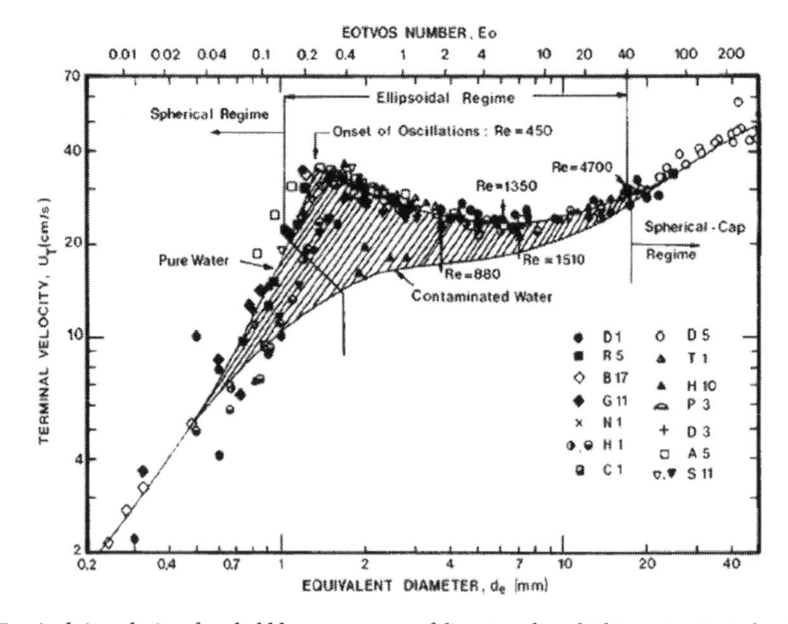

Figure 12.33 Terminal rise velocity of gas bubbles over a range of diameters from [24] in contaminated and pure water. Stokes Law is accurate in the lower left hand region of the drawing.

One interesting observation about these values is that the rise velocity, for relatively small bubbles, can be easy to remember in the units of cm/second, as shown in the figure. For example, a 400 micron bubble will rise 4 cm/sec. Of course, a bubble that has twice the diameter (800 micron) will rise 4 times faster (16 cm/sec). Also, the Figure shown is only for one particular water temperature. For higher or lower temperature, the water viscosity will be different. Nevertheless it is useful to have at least one easily remembered value from which other values can be calculated with ease.

12.3.4 The Bubble / Water Interface

The bubble / water interfacial tension is one of the main determinants for whether the oil droplets will stick to, and perhaps coat the bubble. The bubble / water interface is, in most flotation processes, only a few tens of seconds or a few minutes old. In other words, the bubble / water interface is newly created in every flotation unit. On the other hand, the oil / water interface is already several minutes, if not hours old by the time the oily water reaches the flotation unit. The consequence of this is discussed below.

Multiple Stages of Fresh Bubbles It is well established that there is a dramatic performance enhancement using multiple stages. The main difference between on e stage and another is the introduction of fresh bubbles. To understand this statement it is useful to consider the counter example, which is that of a tall vertical unit that has twice the residence time of any single stage of a multi-stage unit. Such twice-tall units have been built and put into operation. The separation efficiency of these units is almost the same as that of one cell in a multistage unit. Thus, there is something different about putting two stages together each with fresh bubbles compared to stacking two stages one on top of the other and using the same bubbles throughout. Further, a recent entry into the market is the TS Technology vertical flotation unit where in fact multiple vertical stages are employed. Each vertical stage does indeed receive fresh bubble and separation efficiency is similar to that from multiple stages of a horizontal unit with fresh bubbles in each stage. It seems safe to conclude that there is an advantage to adding fresh bubbles in multiple stages. It also seems safe to say that the stages can be configured in either horizontal or vertical configuration.

There are two possible reasons why fresh bubbles might be more effective. The first is that bubbles coalesce relatively rapidly. The number of bubbles present is astronomical. These bubbles are moving and crashing into each other at a high rate. This is what flotation bubbles are expected to do. In a short distance bubbles will coalesce with several other bubbles and their size will increase. As these larger bubbles move rapidly upward, they crash into even more bubbles and reach a relatively large size which is much less likely to capture oil drops.

The second reason why fresh bubbles might be more effective is the presence of surface active compounds in the water. Compounds like fatty acids, and corrosion inhibitors will coat the bubble / water interface making the bubbles repel the similarly coated oil drops. This process is not instantaneous. Thus, it is likely one of the reasons why fresh bubbles are so effective in multi-stage flotation units.

Surfactant Migration to the Bubble / Water Interface Immediately after the introduction of bubbles to the water, the chemical composition of the bubble/water interface will start to evolve [33]. Surfactant molecules, both natural and man-made will diffuse from the bulk phases and from other interfaces to the newly formed bubble /water interface. It is helpful to have a rough idea of the rate at which this occurs. Therefore, the diffusion rate of surfactants onto the bubble / water interface is relevant.

Measurements of the kinetics of adsorption of various surfactant-like molecules to the air / water interface have been made [34]. There were distinct variations in the kinetic of adsorption depending on whether the surfactant was an alcohol, cationic, or anionic, and depending on the pH and temperature. However, under a range of conditions, and for a number of different types of molecules, the kinetics were very fast. Within milliseconds the surfactants diffused from the bulk to the interface. This is one of the main reasons why fresh bubbles are effective at contaminant capture.

12.3.5 Capture Mechanisms and Spreading Coefficient

Capture Mechanisms There are several different ways in which a gas bubble can capture a contaminant particle. Obviously, the particle and bubble must first collide. Having done so, the oil drop may smear across the surface of the bubble forming an interface between the bubble and the water. The interfacial tensions must be favorable in order for this to happen. This is just one example of a capture mechanism. There are others such as:

- Bubble coating by free oil
- Oil droplet and oily-solid attachment to bubbles
- Hydrodynamic capture (dragging of oil drops & solids behind rising bubbles)

Depending on a number of chemical and fluid mechanics variables, one of these capture mechanisms may occur. Probably in many practical situations all of these mechanisms are present to a greater or lesser extent.

Mild Turbulence In most flotation devices there is an advantage to inducing mild turbulence. This induces some horizontal movement in a zig-zag path which increases the change of collisions between the bubbles and particles. In strictly laminar flow, the bubbles and oil drops have a tendency to follow hydrodynamic streamlines which would carry the objects past each other without collision. Turbulence provides a horizontal component to the motion of the oil drops and gas bubbles. Some flotation devices, such as the Wemco, generate a fraction of large bubbles which have the effect of inducing turbulence in the bubble/particle contact zone. Sparging devices tend to produce a vertical bubble pattern with little horizontal bubble movement. Submerged sparger devices can be configured to provide some horizontal movement of the flow.

Hydrodynamic Capture This occurs when the there is capture of oil droplets or solid particles but they not attach or coat the bubble. The lack of attachment or coating may be due to a repulsive energy barrier between the surface of the bubble and the particle surface. Depending on the size ratio of the particle and bubble, and on the size of the bubble, which determines its rise velocity, there may be sufficiently low pressure in the wake of the bubble to trap the particles. This does not necessarily lead to permanent capture and collisions with other bubbles may disrupt the particles. Also, as with any of the capture mechanisms, there is a finite capacity or number of particles that can be captured by each bubble..

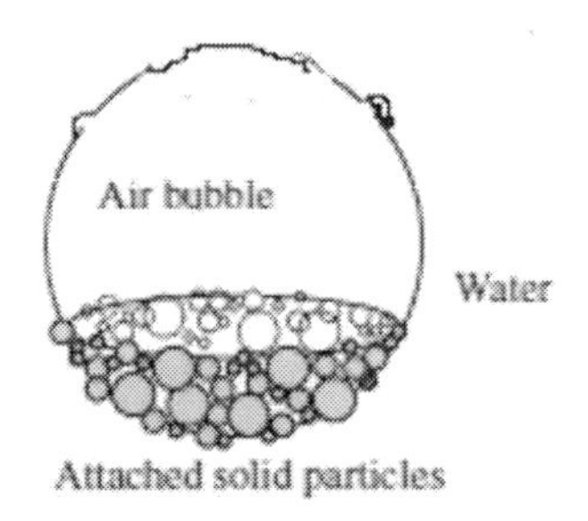

Figure 12.34 A schematic drawing showing the hydrodynamic capture mechanism [22].

Attachment and Spreading One of the capture mechanisms occurs when the oil physically spreads over the surface of the gas bubble [3, 35]. More precisely, this occurs when the oil spreads and forms a film between the bubble and the water. As can be imagined this requires favorable chemistry. Generally speaking the bubble must have a hydrophobic surface. This is illustrated in the figure below. In order for this mechanism to occur, the spreading coefficient must be favorable and there must be a fairly energetic collision between the bubble and oil droplet. This is only one of several capture mechanisms.

The most permanent kind of attachment occurs when the oil forms a film between the water and the bubble surface. This is shown schematically in the figure below.

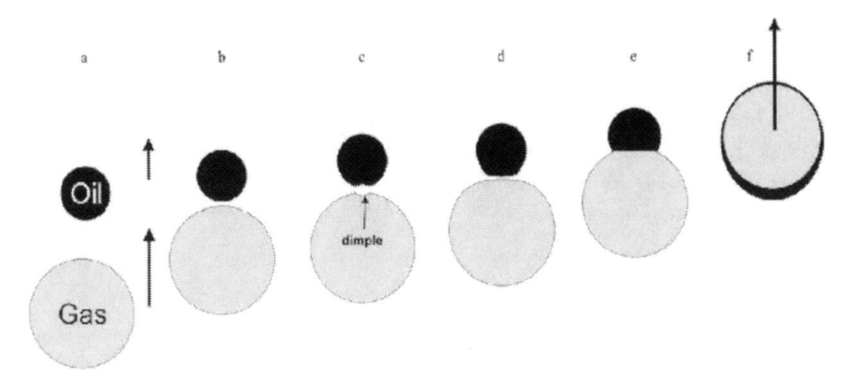

Figure 12.35 One of various capture mechanisms in gas flotation. In this mechanism, the spreading coefficient is favorable.

The size of the oil drop shown here is roughly 10 times larger than realistic. This sequence will be discussed later in relation to the chemistry of the oil/water/gas interaction. For now, it is noted that there are several steps in order for a successful capture of an oil drop.

Spreading Coefficient

The Spreading Coefficient is defined by the following equation [6, 35, 36]:

$$S_o = \gamma_{wg} - \gamma_{ow} - \gamma_{og} \quad \text{Eqn (12.14)}$$

where:

g_{wg} = interfacial tension between water and gas (N/m)

g_{ow} = interfacial tension between oil and water (N/m)

g_{og} = interfacial tension between oil and gas (N/m)

S_o = Spreading Coefficient (N/m)

In order for spontaneous spreading of oil to occur on the bubble surface, the Spreading Coefficient must be positive. Conceptually this can be understood by the following thought process. The line diagram below shows an interface between the three phases. If the oil drop spreads onto the gas bubble, it displaces the gas from the water (reduces the gas/water interface) and creates two new interfaces, the water/oil interface and the gas/oil interface. In order for this process to occur spontaneously the interface thermodynamics must be favorable. For that to be the case, the water/gas interfacial tension should be higher than the sum of the other two interfacial tensions. When this is the case, the Spreading Coefficient is positive and spreading of the oil drop on the gas bubble will occur. From a thermodynamic standpoint, the high water/gas interfacial tension is replaced by the lower gas/oil and oil/water interfaces and the overall Gibbs Energy of the system decreases.

The Spreading Coefficient varies depending on the selection of gas, temperature, production chemicals, concentration and type of naturally occurring surfactants. All of these factors contribute to the Spreading Coefficient. Practically speaking, the interfacial tensions are usually not measured. Instead, a bench top flotation unit, discussed in Section 15.1.8 (Troubleshooting Procedure & Bench Top Flotation Testing), is used to evaluate the tendency of the flotation bubbles to capture the oil droplets. Various chemical additives can be evaluated with such a unit.

Selection of Gas and the Spreading Coefficient Regarding the selection of gas, of the three terms in the spreading coefficient calculation, the most important is the oil/gas interfacial tension [6]. As might be expected, if the gas is a hydrocarbon, then this interfacial tension is small which favors a positive spreading coefficient. This interfacial tension varies depending on the selection of gas, temperature, production chemicals dissolved in the hydrocarbon, concentration and type of naturally occurring surfactants in the hydrocarbon, and the average carbon number of the hydrocarbon. The higher the polarity of the hydrocarbon, the greater the o/g interfacial tension and the more likely that the spreading coefficient will be negative.

12.3.6 Capture Efficiency (E_{capt})

The capture efficiency has two main contributions. The first is the Collision Efficiency which is defined as the ratio of the number of particles that actually collide with a bubble to the number that would collide if the fluid streamlines of the particle were not diverted by the bubble [25, 26]. It is also referred to as the Collision Probability. Another way of defining the Collision Efficiency is as follows. A tube or cylinder can be defined based on the projection of the bubble cross-section along its flow path. The diameter of this cylinder is referred to as the collision cross-section. The volume of the cylinder multiplied by the concentration (number per unit volume) of droplets gives the number of droplets within the cylinder. This is the number of droplets that would collide with the bubble if both the bubble and the droplets continued on their course in a straight line. The collision efficiency is the actual number of collisions divided by the number of droplets in the collision cylinder. It involves complex interactions in the encounter between the fluid, the droplet, and the bubble. Due to its complexity, there are many approximate models that have been developed. Only the basic models that capture the important physical effects are discussed here.

There are two reasons why the Collision Efficiency is important in flotation. The first is that it indicates the importance of bubble diameter relative to particle diameter. It is already obvious from physical intuition that smaller particles are harder to capture than larger particles. Also, large bubbles rise more quickly and push particles out the way without allowing collision. Separation is best when contaminants and bubbles are roughly the same size. The quantitative magnitude of these effects is important. It is particularly important in deciding whether to install a flotation machine with a lot of gas and big bubbles or a machine with small bubbles which may have less gas. This is often the tradeoff that must be made. Quantities such as the Stokes Number and the Collision Efficiency will help in making that determination. That is the subject of this section.

Stokes Number (particle inertia versus viscous drag) It is intuitive to expect that particle collision and capture in a flotation unit will depend on the particle diameter, particle density, bubble diameter, interfacial chemistry, fluid viscosity, etc.. It is also rather easy to imagine that this dependence will depend on details of the fluid flow around these two objects (particle and bubble). Such details are important in understanding flotation but are rather complex from a fluid dynamics standpoint. Thus, there is no generally accepted theory that provides such a separation curve.

The Stokes Number is a dimensionless number that indicates the relative importance for particle inertia versus viscous drag when the particle and bubble collide with one another [25, 27]. As will be shown, particle inertia is not significant in typical upstream oil and gas applications. The Stokes Number also helps in understanding how Collision Efficiency varies with the relevant parameters of the system, i.e bubble diameter, particle diameter, water density and viscosity. As will be shown, the Stokes Number is used to validate different assumptions about particle inertia which are then used to model Collision Efficiency.

Stokes Number is calculated using the following formula:

$$St = \frac{\rho_w u_b}{9\mu_w} \frac{d_p^2}{d_b} \qquad \text{Eqn (12.15)}$$

The Stokes Number was introduced in the literature [3, 27] in order to correlate experimental data of the collision efficiency as a function of bubble diameter, particle diameter, and water temperature (viscosity of the water). For most flotation processes, the Stokes number is much less than unity. For example, consider the case of 200 micron diameter bubbles, and 10 micron diameter oil droplets. The viscosity of water is assumed to be 1 cP (.001 Pa sec), and the density of the water is assumed to be 1000 kg/m3. The rise velocity of the bubble is estimated to be 2 cm / sec. The rise velocity of the oil drop is ignored. The Stokes Number for this case is 0.001. This means that inertia forces are small; particles follow the streamlines which carry them around the bubble; capture efficiency is small. Only a narrow fraction of the collision cross-section results in collision.

Stokes or Viscous Model A simple model for collision probability is due to Gaudin [25, 28 - 31]. It assumes that the flow around the bubble is governed entirely by low Reynolds Number viscous flow. This is the range of Reynolds for which Stokes Law applies. The effect of viscosity is to carry the particle away from the bubble surface.

$Ecoll = 3/2\ (dp/db)^2$

As discussed previously, the Reynolds number for a 100 micron bubble is relatively small. This corresponds to creeping flow in a fluid. This is less of an assumption than it is a more realistic treatment of the situation. However, the assumption that the particle does not have inertia still neglects an important component of the situation.

This model tends to under-predict the Collision efficiency. This is due to the lack of inertia in the particle trajectory. If the particle had inertia, it would have some finite tendency to overcome the viscous fluid streamlines and crash into the bubble.

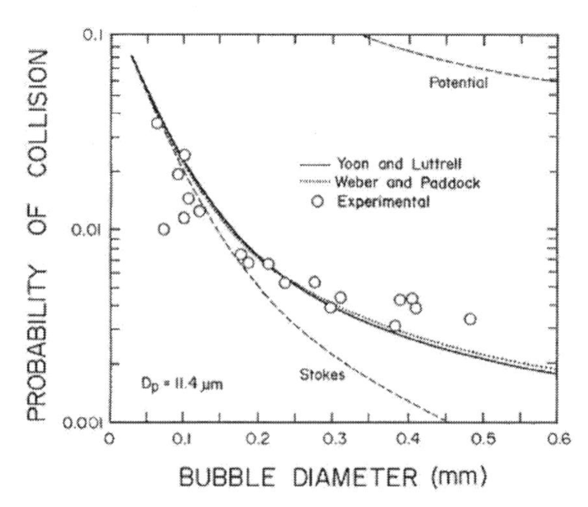

Fig. 1. Effects of bubble size (D_b) on collision efficiency (P_c) under different flow conditions.

Figure 12.36 Probability of Collision as a function of the bubble diameter.

Detailed analysis of Ecoll has been carried out as a function of the ratio (Rp/Rb) for bubbles with and without surface contamination from surfactants [32]. For typical flotation units used in produced water, this ratio ranges from about 0.01 to 0.1. For bubbles that were highly contaminated with surfactant, the values of Ecoll increased from about 0.001 to 0.005. Another study of contaminated bubbles was carried out in [25].

The final term in the equation for Separation Efficiency is the Capture Efficiency. There are three capture mechanisms [18]:

1. The oil drops collide with the rising bubble and the oil coats the bubble. This mechanism is operable when the contact angle between the oil and gas bubble is low (wetting is high). The consequence of this mechanism is a visible oil slick on the surface of the flotation cell (assuming that the surface can be viewed).

2. Oil drops collide with the rising bubble and stick to the surface of the bubble. This mechanism is operable when the contact angle between the oil and gas is high (wetting is low). This mechanism is most efficient when the size of bubble is relatively small since large bubbles tend to move too fast to allow oil drops to merely stick to them.

3. The oil drops and bubbles come to close contact and the oil drop slides underneath the rising bubble, rising with the bubble in its wake. This mechanism does not involve coalescence or sticking. This mechanism is referred to as hydrodynamic capture.

Sylvester and Byeseda discuss the hydrodynamics of oil drop/bubble interaction [37]. Actually, their work is based on theoretical background of bubble interactions with solid particles. They apply this theoretical foundation to bubble / oil drop interactions and provide experimental evidence that such application is accurate. The following figure [38] gives a rough idea of the importance of bubble diameter in collection efficiency over a wide range of contaminant size. Other studies of the effect of bubble size have been reviewed in [26, 39].

It is because of the complexity of the capture process that a detailed predictive mathematical model for flotation will not be developed any time soon. Even if it were developed, it would be difficult to know the precise values of parameters involved. Nevertheless, there is very much to learn about the flotation process from the development of mathematical models. For example, and as will be demonstrated, there are various dimensionless groups that characterize flotation equipment and which can be used to compare one design versus another. The models are also useful in determining the relative importance of gas / water ratio versus gas bubble size. This is useful because it turns out that bubble size is often a trade-off with gas/water ratio. Understanding this tradeoff is useful in selecting the appropriate flotation technology to use in a given application.

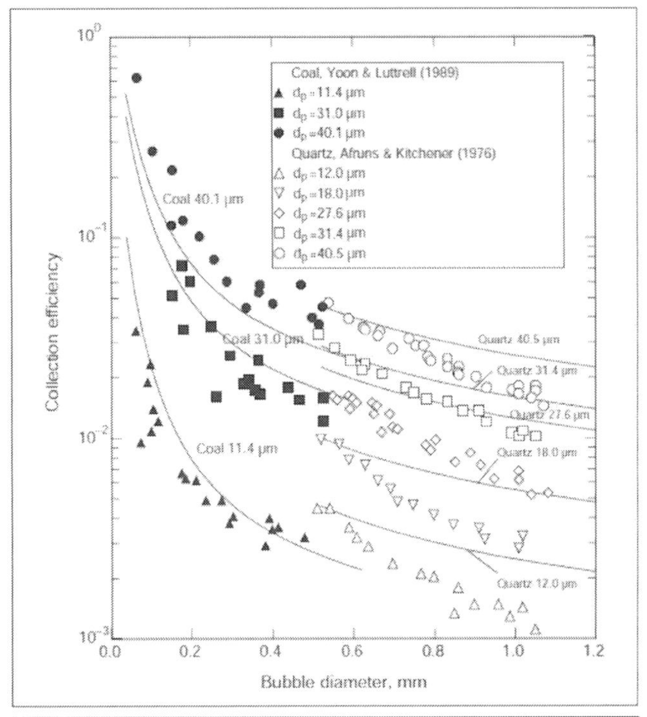

Figure 9.5 Comparison between measured and predicted collection efficiencies. The lines were calculated using Equation 9.14 for each of the two minerals.

Figure 12.37 Collection efficiency as a function of bubble diameter for a number of contaminants [38]

12.3.7 Modeling the Batch Flotation Process

Many laboratory studies have been carried out using oily water in a graduated cylinder. Gas bubbles are introduced at the bottom of the cylinder for a specified period of time after which the volume (or weight) of oil collected at the top is measured and a separation efficiency is calculated. The amount of oil that remains suspended in the water can also be measured in order to verify the material balance. The process is referred to as a "batch" process because the water in the cylinder is not flowing. The same sample of water is used throughout the experiment. In such batch studies, different chemicals can be evaluated, as well as the effect of oil drop size, bubble volume and bubble size. Qualitative differences can be observed. Also, a quantitative assessment of the process can be determined and compared to mathematical models of the process.

Over many years such studies have been carried out in the laboratory and in the field [26, 40]. They have been useful in selecting the right chemicals, in optimizing the process, and in developing an understanding of how flotation works. Data from studies of this type are readily available. Much of the detailed knowledge of gas / oil drop interaction comes from these studies.

The batch flotation model makes use of the general framework presented in Section 12.3.1 (Modeling of Flotation Performance). The model allows a theoretical calculation of oil removal efficiency as a function of typical gas flotation variables such as; gas bubble diameter, gas/water ratio (scf gas/ft3 water), and oil drop diameter. A number of studies have been carried out where the following equation, or one very similar has been used [23, 41 - 43]:

$$E_{sep} = \frac{dn_o}{dt} = F_{coll} E_{capt} = -\frac{\pi}{4} n_o n_b (d_o + d_b)^2 V_{12} E_{capt} \qquad \text{Eqn (12.16)}$$

where:

h = collision frequency (number of collisions/m^3 sec)

n = number density of drops (number/m^3)

This suggests an exponential reduction of oil concentration versus time, as shown in the figure.

$$\frac{d\ln(n_o)}{dt} = -\frac{\pi}{4} n_b (d_o + d_b)^2 V_{12} E_{coll} E_{capt} = -k = -\frac{1}{4} S_b P \qquad \text{Eqn (12.17)}$$

As discussed in ref [43], k is the so-called flotation rate constant. The quantity Sb is referred to as the bubble flux. It is defined as the superficial (calculated) cross-sectional area of flotation bubbles moving into (or out of) a contact cell per cross-sectional area of the contact cell unit time.

Sato et al & Effect of Bubble Size Sato et al. [41] gives experimental data for estimating the capture efficiency when there are large bubbles as with a Wemco. They carried out experiments using a cylindrical tube with a glass frit inserted at the bottom to generate bubbles. The oily water phase was stationary in the sense that it was not flowing in or out of the cylinder. After sparging the oily water for a certain time, the oil remaining in the tube and the oil that was floated to the top was measured.

One of the important findings was that the rate of oil removal depends on the initial concentration of oil to a different extent depending on whether small gas bubbles were used or large gas bubbles. When small bubbles were used, the rate of oil removal did not depend on the initial oil concentration. When larger bubbles were used, the rate of oil removal did depend on the initial concentration of oil to the first power.

This finding implies that there are two different mechanisms occurring in oil removal, depending on the bubble size. For small bubbles, the primary mechanism of oil removal depends strongly on collisions between gas bubble and oil drop. For larger bubbles, this mechanism is also important but a second mechanism also contributes. For larger bubbles, the rate of oil removal depended on the concentration of oil. This implies that collisions between oil drops are important. This later mechanism can be described as oil drop coalescence facilitated by gentle mixing and agitation due to the movement of gas bubbles.

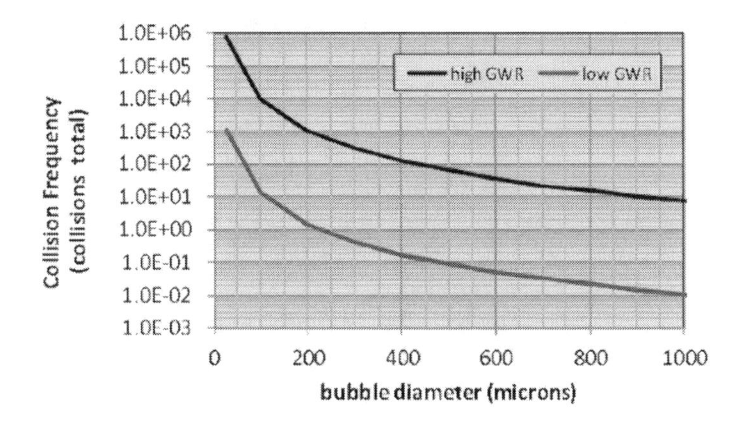

Figure 12.38 The effect of GWR and bubble diameter on the collision frequency.

The kinetic theory of flotation allows the calculation of oil capture. Physically, gas bubbles and oil drops rise at different rates according to Stokes law: V12

The probability of collision is equal the cross sectional area of bubble and drop: π(a1+a2)2, and the concentration of oil drops (n1) and bubbles (n2)

The model allows a theoretical calculation of oil removal efficiency as a function of typical gas flotation variables such as; gas bubble diameter, gas/water ratio (scf gas/ft3 water), and oil drop diameter

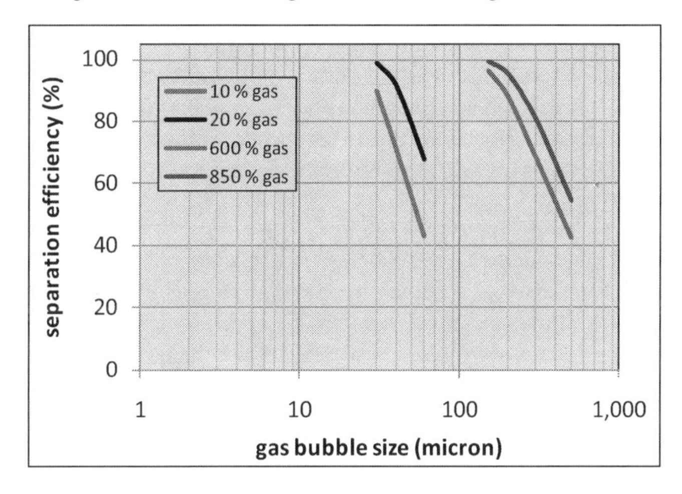

Figure 12.39 Separation efficiency as a function of gas bubble diameter and gas/water percentage. As shown, high separation efficiency can be achieved with large bubbles provided that there is a large gas/water ratio. Conversely, if small bubbles are used, then the gas/water ratio required can be quite smaller. This is due to the greater capture efficiency of small bubbles.

Some equipment deliver large bubbles and lots of gas, and the other equipment produce small bubbles but not much gas. There is no machine that gives small bubbles and a lot of gas. The reason for this is partly related to mechanical configuration and to the cost of generating gas bubbles.

Capture efficiency increases as the bubble size decreases because larger bubbles move too fast to capture the slow moving oil drops. But this effect is less pronounced when chemical is added.

With chemical, capture efficiency is roughly 1/100 for large bubbles. It is 1/10 for small bubbles. An important point to notice is that without chemical, capture efficiency is proportional to 1/d2 while with chemical, capture efficiency is proportional to 1/d.

12.3.8 Correlating Performance of Continuous Flotation Cells

Flotation carried out in a small scale cylinder, as discussed in the previous section, has some similarities but it also has several differences with actual flotation processes that occur in a continuous full scale flotation unit. The first and most obvious difference is that a batch process varies with time whereas a continuous process reaches a steady state. As discussed above, in the batch process, the decrease in oil concentration versus time is an exponential decay. In a continuous process, the oil concentration changes as the fluid moves through the unit, but for any given position within the unit, it is expected to remain essentially constant as a function of time.

There is one very important drawback of these studies however. In a batch flotation study the water is static. The volumetric flow rate of water does not enter into the experimental design because the water has no flow rate. The gas does have a volumetric flow rate. But the gas volume can be expressed in terms of the gas volumetric flow rate, the time of the experiment, and the gas volume per unit cross-sectional area. These are all batch experimental parameters.

The continuous process is more accurately modeled using a Continuous Stirred Tank Reactor model [40]. This is a common approach for chemical engineering continuous processes. This involves a slightly different mathematical formula but the physical interactions are similar to those in the batch process. Nevertheless, the batch process is more commonly used in the lab and quite a bit of information has been gained by published studies. Also, the modeling for the batch process includes all of the important parameters that are important in the continuous process, except the water flow rate.

12.3.9 Sweep Factor

The Sweep Factor is a calculated parameter that describes the rate with which gas bubbles sweep a given volume of water. It is presented here for a few reasons. The first reason is that the Sweep Factor provides a certain degree of conceptual insight into the mechanism of flotation. As the name implies, the Sweep Factor gives a sense of the rate at which gas bubbles sweep through the water.

The second reason is that Sweep Factor occasionally shows up in the literature on flotation and in most instances it is not described nor presented well. This leads to confusion as to what it means. More importantly, it has been used to demonstrate the importance of using small bubbles. There is no doubt that small bubbles have certain advantages. However, both bubble diameter and gas flow have equal importance, as discussed extensively already. When the Sweep Factor is calculated properly, this is shown clearly. The Sweep Factor is presented here following the discussion in Sylvester and Byeseda [37].

The calculation of the Sweep Factor begins with a calculation of the cross-sectional area of a single gas bubble. This is given by:

$$A_b = \frac{\pi}{4} d_b^2 \qquad \text{Eqn (12.18)}$$

where:

A_b = cross-sectional area of a single bubble with average diameter d_b (m^2)

d_b = average bubble diameter (m)

Small bubbles rising at moderate speed through water will remain roughly spherical. Larger bubbles rising rapidly through water will deform. In the present case it is assumed that the bubbles remain spherical. It is noted that the cross-sectional area of a bubble (A_b) is not the same as the collision cross-sectional area. The collision cross-sectional area includes the excluded diameter around the bubble.

It is assumed that bubbles are introduced at the bottom of a cylinder containing oily water and that they rise to the top of the column of water. The volume swept by a single rising gas bubble as a result of traveling from the bottom of the column to the top of the column is given by:

$$V_{sb} = A_b L = \frac{\pi}{4} d_b^2 L \qquad \text{Eqn (12.19)}$$

where:

L = height of the cell (m)

V_{sb} = volume swept by a single bubble (m^3 / bubble)

This is simply the formula for the volume of a cylinder of height L and diameter d_b. For every bubble that is introduced at the bottom, this is the volume that will be swept out once the bubble has risen through the entire column of water. The next step is to calculate the number of bubbles that are introduced to the cell in a given amount of time. This is calculated using the gas volumetric flow rate and the volume of the average bubble, as follows:

$$N_{bt} = Q_{gas} / V_b = Q_{gas} \frac{6}{\pi d_b^3} \qquad \text{Eqn (12.20)}$$

where:

N_{bt} = number of bubbles introduced per unit time (bubbles / second)

V_b = volume of a single bubble (m^3 / bubble)

Q_{gas} = gas volumetric flow rate (m^3 gas / second)

Using SI units, the measure of time is a second. A second is a rather short amount of time to conceptualize the process. Few if any bubbles will travel from the bottom to the top of the cell in one second. However, if say an hour was used, it is easy to see that N_{bt} is the number of bubbles that are introduced to the bottom of the cell and which rise to the top of the cell in one hour. This number is both a function of the time it takes to travel from the bottom to the top, and the number of bubbles introduced at the bottom. The total volume swept out by all of these bubbles calculated as:

$$V_{st} = V_{sb} N_{bt} = \frac{3}{2}\left(\frac{Q_{gas} L}{d_b}\right) \qquad \text{Eqn (12.21)}$$

where:

V_{st} = volume swept out by bubbles in a given amount of time (bubbles / second)

N_{bt} = number of bubbles introduced per unit time (bubbles / second)

V_b = volume of a single bubble (m^3 / bubble)

This last quantity, V_{st}, is the volume swept out by bubbles in a given amount of time. In the case where SI units are used, the given amount of time is one second. However, Q_{gas} can be expressed in volume per minute or volume per hour or any other amount of time. In which case, the parameter V_{st} would be the volume swept out by bubbles in that amount of time. The Sweep Factor is calculated by dividing this bubble volume by the cell volume, as follows:

$$E = \frac{V_{st}}{V_R} = \frac{3}{2}\left(\frac{Q_{gas}L}{V_R d_b}\right) \qquad \text{Eqn (12.22)}$$

where:

E = Sweep Factor (m² gas / m² cell sec)

V_R = volume of the cell (m³)

The Sweep Factor is conceptually the number of times that a volume of gas sweeps out a volume of water in a given amount of time. Again, the unit of time can be chosen as desired. Values of the Sweep Factor are given in the figure below. As shown, the Sweep Factor increases by a factor of ten when the bubble diameter is reduced by a factor of ten. Also, the Sweep factor increases by a factor of ten when the gas flow rate increases by a factor of ten. Also as shown, the Sweep Factor is a very large number for most flotation applications.

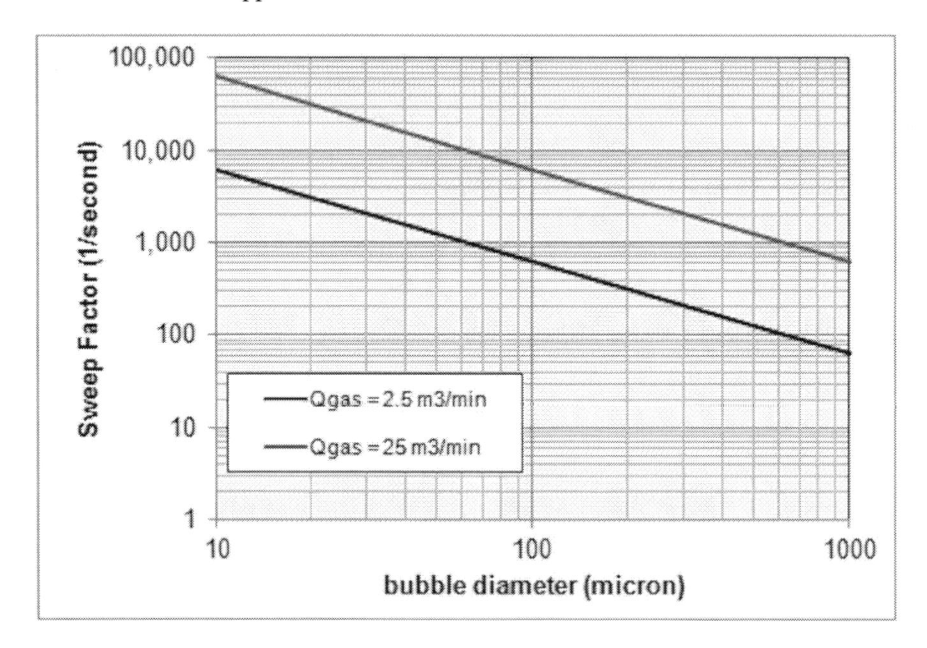

Figure 12.40 Sweep Factor as a function of bubble diameter. Two different gas flow rates are shown. The water flow rate does not have an effect on the Sweep Factor. Thus gas/water ratio is not considered in this model.

While the Sweep Factor adequately illustrates the importance of gas flux, it is somewhat misleading. The misleading aspect has to do with the difference between a static column of water (batch test) versus an actual flotation unit where both gas and water are flowing into and out of a contact cell. Notice that the Sweep Factor is independent of the water flow rate. This is a major drawback since, practically speaking, flotation separation efficiency is strongly dependent on the ratio of gas to water, not just the gas flow rate by itself.

Having a sufficient number of gas bubbles is important in achieving high separation efficiency. A high number of bubbles ensures a high number of oil drop / bubble collisions. However, this alone is not sufficient. There must also be capture of oil drops by the gas bubbles. The Sweep Factor gives no indication of the effectiveness of oil drop capture.

12.3.10 Flux Factor

Flux Factor Three important factors in characterizing flotation equipment are the bubble size, the gas / water volume ratio, and the geometry. Often in the effort to design equipment which optimizes the separation efficiency, at reasonable cost, there is a tradeoff in these factors. This tradeoff can be explicitly understood. As we will see in this section, the design efforts that led to the Unicel, and the VersaFlo demonstrate an understanding of flotation technology in general, and flotation flux factor in particular.

The Flux Factor (N_f) gives an indication of the efficiency with which gas bubbles contact oil drops. It does not take into account the collection efficiency. It only accounts for collision frequency between gas bubbles and contaminant particles. There are several ways to maximize the number of collisions with oil drops. One way is to generate a huge volume of gas. This is the approach embodied in the Wemco design. Another way is to generate a modest amount of gas but to split the gas into relatively small bubbles. This is the approach embodied in the dissolved gas flotation devices. A third way is to generate a moderate amount of gas, and a moderate bubble size but to use a tall narrow contact cell. This is the approach embodied in the Unicel design. Finally, there are units such as the VersaFlo which embody all three approaches.

The Flux Factor (N_f) is a quantity that is defined to help characterize the different design approaches to flotation. The concept behind it is the following. In order to achieve high oil separation efficiency there must be a high number of collisions between oil drops and gas bubbles. In order for there to be a high number of collisions, there must be a large number of bubbles, and the bubbles must cover a high cross sectional area of the water contact cell. The Flux Factor is a measure of the total cross sectional area of bubbles, per unit area of the contact cell.

Contact Cell In the discussion that follows it will be convenient to have a precise definition of the geometry of contact between oil drops and gas bubbles. It is most appropriate to consider a cylindrical volume through which water and gas flow. As will be discussed later, specific types of flotation equipment differ in the size (height and cylinder cross section) of the contact cell but most units that are on the market today have the same basic shape, an upright cylinder. This is true even with the horizontal multistage design. Each cell has a roughly cylindrical volume where active flotation occurs. The corners of such a cell do not contribute significantly to the active volume of flotation. It is just simply cost effective to build rectangular cells that can easily be connected in series. Also, such an arrangement affords a simple path for water to flow from one cell to another. Despite this apparent rectangular geometry, the active volume in a horizontal IGF is cylindrical. Depending on the design of the flotation unit, the water may flow up or down in this contact cell. The gas always flows upward.

Thus, we consider the cylindrical volume to be a general case of flotation geometry.

$$V_c = A_c H_c \qquad \text{Eqn (12.23)}$$

where:

V_c = volume of the cylindrical contact cell

A_c = cross sectional area of the contact cell

H_c = height of the contact cell

These terms are discussed further below.

Contact versus Residence Time In some cases it is necessary to distinguish between contact time and residence time. Contact time is the time that oily water is being swept by flotation bubbles. Residence time is the time that the oily water spends in the flotation cell. The later quantity will include contact time plus whatever time the oily water spends in the flotation device without bubble contact.

The most straightforward way to calculate contact time is to consider the contact cell volume versus the water volumetric flow rate. The contact time is given by the following equation:

$$t_c = V_c / Q_w \qquad \text{Eqn (12.24)}$$

where

t_c = contact time (time that oily water and bubbles are in contact (minutes)

V_c = volume of the contact cell (m3)

Q_w = volumetric flow rate of water (m3/minute)

Bubble Characteristics (area, volume) In this section we define the characteristics of the gas bubbles. Like water drops, gas bubbles have a distribution of diameters. For our purposes here, it is convenient to only discuss the Dv50 value of the gas bubble diameter, rather than the diameter distribution. The reason for this is the need, in this section, to provide relations between bulk averages such as gas volumetric flow rate, gas superficial velocity and number of bubbles. Once these relationships are established, based on the average bubble size, then the intricacies that arise from the fact that gas bubble diameters are actually a distribution of values can be discussed.

The ratio of the bubble surface area to volume, for individual bubbles, is given by the familiar expression.

$$a_b / v_b = 3/2d \qquad \text{Eqn (12.25)}$$

d = Dv50 value of the bubble diameter (m)

a_b = the cross-sectional area of an individual bubble having the Dv50 diameter (m2)

v_b = the volume of an individual bubble having the Dv50 diameter (m3)

Again, this is based on the volume average diameter of the bubbles.

Bubble per Unit Area per Unit Time The number of bubbles that enter the contact cell, during the contact time is calculated presently (N_b). This is given by the volume of gas that enters the contact cell in the contact time ($Q_g t_c$) divided by the volume of each bubble (v_b).

$$N_b = Q_g t_c / v_b = Q_g V_c / Q_w v_b \qquad \text{Eqn (12.26)}$$

where

N_b = the number of bubbles that are swept through the contact cell in the contact time.

This expression contains the gas to water volume ratio (Q_g/Q_w), which is multiplied by the ratio of water volume to bubble volume (V_c/v_b). This later quantity is a measure of the number of bubbles per unit volume of water. The product of these two quantities is the volume of gas divided by the volume of each bubble. Thus, the expression overall gives the expected relation, the number of bubbles that enter the contract volume in the time that the water is treated.

Roughly speaking, the flux of bubbles is the number of bubbles per unit time per unit area. We need only take the previous expression and divide by the cross sectional area of the contact cell. We choose the contact cell cross sectional area since this is the relevant area of interest. The following expression gives the number of bubbles per unit area that enter the contact cell in the contact time:

$$N_b / A_c = Q_g V_c / Q_w A_c v_b = H_c Q_g / Q_w v_b \qquad \text{Eqn (12.27)}$$

The number of bubbles per unit area is the flux of bubbles. It is obviously a relevant quantity. However, an even more relevant quantity is the cross sectional area represented by this flux of bubbles. This is easily obtained by multiplying the flux times the area of each bubble (a_b). This gives the desired quantity, the Flux Factor:

$$N_F = a_b N_b / A_c = H_c Q_g a_b / Q_w v_b = (3\ H_c / 2d)(Q_g / Q_w) \qquad \text{Eqn (12.28)}$$

The final expression on the right hand side is the quantity that we are seeking. It has the height of the contact cell in the numerator. Thus, a tall narrow contact cell gives a higher bubble flux than a short broad cell. This seems intuitively correct. It has the gas volume factor in the numerator. Thus a higher gas volume factor will increase the bubble flux. This also seems intuitively correct. Finally, the bubble diameter is found in the denominator. Thus, as bubble size decreases for a given cell geometry and for a given gas volume fraction, the flux factor increases. This follows the idea that surface area to volume ratio increases with the inverse of the bubble diameter. Thus, this also seems intuitively correct. All things considered then, the Flux Factor should be a reasonable characteristic to use in differentiating between different flotation devices.

The Flux Factor (N_f) gives an indication of the efficiency with which the generated gas bubbles are used to contact oil drops. One of the most important factors in flotation is the number of bubbles. If there is a very high number of bubbles, there is a greater chance that most oil drops will collide with a bubble. Since so few of these collisions result in capture of oil drops, it is important to maximize the number of collisions.

The Flux Factor (N_f) is a quantity that is defined to help characterize the different design approaches to flotation. The concept behind it is the following. In order to achieve high oil separation efficiency there must be a high number of collisions between oil drops and gas bubbles. In order for there to be a high number of collisions, there must be a large number of bubbles, and the bubbles must cover a high cross sectional area of the water contact cell. The Flux Factor is a measure of the total cross sectional area of bubbles, per unit area of the contact cell.

References to Chapter 12

1. K.J. Ives, The Scientific Basis of Flotation, NATO Advanced Science Institutes Series, Martinus Nojhoff Publishers, The Hague (1984).

2. B.K. Parekh, J.D. Miller, Advances in Flotation Technology, Society for Mining, Metallurgy and Exploration, Inc., Colorado (1999).

3. F. Bloom, T.J. Heindel, "Mathematical modeling of the flotation deinking process," Math. Comput. Modeling, v. 25, p. 13 (1997).

4. K. Arnold, M. Stewart, Surface Production Operations, Design of Oil Handling Systems and Facilities, Third Edition, Gulf Professional Publishing, an Elsevier Company, Boston (2008).

5. C.A. Leech, S. Radhakrishnan, M.J. Hillyer, V.R. Degner, "Performance evaluation of induced gas flotation (IGF) machine through math modeling," OTC 3342 (1978).

6. C.H. Rawlins, C. Ly, "Mechanisms for flotation of fine oil droplets," published in Separation Technologies for Minerals, Coal, and Earth Resources, Ed.: C.A. Young, G.H. Luttrell, published by: Society for Mining, Metallurgy, and Exploration. See p. 307.

7. N.K. Shammas, G.F. Bennett, "Principles of air flotation technology," Chapter 1 in Flotation Technology, Ed. L.K. Wang, N.K. Shammas, Springer (2010).

8. L.-S. Fan, R.C. Chen, J. Reese, "Flow structure in a three dimensional bubble column and three-phase fluidized bed," AIChE J., v. 40, p. 1093 (1994).

9. F. Lehr, M. Millies, D. Mewes, "Bubble size distributions and flow fields in bubble columns," AIChE J., v. 48, (2002).

10. M. J. Prince, H.W. Blanch, "Bubble coalescence and break-up in air-sparged bubble columns," AIChE J., v. 36, No. 10, p. 1485 (1990).

11. M. G. Bassett, "The Wemco Depurator System," SPE 3349 paper presented at the Rocky Mountain Regional Meeting of the Society of Petroleum Engineers, Billings (1971).

12. S. Movafaghain, J. Chen, S.S. Wheeler, R.W. Guidry, "Introduction of a new generation of induced gas flotation equipment – A case study," SPE 81135, paper presented at the SPE Latin American and Caribbean Petroleum Engineering Conference, Port-of-Spain (2003).

13. C. Rivet, S. Movafaghain, J. Chen, "Introduction of a dual-cell depurator for FPSO application – theory to practice," SPE 94655, paper presented at the SPE Latin American and Caribbean Petroleum Engineering Conference, Rio de Janero (2005).

14. R. Klimpel, R. Hansen, "Some factors influencing kinetics in sulfide flotation," SME of AIME, Annual Meeting, paper no.: 81 – 14 February (1981).

15. Z. Khatib, "Handling, treatment, and disposal of produced water in the offshore oil industry," SPE 48992 paper presented at the SPE Annual meeting, New Orleans (1998).

16. J.M. Walsh, T. Frankiewicz, "Treating produced water on deep water platforms: developing effective practices based upon lessons learned," SPE 134505, paper presented at the SPE Annual meeting, Florence (2010).

17. C.-M. Lee, T. Frankiewicz, "Developing vertical column induced gas flotation for floating platforms using computational fluid dynamics," SPE-90201, paper presented at the SPE ATCE, Houston (2004).

18. T. Frankiewicz, C.-M. Lee, K. Juniel, "Compact induced gas flotation as an effective water treatment technology on deep water platforms," OTC 17612 (May 2005).

19. L. Jahnsen, E.A. Vik, "Field trials with Epcon technology for produced water treatment," paper presented to the Produced Water Society, Houston, TX (2003).

20. E.A. Vik, L.B. Henninge, "The Epcon CFU zero discharge technology. Case studies 2001 – 2005," Aquateam report No. 04-025 (2005)

21. J.K. Edzwald, "Principles and applications of dissolved air flotation," Water Sci. Tech., v. 31, p. 1 (1995).

22. Z. Huang, D. Legendre, P. Guiraud, "A new experimental method for determining particle capture efficiency," Chem. Eng. Sci., v. 66, p. 982 (2011).

23. A. Ramirez, R.H. Davis, "Microfiltration of fine oil droplets by small air bubbles: experiment and theory," Sep. Sci. Tech., v. 36, p. 1 (2001).

24. L. Parkinson, J. Ralston, "Dynamic aspects of small bubble and hydrophilic solid encounters," Adv. Coll. Interface Sci., v. 168, p. 198 (2011).

25. V. Sarrot, P. Guiraud, D. Legendre, "Determination of the collision frequency between bubbles and particles in flotation," Chem. Eng. Sci., v. 60, p. 6107 (2005).

26. Z. Dai, D. Fornasiero, J. Ralston, "Particle-bubble collision models – A review," Adv. Coll. Interface Sci., v. 85, p. 231 (2000).

27. J. Ralston, S.S. Dukhin, N.A. Mishchuk, "Wetting film stability and flotation kinetics," Adv. Coll. Interface Sci., v. 95, p. 145 (2002).

28. A.M. Gaudin, Flotation, 2nd Ed., McGraw-Hill Book Co., New York (1957).

29. R.-H. Yoon, "The role of hydrodynamic and surface forces in bubble-particle interaction," Int. J. Mineral Proc., v. 58, p. 129 (2000).

30. Z. Dai, S. Dukhin, D. Fornasiero, J. Ralston, "The inertial hydrodynamic interaction of particles and rising bubbles with mobile surfaces," J. Coll. Interface Sci.., v. 197, p. 275 (1998).

31. T.J. Heindel, F. Bloom, "Exact and approximate expressions for bubble-particle collision," J. Coll. Interface Sci., v. 213, p. 101 (1999).

32. Z. Huang, D. Legendre, P. Guiraud, "Effect of interface contamination on particle-bubble collision." Chem. Eng. Sci., v. 68, p. 1 (2012).

33. S.S. Alves, S.P. Orvalho, J.M.T. Vasconcelos, "Effect of bubble contamination on rise velocity and mass transfer," Chem. Eng. Sci., v. 60, p. 1 (2005).

34. A.M. Posner, A.E. Alexander, "The kinetics of adsorption from solution to the air / water interface. Part II. Anionic and cationic soaps," J. Coll. Sci., v. 8, p 585 (1953).

35. C. Grattoni, R. Moosai, R.A. Dawe, "Photographic observations showing spreading and non-spreading of oil on gas bubbles of relevance to gas flotation for oily wastewater cleanup," Coll. Surf., v. 214, p. 151 (2003).

36. R. Moosai, R.A. Dawe, "Gas attachment of oil droplets for gas flotation for oily wastewater clean-up," Sep. Pur. Tech., v. 33, p. 303 (2003).

37. N.D. Sylvester, J.J. Byeseda, "Oil/water separation by induced-air flotation," SPE J., p. 579 (Dec 1980); and SPE-7886 (1980).

38. R.P. King, C.L. Schneider, E.A. King, "Chapter 9: Flotation," in Modeling and Simulation of Mineral Processing Systems, 2nd Ed. Society for Mining, Metallurgy and Exploration (2012).

39. P. Diaz-Penafiel, G.S. Dobby, "Kinetic studies in flotation columns: bubble size effect," Min. Eng., v. 7, p. 465 (1994).

40. J.B. Yianatos, "Fluid flow and kinetic modeling in flotation related processes. Columns and mechanically agitated cells – A review," IChemE, v. 85, p. 1591 (2007)

41. Y. Sato, Y. Murakami, T. Hirose, Y. Uryu, K. Hirata, "Removal of emulsified oil particles by dispersed Air flotation," J. Chem. Eng. Japan, v. 13, p. 385 (1980).

42. M. Loewenberg, R.H. Davis, "Flotation rates of fine, spherical particles and droplets," Chem. Eng. Sci., v. 49, p. 3923 (1994).

43. L. Mao, R.-H. Yoon, "Predicting flotation rates using a rate equation derived from first principles," Int. J. Mineral Proc., v. 51, p. 171 (1997).

44. D.B. Rochford, "Design and operation of a hybrid gas flotation system," Doc. Ref. No.: 1638e, Britoil Report, Aberdeen (1986).

45. W.T. Strickland, "Laboratory results of cleaning produced water by gas flotation," Society of Petroleum Eng. J., p. 175 – 190 (1980).

CHAPTER THIRTEEN

Tertiary Equipment

Chapter 13 Table of Contents

13.1 Introduction.....209

13.2 When is Tertiary Treatment Required?.....212

13.2.1 Filtration for Water Flood and Injection.....212

13.2.2 Filter Cake Permeability and Barkman-Davidson Test.....216

13.2.3 Comparison of Injectivity with Deep Bed Filtration.....218

13.3 Tertiary Treatment Mechanisms and Equipment Types.....222

13.4 Straining Filtration.....225

13.4.1 Wedge-Wire Filtration.....226

13.4.2 Cartridge Filtration.....226

13.5 Cake or Pre-Coat Filtration.....230

13.6 Deep Bed Filtration:.....232

13.6.1 Sand and Multi-Media Filters.....240

13.6.2 Nut Shell Filtration.....241

13.6.3 Clay Media Filtration.....248

13.6.4 Activated Carbon Filtration.....248

13.7 Cross-Flow Filtration.....261

13.7.1 Microfiltration & Ultrafiltration.....262

13.8 Ion Exchange.....266

13.8.1 Boron Removal.....270

13.9 Media Coalescence.....271

References to Chapter 13.....276

13.1 Introduction

In this chapter, tertiary treatment technologies for produced water are discussed. Several important technologies are discussed, all of which depend on the flow of produced water through or past some form of media. Such media include but are not limited to activated carbon, cartridge filters, sock filters, walnut shells, anthracite and garnet, coalescing media, absorption media organo-clays, oleophilic polymers), cross-flow membranes, ion exchange, desalination membranes, and various other forms of media. Other technologies are also discussed which satisfy one or more specialized separation requirements. They all have strengths and weaknesses.

The technologies discussed here are typically located in a process downstream of primary (separators and hydrocyclones) and secondary (flotation) technologies. The intention of tertiary treatment is to remove oil, solids, or oily solids down to a few parts per million. The mechanisms of separation include straining, dead-end filtration, cross-flow filtration, coalescence, and ion exchange. In municipal and industrial water treatment, the term "tertiary treatment" is commonly used [1]:

> "Removal of residual suspended solids (after secondary treatment), usually by granular medium filtration or micro-screens."

In the definition above, "suspended solids" could be replaced by "suspended solids, liquids (such as oil), organics or colloidal material, etc.."

Primary and secondary water treating for oily water involves just a few technologies, and essentially two or three separation mechanisms. Primary separation is driven by the density difference between the water and the contaminant. Secondary (flotation) is based on collisions between bubbles and contaminant particles. Tertiary treatment involves contact of the produced water with some form of media. Becoming knowledgeable about tertiary treatment requires an in-depth understanding of several technology types, separation mechanisms, and an even larger number of individual technologies. This is challenging. Nevertheless, it is becoming increasingly necessary for water streams in the oil and gas industry to be treated with some form of tertiary technology. Even when primary and secondary technologies are working properly, there may be a need to remove one or more constituents to levels that cannot be achieved otherwise.

For most of the equipment described in this chapter, with some exceptions, it is assumed that primary and secondary treatment has already been applied. In the context of this discussion, such treatment steps are referred to as "pre-treatment." Some tertiary technologies perform quite well with high contaminant loading and pretreatment is not required. For the most part though, the technologies presented here do not perform well when there is a high loading of oil or oily solids. For example, specialized resin beds for removing mercury become quickly fouled if pre-treatment has not removed the fouling components such as oily solids.

There is a general rule of thumb in equipment design. To prevent fouling – eliminate internal surfaces. The most broadly applicable and successful water treating equipment (separators, hydrocyclones, flotation units) in the upstream industry have very simple internals. The lack of internal complexity reduces fouling and is one of the reasons why such technologies are so widely applied. These technologies do seem like the silver bullets for every application. But these technologies only go so far. On many platforms, some additional treatment is required beyond primary and secondary treatment.

There is another rule of thumb in equipment design. To enhance separation performance – add internal elements. In fact, there are several technologies where the main goal of the design is to maximize contact between the media and produced water. That contact is what provides the separation efficiency.

Obviously, these two rules of thumb (prevent fouling by minimizing use of internal devices / enhance performance by increasing the surface area of the media) are at odds with each other. If fouling is a potential problem the last thing that you should do is add media. This is one reason why a universally applicable, robust, and multipurpose tertiary treatment technology for produced water does not exist. Regarding tertiary media, there is no silver bullet. In some situations, fouling is the overriding concern and tertiary treatment is not applicable. However, there are many locations where fouling can be eliminated, or managed, and tertiary treatment technology is just the thing that solves a problem. As will be discussed, media filtration has been used effectively for years in certain upstream oilfield water treatment applications. Whereas primary and secondary treatment have broad applicability, tertiary treatment requires a much deeper, location by location understanding of water quality, fouling tendency, and tertiary treatment options.

One of the complications that must be addressed when considering tertiary treatment is the weight and space required. The media itself has a certain weight. In normal operation, the media is flooded with produced water and the superficial velocity is rather low such that the volume of media and water can be relatively high. This adds considerably to the total weight of the equipment. Many tertiary technologies generate a waste stream that must be recycled or disposed. With some technologies the waste stream is continuous. With other technologies the so-called backwash is the waste stream which must be handled. Some tertiary technologies require disposal of the media which is an added cost and complication. All of these factors must be taken into account for economical application.

A further complication with tertiary treatment is the space and weight that are typically involved. The media itself has a certain weight. In normal operation the media is flooded with produced water and the superficial velocity is rather low such that the volume of water can be relatively high. This adds considerably to the total weight. Many tertiary technologies generate either a waste stream that must be recycled or disposed of. All of these factors add to the weight and space requirement. Some tertiary technologies require disposal of the media which is an added cost and complication. Thus, only a small number of technologies are economical for offshore application.

Those tertiary treatment technologies that are applicable offshore are typically designed to operate at their upper limit of flow rate, lower limit of residence time, and lower limit of reject or waste volume. This is done to economize on weight and space and make the unit economical for offshore application. However, it does decrease the margin of error and increase the precision required in selecting, designing, integrating it into the system, and operating it properly.

Since two thirds of all oil and gas that is produced in the world is produced onshore, most water treating specialists must be knowledgeable in both onshore and offshore applications. Most onshore applications of produced water will involve disposal injection. Those that do not will likely require tertiary treatment for surface or ground water discharge. All things considered, there is almost no way for a water treating specialist in the oil and gas industry to escape the need for a detailed knowledge of tertiary treatment technologies.

Industrial & Municipal Filtration: It is worthwhile to note that by the time oily water has been treated through primary and secondary separation processes, the characteristics of its suspended material begin to converge to that of food and chemical waste water streams that have been treated by appropriate secondary processes. In other words, waste water streams in oil and gas, pulp and paper, and the food and beverage industries, tend to be similar by the time they reach the tertiary stage.

Of course, this is an oversimplification since waste water streams within such broad industry categories do have wide variation. Nevertheless, at a high level of discussion, by the time secondary treatment has been applied, the character of these waste water streams tends to be similar. It is with this perspective that the material in this chapter was developed. The discussion presented in this chapter is based on the terminology, classification, and literature of the industrial water treatment industries, particularly those that deal with oil, grease, fat, and oily solids. The main advantage of appealing to these other industries for material is that there is a large literature to draw upon. For example, in the oil and gas literature, there are perhaps five papers on walnut shell filtration. In the chemical and food water treatment literature, there are hundreds of papers on deep bed and granular filtration, of which walnut shells are discussed as one particular example.

Seawater Filtration for Water Flood: It is instructive to compare filtration applied to seawater for water flood application to that required for produced water, i.e. for oil and oily solids separation. Filtration of seawater for water flood is mostly a case of removing non-oily solids. Seawater does usually contain some suspended organics but typically those organics are not sticky in the way that moderate to heavy oil sticks to solid particles. When seawater does become contaminated with sticky organics (slime) it is usually short-lived and seasonal.

In 1985 Matthews [55] reported the results of an evaluation of four different filter systems for seawater. Seawater is typically contaminated with phytoplankton which is produced in the upper one hundred or so feet of the sea. The solids are composed of diatoms, coccolithophores, and dinoflagellates. It is important to recognize that seawater filtration for water flood typically involves large volumes of water containing a low concentration of suspended solids with a rather broad particle size range. Typically, particles larger than 10 to 20 microns must be removed in order to prevent plugging of the reservoir. For example, in the application reported [55], the particle size range was from 1 to 40 microns. The organic content of the seawater was low. In terms of particle counts, the concentration of suspended solids, as measured on a 0.45 micron filter paper, was between 26,000 N/mL to 34,000 N/mL, where N is the number of particles. Most of these particles were in the range of 1 to 3 micron. These small particles do not typically need to be removed since they do not cause reservoir plugging. As will be discussed, produced water can contain 100 to 1,000 times higher particle counts, particularly in shale produced water. Also, small particles in produced water can cause reservoir impairment due to their sticky nature and tendency to bridge across pore throats and eventually cause plugging. Thus, seawater filtration is a case of large volumes of low contaminant concentration with a moderate number of particles in the 1 to 3 micron range and a relatively low concentration of particles in the 3 to 20 micron range which must be removed.

Matthews reported on the performance of four filtration technologies:

- Two stage back washable cartridge filter
- Back washable fine mesh membranes of synthetic fibers
- Back washable dual media anthracite and garnet
- Precoat filter

All of the filter systems has strengths and weaknesses. Separation efficiency was best for the synthetic fibers and the precoat filter. The cartridge filter had poor separation efficiency. The dual media had good separation. The synthetic fiber media had low footprint but required four hours per backwash. The precoat filter was sensitive to flow and pressure variation. Back flow caused the precoat to come off the support material which then required the initiation of a new regeneration cycle. The author suggested that the final selection of which system to use would depend on each location and the importance of the various strengths and weaknesses identified. As discussed below, these systems have also been considered for produced water treatment.

13.2 When is Tertiary Treatment Required?

The need for tertiary water treatment technologies in the oil and gas industry continues to grow. The obvious reason for this is that tertiary treatment is required when primary and secondary treatments are incapable of delivering the required quality. This may be due to a high loading of small diameter oil droplets or solid particles, due to chemical stabilization of an emulsion, or due to some design or integrity problem with the upstream equipment or process. In the oil and gas industry there is a wide range of water quality requirements. Overboard discharge to the sea is one disposal method that requires relatively clean water. Another example would be injection of produced water into a hydrocarbon reservoir (PWRI – Produced Water Reinjection). As discussed below, there are various reasons why this is, or should be, an example where relatively clean water is required. Produced water reuse for steam flood, and hydraulic fracturing of shale are other examples.

Filtration as Pre-Treatment for Removal of Specific Contaminants: Tertiary treatment may be required as a pre-treatment for some other specialized technology used to remove a specific contaminant. Such contaminants may include a heavy metal (mercury, zinc, etc), or a toxic (H_2S) or radioactive substance (H_2S, mercury sulfide, mercury, radium, etc). Specialized technologies for this purpose might include micro- and ultra-filtration, nano-filtration (sulfate removal), reverse osmosis (RO), and ion exchange resins. Most of these specialized technologies are prone to fouling when used to treat produced water. For this reason, pre-treatment requirements, ahead of the specialized treatment technologies, are typically stringent. An example would be the use of reverse osmosis for the removal of a dissolved ion such as boron or radium in an oily waste water stream. In that case, various tertiary treatment systems are required such as fine filtration, nut shell filters, or other deep bed filtration, followed by activated carbon application, and finally followed by the RO itself. Another example might be the removal of ionic mercury from a gas condensate oily water stream. Again, it may be necessary to use activated carbon to remove the last remnants of condensate before passing the water through an ion exchange resin bed for the specific removal of ionic mercury. There are many other examples that can be sited.

13.2.1 Filtration for Water Flood and Injection

Injectivity of produced water is becoming a significant economic issue in the Permian basin. The term, "injectivity" is well defined for a disposal injection well. The injectivity Index is calculated using the formula:

$$II = \frac{Q}{P_{bhi} - P_e} \qquad \text{Eqn (13.1)}$$

Where:

- Q = volumetric flow rate (Stock Tank Barrels/day – defined in the Appendix)
- P_{bh} = bottom hole pressure (psia)
- P_e = far-field reservoir pressure (psia)

There are several factors that impact the Injectivity Index including the water quality or cleanliness. The cleaner the water the lower the required injection pressure, and the less frequently the well must be shut down for stimulation. Stimulation has direct costs in terms of chemical, manpower, fuel, and equipment. A less direct but significant cost is the handling of produced water while the injection well is shut down for the stimulation job. If trucking is required to move the water to another site, then costs escalate quickly. Another cost that is beginning to be recognized is the cost associated with capacity limitation. When an injection zone has been injected with poor quality water over a few or several years, the pore space becomes blocked and the available capacity of the zone can be dramatically reduced. When this occurs, the injection zone must be abandoned. Again, this is a considerable cost.

Thus, the closely related issues of produced water cleanliness and injectivity are beginning to be recognized for shale developments, at least in the Permian. In this paper, results from sampling, analysis, and field testing in the Permian will be presented and discussed in the context of the large body of published knowledge available on water injection. Results found in the Permian will be compared with results chronicled over decades in various parts of the world for various types of injection projects and various types of formations.

Regulations for Disposal Injection: The Underground Injection Control (UIC) program of the US EPA defines different types of wells depending on the use and fluid types to be injected. The overall goal of the program is to protect ground water from contamination. A Class II well is used exclusively to inject fluids associated with oil and gas production. States can, and many states do provide guidelines and regulations for the permitting of Class II disposal wells.

In New Mexico the UIC Program Manual [48] states that all injection wells are subject to a surface injection pressure maximum. In general, the permitted injection pressure is limited to 0.2 psi / ft, to the uppermost perforation. This means that the surface injection pressure, for an injection target at a depth of 1,000 ft, could be as high as 200 psi. Higher surface pressure is allowed provided that a step-rate test or other test is documented in order to prove that injection will not initiate fractures in the confining rock and potentially contaminate nearby aquifers. This test involves measurement of injection flow rate, topsides and downhole pressure at progressively greater rates of injection [42, 43]. The test is designed to measure the pressure at which reservoir fracturing will occur.

In Texas, in order to operate a Class II injection well, the operator must obtain a permit from the Texas Railroad Commission [53]. The intent of the permitting process is to ensure that sources of fresh water are protected from contamination. The maximum surface injection pressure is limited to 0.5 psi / foot of depth to the top of the disposal interval. Higher pressure is allowed provided that a step-rate test or other test is provided to prove that injection will not initiate fractures in the confining rock. Guidelines for step rate testing are available [49].

Injectivity Impairment Mechanisms: There are several mechanisms by which contaminants can impede flow through the formation. It is worthwhile to understand these different mechanisms since it may be necessary to remediate the well and improve the injection water quality at some point in the life of the well. There are various ways to improve water quality such as elimination of corrosion, scale inhibition, bacteria control, and various water treatment methods such as coagulation / flocculation, flotation, filtration, etc.. Understanding the impairment mechanism will allow a fit-for-service low cost strategy to be developed.

As discussed by Eylander [34], the rate of impairment of a water injection well is a function of the contaminant particle size distribution, the concentration of contaminant solids, surface characteristics of the solids and formation rock, presence of oil with the solids, flow rate of water, and the pore size distribution of the formation. A suspension of larger particles at higher flow rates will lead to impairment at lower concentration. Injection reservoirs that have low permeability to start will impair more rapidly. However, smaller particles also can cause problems, albeit over a longer period of time and perhaps requiring a greater concentration. In the case of shale produced water, where the contaminant particles tend to be small, the plugging mechanisms of small particles are of greater interest here.

The extent to which the particles stick to the reservoir rock and to each other as well as the extent to which they form bridges are all related to the surface characteristics of the contaminant particles and the reservoir rock, as well as the other characteristics such as pore size and particle size and concentration. The point being that the characteristics of the particle surface can play an important role in allowing particle bridges to form.

Much of what is known about the role of fines in blocking pores and leading to impairment comes from studies of producer wells [54]. Injection wells are usually less susceptible to fines-migration problems than production wells, because the fines being generated are pushed away from the well-bore, leading to less severe impairment in the near-wellbore region and therefore relatively small losses in injectivity. There are cases however when a large concentration of fines causes injectivity problems in injection wells over extended periods of time.

Size Ratios: In the discussion that follows, three size ratios are discussed. They define the relationship between the pore throat diameter (*dp*), the contaminant size (*dc*), and the grain size of the particles that comprise the formation rock (*dg*). The ratios of these quantities can be used to estimate the tendency of contaminant particles to plug the pores.

dc / dp Ratio of contaminant size to pore throat diameter. This is the most important ratio since it determines the tendency of the contaminant to plug the pores. Large contaminant particles flowing through small pores can easily cause plugging. Smaller contaminant particles may infiltrate the pores. Even smaller particles can form bridges which may eventually block the pores.

dp / dg Ratio of the pore diameter to the sand grain diameter. This ratio depends on the shape and packing of the formation rock grains. Spherical grains that are closely packed have a ratio of *dp / dg* = 0.15. This means that the diameter of pore throats between sand grains is 15 % of the sand grain diameter. In the case where pore diameter has not or could not be measured, empirical estimates can be made for this ratio which can then be used to estimate pore throat diameter.

dc / dg Ratio of the contaminant size to reservoir grain size. Often the contaminant size can be easily measured, and the grain size is known. This ratio, together with *dp / dg* can be used to estimate *dc / dp*, which is the most important ratio of the three ratios.

Geometric Matching Criteria: Various rules of thumb have been developed to simplify the selection of water treatment system design. One of the earliest rules of thumb is the well-known one third / one seventh rule [34, 35, 44, 47]. The idea is to eliminate the particles that are one third (33 %) the size of the median formation pore throat diameter or larger (eliminate $dc / dp > 0.33$), and to significantly reduce those particles that are between one seventh (14 %) and one third (significantly reduce $0.14 < dc / dp < 0.33$). This approach is broadly referred to as geometric matching. The thinking is that particles larger than one third the pore diameter will deposit on the surface of the porous formation rock forming what is referred to as an external filter cake. This filter cake causes a significant flow restriction. Thus, essentially all of the particles larger than one third of the pore diameter should be removed. Smaller particles will invade into the porous formation rock and, depending on their size, can stick to formation grains, form bridges across the pore throat, and form an internal filter cake. This internal filter cake is thought to be formed by particles that are larger than 14 % of the pore diameter, and smaller that the previously mentioned 33 %. The depth of the internal filter cake will depend on various factors but is generally in the range of several centimeters from the porous rock face. The conventional thinking, which is discussed below, is that particles smaller than 14 % of the pore diameter will remain suspended in the injection water as it infiltrates the porous rock and will not plug the formation pores.

Pautz and Crocker [47] carried out laboratory measurements using sandstone cores and various sizes and types of contaminant particles. For the most part, their results supported the rule of thumb. The

1/3 rule was verified. However, they did find inconsistent results for the 1/7 rule ($dc/ dp = 0.14$). Some reduction in permeability was seen even for such small particles.

As discussed by Eylander [34], when this approach is applied in the field the success rate is more or less random. In other words, when this approach is applied, some wells have relatively long life and others do not. The geometric criterion ignores the impact of the type of solids, surface chemistry of the contaminant and formation grain, and type of formation material. When there is a strong attraction between contaminant particles and formation grains, there is a stronger tendency for sticking, bridge formation, and formation of a filter cake within the porous media.

Injection Velocity: Measurements of particle deposition and bridge formation of small particles has shown a dependence on not only particle size but also on injection velocity as well. Pore throat blockage has been observed for systems with dc/dp = 0.07, provided that the flow rate was no greater than 0.033 cm/sec [44]. In other words, high velocity flow (0.17 m/sec) prevented the formation of bridges for systems with dc/dp < 0.14. However, at low velocities (0.033 cm/sec), smaller particles dc/dp = 0.07 would cause bridging and pore throat blockage.

Estimating Pore Size based on Grain Size: In the discussion that follows, various laboratory and field tests are reported. For many of these reports the pore diameter of the formation rock has not been measured. Despite the fact that pore diameter is an important variable, it is sometimes not known. In the data that we collected for Permian fields, pore diameters are not available. More often, grain size and contaminant size are available and can be used to give a semi-quantitative comparison of one set of data to another.

In a geometrically perfect system of hexagonal close pack (HCP) spheres there is a well-established relation between the sphere diameter (grain diameter) and the effective pore diameter [39]. In the HCP system, the effective pore diameter is the diameter of contaminant spheres that can traverse the pore space of the spheres. The effective pore diameter of this system is $dp / dg = 1/6.5 = 0.15$, where dg is the diameter of the grain particles. Although this is an idealized system it does give us a useful order of magnitude or frame of reference for the relation between pore diameter and grain diameter. Wu [45] reported the results for a close-packed perfect tetrahedral system as $dp / dg = 0.23$. He also carried out measurements of the relation between soil particle sizes and pore diameters and found values very close to close-packed perfect tetrahedral system. As will be shown, this order of magnitude analysis for pore diameter ($dp / dg = 0.15$ to 0.23) will be sufficient for our purposes. For simplicity, the average of these values will be used (dp / dg = 0.19). Using this value, the pore throat diameter can be estimated on the basis of the grain diameter.

Applying these relationships to the size ratios defined earlier, the following table can be generated. It must be emphasized that this table is not quantitatively accurate. It is only a rough guide that should be used in making order of magnitude estimates when data are missing.

Table 13.1. Qualitative guide to size ratios for contaminant particles, pore throat diameters, and grain sizes. Intended for rough order of magnitude comparisons.

dc / dp	dc / dp	dp / dg	dc / dg
1 / 7	0.143	0.15	0.021
	0.200	0.15	0.030
	0.250	0.15	0.038
	0.300	0.15	0.045
1 / 3	0.333	0.15	0.050

By applying the relationship that the pore diameter is equal to the square root of permeability, in microns, the following table is generated [51].

Table 13.2. Relationship between permeability, pore size, and range of particle sizes that would cause plugging [51].

Permeability, mD	Pore size, microns	Plugging range, microns
100	10	1.4 to 3.3
250	15.8	2.2 to 5.2
500	22.4	3.2 to 7.4
750	27.4	3.9 to 9.1
1000	31.6	4.5 to 10.5
1500	38.7	5.5 to 12.9
2000	44.7	6.3 to 14.9

13.2.2 Filter Cake Permeability and Barkman-Davidson Test

Solids suspended in produced water may be treated using some sort of filtration media, or porous material where a filter cake may buildup. The formation of a filter cake is an absolute certainty in dead end filtration. A filter cake may also build up on the formation face in an injection well or within a target injection reservoir. Filter cake formation is also a possibility in the proppant pack of a fractured shale production zone. Filter cake permeability is an important characteristic. It is a characteristic of the solids concentration, size distribution, composition, and the surface charge of the suspended solids particles. All of these factors can influence the permeability of the cake that can form. The Barkman-Davidson test was devised as a standard method to measure the permeability. It is described below and by the following equation.

$$\Delta P = \mu(r_c h_c + r_s h_s)q \qquad \text{Eqn (13.2)}$$

Where:

ΔP = pressure drop across a porous Millipore filter paper (kPa)

μ = viscosity of the water (Pa sec)

r_c = filter resistance of the filter paper ($1/m^4$)

h_c = height (thickness) of the filter cake (m)

r_s = filter resistance of the filter paper ($1/m^4$)

h_s = height (thickness) of the filter cake (m)

q = volumetric flow rate through the filter paper and filter cake (m^3/sec)

Filter cake permeability (Barkman-Davidson) Test: A slight improvement to the above rules of thumb was developed by Barkman and Davidson [36]. According to them, impairment is thought to occur by the following mechanisms:

1. The solids form a filter cake on the face of the formation (wellbore narrowing).
2. The solids invade the formation, form bridges, and eventually form filter cake (invasion).

3. The solids become lodged in the perforations (perforation plugging).

4. The solids settle to the bottom of the well and decrease the injection zone height (wellbore fill-up).

The Barkman-Davidson impairment mechanisms can be interpreted in terms of the one third / one seventh rule to obtain a more precise definition of the impairment mechanisms [37]:

1. When $dc / dp > 1/3$, the contaminant particles bridge the pore entrances at the formation face to form an external filter cake.

2. When $dc / dp < 1/3$, and $dc / dp > 1/7$, the contaminant particles invade the formation and are trapped to form an internal filter cake.

3. When $dc / dp < 1/7$, the contaminant particles cause no formation impairment because they are carried through the formation by the infiltrating water.

Over the years there have been extensive measurements published on whether or not these mechanisms are realistic in predicting injection well impairment in the field. Eylander [34] found that the one third / one seventh rule is not realistic. Van Oort [37] found it is. Both groups however argue in favor of on-site filtration tests using core samples from the injection well. These relationships are shown schematically in the figure below.

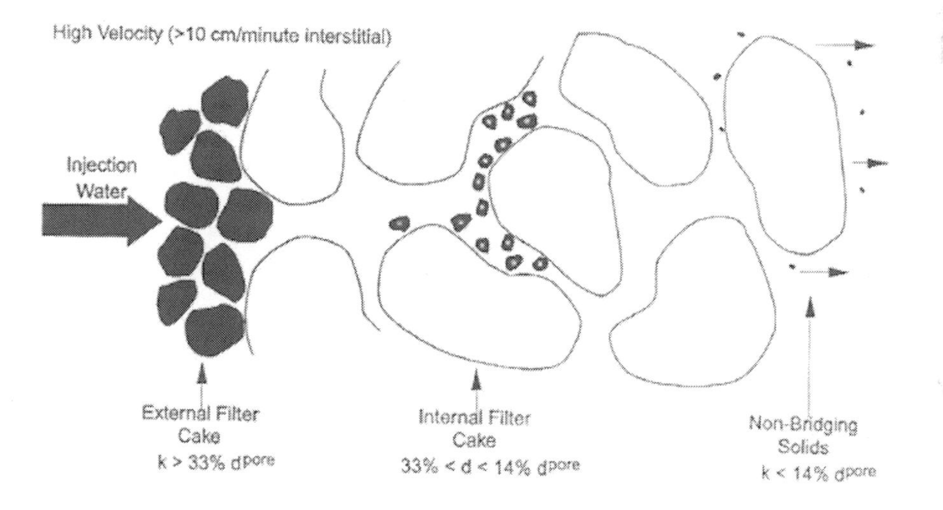

Figure 13.1. Schematic diagram showing penetration of solids particles as a function of particle size and pore throat diameter [52].

Injectivity Tests of Core Samples: The drawbacks of simple rules of thumb, and of the Barkman-Davidson test are rather easy to point out. Developing a better method, suitable for field application, is a lot harder. The rules of thumb are too simplistic to be of widespread applicability. The Barkman-Davidson test only investigates one mechanism (pore blockage from filter cake buildup), and it only tests the water quality and not the formation rock features.

The Role of Fines in Injectivity Impairment: Amaefule et al. [54] give a mechanistic description of fines migration and pore throat plugging in producer wells. Muecke [40] constructed a view cell that allowed direct microscopic observation of fine particles flowing through porous media. The objective of his study was to understand the mechanisms behind fines migration in porous sandstone formations. In general, formation fines are composed of clay, feldspars and even fine sand particles. In a producing well zone such fines move with the fluid as it flows toward the well and tend to accumulate and cause flow reduction in the near wellbore region. Muecke's observations showed that fine particles can form bridges at pore restrictions. Once bridges form, they become effective traps for the collection and buildup of particles that follow. The higher the concentration of contaminant particles in the injection water the more bridges that form and the greater the percentage of fine particles that participate in bridge formation. In other words, the greater the fines concentration, the greater the tendency to bridge. Higher flow velocity reduces the fraction of particles that participate in bridges. It also reduces the number of bridges by reducing the formation of bridges at the relatively larger pore openings.

Fines are much smaller than typical sandstone grain sizes found in moderate permeability (100 to 1000 mD) injection wells. The size of fines range from a few microns to 37 micron. The upper size limit is somewhat arbitrarily but is agreed to be the size of particle that passes through a 400 mesh screen. Muecke studied porous media composed of grains that ranged in size from 400 to 850 micron (20/40 mesh). The fines ranged in size from a few microns to 10 micron. Thus, the *dc* / *dg* ratio ranged from 0.001 to 0.02. Comparing these values to those in Table 13.1, and keeping in mind that Table 13.1 is only intended for approximate guidance, the fines sizes studied by Muecke are at the smaller end of the *dc* / *dp* ratio.

Gabriel and Inamdar [38] studied the behavior of fines in Berea sandstone formations in the laboratory. Berea sandstone is generally reported to have pore throat diameters in the range of 10 micron. The fines diameter ranged from 1 to 37 micron. They considered mobilization, migration and plugging from both a chemical and mechanical standpoint. From the chemical standpoint they applied the DLVO / electric double layer concepts to explain adhesion of fines to sandstone grains. These concepts were used to explain observed differences in behavior from one brine to another, and the effectiveness of inorganic polymers in reducing migration of fines, and generally reducing the tendency of fines to cause permeability reduction. From the mechanical standpoint they identified a critical velocity of flow for fines entrainment, which was found to be highly dependent on the surface properties and pore structure of the porous medium, as expected.

Gruesbeck and Collins [39] performed a number of flow studies using packed columns of unconsolidated sand with grain diameters ranging from 840 to 2,000 micron. The sand particles were mixed with fine particles of calcium carbonate. In this case the carbonate particles represented fines that could become entrained in the flow and form bridges and eventually plugs. The sand particles of course represented an unconsolidated formation sample. The calcium carbonate particles had a mean diameter of eight micron. Thus, *dc* / *dg* ranged from 0.01 to 0.004. Again, the pore size distribution was not measured. However, given the small values of *dc* / *dg*, it is likely that *dc* / *dp* is somewhat smaller than 1/7. Thus, plugging would be a function of deposition and bridging of particles. Gruesbeck and Collins found that below a superficial velocity of 0.13 cm/sec there was essentially no mobilization of the contaminant particles. Above this flow rate, the fraction of particles mobilized was proportional to the superficial velocity.

13.2.3 Comparison of Injectivity with Deep Bed Filtration

Comparison with Deep Bed Filtration: Generally speaking, filtration of fine solids using a deep bed of media has a lot in common with fines migration in producing wells, and with injection well impairment. Each process has significantly different objectives but all three are fundamentally dependent upon the surface / interfacial chemistry of particles, particle mobilization, particle capture, and the flow of suspended particles through porous media. In the case of deep bed filtration, the objective is to capture the suspended particles during the filtration cycle and to release the particles during the backwash / media cleaning cycle. In the case of an injection well the objective is to maximize the flow rate of the contaminated fluid without allowing the particles to adhere to the porous formation and cause blockage of the pore space. In the case of a producing well the objective is to maximize the flow rate of the production fluids without mobilizing fine particles into the flow where they can subsequently cause blockage and permeability decline. As early as 1982 [39] the common features of these processes was recognized in the oil and gas industry.

The Gruesbeck and Collins experiment [39] was meant to address the question of fines mobilization in a producing well. However, as pointed out by the authors, the chemistry and the physics of the situation suggest that the measurements are applicable to deep bed filtration as well. One of the fundamental challenges in deep bed filtration is to retain captured particles within the media and not entrain (or "mobilize") them in the fluid flow. It stands to reason that there is a critical flow rate above which captured particles would be mobilized. In fact, Gruesbeck and Collins found that below a superficial velocity of 0.13 cm/sec there was essentially no mobilization of the contaminant particles. Above this flow rate, the fraction of particles mobilized was proportional to the superficial velocity. In deep bed filtration using sand as the media it is well known that the flux must be maintained below 2 gpm/ft^2 in order to retain the contaminant particles in the filter media. Above this value, separation efficiency decreases as a result of decreased particle capture as well as increased mobilization of captured particles. Flux is equivalent to superficial velocity. When the sand flux is converted to metric units, the critical superficial velocity is 0.13 cm/sec in agreement with the measurements made by Gruesbeck and Collins. It is perhaps fortuitous that there should be such good agreement. Nevertheless, it is gratifying that two such different approaches should give close agreement and it does underscore the fundamental relationships involved in well impairment, productivity decline, and deep bed filtration and the need to remove small particles due to bridging. It worthwhile to note that Gao [50] has given a review of studies related to particle retention for both reservoir media and deep bed filter media.

<u>Surface Characteristics of Rock and Contaminant:</u> The surface characteristics of the injected contaminant and the formation rock play an important role. If the contaminant sticks to the rock it will generally impede flow. Sticking tendency, otherwise known as adsorption tendency is an important characteristic. Small particles can either be washed into the formation and carried with the injected water with little impact on the fluid flow. Or they can stick to the formation particles and cause other contaminant particles to stick and begin a buildup of material. This can happen at a multitude of locations and eventually cause significant impairment. Surface characteristics can also cause the wettability of the rock to change. Once there is a change in the wettability, there is then a change in the relative permeability. This usually reduces the permeability. Fine particles can either be present in the injection water or can be generated by an incompatibility between the injected water and the water originally in place. The incompatibility can result in mineral precipitation in the regions where the incompatible water mix. These particles can cause subsequent blockage. Or scale can precipitate on water-wet rock and buildup on the rock surface and eventually block off the pores.

<u>Matrix versus Fractured Injection:</u> As discussed by Bansal [42], when water is injected above parting pressure (fractured injection), these mechanisms change somewhat. Above parting pressure, larger particles and much greater concentrations are required to cause impairment. Above parting pressure, when blockage begins to take place, new fractures are opened up due to the pressure differential supplied by the injection pump and the blocked pores are then bypassed by the flow. This can eventually lead to out-of-zone injection. Depending on the details, this can lead to surface breakthrough, injection into other well zones such as producing wells, etc.. In the case of a waterflood, this can lead to bypass of oil and low sweep efficiency. This will negatively impact oil recovery. Thus, even when injecting above parting pressure, there are a number of detrimental processes that result from injecting excessively contaminated water.

There is another mechanism that can cause significant impairment above parting pressure. If the contaminants form blockages near the injection wellbore it may be impossible for the injected water to establish large fractures in the first place. Near wellbore plugging can be the result of rapid and impermeable filter cake formation due to small particles that have a surface affinity to the reservoir rock. In this case, impairment can be rapid (within a few months) and can require frequent stimulation.

In the Gulf of Mexico offshore environment, Morgenthaler and co-workers [41] found rapid injectivity decline in a field of five injection wells. The target reservoir had high permeability (700 to 1,200 mD) and moderately large grain size (80 to 100 micron). The injection water was sourced from the open sea and had contaminant particles of 2.5 to 3 micron mean particle size with concentration in the range of 10 ppm. By most measures this would be considered to be clean water. Various causes of injection impairment were investigated. Based on compatibility studies of the formation water and the injected seawater, filtration studies, various forms of acid stimulation, scale inhibitor treatments, and other field studies it was concluded that the injection well impairment was due to the contaminant particles in the injection water which reduced the permeability in the near wellbore region. Conventional particle trapping due to size exclusion was not observed. Also intrusion of contaminant particles into the pore structure did not seem to occur either. Instead particle retention on the formation rock due to strong surface forces, followed by bridging and formation of a tight filter cake due to accumulation of particles was thought to be responsible for rapid decline of permeability in the near wellbore. The specific surface forces involved were not identified.

Filtration for Water Injection Applications – Oman: It is now well established that the cleaner the water, the less pump energy is required to inject it. This is always true without exception. The only variation from one location or project to another is the degree to which the statement is true. This is true for injection disposal or water flood injection. If the injection target is a tight reservoir and the water is fairly dirty, then cleaning the water will have a dramatic effect on the pump energy requirement. Also cleaner water will reduce fracture propagation and will lead to more efficient oil sweep and production.

Experience in Oman gives a striking example of the long term detrimental effects of injecting dirty water, propagating fractures, and allowing water to bypass the oil in the reservoir. In the early 1990's Oman was producing 100,000 m3/d of crude oil with 140,000 m3/d of co-produced water [2]. This represents a water cut of roughly 60 percent. By the turn of the century, the water cut had risen to roughly 80 %. As discussed in [3], by 2006 oil production had only increased to120,000 m3/d while water production had increased to 680,000 m3/day. This represents nearly 85 % water cut and a nearly 5 fold increase in water production in 15 years. By 2010, in the author's experience, water cut had reached over 90 % and was still rising.

By 2007 or so, some steps were being taken to reverse the trend, cleanup the water, and reduce the water handling capacity required of the facilities. Flotation was implemented in roughly a half dozen locations. In the Haima West field in Oman, a field trial demonstrated the close relationship between the injection water quality and fracture length. An improvement in water quality resulted in a reduction in fracture propagation which led to an increase in oil production of between 50 and 100 % for the pilot wells. In the Al-Khlata field, significantly higher injection rates were obtained when tertiary water treatment was installed, and injection water quality was improved.

Filtration for Water Injection Applications – North Sea: Water flood using seawater or produced water is, and has been extensively applied in the North Sea. In [4] a team of BP research scientists give a technical and historical review of North Sea experience in seawater injection up to 1991. The fields that were studied were: Forties (offshore UK), Magnus (offshore UK), and Ula (offshore Norway). Several important observations were made. One of the important points is that water quality improved as it progressed though the topsides filter system but the water quality at the bottom of the injection well was worse than the untreated raw water on the topsides. The filtration systems seemed to be working well. Immediately downstream of the filters the water had high quality. The water entering the wells was relatively clean. However, by the time the filtered water reached the bottom of the injection well, the concentration of particles exceeded the concentration upstream of the fine filters. Presumably corrosion, microbial action, and mineral precipitation caused the solids content at the bottom of the injection well to deteriorate. This was found in all cases studied. The conclusion of these measurements was that fine filtration was pointless

But this finding raised other questions such as, what is the impact of poor water quality on injectivity, and what should be the strategy for filtration of seawater going forward. It was noted that by-passing the fine filters had already become a trend in the North Sea even before the measurements in [4] were made. In doing so, operators claimed that there was no detectable change in injectivity. The original specification on many of the injection wells was to provide fine filtration to remove 95 % of particles greater than 5 micron, and 100 % of particles greater than 10 micron. In the Ula field, the requirement was to remove 98 % of particles greater than 2 micron. Both of these specifications are rather stringent, and the filtration systems were probably somewhat troublesome to maintain.

Based only on the criterion of injectivity there seemed to be no point to fine filtration. This prompted a review of the assumptions and modeling that had led originally to the fine filtration specifications. One of the main assumptions was that clean water was required to maintain injectivity at reasonable pump pressures. It was thought that fracturing pressures would be high and expensive to attain from the standpoint of pump pressure, capacity and power requirement. The issue of fracturing was addressed in detail [4]. One of the findings was that the cool temperature of the seawater had not been adequately taken into account in core fracturing studies and that it was responsible for much greater fracturing tendency than previously thought. The new understanding was that cool water caused the formation to contract and reduce the compressive stress which allows fractures to propagate at significantly lower pressures than for warmer water. This effect is referred to as thermally induced fracturing. Thus, the injection water was fracturing the reservoirs at much lower pressures than anticipated without apparent loss of injectivity, and without apparent loss of pressure maintenance. Within a few years, in the North Sea fine filtration of seawater for improved injectivity was a thing of the past.

However, there is an important issue that has already been highlighted in relation to the Oman experience. Oil by-pass is much greater in a fractured reservoir than in one that is not fractured. Water injection under conditions that do not fracture the reservoir is also known as matrix injection. The main objective of water flood is to recover more oil. Oil recovery is measured both in terms of immediate production and ultimate recovery. While immediate production and pressure maintenance may not appear to suffer from reservoir fracturing, ultimate recovery always will decrease as the extent of fracturing increases. Also, early water breakthrough can occur in fractured reservoirs.

Also, it must be pointed out that in Produced Water Reinjection (PWRI) and Chemical Enhanced Oil Recovery (cEOR) the composition of the solids is significantly different than for seawater. The former tends to stick to the formation pores causing bridging at much lower concentrations and for much smaller size particles.

System Design for Water Injection Applications: In all of these examples, the design strategy for tertiary treatment systems is rarely a simple matter of selection, capacity determination and detailed design. System upsets and the risk of inadequate upstream design must be taken into account. This requires flexibility of design and a good understanding of the importance and role of the chemistry of the waste water and how the chemical treating system can be used effectively to enhance the performance of the specific tertiary treatment system.

For onshore applications, coagulation, in some cases followed by flocculation chemical treatment, followed by sedimentation is often used upstream of filtration. With very high flows and high suspended solids loading, a clarifier should be considered upstream of the filtration system. Settling is inexpensive and, with proper chemical treatment, can eliminate most large particles before further treatment.

13.3 Tertiary Treatment Mechanisms and Equipment Types

There is an overwhelming number of tertiary treatment methods and technologies. It helps a great deal to categorize them. At the highest level, tertiary treatment is differentiated into the following broad categories:

- Straining (size exclusion, dead-end)
- Cake Filtration
- Deep Bed Filtration
- Cross-Flow Filtration
- Ion-Exchange
- Media Coalescence

Table 13.3 Characteristics of granular media and leaning (backwash) possible

1. Media Filteration		Comments
1.1 Rapid Filtraton		
	a. Gravity filters	5-12 m/h
	b. Pressure filters	5-30 m/h
	c. Up flow filters	3-6 m/h course to fine filtration, disadvantage: break-through
1.2. Moving bed filteration (continuous filtration)		14-18 m/h (water), 2-20 mm/s (sand)
1.3. Dry filtration		7 m/h (gravity), 10-18 m/h (pressure), air + water in media for continuous aeration (e.g. NH4 > 3 mg/l)
1.4. Slow sand filtration		0.1-0.3 m/h, gravity filters, no backwash
2. Fiber Filtration		
2.1 Cartridge filters		range 1-10 micron, application: pre-treatment reverse osmosis
2.2 Sock filters		range 100-150 micron
3. Membrane Filtration		
3.1 Cross flow filtration	MF, UF, NF, RO	
3.2 Dead end filtration	MF, UF	most commonly applied for MF and UF (also sub-merged systems)
3.3 Semi dead end filtration	UF	permeate backwash without chemicals (every 15 min)
4. Screen Filters		
4.1 Micro sieves		rotating screens, range 20-200 micron (surface water pre-treatment), cleaning system
4.2 Screen filters		range 10-3500 micron, automatic self cleaning filters
4.3 Disc filters		range 10-200 micron, automatic self cleaning filters

These various categories of tertiary treatment technology have one major characteristic in common. They utilize some form of media. In the discussion below, each of these technologies is discussed together with the type of media that is involved. The figure below, shows schematically the four different types of filtration.

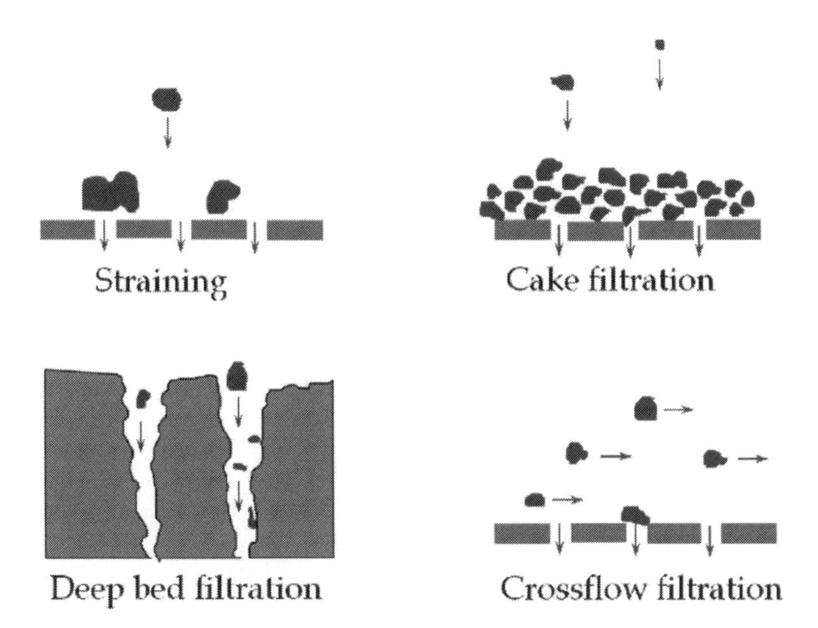

Figure 13.2 Schematic diagram showing different modes or mechanisms of filtration.

Straining (also referred to as dead-end filtration) is a basic filtration mechanism that relies on size exclusion. The media is designed and selected such that the particles to be separated are simply too big to fit through the openings in the media. Obviously, the size and shape of the particles is important, as well as the size and shape of the openings in the media. Both of these characteristics will typically have a distribution.

Cake Filtration Cake filtration is based on a strainer, plus the addition of a cake material, or pre-coat material such as diatomaceous earth or Pearlite. The cake material is also referred to as filter-aid. It is added upstream of the strainer. As can be easily imagined, it greatly increases the separation of contaminants. As contaminant material becomes trapped on or in the cake, the pressure drop across the cake rises. Periodically the cake material must be swept off the strainer, and new cake material must be deposited. The waste that is generated includes the separated contaminant together with the cake material itself. In some cases, the cake material can be regenerated. When this is true, it generally requires specialized offsite facilities.

Deep Bed As the name implies, deep bed (or depth filtration) involves passing the waste water through a bed of granular or fibrous media where a number of different particle capture mechanisms are generally at work, including particle trapping, adsorption, or absorption. Typically, it is important to understand which of these mechanisms is operative because they differ in the extent to which chemical addition, lowering the flow rate, and changing the media particle size has an effect on the separation performance.

The figure below gives optimal configurations, not on the basis of particle size, but instead on the basis of the quantity of solids in the feed stream. It's obvious that if some 1% solids are present in the water, this could easily be called sludge.

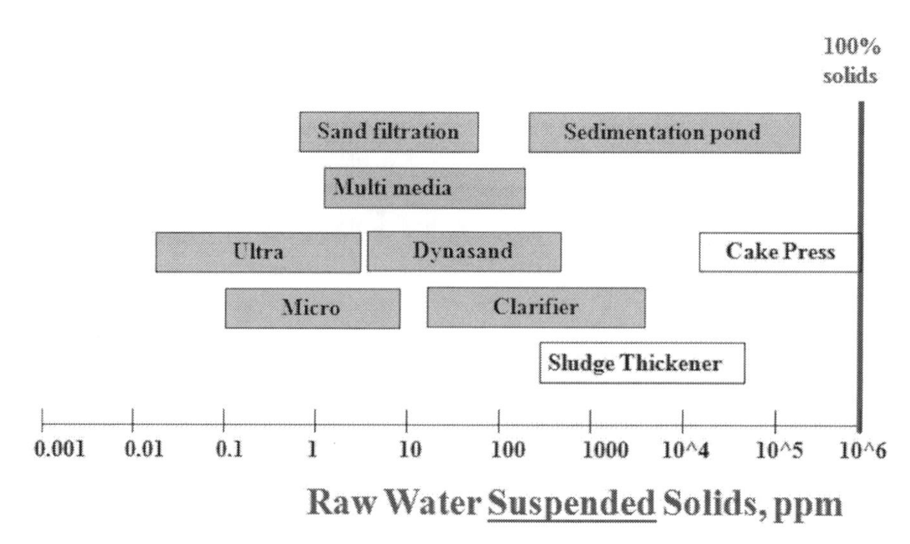

Figure 13.3 Different configurations of filtration as a function of the suspended solids loading.

Cross-Flow Filtration As the name implies, cross-flow filtration involves flow of the waste stream across a filtration material with effluent passing through the material and the contaminants remaining in the reject stream. The important variable is the trans-membrane pressure (TMP). There are four technologies that utilize cross flow: Micro-Filtration (MF), Ultra-Filtration (UF), Nano Filtration (NF), Reverse Osmosis (RO). Each of these technologies is characterized by a size cutoff. There are many materials and configurations that are employed. These include: spiral wound membranes, hollow fiber membranes, ceramic membranes, to name only a few.

Ion Exchange: Ion exchange is a technology that utilizes media with unique surface chemistry to separate an unwanted ion from the waste water. A transfer of ions occurs. An ion from the waste water displaces an ion from the media. The ion from the waste water gets adsorbed onto the media. The ion from the media gets dissolved in the waste water.

The process is used most commonly when ions such as Ca^{+2}, Mg^{+2}, B(OH)x, Hg^{+1}, Sr^{+2}, Ra^{+2}, etc, need to be removed. The most common exchange ions are Na^{+} and H^{+}. In most cases, these exchange ions are acceptable. Obviously, the ion exchange media will have a finite capacity for exchange and must be periodically recharged. Without going into details here, this is done by washing with a concentrated solution of the Na^{+1} or H^{+1} exchanged ions.

Coalescing Media: There are two types of coalescing media used in the oilfield, vapor phase and liquid phase. A vapor phase coalescing media might be used to coalesce a fine mist upstream from a compressor to protect it from having droplets impacting the turbine blades. Since we are dealing with produced water, this discussion concerns liquid / liquid coalescing where small oil droplets in the produced water are coalesced to aid in their separation.

Coalescing media is not entirely unique to the oil and gas industry; however it finds greatest application there due to the need to remove oil droplets suspended in water. The strategy behind coalescing media is to promote coalescence between oil drops, which creates a smaller number of larger drops. The larger drops are then easier to separate. As discussed in other parts of this book, coalescence can occur naturally due to mild turbulence. It is greatly facilitated by the presence of media having high surface area and a hydrophobic surface.

One of the great advantages of coalescing media is that there is no waste stream. There is only an oil stream and an effluent water stream. In most cases, the oil stream is more accurately referred to as an oily water stream. But in some cases, the water cut of the oil stream is quite low. In most cases, the oily water stream can be routed to the dry oil tank. One of the disadvantages of coalescing media is the tendency of sticky crude oil, or crude oil with a high solids content to foul the media. This is discussed below.

13.4 Straining Filtration

Straining filtration relies on size exclusion. The filtration or straining material used can be as simple as a screen, or may be a deposited fiber mat, woven fiber mat, or may be more complex such as an array of wedge-shaped wires. For this reason, this mechanism is also sometimes referred to as "mechanical straining," or as "size exclusion filtration." It is also sometimes referred to as "surface filtration" in order to differentiate it from depth filtration which is described below.

When considering the filtration of non-spherical shaped particles, it must be kept in mind that a long slender particle shape may allow a particle to pass through the filter. In that case, it would be misleading to characterize the particle using a single value of particle diameter. Instead, a shape factor must also be used. In the case of oil drops, the deformability may influence whether the drop passes through the opening or not. Low API, high gravity oil generally has higher viscosity and is less flexible. High API, low gravity oil is more deformable, has a lower viscosity, and tends to pass through screen filters. The most common occurrence of this situation is in sampling and analysis of an oily water stream where a 0.45 micron filter is used to differentiate between suspended and dissolved oil.

For most straining filters, a filter cake will build up. The pressure drop across the cake will add to that across the strainer (screens or fabric) itself. As a filter cake builds up, the separation efficiency increases since the filter cake is a more effective filter material than the strainer. In some cases, this effect is required to achieve the desired separation. In other cases, this effect is not desirable, and steps are taken to periodically back flush or otherwise remove the buildup of filter cake and filter residue.

The volume of waste that is generated depends on the type of straining filter media that is employed. In the case of a simple screen or wedge wire screen, the media is periodically backwashed. In this case, the waste is a concentrated suspension of contaminant together with the backwash water. In the case of a fiber mat, such as used in a cartridge filter, the filter itself must be replaced once it has reached its dirt holding capacity.

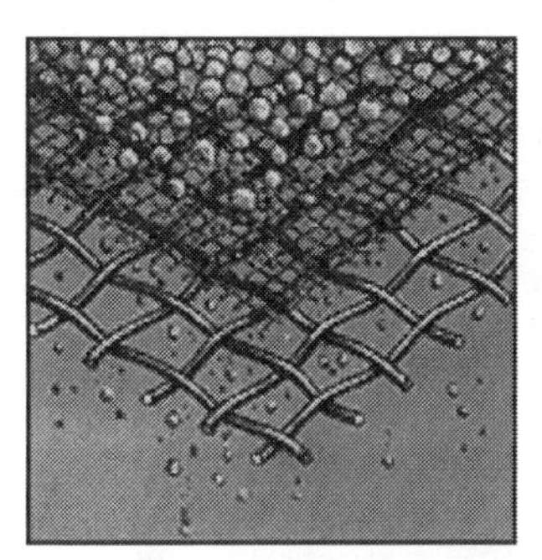  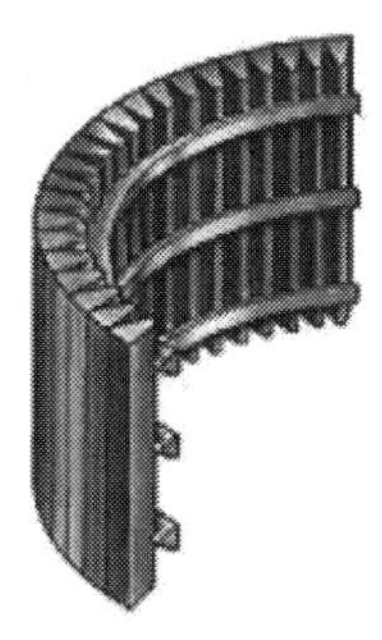

Figure 13.4 Photograph and schematic diagram of a wedge wire filtration media

13.4.1 Wedge-Wire Filtration

Backwash wire-based filtration systems are common throughout the industry. They minimize the volume of liquid reject compared to deep bed filtration. In deep bed filtration systems, a considerable volume of backwash water is required to lift the media and separate the contaminant particles from the media. Also, wire-based systems do not require the operator time and do not generate the large volume of waste compared to expendable media such as cartridge filters.

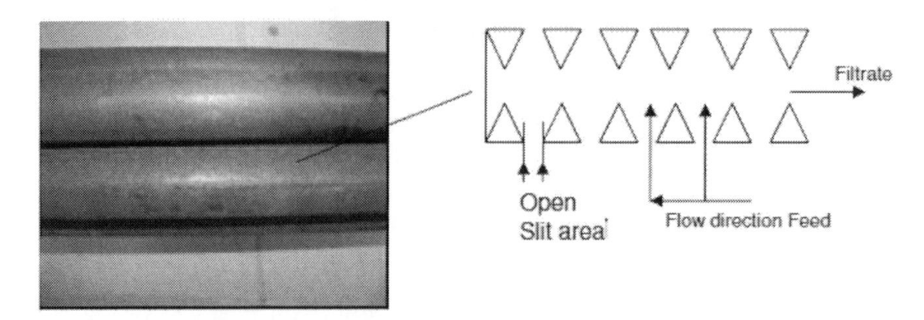

Figure 13.5 Photograph and schematic diagram of a wedge wire filtration system

A particularly effective design is that of wedge wires. As the figure shows, the wedge wires are triangular, or approximately triangular in cross-section. The waste water is pumped in the direction of expanding gap. This ensures that the highest hydrodynamic forces are located at the entrance to the strainer. If this is referred to as the outside of the strainer, then in backwash cycle, the flow is from the inside to the outside. In backwash mode, the hydrodynamic forces increase as the backwash water passes through the mesh. This helps to dislodge any particles that have become lodged in the mesh itself. Also, the fluid flow rate is designed to be somewhat higher in backwash mode, for the same reason. Thus, any particles that have been lodged would hopefully become detached in backwash given the fact that the hydrodynamic forces are greater during backwash than during operation.

Nevertheless, in some applications, sticky solids or solids with sharp edges can get lodged in the gap between the wires. Such particles lead to damage of the wire which increases the gap width and increases the cut-off size. In some cases, the material can become irreversibly lodged in the gap. Over time, this can lead to high pressure drop across the device since the available area for water flow is reduced.

These problems have led to an evolution in the design of the backwash system. Today, a number of manufacturers have developed high pressure sprays with high backwash jetting action. This has improved the range of fluids that can be successfully treated with this technology. Overall this technology has been successful. Today, it is common for suppliers to offer gap width in the range of say 10 to 25 microns. This is an incredible achievement. Nevertheless, it is still a fact of water treatment in the oil and gas industry that even smaller gap widths are needed for some applications.

13.4.2 Cartridge Filtration

A typical cartridge filter consists of a hollow cylindrical filter element, or cartridge. The feed water normally passes from the outside of the cartridge through the filter medium into the center of the element, rather than from the inside to the outside. This is done in order to maximize the probability of particle capture while minimizing the pressure drop for a given particle loading.

The water that enters the cylindrical center of the element is referred to as the clean filtrate, or simply, the filtrate. The filtrate leaves from the center of the cartridge. The separated waste stays within the cartridge filter material and the cartridge is then disposed of, once it has become fully loaded with waste material.

Cartridge filters are available down to a rating of less than a micron. Depending on the design and manufacture, cartridge filters can provide a sharp separation of particles. In other words, some cartridge filters provide a precise cutoff of particle sizes. This is discussed in further detail below.

Cartridge filters are referred to as "expendable media." When the cartridge filter has reached its particle holding capacity, it must be removed from the system and replaced. The spent cartridge filter will then be taken to a waste facility for disposal. This is both an advantage and a disadvantage. It is an advantage due to the fact that there is no waste stream. Thus, the process configuration for a cartridge filter is not required to accommodate a reject or recycle stream. It is a disadvantage in the sense that it generates a waste that must be transported and disposed. Depending on the particular application, the overall advantages and disadvantage of these characteristics will determine the suitability of using cartridge filters. In many applications, cartridge filters are used as a "guard bed." In that case, the equipment upstream of the cartridge filter is intended to provide most of the required separation of waste contaminants. The cartridge filter then is only intended to provide a final barrier in case of process upsets and excursions. In that application, the loading of the cartridge filter is intended to be minimal and hence change-out frequency is intended to be low. Cartridge filters are effective in providing a process barrier.

In some cases, cartridge filters may be back-washed or cleaned and put back into service. However, when this is done, the filtration characteristics of the filter are irreversibly changed. The recycled filter loses some if not most of its separation ability and particle holding capacity.

Cartridge Filter Media: There are many different manufacturers of cartridge filters. Materials used range from glass fiber, pleated paper, polypropylene, woven metal mesh, woven wool or cellulose fibers, sintered metals and molded fibers. The filter media can be formed or organized in different ways including woven, highly structured and organized, or can be somewhat random as would be the case for a fiber mat. The filtration material can be wrapped around a hollow permeable tube or can be formed into pleats. The material can also be formed into circular leafs which can be stacked on top of each other.

The combination of material used, and the way in which the material is formed will determine the mechanism of filtration and the details of separation performance. The filtration mechanism may employ surface filtration only or the media may be designed to give a depth filtration effect. Pleated materials are most suitable for solids removal from water as they offer higher surface area and hence higher flow rates and higher dirt holding capacities per cartridge.

Dirt Holding Capacity: Dirt Holding Capacity is a fundamental property of any filtration media. It is useful to know for back washable multimedia since it will determine the time between backwashing. It is particularly important for cartridge filters or any filtration media where the media is consumed since it is a main determinant of media consumption. The DHC of cartridge filters varies depending on the micron size, the manufacturer, and the media type and can range from a few lbs per cartridge to 30 lbs/cartridge or more.

Housing: Individual cartridge filter elements are contained in a vertical or horizontal pressure vessel housing. The housing typically has a single feed line, and a single effluent line. It will also have the necessary instrumentation and safety equipment such as a pressure relief valve and line. As mentioned above, there is no waste or reject stream.

The housing must be designed with operability features that make it relatively easy for operators to take the unit offline, open it, replace some or all of the filter elements, close the unit, and put it back online. In many installations, the cartridge filters are bundled together in a basket that allows all of them to be replaced in a single operation. Also, in considering the storage of the spent elements some thought must be given to the possibility that pyrophoric iron sulfide may form which can spontaneously ignite. The chemistry of iron sulfide was discussed in Section 2.7.4 (Iron Compounds). Handling of pyrophoric iron sulfide is discussed in Section 18.1.3 (Iron Sulfide).

Applications: There are various uses for cartridge filters in oil and gas applications. In seawater filtration, cartridge filter systems capture particles down to a well-specified diameter. This is beneficial in order to protect the injection reservoir where particles with a diameter above a specific value may plug the pores. Particles below this diameter are able to pass through the pores. In this application, the cartridge filter pore size, and the dirt holding capacity (DHC) are the main variables of interest.

For the purpose of water injection, cartridge filters are often used to guarantee the removal of certain size particles in the injection water, which is crucial for matrix injection. Cartridge filters are potentially expensive when the suspended solids loadings are high in combination with a large flow. Such a filtration system is usually present as a kind of last line of defense, i.e. other filter systems are removing the bulk of the suspended solids load upstream of the cartridge filter. There will normally be a pressure gauge on the inlet and outlet of these solids filters. The pressure drop will increase as more solids accumulate in the filter. At a given point for a filter type, the operator will need to change out filter elements.

Replacement of cartridge filters is a major concern of operating units. Such replacement requires staff time and transportation cost to handle the waste material (spent cartridges). For this reason, the system design philosophy is often focused on minimizing the frequency of cartridge filter replacement. When this is the case, the cartridge filter system is referred to as a "guard bed." The upstream equipment is selected and designed in order to minimize cartridge filter replacement.

At the other extreme, cartridge filtration is also used for high fouling or sticky contaminants. In this application, cartridge filters are expendable media, rather than guard bed media. Pore size and DHC are still important criteria but also compaction tendency must be minimized. Compaction tendency is the tendency of the internal surface to become stuck together. When this happens, effective surface area decreases and DHC decreases in proportion. Currently, there are no analytical tests that can be run to determine if this will occur. The best way is to run a pilot study. At intervals in the pilot study, cartridges are taken out of service and DHC is measured. Also, the internal pleats and surfaces are inspected. The internal surfaces are pealed apart to determine if the filtered material is uniformly distributed versus caked on some small section of the filter cartridge. If excessive caking has occurred, it may be beneficial to replace the cartridges with a brand that has less surface area. Since surface area adds to cost, then lower surface area may actually make sense particularly when compaction occurs. Depending on the manufacturer and model, cartridge filters are capable of removing oil drops and solid particles to a small diameter. For injection into a reservoir, a typical specification is 98 % removal of all particles down to 2 microns.

Nominal versus Absolute Rating: Cartridge filters are rated according to the particle size that the filter will trap. They are given either a "nominal rating" or "absolute rating". Nominal rating refers to the filter's ability to trap a nominated minimum weight percent of solid particles greater than a specified size (e.g. 95% of 2 micron). It gives some indication as to the amount of filtration that a specific filter may provide. Absolute rating refers to the largest particle that will pass through the filter under laboratory conditions (e.g. 99.98% of 1 micron). A nominal rating indicates the median particle diameter that is removed. These are two very different quantities. For a given nominal rating, a percentage of particles larger than the nominal rating will pass through the filter.

Beta Ratio: Another measure of a filter's ability to remove contaminants is the beta ratio, which is defined as the number of particles of a specified size (and larger) upstream of the filter relative to the number of particles of the same size downstream of the filter.

Beta ratio testing or multi-pass testing is a generally accepted test method to characterize the performance of cartridge filters. It provides the filter manufacturer and user with an accurate and representative comparison between different filters. Multi-pass testing uses a specified contaminate, of known sizes, added in measured quantities to the feed water which is pumped continuously through the filter. Samples of the feed and effluent are taken simultaneously at timed intervals. The concentration of particles is determined by automatic particles counters. From these measurements the Beta ratio (β) is calculated. The equation which defines β is given as [see page 43 of ref 5]:

$$\beta_X = \frac{N_U}{N_D} \quad \text{Eqn (13.3)}$$

Where:

bx is the beta ratio for contaminant particles of diameter x

Nu is the number of particles larger than x mm per unit of volume upstream

Nd is the number of particles larger than x mm per unit of volume downstream.

As shown, beta is the ratio of the number of particles of a particular size in the upstream flow by the number of particles of the same size in the downstream flow. Filters with a higher beta ratio retain more particles and have higher efficiency. Obviously, β must be reported as a number for a given particle size. The separation efficiency can be calculated from the above equation as:

$$E_X = 100\frac{\beta_x - 1}{\beta_x} \quad \text{Eqn (13.4)}$$

Where:

E_x = separation efficiency for particles of diameter x (percent)

β value	Separation Efficiency
200	99.5
1000	99.9
5000	99.95
10000	99.99

As a rule of thumb, a beta value of 1000 (three zeros) has a separation efficiency with three nines (99.9 %). A beta value of 10,000 (four zeros) has an efficiency with four nines (99.99 %). Some manufacturers provide a relation between absolute rating and beta rating. For example, some suppliers define the absolute rating of their filters as being that particle diameter for which the beta rating is equal to 1000. This gives a precise meaning to the term, absolute rating since in this case it is the micron size for which 1 particle passes to the effluent for every 1000 particles in the feed. Not all manufacturers do this however.

Pressure Drop A typical pressure drop across a clean cartridge filter is on the order of 0.2 to 0.4 bar. The pressure drop increases as the cartridge becomes loaded with contaminant. Cartridges are generally changed out when the pressure drop increases to the range of 1.4 to 1.6 bar.

Filter Models and Designs: There is a wide variety of screen, sock and cartridge filters with new designs being introduced each year. Disposable filters can be used with water that is significantly contaminated with oil and grease. For these filters fouling is only a concern to the extent that it limits the capacity of the filter. Most of the back washable designs are used in industrial and municipal treatment where there is little or essentially no oil and grease. Some of the systems listed below can be used for limited oil and grease applications. When that is the case, the pore size is generally restricted to larger values, and the pressure drop across the filter is generally limited to smaller values. Both of these limitations help to prevent irreversible fouling.

- **Vertical Tubular Backwashing Cartridge Filters:** There are at least a few manufacturers of vertical tubular filter systems. The filters can be made of wire mesh and or non-woven fabric media. A number of vertical tubes are arranged around and sealed to a distributor pipe. central cleaning valve. Flow is usually from the outside of the tube to the inside. Backwashing can be configured in the same direction, or in the opposite direction (preferred). Most manufacturers claim that such filters will remove particles larger than 2 microns at 95% separation efficiency. However, these results are generally for non-oily particles. For oily-solids, the filter pore size is generally restricted to much larger size in order to avoid fouling.

- **Acid Regenerated Cartridge Filters**: The system comprises multiple filter vessels arranged in parallel. Each vessel contains two filtration stages in series with water flowing in the upward direction. The first stage contains cartridges having a higher removal rating than the second, which therefore acts as a final polishing stage. Cartridges are located in tubeplates in the conventional manner. Filter cleaning is achieved using sulphuric acid. The acid dehydrates the organic material adhering to the cartridge element producing free carbon which is subsequently flushed to drain. Inorganic components will also be attacked by the acid and reduced in size. A gas cyclone installed into the vessel drain system prevents acid droplets being vented into the atmosphere. After a predetermined time, the filter is flushed several times with seawater and returned to duty.

- **Back Washable Cartridge Filter:** it has two stages of filtration. The first stage has a 30.5 cm diameter stainless steel disk with pore size of 80 microns. The second stage was cylindrical filter mat of needled polypropylene felt (nominal pore size of 3 microns) 1.3 cm thick and fitted to a perforated basket. Both stages were regenerated by backwashing with filtered water through a rotating arm passing segments of the filter elements. The filters were re-generated at preset 30 minutes interval unless the pressure drop exceeded 3.5 psi before that time. When regeneration becomes continuous, the filter cartridges have to be replaced.

- Performance: best performance during normal operation is around 80% and 90% removal of particles larger than 1 micron and 2 microns respectively. At worst it allows the majority of 1 micron particles to pass through. Backwashing was found to be not completely effective.

13.5 Cake or Pre-Coat Filtration

Cake filtration is based on a strainer, plus the addition of a cake material, or pre-coat material such as diatomaceous earth or Pearlite. The cake material (media) is added upstream of the strainer. As can be easily imagined, it greatly increases the separation of contaminants. A typical pre-coat filter consists of a pressure vessel housing with vertically mounted leaves or candles composed of a course metal mesh. The leaves are used to support a finer mesh material fabricated from woven fabric or stainless steel. The pore size or opening in the finer mesh is typically on the order of 20 to 100 micron. Prior to putting the system into service, a slurry of media is injected upstream of the leaves into a circulation water stream. Circulation continues until all of the media has been injected. The fine mesh material captures the media and allows the formation of a bed or cake of media. Usually diatomaceous earth or Perlite is used as media. As contaminant material becomes trapped on or in the cake, the pressure drop across the cake rises. Periodically the cake material must be swept off the strainer, and new cake material must be deposited.

When the differential pressure across the filter reaches a pre-set value, the unit is taken offline and the accumulated filter cake is removed by back-washing or by water spray or some combination of both. Once the media is removed, the cycle is repeated starting with the circulation of water containing a fresh slurry of media. The waste that is generated includes the separated contaminant together with the cake material itself. In some cases, the cake material can be regenerated. When this is true, it generally requires specialized offsite facilities. This media is generally less expensive than the equivalent of disposable cartridge or sock filters. The strainer material must also be replaced periodically though not nearly as often as the media. In general, the operation of a pre-coat unit is more complicated, which makes it less economical for very small flows.

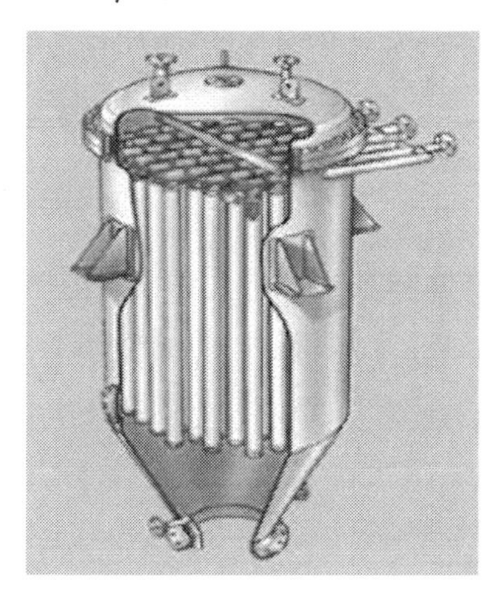

Figure 13.6 typical vertical pipe configuration for pre-coat filtration system.

The following design guidelines can be used:

flux: 1.2 to 2.5 $(m^3/hr)/m^2$

bulk density of diatomaceous earth:	dry powder:	110 to 240 kg/m^3
	wet filter cake:	240 to 320 kg/m^3
bulk density of Perlite:	dry powder:	60 to 90 kg/m^3
	wet filter cake:	110 to 130 kg/m^3
typical pre-coat coverage:	100 to 150 kg/100 m^2 of filter area	

Table 13.4 Applications of Pre-Coat filtration.
All of these applications involve schmoo-like fluids with pre-coat systems.

Opeartor	Total Flow (m^3/hr)	Filter Area (m^2)	Specific flow m^3/m^2.hr
BP UK	21.7	13	1.67
Total Fina Elf	57.5	36.4	1.58
Sirte Oil	109	62.4	1.75
Nerefco	90	62.4	1.45
Agip Lybia	206	109.2	1.88
Agip Lybia	132.6	79.3	1.62

13.6 Deep Bed Filtration:

As the name implies, deep bed (or depth filtration) involves passing the produced water through a bed of granular or fibrous media. Media particles are generally larger than the contaminant particles in the produced water. A number of different contaminant capture mechanisms are involved, including particle trapping, adsorption, or absorption, as discussed below. It is important to understand which of these mechanisms is operative because they differ in the extent to which a change in chemical treatment, flow rate, and or media will impact the separation performance. Often a combination of mechanisms is involved. Walnut Shell Filters are the most common form of deep bed filtration used for produced water. They are discussed in this section. Prior to the discussion of Walnut Shell Filters background on deep bed filtration and other media types are given.

Design and Operating Variables: There are many design and operating variables in deep bed filtration [1, Section 6.8]. They can be categorized as follows:

1. media type, composition, size, shape, surface charge, capture mechanisms
2. bed depth, cross-sectional area
3. flow rate, flux, hydraulic residence time
4. head loss
5. up flow versus down flow
6. backwash process, water volume, duration
7. redundancy
8. process configuration

Each of these is discussed below. The size and performance of the media are the main variables that determine the equipment size, weight, and performance of deep bed media, in general. The separation performance of the media is often a tradeoff between the desire for a compact design and the need to meet certain performance specifications. The following discussion helps to explain the role of the various design variables.

Media Type, Composition, Size: The composition of the media is obviously one of the most important design variables. There are many different materials that can be used. If an oleophobic material is used then adsoption / absorption will not be one of the main capture mechanisms (see below). The oil will not stick to the media surface. If an oleophilic material is used, there may be greater capture of oily contaminants but fouling, mud balls, and poor dirt holding capacity may be a consequence.

The following table gives the relation between mesh size and media particle diameter. As shown, media particles tend to be much larger than the contaminant particles that are removed. This implies that straining will not be a major separation mechanism. On the other hand, straining through a filter cake of contaminant particles that has built up over some time is a common mechanism, as will be discussed.

Table 13.5 Relation between mesh size and media particle diameter

US Mesh	mm diameter
4	4.7
8	2.4
12	1.7
20	0.8
35	0.5
50	0.3

Table 13.6 Common bed materials are listed in the table below

Media	Characteristics
sand	inexpensive default media for large scale filtration
anthracite	suited for sand / anthracite dual media filtration
granular activated carbon	absorption of non-polar compounds (oil)
garnet	high density, can be used in multimedia filtration
magnetite	high density, can be used in multimedia filtration
lime stone	used for remineralization
dolomite	used for remineralization
plastic	low density, can be used in multimedia filtration
pumice stone	low density, can be used in multimedia filtration
nut shell	suited for removal of oil
greensand	suited for iron and manganese removal

Capture Mechanisms: There are several possible capture mechanisms that depend on hydrodynamics and the surface characteristics of the contaminant and media [6], [7, p. 261]; [8, Chapter 9]. Generally speaking, contaminant particles, to a greater or lesser extent, follow a streamline in the flow. This tends to carry the contaminant particle around the media. There are various mechanisms that cause the particle to deviate from the streamline and get captured. First of all, the flow is laminar which means that turbulent fluctuations are absent. However, Brownian motion of small particles will cause a somewhat random deviation from the streamline which can lead to collision with the media. This is particularly true for small particles, in the range of a few microns in diameter or less.

It is important to note that media beds do not operate by surface filtration. Surface filtration is the separation mechanism where the particles or droplets to be filtered from the water become wedged in the spaces between the grains of media. Surface filtration in a deep bed filter is undesirable as it will quickly result in blockage of the top layer of the bed and the capacity of the subsequent layers of the media bed will not be effectively utilized. The openings or passageways between the granular media grains are relatively large compared to the suspended particles, allowing the particles to pass into the bed of the filter. Inside the bed, the suspended particles become trapped by first intercepting, then attaching to the media. Thus the suspended particles and droplets in the feed stream are removed within the body of the media bed, not on the surface of the bed.

In general, most media types have surfaces that are negatively charged when submerged in water. Most contaminant particles and oil droplets also have negatively charged surfaces. Surface attraction forces, including hydrogen bonding, charge transfer complex formation, and van der Waals forces may cause the particle to adhere to the media surface but these forces are very short range. The particles are likely to be repelled by the media surface charge before they are drawn to the surface by the short range attractive forces. Charge-charge repulsion occurs at relatively long distances (tens of nanometers) compared to surface attraction forces which are very short range (few nanometers). Therefore, one strategy for improving particle capture is to add coagulation or flocculation chemicals. These chemicals modify the surface of the contaminant and media thus making the interaction with contaminant more favorable. These chemicals can cause particles to flocculate into clusters which improves the probability of capture.

Over the years, chemical treatment of media to modify its surface properties has been tried. The effect was initially positive but not long lasting and therefore not economically viable [9]. Chemical treatment where used is applied to the produced water rather than the media surfaces.

Particles can become trapped in the interstices between particles. Gravity, as well as particle inertia may cause a particle to deviate from a streamline and become captured. Straining, as a result of filter cake build up is an important mechanism. This occurs when some particles have already been captured and have begun to build up a cake of material with very fine pores. This can create a matrix of material on the media surface that becomes effective at particle entanglement and capture. The cake of previously captured material then becomes an effective filter.

As mentioned, these physical mechanisms are different from each other and it is generally worthwhile to perform at least some simple bench top laboratory tests to determine which mechanisms are likely to be most operative, and thus how best to optimize the process. Such tests can be carried out using a clear plastic packed column on a bench top. The main strategy in conducting the tests is to test various chemical additives, adjust the pH, salt content, and evaluate separation performance and the extent to which the contaminant particles penetrate the media bed versus the formation of a filter cake at the top surface of the media (in down-flow). This is useful information to have before proceeding to field trials and field implementation. Also, it must be kept in mind that the capture mechanism can change depending on the zone (height) within the bed, and the time since the last backwash. There is a tendency for particle / media capture to be the dominant mechanism in the early stages which is then followed by filter cake formation and capture as the overall level of contaminant builds up. If filter cake capture is the dominant mechanism throughout the life of the media, then fresh media will require a break-in time during which separation efficiency will increase to the desired level. This is the case with cake or pre-coat filtration using, for example, diatomaceous earth.

As discussed presently, capture is typically not a permanent process. As the time between backwash proceeds, there is a tendency for captured particles to break off from the media and become entrained in the water flow. The bed depth should be deep enough that these particles are recaptured. The re-capturing of entrained particles is one of the parameters that determines the bed depth in the design of deep bed filters. As time proceeds further, the depth at which these particle clusters shear off and are re-captured extends deeper and deeper into the bed. This eventually requires a backwash or there will be break-through of contaminant.

Bed Depth: Media bed depth is selected on the basis of two key objectives. One objective is to prevent wash out of contaminant. Wash out occurs when the bed height is too short to adequately capture the contaminants even though the porosity, capture mechanisms, and flux would suggest that capture is likely. The bed depth, and the superficial velocity together determine the Hydraulic Residence Time (HRT). The longer the HRT, the greater the probability of capture. The second objective is to achieve a certain Dirt Holding Capacity (DHC). Having a high DHC will extend the run time between back-washing and thus reduce the volume of waste water. The optimal bed depth is often selected on the basis of these two objectives, plus a number of tradeoffs or drawbacks that come with increasing bed depth. One such drawback is that a deep bed adds to the weight and space of the unit. It does so both directly and indirectly. The direct impact is simply that the bed is deeper and therefore occupies more space and requires a larger vessel. Indirectly, a deep bed requires more vessel height in order to obtain good fluidization for adequate back wash. Also, a deeper bed causes a greater pressure drop across the bed. Thus, additional feed pressure (hydrostatic head) may be required to force water through the bed particularly as it becomes loaded with contaminant.

Bed Cross-Sectional Area & Flux: The flux through the bed is defined as the flow rate of contaminated water per unit cross-sectional area of the bed. Flux is numerically equal to the superficial velocity. They are typically expressed in different units, as shown below.

Flux, bed depth, and media type and size etc. are the main variables that determine the separation efficiency. Of course, it would be desirable to design and operate a unit for a high flux. This is not always possible since performance will decrease as the flux is increased. This reduction in performance may be compensated, at least partially by increased bed depth, but there are drawbacks to this, as discussed.

A filter should not be operated above its design flow rate. Operating a filter at capacities above the design flow rate will raise the velocity though the open pores in the media bed. This will tend to detach the contaminant particles that had previously been captured in the media bed and carry them through the bed into the effluent, resulting in a deterioration of the filtrate quality and a shortening of the length of the filtration cycle.

It should be noted that the length of the filtration cycle will decrease dramatically with increasing flow rate as the increasing flow rate has a double impact. Firstly, forcing a higher flow rate through the filter will itself result in a higher pressure drop. Secondly, the higher flow rate will mean that the rate of solids entering the filter will be higher, blocking the filter more rapidly. Thus a doubling of the flux through the filter can lead to a fourfold reduction in the length of the filtration cycle for a differential pressure limited filter.

Flux is calculated using the following formula:

$$F = Q / A_B \qquad \text{Eqn (13.5)}$$

Where:

F	=	flux of water through the bed (m3/m2 sec)
Q	=	volumetric flow rate (m3/sec)
A_B	=	cross-sectional area of media (m2)

The superficial velocity is numerically equal to the flux, but has different units.

$$V_S = Q / A_B \qquad \text{Eqn (13.6)}$$

Where:

Vs = superficial velocity (m/sec)

Hydraulic Residence Time The HRT is not a design variable per se since it is uniquely determined by the bed depth and the superficial velocity. Once values of those variables are selected the HRT can be calculated. HRT is a useful variable since it gives a rough understanding of the level of compactness of a unit compared to other technology and compared to other designs. The Hydraulic Residence Time (HRT) is defined as:

$$t_H = H_B / V_S \qquad \text{Eqn (13.7)}$$

Where:

t_H = Hydraulic Residence Time (sec)

H_B = Bed depth or height (m)

Typical HRT values for Walnut Shell Filter systems are on the order of 20 to 60 seconds. Primary separators have a water residence time of a few to several minutes. Hydrocyclones have a residence time of about 3 seconds. A multistage horizontal flotation unit has a residence time of 4 to 6 minutes.

Up-Flow versus Down-Flow Most deep bed filters that are intended for the removal of oil are designed in a down flow configuration in order to improve oil capture. In a down flow configuration, the buoyancy of the oil droplets tends to increase their hydraulic residence time (HRT) in the bed which improves the likelihood of capture. In an up flow configuration, the Stokes Law upward flow of oil droplets is added to the main flow rate of water through the bed. This reduces the HRT and increases the likelihood that the droplets will get pulled away from the media and into the main flow which eventually would lead to discharge with the effluent water.

Captured Solids Distribution: In a down-flow media bed, it is desirable to have larger media particles on the top and smaller media particles at the bottom. This reduces the blinding-off effect that occurs when a sediment layer or filter cake forms at the top of the media. Blinding-off occurs when a tight, relatively impermeable filter cake builds up and causes a large pressure drop over a relatively thin section of the bed [7, p. 168], [1, p. 263]. One of the design objectives is to ensure that the particles are trapped throughout the bed and not just in the top-most layer. If straining occurs in the top-most layer, then head loss will climb quickly without significant capture of contaminant throughout the bed. Eventually, the head loss will match the available head and flow will stop, or perhaps a backwash sequence will be triggered. Thus, the media size, together with the surface properties and chemical addition must be selected in order to minimize surface straining and maximize the capture of particles throughout the depth of the bed.

Inlet Contaminant Concentration: Media filters are not tolerant of high concentrations of solids in the feed. High solids concentrations have a similar surface filtration effect due to bridging. A number of solid particles can wedge together, effectively bridging over and blocking the pores in the media bed. The solids concentration in the feed should typically be less than 50 mg/L.

Contaminant Particle Size: Media filters are typically capable of removal of 85% of particles 2 μm or larger and 90-95% of particles 5 μm or larger. However, the performance of a filter must be considered in relation to both particle removal efficiency and overall filtration efficiency. Particle removal efficiency is a measure of how efficiently a filter removes a particle, regardless of the throughput through the filter. Filtration efficiency takes the flow rate through the filter into consideration.

For example, large solid particles entering a media filter will be quickly removed in the uppermost layer of the media bed, leading to surface filtration. The surface layer will be very efficient at removing incoming particles; hence will have high particle removal efficiency. However, the actual flow rate through the surface layer will become restricted as contaminant particles accumulate, resulting in low overall filtration efficiency.

This behavior is illustrated in figure below. It can be seen that above a certain ratio of particle diameter to media diameter, surface filtration effects will dominate and although the particle efficiency will continue to rise, the overall filtration efficiency will fall.

Media Grain Size: Thus, for a given media grain size, there will be an optimum feed particle size that will be efficiently filtered without inducing surface filtration. As an indication, for a media bed of spherical particles, the largest particle that can pass into the bed has a diameter of about one seventh (1/7) of the diameter of the media particles. Thus, the feed to a filter consisting of 0.5 mm sand particles should not contain particles larger than 70 microns. In general, it is recommended that particles larger than 100 μm are not allowed to enter the filter. This particle size corresponds to the size of solid particle or oil droplet that is readily removed by relatively simple upstream equipment such as a plate pack interceptor.

The Bed Depth / Media Grain Diameter Model: As discussed by French [10], one of the key design parameters for deep bed filtration is the ratio of the depth of the bed (L) to the media grain diameter (d). Models of deep bed filtration show that the greater the number of filter grains (collectors), the better the filtrate quality. Iwasaki [11] presented a mathematical model that describes the removal of particles based on solids influent concentration, media grain diameter (d), and bed depth (L). According to his equation, as the L/d ratio increases the removal of particle efficiency increases. This parameter (L/d) can be increased by increasing L or decreasing d. The model was first validated by Wegelin et al (1986). Regardless of the capture mechanism (Brownian motion, inertial impaction, sedimentation, straining or diffusion), the model generally holds true.

However, the Iwasaki model does have limitations. When the media grain size is too small, the clean filter head loss is impractically high, and filter run times are too short, typically due to surface blinding of the fine media. For example, when testing filter sand of 0.5 mm size at 24 inch depth (an L/d of 1,220) the clean head loss may be acceptable but the filter surface blinding tendencies may limit the applications for this configuration. When using deep bed filters such as those 6 feet deep with 1.5 mm sand media (an L/d of 1,218), the effluent quality may be similar to that obtained from use of finer filter sand and the clean head loss will be reduced. However, the implementation of beds 6 feet deep has both significant installation and operational costs and may be impractical. Thus, selection of these parameters must be made with the particular application taken into account. Nevertheless, the general consensus is that an L/d ratio of 1200 is a good starting point.

Media Size Distribution and Stratification: By arranging the course media at the top and fine media at the bottom, a more gradual capture of contaminant particles can occur throughout the bed. Large contaminant particles are captured in the top layer and removed from the water. Another way of saying this is that the coarser top layer will provide interstitial spacing between media particles to capture larger contaminant particles solids. The selection of the top layer is critical to achieving acceptable Unit Filter Run Volumes (UFRV) thereby providing reasonable filter throughput between backwash events. By utilizing the correct media grain size at the top of the filter bed the depth of penetration of solid contaminant particles is enhanced. This improves the overall Dirt Holding Capacity (DHC) of the bed. Finer particles and finally the smallest particles are captured toward the bottom of the bed where they do not create a large pressure drop.

While this is most desirable it can only be achieved with a limited number of media types. The problem is that backwashing is carried out with upward flow from the bottom of the bed. This provides separation of media so that it releases the trapped contaminant particles. This is best achieved when the media is fluidized during backwash. The upward flow will naturally drag the contaminant particles to the discharge, which is the objective of the backwash sequence. It will also naturally drag and suspend the small diameter media particles to the top of the bed and will allow the larger particles to settle further down. This is a natural consequence of Stokes Law. Chapter 11 of reference [12] discusses the fluid mechanics of backwash. If the media particles have a range of diameters, and all the particles are of the same density, the smallest particles will settle on top and the largest particles will settle on the bottom. This is just the opposite of the desired situation. For this reason, multimedia filtration systems have been designed. Smaller, more dense media material is used together with larger less dense media material. Upon fluidization, the larger particles migrate to the top due to their low density and the smaller particles migrate to the bottom due to their higher density.

Head Loss: Backwash is initiated by one of three situations: (1) timer, (2) head loss (pressure drop through the bed), or (3) breakthrough of contaminant. During filtration, contaminant builds up in the bed which restricts the flow of water and raises the pressure drop across the bed. This requires a source of pressure such as a pump of hydrostatic head in order to sustain flow. The greater the source of pressure, the longer the bed can run before backwash is required due to head loss. Due to this, media is selected in part due to head loss. Very fine media particles will give rise to greater head loss. Course media will have less head loss associated but will not be as effective as a filtration media. Therefore, head loss is an important design consideration. It is worthwhile to consider the mathematical relation between media bed characteristics and head loss as given by the Kozeny equation [7, see p. 167]:

$$h = H_B \frac{k\mu}{g\rho} \frac{(1-\theta)^2}{\theta^3} a\upsilon \qquad \text{Eqn (13.8)}$$

Where:

h	=	hydraulic head loss across the bed (m)
H_B	=	bed depth or height (m)
k	=	dimensionless parameter (usually set equal to 5)
m	=	viscosity of water (Pa-sec)
g	=	gravitational constant (9.81 m/sec^2)
r	=	density of water (kg/m^3)
q	=	bed porosity (dimensionless)
a	=	surface area / volume ratio for average media particle (1/m)
u	=	superficial velocity (m/sec)

This equation shows the main relationships between the design variables. First, as the superficial velocity increases, the head loss increases as well, as expected. The faster the water flows through the media bed, the greater the head. Or conversely, the greater the pressure drop across the bed, the higher the water flow rate. This is a design trade-off. If high flux is desired in order to reduce weight and space, then hydraulic head loss will be higher and a larger pump may be required.

As the bed porosity decreases the head loss increases, as expected. This is another design trade-off. If small media particles are desired in order to improve separation efficiency, then hydraulic head loss will be higher. Similarly, the deeper the bed, the greater the hydraulic head loss, as expected. Again, this is another design trade-off.

The above equation is given here only for the purpose of discussion since application to an actual design is somewhat problematic. The porosity and surface area / volume ratio of the media varies across the height of the bed, and with time as contaminant builds up. Both of these effects are difficult to quantify accurately. Nevertheless, it does provide a conceptual relation between the various design variables.

Backwash Design: The weight and space for a given deep bed filter design depends upon two important factors: the flux, and the auxiliary equipment required to operate the backwash. Also, the efficiency of a deep bed filter is linked to the efficiency of the washing procedure. There is no other variable that is as important as the variables related to the backwash procedure, equipment required, and frequency of backwash.

There are typically four main steps in backwash:

1. Partial emptying: discharging the water above the filtration bed.

2. Air scour and wash: During this step, the sludge retained in the filter bed is separated from the media particles.

3. Backwash with water only: the aim of this step is to remove all dirty water over the media and to replace it with clean water. This ensures that upon re-start clean water emerges rather than a slug of dirty water.

4. Filling: the backwash water outlet is closed at the end of the backwash step allowing the filling of the filter with raw water.

For many deep bed filter designs the backwash duration is in the range of 17 to 20 minutes. The frequency of backwash depends on the solids removal rate (difference between feed water contaminant concentration and discharge water), and the dirt holding capacity of the bed. The dirt holding capacity of the bed is really a measure of how much contaminant a bed can hold before it exceeds the maximum allowed hydraulic pressure drop.

As an example, if a bed can hold 30 kg of a certain type of contaminant, the feed water flow rate is 6,555 m3/day, and the feed concentration is 20 mg/L, while the effluent concentration is 2 mg/L, then the frequency of backwash is given by:

$$t_B = \frac{30 \text{ kg}}{(0.02 \text{ kg/m}^3 - 0.002 \text{ kg/m}^3) \text{ x } 6{,}555 \text{ m}^3/\text{day}} = 0.25 \text{ day} \qquad \text{Eqn (13.9)}$$

This calculation indicates that there will be a backwash 4 times per day. As mentioned already, the dirt holding capacity is the amount of contaminant that the bed can hold before it reaches or exceeds the upper limit of pressure drop across the bed. If this upper limit were increased, by adding pumping pressure for example, then the dirt holding capacity would increase and the backwash frequency would decrease.

Process Configuration: In onshore applications, coagulation chemical treatment followed by sedimentation is often used upstream of deep bed filtration [1]. This is particularly true when the produced water contains high concentrations of contaminant. In that case, the upstream equipment is used to remove the bulk of the contaminant concentration to prevent overloading the deep bed. With very high flows and high suspended solids loading, a clarifier should be considered upstream of the filtration system. Settling is inexpensive and, with proper chemical treatment, can eliminate most large particles before further treatment. The deep bed then provides deep removal of the fine material. Without the upstream steps, the deep bed will be overwhelmed with contaminant loading and will not perform well to remove the fine material. In an offshore installation, Walnut Shell Filtration is installed downstream of a flotation unit, for similar reasons.

13.6.1 Sand and Multi-Media Filters

Slow Sand Filters Slow sand filters operate at slow flow rates, 0.1 - 0.3 meters per hour (0.1 – 0.3 m^3/ hour m^2). The top layers of the sand become biologically active by the establishment of a microbial community on the top layer of the sand substrate. These microbes usually come from the source water and establish a community within a matter of a few days. The fine sand and slow filtration rate facilitate the establishment of this microbial community. The majority of the community is composed of predatory bacteria that feed on water-borne microbes passing through the filter.

One of the distinguishing features of slow sand filters, which distinguish them from rapid sand filters, is that they are not designed to be backwashed. Another feature is that the primary mechanism of filtration is due to the action of the microbial community. The microbial community forms a layer called the Schumtzdecke layer. It and can develop up to 2 cm thick before the filter requires cleaning. Once the Schumtzdecke becomes too thick and the rate of filtration declines further it is scraped off, a process done every couple of months or so depending on the source water. Once this has been carried out, the slow sand filter will not be fully functional for another 3 to 4 days until a new Schumtzdecke has developed, although this procedure can be speeded up by seeding the filter with bacteria from the removed Schumtzdecke. Slow sand filtration is extremely good at removing microbial contamination and will usually have no indicator bacteria present at the outlet. Slow sand filters are also effective in removing protozoa and viruses.

Slow sand filters require low influent turbidity, below 20 NTU and preferably below 10 NTU. This means that efficient pretreatment is required to ensure that the filters do not become overloaded. Slow sand filters can cope with shock turbidities of up to 50 NTU, but only for very short periods of time before they block. The sand used in slow sand filters is fine, thus high turbidities cause the bed to block rapidly and necessitates more frequent cleaning and therefore greater time out of action. Nevertheless, slow sand filters are still used in London and were relatively common in Western Europe until comparatively recently and are still common elsewhere in the world. The move away from slow sand filtration has largely been a function of rising land prices. Where this is not the case, slow sand filters still represent a cost-effective method of water treatment.

A sand filter is suitable for even higher suspended solids loadings, at relatively medium to high flows. If extreme suspended solids loadings are to be expected, the Dynasand filter is preferred, because of the continuous backwashing capability. If the flows are very large, it will become attractive to install horizontal sand, dual media or multi-media filters. The horizontal alignment makes the filter relatively cheaper than the vertical cylindrical sand filters.

Rapid Sand Filters: Rapid sand filters work at much higher rates of flow (up to 20 m^3/ m^2 hour) and rely on physical removal of suspended solids. Although rapid sand filters achieve some reduction in microbial populations in water as it removes particles to which bacteria are attached, it is not a biological treatment and the use of a disinfectant is vital to ensure that bacteria in the water have been inactivated. Rapid sand filters require frequent cleaning (daily or twice daily) which is achieved through backwashing with clean water. Cleaning takes relatively little time and the filters can be put back into operation immediately. However, sand is relatively heavy and requires relatively high backwash energy in order to fluidize the bed and release the contaminant particles. Also, sand has a higher affinity for oil than most deep bed media types. Thus, there is a greater tendency for fouling, mud ball formation, and sticking of asphaltene crude to the media.

Rapid sand filters are far smaller than slow sand filters and are commonly employed in 'batteries'. The rapid flow rate through these filters means that demand can be more easily met from smaller plants. Rapid sand filters do not require low influent turbidities, as they are essentially a physical treatment process, although higher suspended solids loads will result in more frequent cleaning. Backwashing is usually rapid, and filters are not out of commission for more than a matter of minutes. Cleaning and operation can be largely mechanized, and air scour is commonly employed to make backwashing more effective. With the small land requirement, several rapid sand filters can be accommodated in small area and thus it is easy to maintain capacity to meet demand when filters are being cleaned.

Multi-Media Deep Bed Filtration In a down flow configuration it is possible to use two media types and still have large media on top and small media on the bottom. This requires that the smaller media have a higher density than the larger media. The top media should have lighter density and larger diameters. If the density is small enough to overcome the effect of surface area, this material will end up on top after backwash. The bottom material should have higher density and smaller media grain diameter. If the density is high enough to overcome the effect of surface area, then these smaller grains will end up at the bottom of the bed after backwash. Media combinations such as anthracite / garnet, and anthracite / silica sand are examples.

Multimedia filtration removes colloidal and larger particles from water by passing it through a porous filter media that may consist of sand, anthracite and garnet. Suspended solid particles from less than 1 micron to 50 micron and larger may be removed although they are typically capable of removal efficiency in the order of 85% of particle 2 microns or larger and 90-95% of particles 5 microns or larger. As a guide, particle size of the medium often used is as follows;

- Anthracite 0.8 to 2.0 mm
- Sand 0.4 to 1.0 mm
- Garnet 0.2 to 0.4 mm

The selected configuration will be influenced by the operating philosophy and the required injection water quality. Typically, the source pump (seawater lift) feeds into the coarse strainer to remove large particles. The filtered water may then be passed through a multimedia filtration system to reduce the total suspended solid to meet the requirements any further filtration levels or technology to deliver the required water quality [5].

Media filtration packages are normally comprised of a number of smaller units which have a means of flow sharing between the units to ensure even flow and that the maximum design flow is not exceeded through any single unit. A separate coarse filter (100 microns) may be required in the backwash water supply to prevent blockage of the packed bed support screens. A supply of air for scouring of the packing during the cleaning operation is also required [1, 8].

13.6.2 Nut Shell Filtration

Nut shell filters are used in produced water applications. It is a form of deep bed media filter. They are mostly deployed onshore due to relatively large weight and space requirements. However, some units have been installed offshore. In general, they may be composed of crushed black walnut shells or pecan nut shells, or a combination of the two. The black walnut shell has characteristics that make it most suitable. These are discussed below. The hydrophilic surface ensures that sticky oil drops and solids are less prone to stick to the surface of the media. This is an advantage because once a high viscosity sticky material has become adherent to a surface it is difficult to remove. Black walnut shell filters are often used for heavy oil, particularly steam floods where the water for steam generation is recycled produced water. The low API gravity of the oil does not seem to have an adverse impact on the separation efficiency. Low gravity oil does of course have a greater tendency to foul which must be taken into account in the design of the back-wash system.

Oil drops and oil-coated solid particles can be effectively removed in a walnut shell filter. Water-wet particles tend not to be removed. There are many variations of backwash procedure available from the different manufacturers. In all cases, the objective of backwash is to lift the media particles and separate the media which releases the contaminant that had been trapped in the gap between the particles.

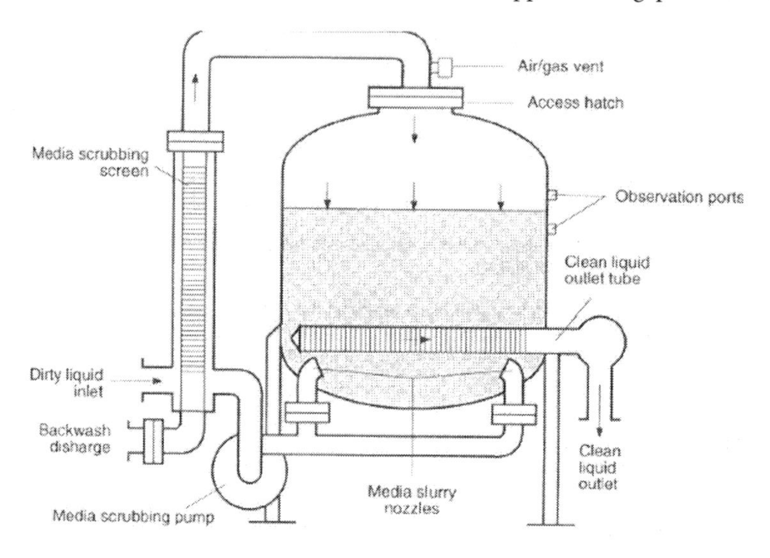

Figure 13.7 Silverband Vertical pipe configuration for Walnut Shell Filter system.

After a period of operation, the effluent water quality will begin to deteriorate. When this occurs, the system must be back-washed. There are other advantages, but at this time, we will only discuss the above two and further we must discuss the manufacturers of this type of equipment.

There are a few designs that have historical and practical significance and are therefore discussed here. They represent stages in the evolution of filter design. They are primarily distinguished by the method of backwash. They are:

1. Filtra Systems (Hydromation Division) [13]
2. Wemco Silver Band
3. U.S. Filters.

Both Filtra systems and U.S. Filters use crushed black walnut shells while Wemco use a mixture of pecan and English walnut shells.

Kashaev and Lee have given a description of the historical development of Deep Bed Nutshell Filter evolution [9]. Over several decades the basic media bed filtration system has remained relatively unchanged. However, the design of the backwash system has gone through several major modifications and improvements. These are discussed below.

Nut Shell Media Properties and Advantages Another advantage of the Black Walnut Shell is that its specific weight is almost the same as water. Therefore it is a great advantage in that the shells can be fluidized and with the aid of the high velocity nozzles, all the water contained in the vessel along with the oil and solids are washed out of the vessel through the scrubber tube and back through the pump to the base of the vessel through the high velocity nozzles. This procedure is known as the Agitation Cycle and it ensures that all the media, dirt and oil is in a state of flux.

Wemco Silver Band Filter This is a down flow filter. The media is a combination of ground pecan shells and walnut shells. During backwash the media bed is fluidized within the vessel. There is a fluidization pump located on the top of the vessel. The pump fluidizes the media and circulates it through the vessel. There is a screen that extends down from the top of the vessel that separates the contaminated water from the nut shells. The contaminated water passes through the screen. Dirty, unfiltered water flows into the vessel continuously replacing the backwash water that passes through the screen which is considerably dirtier than the feed water. The screen ensures that only a minimum of media is discharged during backwash. Backwash times are reported to be about 14 minutes. The design flux is about 13.5 to 15.0 gpm/sq ft [14]. The manufacturer claims 98 % removal of particles larger than 2 micron. The feed concentration should be 100 mg/L or less.

Hydromation Filter Hydromation was the pioneer of the use of Black Walnut shell media, starting in the late 1950's. Various generations of design we developed. The most popular design was first tested at the Chevron production Facility at Sacroc in West Texas in 1979. The Hydromation Company was sold to Filtra Systems in the mid 1980's.

The Hydromation Filter is down flow with the inlet water being deflected by a splash plate and the outlet water passing through a screen which prevents the media from leaving the vessel. A wide range of flux are stated for this type of filter. When used for seawater filtration, the flux can be as high as 22 gpm/sq. ft. In this application, a polyelectrolyte flocculating agent (filter aid) is often used. These chemicals are not typically used for produced water. For produced water from heavy crude oils it can be as low as 6 gpm/sq ft. However, for most produced waters the optimum rate is about 12 gpm/sq ft. At this rate depending on the influent quality of the water, it is possible to achieve effluent qualities of 1 mg/l solids and 5 mg/l oil. There are however many factors to weigh before these limits can be reached and this point will be discussed later.

The filter has a scrubber tube outside the vessel whose top is connected to the top of the filter vessel. The bottom of the scrubber is connected to a backwash pump whose discharge is connected to the high velocity nozzles in the underside of the vessel. Inside the Scrubber Tube is a screen with a slot opening of 18 thousandths of an inch plus 2. This means that the slot is never narrower than 18 thousandths but can be as wide as 20 thousandths. Inside the vessel is a similar screen as mentioned earlier that has a slot opening of 16 thousandths minus two, meaning of course that the slot is never bigger than 16 thousandths but can be as narrow as 14 thousandths. Most of the time this screen is known as a Johnson Well screen, bur it is manufactured by many other organizations.

After the Agitation step is complete, which takes normally about 3 minutes, a valve at the bottom of the scrubber tube opens. This valve is called the backwash valve and when it opens, the solids and oil are drawn across the screen and the walnut shells are retained in the system.

We have now to consider the dynamics of the system and it should be realized that the filter feed pump with this type of filter, never stops running. During the filtration cycle of course, the pump is required to pressurize the system. In backwash, a second pump is in operation circulating the system as told in the agitation cycle. The filter feed pump discharge is situated on the inlet of the backwash pump and while the system is in backwash, the filter feed pump is making up for that water that is lost across the Scrubber Screen. This step is called the backwash step and usually lasts for 25/35 minutes, depending on the contamination in the system.

With respect to the pumping rates of this type of system, it can be approximated that the backwash pump rate is roughly ten times greater than the filter feed rate. However, herein lies another advantage as the backwash pump rate is not the backwash rate as is the case in all other types of media down flow or up flow filters. The Backwash water rate is 0.6% of the daily rated flow, per backwash, over the 25/30 minutes, of the backwash.

An example would be: For a 10,000BWPD filter the backwash total volume is 0.6% of 10,000 = 60 barrels =2520 gallons. If the filter backwashes twice a day then the total backwash per 24 hours would be 120 barrels, for three backwashes, 180 barrels and so on. For a conventional filter the rate is some 10 times higher and hence there is a distinct advantage in the volume of backwash water that has to be handled.

At the end of the backwash cycle, the media is all over the system and now it has to be returned to the vessel. With both pumps running and by a system of valve changes, the water flow bypasses the filter vessel allowing the walnut shells to settle in the vessel. This takes approximately 30 seconds and then there is another valve change and the flow is reversed to allow all the pipework to be flushed back into the vessel and the suction is taken from the filter outlet back to the backwash pump suction. This operation is called the media purge and takes approximately 3/5 minutes. At the end of this time, again there is another valve change and the backwash pump stops while the filter goes back into the filtration mode.

The backwash pumps are not very efficient as there has to be a 100 thousandths gap between the impeller and the pump casing to ensure that the shells are not crushed. This brings forth another advantage as the shells never need to be changed out, only the degradation replaced every year. The manufacturer lists this degradation as 5% per year but in fact it is nearer 10/15%. The cost however is minimal as the price of Black Walnut Shells in the USA is 25/30 cents /lb. depending on the quantity that is purchased.

The main advantages of this model of Walnut Shell Filter are low backwash volume, water-wet media such that backwash is more effective and there is minimal oil retention, and low cost media. Towards the end of the patent life for the Hydromation Filter, a few alternative designs were introduced to the market. The Silverband Company developed a particularly successful design. Later, the designs and engineering for the Silverband Company were purchased by Wemco whose full title is 'The Western Engineering and Mining Company'.

One of the original patents related to the Hydromation Filter design was a process patent that covered the use of Black Walnut Shells. The existence of this patent prompted the use of the English Walnut and Pecan Shell mixture. From this point on the Silverband will be referred to as the Wemco Filter and although the Black Walnut Shell patent has long since run out, Wemco have felt more comfortable with the two-shell mixture and continue to use it.

The main change in the Wemco filter was the thought that the Hydromation Filter was too complicated and used far too many valves. Also, there was the thought that it is not necessary to take the media outside the vessel during the backwash process and that the whole process could be carried out inside the vessel. This gave birth to the Silverband design. The scrubber tube is on the inside of the vessel instead of being attached to the outside. The Backwash pump is mounted on the top of the vessel and this gives rise to the first advantage of this type of filter. There are not as many automatic valves required to operate this type of filter as opposed to the Hydromation Filter. Another change in this filter which acts as a support for the Scrubber tube is the use of a flat plate screen to hold the media in the vessel. The principle is the same however and the scrubber tube slot is bigger than the clean or media retention screen slot.

It was stated that all the walnut shell filters are high rate down flow and Wemco differentiate between filters that are not filtering oil and solids against those that are. The non-oil filters have a larger flow rate and the specifications are not so stringent if the filter is not being purchased by an oil company. The filter has an inlet and outlet and requires a pump or pressure to supply the force to move the fluid through the filter.

- Hydromation and Silver Band
- Nut shells have high porosity which traps the oil drops
- They are hydrophilic and therefore not wet by "clean" oil, thus the oil can be removed, and solids regenerated by back-washing with water or steam
- Inlet concentrations to 200 ppm with outlet in the range of 50 ppm
- Major drawback: easily fouled by waxes, asphaltenes, or other sticky hydrocarbons. This reduces backwash effectiveness.

Figure 13.8 Different media types

Comparative Study of Media Types Six media types were studied: English Walnut, Pecan, Black Walnut, Silica (quartz), Anthracite, Garnet [15]. Each media type was sieved using 12 / 20 mesh. This would have produced a grade of media particles with diameter less than 1,700 micron and greater than 800 microns. Two media types were sieved with different size. Garnet was sieved at 14 mesh and anthracite was sieved at 10 mesh.

Surface Wetting: A Rame-Hart goniometer was used to measure contact angle between the media and various liquids including water, dodecane and Mobil Velocite # 6 [15]. There is significant confusion in the early literature about the wetting characteristics of various media types. Direct measured results are shown in the Table below. In general, wetting of a liquid on a solid is defined by a contact angle of 90 degrees or less. A contact angle near 0 degrees indicates full wetting. A contact angle near 180 degrees indicates no wetting. According to the results in the table, all three liquids are able to wet all of the media types in the study. This is not particularly unusual. The confusion sets in when the general statement is made that "the media is wetted by water." This is a true statement, but most readers would interpret this to mean that the media is "preferentially wetted by water," which is not correct. The media is wetted by both oil and water, but it is preferentially wetted by oil. The media bed is always flooded with water prior to being put into service. This pre-wets the media with water making it less likely to foul. It also purges air out of the bed which can reduce the effective media bed volume if not removed. Pecan and Black Walnut have the least preference for oil which is perhaps why it is generally remarked that backwash of these media types is most effective at removing the oil. English walnut and anthracite standout somewhat as having a greater preference for oil, compared to the other media types. Besides those two, there does not appear to be a great difference in the wetting characteristics of one media type versus another.

Oil Retention: Oil retention was also measured and reported as grams of oil retained per gram of dry media [15]. The tests were carried out with dry media and with water-wet media. The media is contacted with oil for a certain time. The oil is then gently removed and the amount of oil that remains on the media is determined gravimetrically. Oil retention is a partial indication of the ability of the media to capture and hold oil. The oils ranged in API gravity from 23 to 58. The higher API (lower density) oils showed the greatest oil retention. For the high API oil, values ranged from 40 to 60 percent for both dry and water-wet media. For the heavier oils, there was much less oil retention with the dry media retaining between 20 to 30 percent, and the wet media retaining between 10 and 20 percent. Black Walnut, pecan, English Walnut and anthracite behaved similarly. The garnet and silica had significantly lower retention, in the range of 2 to 15 percent. WSF beds have been shown to not only capture oil droplets but they also provide effective coalescence of the oil droplets into larger droplets [13].

Separation Performance: Walnut shell filter systems remove oily solids by a particle capture mechanism, and they remove oil droplets by a combination of capture and coalescence. Smaller droplets collide with media particles and attached to the media particle. The oil droplet may then coalesce with an existing oil film of the media particle or can coalesce with another oil droplet as it collides with the media. The coalescence mechanism is similar to that described for Corrugated Plate Interceptors (CPIs). Separation performance for oil in water is shown in the figure below.

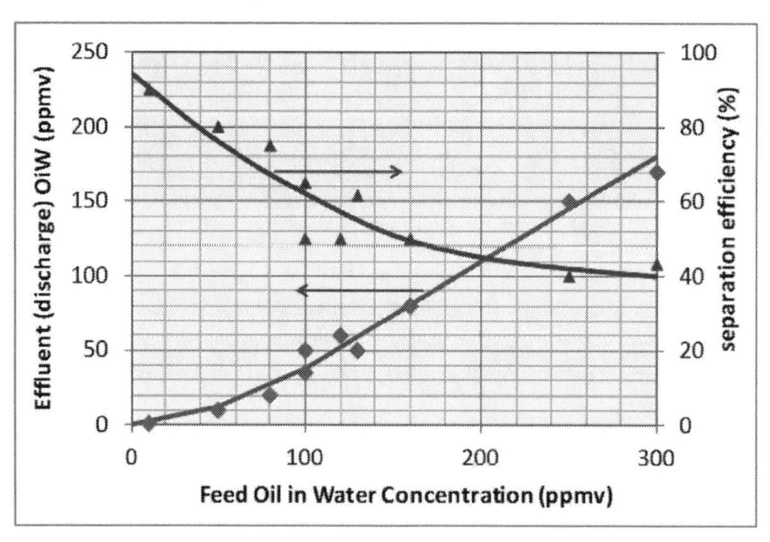

Figure 13.9 Effluent concentration and separation efficiency as a function of the inlet water quality. Note the significant decrease in separation efficiency as the oil concentration of the inlet water increases.

Attrition Tests: Attrition tests were also carried out. Attrition is a measure of the amount of media that will be lost due to fluidization and collision of media particles which occurs mostly in the backwash stage. Anthracite had a high attrition while Black Walnut had the lowest attrition of the media studied. This is generally confirmed by performance tests and observations of systems operating throughout the industry.

Flux Required to Operate and to Backwash / Fluidize the Media Bed: The flux required to fluidize the bed was studied. Garnet and silica required the highest flux, in the range between 12 and 25 gpm/ft². These two media types had the highest density. English Walnut, Black Walnut, pecan, and anthracite had essentially the same flux requirement for fluidization, in the range of 3 to 4.7 gpm/ft².

Pressure drop as a function of flux was measured for bed depth of 72 inch in a downward flow configuration. Silica and anthracite had the highest pressure drop, in the range of 0.5 psi for a flux of 12 gpm/ft². All other materials had only about 0.1 to 0.2 psi for a flux of 12 gpm/ft².

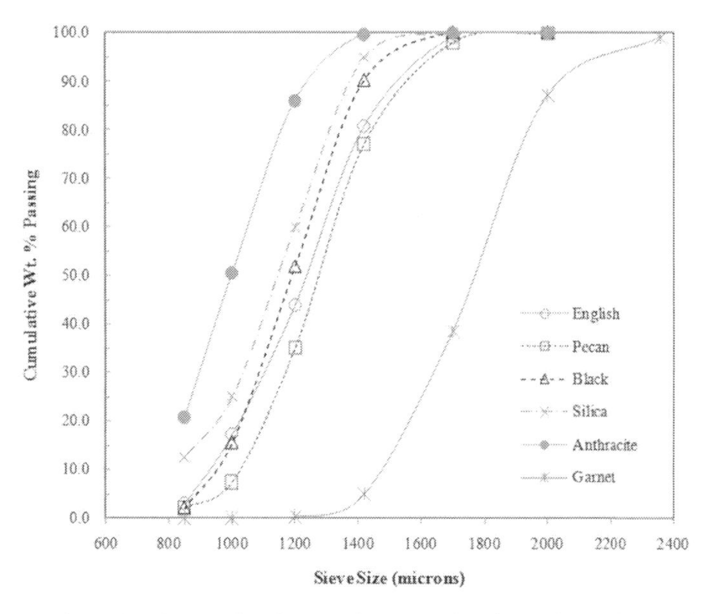

Figure 13.10 Cumulative distribution of media particle sizes used in the Rawlins and co-workers study [15].

The true and apparent specific gravity of the media particles is reported in the Table below. The void percentage is calculated based on the percent difference between the true and apparent density.

Table 13.7 Density and porosity of media used in the Rawlins study [15]. Measured contact angle for media used in Rawlins study [15]. Contact angle of 0 means that the media is fully wetted by the liquid. Contact angles in the range of 23 to 51 are still in the range considered to be wetting but are less wetting than the hydrocarbons.

Media	Specific Gravity		Void %	18 MΩ Water	Dodecane	Mobil Velocity #6
	Apparent	True				
English	1.34	1.44	7.1	51	0	0
Pecan	1.34	1.46	8.3	23	0	0
Black	1.39	1.45	3.8	31	0	0
Silica	2.63	2.64	0.4	35	0	0
Garnet	4.14	4.17	0.6	35	0	0
Anthracite	1.60	1.67	4.5	44	0	0

Typical WSF Configurations Essentially all WSF systems are configured for downward flow of contaminated water [15]. The depth of the media bed varies widely from one manufacturer to another and can range between 32 to 72 inch. The typical flux is in the range of 10 to 16 gpm/ft². The initial pressure drop of a clean bed is in the range of 0.5 psi. With time, and with contaminant removal and buildup within the bed, the pressure drop across the bed rises to the range of 15 to 20 psi. Backwash is typically initiated based on a pre-set pressure drop in this range. During backwash, the flow rate of feed water is decreased but not stopped altogether. A portion of the feed water flow (contaminated) is run through the vessel in order to sweep out the contaminant particles. The main objective of backwash is to fluidize the media such that particles trapped between and on the surface of the particles becomes detached and is be discharged.

13.6.3 Clay Media Filtration

Clay: Clay media manufacturers claim up to 70 pounds of dissolved and/or dispersed hydrocarbon per 100 pounds of media before becoming saturated. Laboratory and field experience by ETT [16] usually indicates 40 –50 pounds of oil removed per 100 pounds of clay media.

Should a water containing 100 mg/L hydrocarbon content be produced for 24 hours at the rate of 1,000 Bbl/day, the total hydrocarbon weight in water for the day would be approximately 35 pounds, assuming that the oil specific gravity is 0.90. So a filter containing 2,000 pounds of activated clay media would provide removal for 22 - 28 days, assuming 40 – 50 pounds of hydrocarbon removal per 100 pounds of clay media.

Field and laboratory experience with clay type absorption media indicate that the effluent quality water from these type beds spikes sharply upward in oil content when the media becomes spent. Therefore, when using clay media it is common to follow the clay media with GAC filtration.

Clay media bulk cost/lb quotations to ETT are approximately four times higher than those of GAC. So while there is an adsorptive rate/lb and thus smaller size advantage for clay media, there is a strong economic advantage for GAC media. Since there is no proprietary advantage to ETT for either type media, and since clay media are usually followed by GAC, ETT has been chosen as the final polishing step for flow back water treating and special produced water problems.

13.6.4 Activated Carbon Filtration

Activated carbon filtration is an important tool for polishing oilfield waters. It is referred to as a "consumptive" media because it removes constituents up to a finite capacity and then the activated carbon must be removed and replaced with fresh activated carbon. This can be costly. Nevertheless, it is so effective at contaminant removal that it is often used as a final stage for treating produced water and flow backs. When this is the case, the upstream equipment must be selected and designed in order to minimize the consumption of activated carbon. In that application it is referred to as a "final barrier."

Separation Performance: Activated carbon is effective in removing organic compounds from water. Such organics include the saturated and unsaturated hydrocarbons, BTEX, phenol, PAHs, as well as some resins and asphaltenes. It has essentially no separation ability for ionic compounds and constituents such as low molecular weight organic acids in the ionic form (pH 5 and above), and the ions of mineral salts. It is also not effective for short-chain alcohols, such as methanol. It will not effectively remove glycols and amines from aqueous streams, although for the longer chain alcohols it does have some affinity. In produced water streams that contain high concentrations of methanol (percentage concentration), activated carbon is still effective at removing hydrocarbons. The absorptive capacity of carbon for several individual compounds is given below.

The main mechanism of separation is absorption due to nonpolar van der Waals forces. Compounds such as hydrocarbons that have limited solubility in water are readily adsorbed onto activated carbon. As a rough rule of thumb, separation effectiveness is inversely proportional to the solubility of the constituent in water. However, this statement must be applied with caution because there are exceptions to this rule.

Manufacture of Activated Carbon: Activated carbon is manufactured from a variety of natural materials including coal, lignite, wood, peat, and coconut shells. The manufacturing process has several steps but the two most important steps are carbonization and activation. In the carbonization step, pyrolysis is used to release volatile components, and to form a low density carbonaceous material. Pyrolysis is essentially high temperature (600 to 900 C) treatment in the absence of oxygen. In the specific case of activated carbon manufacture, the pyrolysis step generates a gas rich in hydrogen, which escapes from the char leaving a porous structure mostly composed of carbon. The resulting carbon char has relatively high porosity from the evolution of the gases. However, the pore structure is relatively closed, with low permeability, and is not optimized for waste water treatment applications.

The second step of manufacture is the activation step where the pore structure is created, and permeability is increased. This is accomplished by subjecting the carbonized material to a partially reducing atmosphere, at somewhat higher temperature (600 to 1200 C), and in the presence of steam. This is called the water-gas shift reaction. This step generates gases rich in (H_2, CO_2, etc). This further increases the porosity, reduces the density, and most importantly creates a permeable pore structure and very high surface area. This is the basic process, from which there are numerous variations.

Although the term "activated carbon" is used frequently in the industry, it is important to recognize that there are a large number of different types of activated carbon on the market with widely varying properties and effectiveness. Different raw materials produce different types of activated carbon which vary in hardness, density, pore and particle size, surface area, extractable material, ash, and pH. These variations in properties make certain raw materials preferable than others. The Table below gives some information on the different types of activated carbon available. As shown, there is a wide range of properties, depending on the specific type of activated carbon.

Table 13.8 Different Types of GAC and the Parameters that characterize them [17].

Parameter	Bituminous	Sub-Bituminous	Lignite	Nut Shell
Iodine No.	1,050	850	600	1,000
Molasses No.	235	230	300	---
Abrasion No.	85	75	60	97
Bulk density (lb/ft^3)	27	25.5	23	29.5
Capacity (lb organic removed/100 lb GAC)	26	25	13.8	---

Note that if crude oil is being removed, then the lignite would actually have a higher capacity than the bituminous coal, because of its greater pore volume. Iodine and Molasses numbers are measures of pore size distribution. Iodine number is a relative measure of the number of small pores (in the range of 10 to 20 Angstroms) which are available for adsorbing small molecular weight organics. It is reported in units of milligrams of elemental iodine adsorbed per gram of GAC. The Molasses number is a relative measure of the number of larger pores (in the range of 28 angstroms) pores which are available for adsorbing large molecular weight organics. Abrasion number is a relative measure of the degree of particle size reduction after tumbling with a harder material. No reduction is rated with a value of 100. Complete pulverization is rated 0.

Activated carbon with smaller pores will have larger surface areas. However, much of this surface area might not be available, if the target contaminant compounds are higher molecular weight. For example, a coconut shell carbon might have 1,100 sqm/g of surface area but would perform poorly adsorbing crude oil fractions from water. A lignite-based carbon with only 600 sqm/g of surface area would be much more effective adsorbing crude oil fractions because of the much larger pores in this grade of activated carbon make the surface area available to crude oil fractions.

For illustration purposes only, a typical activated carbon will be defined here as having in excess of 500 m^2 surface are per dry gram of material. As discussed elsewhere, in surface area determinations, it is always important to specify the method used in the measurement. In this case, surface area is measured by generating an adsorption isotherm using carbon dioxide at room temperature. Activated carbon with surface area in the range of 1,000 m^2/dry gram is also available.

Activated carbon is packaged in two major forms, either as bulk material for use in a media bed vessel, or in a cartridge canister. When packed as bulk material, it is often referred to as granular activated carbon (GAC).

GAC (Granulated Activated Carbon): There are numerous types of GAC used for water treatment. The dominant GAC used in the oilfield is bituminous coal based, although lignite carbons are widely used for amine and glycol treatment systems to remove oils and other larger molecule contaminants. The reason for the dominate position of the bituminous coal carbon is that the commercial thermal reactivation facilities are set up to process this type of carbon. Lignite carbons are reactivated at lower temperatures, so they cannot be combined with bituminous coal-based carbons when reactivating. The rest of this discussion on activated carbons will refer to bituminous coal types.

Amount of GAC Needed: The amount of GAC that is brought to a site will depend on the volume of water to be treated, the contaminant concentration in the water, and the number of days that the operator wants the GAC to last before it has to be changed out. Alternatively, this can be calculated as the dry mass of contaminant in the water. Field and laboratory experience indicates that as the GAC becomes saturated (spent), effluent water quality from the carbon bed gradually increases in oil content until it exceeds limits.

Manufacturers claim up to 30 pounds of contaminant removal per 100 pounds of GAC media before becoming saturated. Field experience reported by ETT [16], a service company that uses GAC, indicates that GAC can remove 20 – 25 pounds of oil per 100 pounds of GAC media. This is equivalent to 20 – 25 kg of oil per 100 kg GAC. For example, a produced water stream of 10,000 BWPD (barrels of water per day), with 10 mg/L hydrocarbon content will, over a 24-hour period, contain 15.9 kg of hydrocarbon. For the sake of calculation, a GAC filter containing one mt (metric ton = 1,000 kg) will absorb, say 200 kg of hydrocarbon (between 20 and 25 wt %) before it becomes spent. If it is assumed that the GAC removes 100 % of the hydrocarbon, such that the effluent water contains 0 mg/L hydrocarbon, the GAC filter will last roughly 12 to 13 days (200 kg / (15.9 kg/day) = 12.6 days).

Using Reactivated GAC When oilfield GAC is exhausted, it may be disposed of or reactivated. Reactivated GAC is typically sold back into the market at substantial discounts versus virgin GAC, so there may be significant cost savings to using reactivated GAC. The most common method of reactivating loaded GAC is through thermal reactivation, typically in a rotary kiln or multi-hearth furnace. For this discussion only bituminous coal based GAC is reviewed, since this is the dominant grade used in the oilfield. The reactivation process is similar to the initial activation process. The rising temperature drives off volatile material and carbonizes any material that remains. As the temperature elevates in the presence of a reducing atmosphere (steam) the water-gas shift reaction begins to generate new pore volume. A well activated GAC will have slightly less total surface area and slightly more total pore volume, which should be suitable for most oilfield applications removing hydrocarbons.

While a well reactivated GAC might generate some cost savings, a poorly reactivated GAC may actually create higher costs. The cost of transportation and more frequent GAC change outs will quickly erode any cost savings from using reactivated GAC. To avoid this problem, it is important to establish a specification on the GAC quality. An example of a quality specification for reactivated GAC is in Table 13.9

Table 13.9 Sample Reactivated GAC Specification

Iodine Number	800 (min)
Density	0.58 g/ml (max)
Moisture	8% (max)
Dust	<1.5%

Since reactivated GAC is likely to be more variable in quality than virgin GAC, it is a good idea for the operator to do spot checks. The easiest property to check is the density. The manufacturer should be reporting results using the ASTM D 2854 Vibrating Feed Apparent Density test. The operator might not have the laboratory apparatus to run this test. An option would be to use the Poured Apparent Density (PAD). This is simply pouring a sample of the reactivated GAC into a 1,000 ml graduated cylinder and weighing the carbon. The PAD result is always a lower number than the VFAD test. The measured density should be within 15% of density claimed on Manufacturer's QC report. The following table gives guidance on the suitability of reactivated GAC density by the PAD test.

> 0.60 g/cc	Reject
0.57-0.60 g/cc	Marginal quality
< 0.57 g/cc	Acceptable quality

PAC Powdered Activated Carbon (PAC) is also commercially available. It has a small particle size (mean of about 24 micron). It is typically added to a waste water stream. It flows with the waste water thus achieving a relatively high residence time without requiring a separate unit operation for residence time or contact. However, it does at some point need to be removed. This can be accomplished by filtration or settling. If these steps are combined with some other process, then use of PAC has minimal process requirements. In addition, in this manner it can be used in adjustable concentration, or not at all until needed. This makes it a convenient and flexible treatment material. The ZIMPRO [18] Wet Air Oxidation process uses PAC in the activated sludge system very effectively for treating complex wastewaters.

Factors that affect performance Flow rate is obviously one of the major factors that has an effect on performance. The key to achieving the desired performance is to have sufficient residence time to allow diffusion of the organic molecules into the pores of the GAC. Most GAC used in the oilfield is 8 x 30 mesh, although occasionally 12 x 40 mesh may be used. The smaller 12 x 40 mesh will provide faster adsorption and shorter mass transfer zone. This is not typically an issue in a properly designed GAC adsorption system for hydrocarbon service. It does become an issue when target compounds are difficult to adsorb or very low treatment levels are required, i.e. ppb and ppt levels.

Temperature may or may not have an effect on performance. There are two effects which act in opposite directions. Higher water temperature will decrease the water viscosity and increase the diffusion rate of the organic constituent. If this mechanism is dominant, then performance will improve with higher temperature. However, higher temperature will also decrease the effective adsorption coefficient of the organic molecules on the GAC. If this mechanism is dominant, then higher temperature will reduce the performance.

The effectiveness of activated carbon increases with lower temperature. This is due to the fact that solubility of hydrocarbon generally decreases as temperature goes down. For similar reasons, activated carbon effectiveness for weak acids increases as pH goes down. With the added number of protons available at lower pH, the organic acids become protonated and less polar. A similar situation exists in the removal of alkaline components. At higher pH the alkaline components are less soluble and therefore they have a higher affinity to the carbon.

As concentration of contaminant in the feed water increases, leakage into the effluent can also increase. The upper limit of feed water contamination for continuous GAC treatment is in the range of 10 to 20 mg/L. Some GAC absorbers are designed only for use during excursions caused by upsets or equipment problems upstream. In this case, contaminant levels might be several hundred parts per million for brief periods. The GAC will provide good treatment at these high levels, but carbon exhaustion will be rapid. GAC would rarely be cost effective at even 20 ppm of contaminant in the influent, unless it is a guard bed for excursions. You would never use GAC to treat a process stream of 100-200 ppm on a continuous basis.

The removal of organics is enhanced by the presence of hardness in the water. Thus, GAC units should be placed upstream of ion exchange. Since ion exchange media is more expensive than GAC, this is prudent as well in order to protect the ion exchange from fouling.

Performance evaluation and determination: Since the mechanism of separation for activated carbon is adsorption, it is expected that the effectiveness will be compound-specific. In other words, activated carbon will be more effective for some compounds than others, on the basis of the chemical properties of the contaminants and the solution. This is typical of technologies that depend on adsorption but is significantly different from those that depend on particle trapping or size exclusion. Due to this chemical-related separation mechanism, it is prudent to carry out a preliminary separation test, particularly when specific toxic compounds must be removed. An adsorption isotherm is the most common performance measure for activated carbon. It is relatively easy to generate and there is a wealth of data available in the literature for validation [19, 20].

An adsorption isotherm is generated in a batch test using method ASTM D3860. A specified quantity of waste water is treated in a beaker with varying amounts of finely ground powdered carbon. The quantity of carbon is determined gravimetrically. The carbon and water are mixed well in order to achieve an equilibrium value (i.e., one that does not change with time). The waste water is then sampled, and the concentration of contaminants is determined. The mass of contaminant adsorbed is calculated and plotted on a log-log plot of adsorption capacity (mg contaminant / gr of carbon) versus the concentration of unadsorbed contaminant in the solution. A straight line is typically evident.

For PAC, the Freundlich equation is used to model the results. PAC usually has slightly greater adsorption than GAC.

$$y = x/m = kC_f^{1/n} \qquad \text{Eqn (13.10)}$$

Where:

y	=	adsorptive capacity (mg contaminant / gr of carbon)
x	=	mass of contaminant adsorbed (mg contaminant / L solution)
m	=	mass of activated carbon (gr activated carbon / L solution)
k	=	intercept parameter
C_f	=	concentration of un-adsorbed contaminant in solution (mg/L)
n	=	slope parameter

Table 13.10 Parameter values for the Freundlich equation for calculating the adsorption isotherm for PAC [19, 20]

	k	1/n
aniline	25	0.322
benzene	1.0	1.6
benzene sulfonic acid	7	0.169
benzoic acid	7	0.237
butanol	4.4	0.445
butyric acid	3.1	0.533
chlorobenzene	40	0.406
nitrobenzene	82	0.237
phenol	24	0.271
trinitrotoluene	270	0.111
toluene	30	0.729
o-xylene	14.6	0.36
ethyl benzene	53	0.79

The calculation procedure is a bit non-intuitive. This is because the independent variable in the equation is the concentration of the contaminant left in solution after contact with the carbon. It would be more intuitive to choose the initial dose concentration as the independent variable and to use the equation to calculate the final concentration after contact with carbon. But that is not how the model was developed. Therefore, some explanation regarding how to use the model is worthwhile:

1. pick a value of C_f, the final concentration of contaminant (mg contaminant / L solution)
2. use Equation (13.8) to calculate the absorptive capacity of the carbon y (mg contaminant / gr carbon)
3. pick an amount of carbon m (gr carbon / Liter of solution)
4. calculate x, the amount of contaminant absorbed per liter of solution (mg contaminant / Liter of solution). This is calculated as: $x = my$.
5. calculate C_0, the initial concentration of contaminant present before contact with carbon (mg contaminant / L solution).

The equation for calculating the initial concentration of contaminant is given by the following equation. This equation is equivalent to following the procedure above.

$$C_0 = x + C_f = mkC_f^{1/n} + C_f \qquad \text{Eqn (13.11)}$$

The next step in the calculation is to determine the final contaminant concentration as a function of the feed contaminant concentration. This will require a specification of an amount of carbon. A convenient basis for calculation is 1 liter of contaminated water. A certain amount of carbon is added to the 1 liter of contaminated water. The amount of carbon that is added is m. Once this is specified, then x (the amount of contaminant absorbed) can be calculated as x = ym. Then $C_0 = x + C_f$.

In the table below, the experimental values reported by Dobbs are used again. Roughly the same initial concentration of contaminant (C_0) are used for each of the tests. The amount of carbon (m) used is varied from one test to another. Following the calculation procedure above, the amount absorbed (x) and the percentage removal are calculated.

Table 13.11 Example of calculation using Eqn (13.11) to calculate the percentage of removal of contaminant.

Cf (mg/L)	x/m (mg/gr)	m (gr C/L)	x (mg/L)	Co (mg/L)	% removal
14.2	69.8	0.096	7	21	32
10.3	41.7	0.191	8	18	44
6.2	18.5	0.573	11	17	63
5.8	16.7	0.951	16	22	73
4.0	9.2	1.91	18	22	81

Adsorption isotherms can be used for a preliminary estimate of the adsorption effectiveness of activated carbon for certain constituents. It is useful to have some initial estimate. The adsorption isotherm can also be used for preliminary design. Detailed design should either be flexible to account for potential differences with lab results or actual field trials should be carried out.

Operating considerations: The effectiveness of activated carbon increases as the temperature goes down. This is due to the fact that solubility of hydrocarbon in water generally decreases as temperature goes down. For similar reasons, activated carbon effectiveness for weak acids increases as pH goes down. With the added number of protons available at lower pH, the organic acids become protonated and less polar. A similar situation exists in the removal of alkaline components. At higher pH the alkaline components are less soluble and therefore they have a higher affinity to the carbon.

Performance of GAC in a Single Component System: Rarely in the real world are there one component systems. An example of a real-world mixture might be a condensate / water mixture where the condensate is composed of BETX. In this case, there are four target compounds that need to be removed. When this stream first starts going into a fresh GAC bed, all four compounds are adsorbed and none are found in the treated effluent. However, after the bed has been in service for some time, bands of adsorbates form. The more strongly adsorbed Ethylbenzene begins to replace the less strongly adsosrbed BTX compounds. This results in a segregation of adsorption bands, see Fig 13.12 for a simplified example of how these adsorption bands pass through a GAC column.

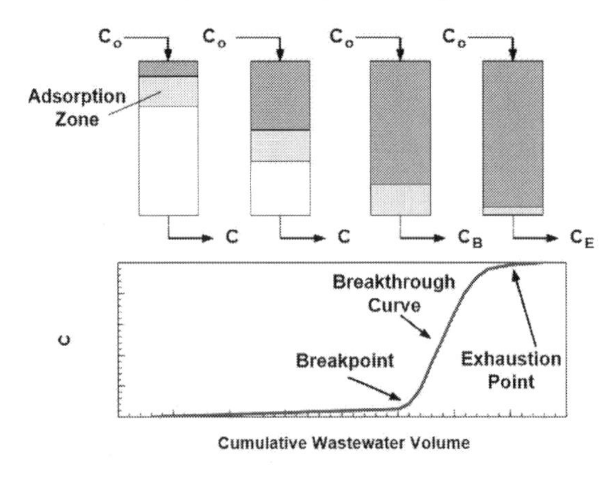

Figure 13.11 Flow diagram of zones in a GAC bed[3]

As the mass transfer zone moves through a carbon bed and reaches its exit boundary, contamination begins to show in the effluent. This condition is classified as "breakthrough" and the amount of material adsorbed is considered the breakthrough capacity. If the bed continues to be exposed to the water stream, the mass transfer zone will pass completely through the bed and the effluent contaminant level will equal the influent. At that point, saturation capacity is reached. The saturated capacity is that which is represented by the isotherm. To take full advantage of the adsorption capacity difference between breakthrough and saturation, several carbon beds are often operated in series. This allows the mass transfer zone to pass completely through the first bed prior to its removal from service. Effluent quality is maintained by the subsequent beds in the series.

EBCT (Empty Bed Contact Time): When sizing an activated carbon system, it is necessary to choose an appropriate contact time for the wastewater and the carbon. EBCT (empty bed contact time) is the terminology used to describe this parameter. EBCT is defined as the total volume of the activated carbon bed divided by the liquid flow rate and is usually expressed in minutes. The appropriate EBCT for a particular application is related to the rate of adsorption for the organic compound to be removed. While this rate will vary for individual applications, experience has shown that for most low concentrations of dissolved hydrocarbon organics, an EBCT contact time of 10 to 15 minutes is normally adequate. This contact time may be provided by multiple carbon beds.

Performance of GAC in Complex Systems: Rarely in the real world are there one component systems. An example of a real world mixture might be a condensate / water mixture where the condensate is composed of BETX. In this case, there are four target that need to be removed. When this stream first starts going into a fresh GAC bed, all four compounds are adsorbed and none are found in the treated effluent. However, after the bed has been in service for some time, bands of adsorbates form. The more strongly adsorbed Ethylbenzene begins to replace the less strongly adsosrbed BTX compounds. This results in a segregation of adsorption bands, see Fig 13.14 for a simplified example of how these adsorption bands pass through a GAC column.

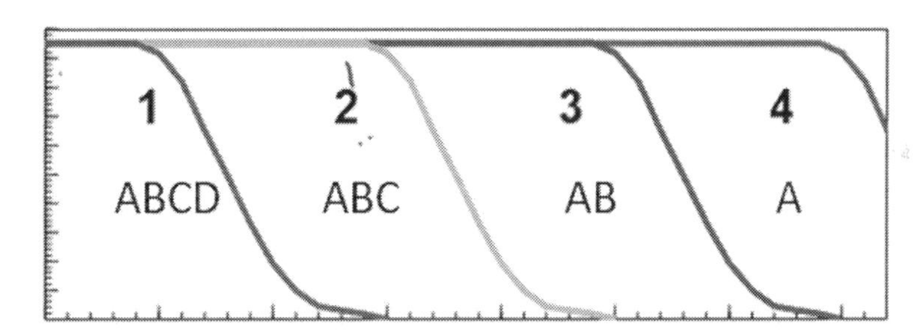

Figure 13.12 Adsorption wave bands. x-axis is time. y-axis is concentration

In this example BETX would be represented, as follows:

A. Benzene

B. Toluene

C. Xylene

D. Ethylbenzene

In zone 1 all four target compounds are being adsorbed. In zone 2 virtually all the ethylbenzene has been adsorbed. By the time the adsorption wave front reaches zone 4, benzene is the only target compound still in the effluent water. All of the other compounds have been removed.

This is similar to what occurs in a high pressure liquid chromatograph (HPLC) column. For example, if a sample of water containing BETX were put into a HPLC, the results of the printout would show four peaks, one for each of the BETX compounds. The least strongly adsorbed benzene would be the first peak followed by peaks for toluene, xylene and finally ethylbenzene. A similar phenomena occurs in a GAC column. A properly designed GAC treatment system will typically use multiple stages. While this adds to the capital cost, it reduces the operating cost. It is desirable to allow the first adsorber on line to become completely exhausted, while the following adsorbers continue to polish the process stream. The GAC in the first adsorber is then replaced with fresh GAC. After proper conditioning, the fresh adsorber is put on stream last in line.

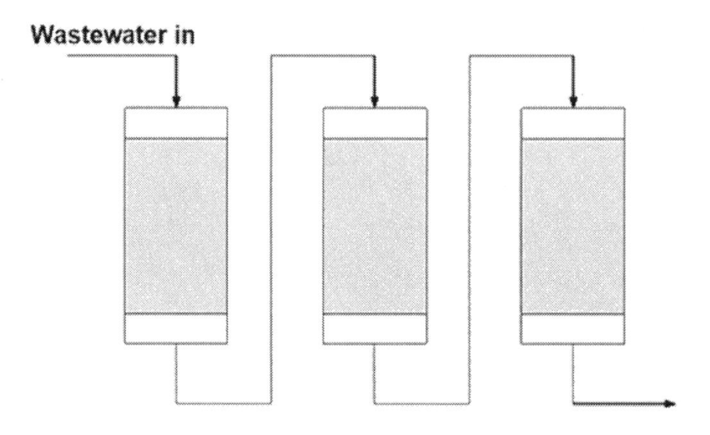

Figure 13.13 Multistage GAC system.

Design of a GAC Vessel: The following design considerations pertain to produced water applications [21]. Typically a GAC vessel is designed with a 2H:1D ratio. The actual GAC bed fills about 65% of the volume between the top and bottom distributors. This leaves adequate room for up to 50% bed expansion during backwash. Most liquid phase GAC adsorbers are operated in down flow mode. There are times when up-flow operation is more desirable, as discussed below.

The design of the distributors is important, so that the operator gains maximum utilization of the GAC bed. The best designs have a lower distributor "nested" in the bottom dished head of the pressure vessel, see the figure below. This approach avoids wasting adsorption capacity with too much GAC bed below the distributor. It also eliminates the need for a support bed or gravel in the bottom of the column. The top distributor can be a simple design of a screened hub. It is not critical to have a distributor design on top covering the full area of the adsorber, because the open volume above the GAC bed acts, as a plenum. Typically the backwash rates will be higher than the process flow rates, so piping and nozzles should be designed to accommodate the backwashing flow rates.

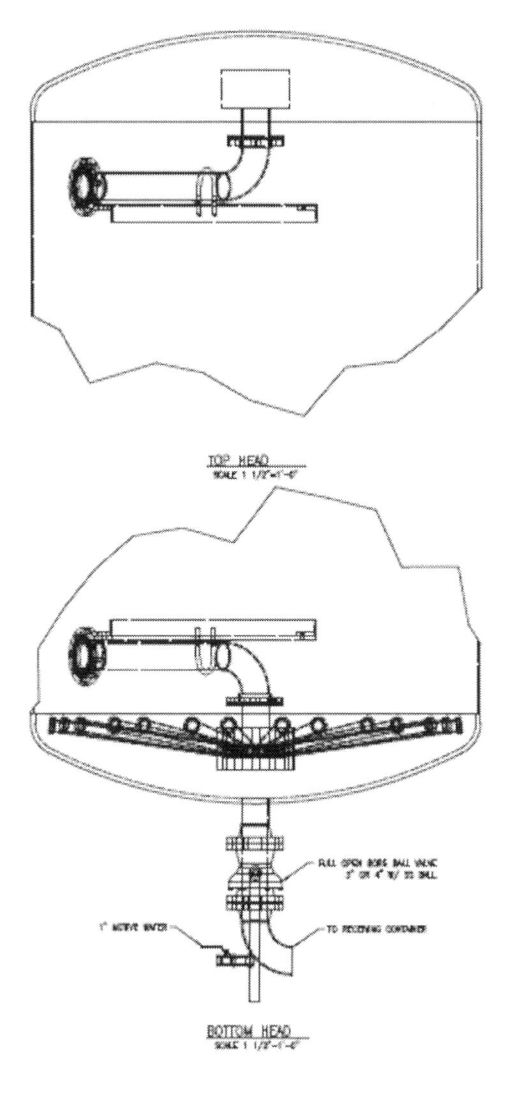

Figure 13.14 GAC Vessel Design Features4

It is a significant operational advantage to have adsorbers designed so they can be hydraulically unloaded. This means installing a GAC slurry discharge line in the middle of the bottom dished head, see Fig 13.14. For adsorbers with over 2,000 lb GAC bed capacity, this line should be 3 inch (minimum) or 4 inch (preferred) discharge. The bottom GAC ball valve must be a full open throat design. A 1 inch line for motive water should come into the back of the discharge elbow. This motive water will keep the GAC moving through the discharge line during GAC transfer.

In cases where there is a high level of suspended solids, i.e. >20 ppm, it might be advantageous to operate the adsorber in up-flow instead of down flow. When operating in down flow configuration, suspended solids will be captured in the GAC bed. This will lead to high pressure drop and channeling through the GAC bed, requiring frequent backwashes and managing backwash water. If solids can be tolerated in the effluent, i.e. the water goes to a settling pond before outfall, then up-flow configuration has big advantages.

In up-flow these solids are temporarily captured in the bottom of the GAC bed. As pressure drop starts to build, the bed will "burp" and allow these solids to pass through. This burping action relieves the pressure drop without requiring backwashing. It also prevents flow channeling. This is an especially effective way of treating a water stream that contains bio-solids. The bio-solids will not hurt the carbon. They will pass through during the burp slugs.

Operating Considerations Most people are familiar with sand filters for filtering solids. These filters work along the cross sectional area of the top of the sand bed. Activated carbon is an adsorbent. It works through contact time with the fluid, so the whole bed is employed in removing target organic compounds. This means it is the volume of the bed and its contact time at a given flow rate that determines performance. Flow capacity guidance is provided in the Table below, based on this minimum EBCT (Empty Bed Contact Time – see above).

The flow rate through a GAC bed is determined by the treatment target. If the treatment goal is to remove a few parts per million of oil from water, then this application can generally be met with about 7-8 minutes of EBCT. When polishing target organics to very low levels (well below 1 ppm), the normal contact times below might need to be doubled. In the case where the GAC bed is simply a "guard bed" to catch occasional spikes in concentration, then the flow rates can be increased substantially. Of course, at higher flow rates, more organic material is carried into the bed, which consumes available adsorption capacity more quickly. Also at higher rates, the percent of target organic removed will be reduced

This table is a general guide for routine removal of oil based compounds typically encountered in upstream operations.

1. Minimum bed depth: one vessel diameter above the lower distributor
2. Normal EBCT: 7.5 minutes
3. Guard bed: EBCT: 4 minutes

Preparing a GAC Bed: Carbon fines are generated during transport and handling of activated carbon. These fines must be washed out before the carbon bed is put into service. Prior to filling a liquid phase adsorber with fresh bulk GAC, the vessel should be filled half full with water. The GAC is then poured in from the top of the vessel. The water cushions the GAC as it is poured into the vessel preventing generation of fines. Typically, the GAC should not be filled to over 65% of the height between the lower and upper distributors. This allows room for bed expansion during back wash of the GAC bed.

If time is available, it is desirable to let the GAC soak for 24 hours to allow the air in the pores to be displaced by water. After the GAC is loaded, the vessel should be filled the rest of the way. The carbon fines in the bed will be washed out by gradually increasing the back flush water rate (fed through the bottom distributor). This step accomplishes two important goals. It removes fines from the GAC bed, which could inadvertently wash through the bed during operation and create false readings when samples are taken. The most important reason is to flush out the air trapped in the bed. If this is not done, it is possible to have part of the GAC bed "air locked" leading to channeling through the bed and premature bed exhaustion.

Loading, Unloading and Conditioning GAC: Carbon fines are generated during transport, handling and loading of activated carbon. These fines must be washed out by properly conditioning the bed before the carbon bed is put into service. The generation of carbon fines during loading can be kept to a minimum. GAC can be loaded as a slurry (hydraulic loading) or dry loaded. Usually hydraulic loading is done for larger quantities of 5-20 tons, but smaller volumes can also be transferred this way. The advantage of wet loading is it is generally faster, less labor intensive, and much less dusty. Here are considerations for dry loading and hydraulic unloading.

Simplified Procedure for Dry Loading GAC:

1. Fill the vessel half full with water

2. GAC is typically supplied in 1,000 lbs. bulk bags, but might also come in 25 kg paper or plastic bags loaded onto pallets. Assuming GAC is delivered in bulk bag packaging, use suitable lifting equipment to lift GAC to a position immediately over the top hatch

3. Discharge the GAC into the vessel

4. Set process valves into back flush configuration. Start slowly adding back flush water until the level approaches the top of the vessel. Notice large volumes of gas bubbling to the top. This is normal; it is the air from the pores of the GAC, which is being displaced by water

5. Let the GAC soak for at least 24 hours, if the operation has the time.

6. After soaking the GAC restart the back flush water, slowly at first. You might hear noise coming from the vessel as air trapped in the bed surges out in two phase flow with the back flush water

7. Gradually increase the flow of back wash water to the predetermined back wash rate. This ramp up of back flush flow rate should occur over 10-15 minutes. If the back flush rate is ramped up too fast, then the GAC bed could be lifted, potentially damaging the upper distributor

8. When the back wash water starts running mostly clear (few entrained fines) then the back wash can be stopped

9. Slowly reduce the back flush flow rate over two minutes until flow is fully shut off. By reducing the back flush rate slowly, it will allow the bed to segregate by particle size. This means more of the larger particles are on the bottom and more of the smaller particles are at the top. This helps to maintain the mass transfer zone in the bed, if the bed must be backwashed again during operations.

10. The new GAC bed is now ready for use. Reposition process valves for normal downflow operation. Once a GAC bed has been conditioned, it is best to leave the vessel filled with water; wet GAC in the presence of air is quite corrosive

Hydraulic Unloading of GAC, Simplified Procedure Typically GAC can be hydraulically unloaded from a vessel more quickly and much more safely than vacuuming or shoveling the spent media out. Hydraulic transfer is also a much cleaner media transfer option. The following procedure can be used.

1. Position a dumpster or other receiving vessel in the immediate vicinity of the adsorber

2. A screened drainage pipe should be placed in the bottom of the receiving vessel to allow water to be drained from the dumpster without losing GAC

3. Attach a transfer hose to the line coming out of the bottom of the dished head on the adsorber. The discharge end of the transfer hose should them go into the receiving vessel/dumpster

4. Be sure the adsorber is full of water. Put at least 2 bar of air or water pressure on the adsorber

5. Start the motive water flow, then open the bottom discharge valve; the GAC slurry will begin to flow to the receiving vessel.

6. Maintain pressure on the adsorber during the transfer operation; usually 2 bar is adequate unless the slurry is being pushed a long distance or to a point at a higher elevation than the discharge line.

7. It is important to keep the GAC slurry flowing. If the slurry flow stops, then it could become difficult to restart the flow

8. Continue until the vessel is empty; the GAC "slug" will usually end rather abruptly as the remaining water in the vessel rushes out

WARNING: Activated carbon (especially when wet) can deplete oxygen from the air and dangerously low levels of oxygen may result. When workers enter a vessel containing activated carbon, procedures for potentially low oxygen areas should be followed.

Conditioning of the Bed The most frequent cause of underperforming GAC is poor conditioning of the beds before putting them into service. Conditioning a carbon bed is accomplished through pre-soaking then backwashing the bed. It is more important to condition GAC beds than any other filtration media. Virtually all solid media filters will trap some air in the bed, but most of this air is quickly released during operation. However, in the case of a GAC bed, the trapped air can seriously degrade filter performance. Activated carbon has extensive pore volume holding quite a lot of air. About 40% of the volume in a GAC bed is the inter particle volume. Another 40% is the pore volume in the granular carbon. Only about 20% of the bed volume is the carbon skeleton. Even more gases are adsorbed onto the surface of the carbon. The presoaking and backwashing of the bed removes most of this air. About 90% of the air is removed after 24 hours of soaking. If this is not done effectively, then the air will often be trapped in the GAC bed resulting in flow channeling and premature bed exhaustion.

If there are suspended solids in the influent stream, then the GAC bed will eventually exhibit high pressure drop. A GAC bed should not be operating at greater than 15 psi pressure drop across the bed. When high pressure drop develops, the GAC bed needs to be backwashed. This backwash does not regenerate the media, nor desorb the contaminants. However, backwash is still necessary in order to reclassify the media, eliminate any channels that may have developed, and to remove the suspended matter from the bed.

Application in oily water and Fouling Activated carbon is also effective in separating dispersed hydrocarbons from water. This is reasonably expected from consideration of the mechanism of separation. However, the cost and capacity of activated carbon for removing dispersed hydrocarbons generally limits its use for this application. Like most filtration systems however, there can be an application in a guard bed situation.

If activated carbon is chosen as a guard bed material, it will indeed provide a highly effective barrier to passage of organics and dispersed hydrocarbons. In this application, it is necessary to design the upstream water treatment system to minimize the hydrocarbon content to the activated carbon in order to minimize the cost of carbon consumption.

Disposal: Once the carbon becomes saturated it must be either regenerated or disposed. While reactivated GAC is often used in the oilfield, it is not typically practiced for spent GAC in the oil and gas industry. This is due in part to the availability and relatively low cost of virgin and reactivated GAC grades. Typical disposal of spent activated is to a landfill or to incineration. Most oil and gas companies that use activated carbon do so through a use and waste disposal / management agreement with a services provider who assumes responsibility for the permitting and disposal of the spent material.

13.7 Cross-Flow Filtration

As the name implies, cross-flow filtration involves flow of the waste stream across a filtration material with effluent passing through the filter membrane and the contaminants remaining in the reject stream. The objective of this design is to promote passage of clean water through the membrane while passing the contaminants from the inlet to the outlet without allowing them to build up on the membrane. One of the important variables is the pressure required to push water through the membrane. This is referred to as the trans-membrane pressure (TMP).

Most cross-flow membranes used for produced water treatment were originally developed for use in industrial and municipal applications. In those applications, one of the important contaminants is referred to as NOM (Naturally occurring organic matter). NOM are typically negatively-charged small diameter suspended solids. The composition includes tannins, lignin's, water soluble humic acid compounds or fulvic acid compounds resulting from the decay of vegetative matter.

Each of these technologies is characterized by a size cutoff. There are many materials and configurations that are employed to provide this. Choice of module design is important. Considerations include: concentration polarization, cleaning, surface area / volume ratio, and isolation and repair of damages or leaking elements. At sufficiently high pressures, a contaminant layer forms on the membrane. This layer is caused by concentration polarization. It typically reduces the flux across the membrane, sometimes significantly. Removing this contaminant layer and maintaining flux is an important consideration in membrane design. The thickness of this layer can be reduced by increasing the Reynolds number of the feed water through the membranes. Module design includes: spiral wound (thin film), hollow fiber, tubes, plate and frame.

Spiral Wound: Spiral wound membranes are composed of layers of material (support, channel material, membrane material) that are rolled into a cylindrical cartridge. The objective is to maximize the surface area of membrane material without creating a restrictive path that is prone to fouling.

Hollow Fiber: The membrane fibers are typically oriented in the direction of the major axis of the vessel. Pressure is applied to the fiber bundles from the outside and the permeate flows through the fiber to the inside and along the fiber to the effluent discharge. Since the feed water flow is on the outside of the tube bundle, it is easier to achieve a high Reynolds Number on the feed side of the membrane, at least for the fibers near the outside of the bundle. A fiber bundle will have a high surface area to volume ratio. Cleaning is usually accomplished by an air scour on the outside of the fiber bundle.

Tubular Membrane: Feed water flows into one end of the tube. Permeate flows from the inside to the outside of the tube. The tubes can be installed in parallel or in series. This configuration is relatively tolerant to suspended solids without fouling. A disadvantage is that the surface area is small compared to other configurations for a given volume or space.

Plate and Frame Membrane: Membranes are configured in a stacked array with spacers and membrane supports in between the membrane elements. An advantage is that the surface area is large compared to other configurations for a given volume or space.

Capillary: Capillary modules have a large number of membrane capillaries with small diameters (0.5 to 1 mm). The feed is passed through the center of each capillary and the filtrate permeates the walls of the capillary tubes. In this configuration it is difficult to achieve a high Reynolds Number in order to prevent concentration polarization and buildup of contaminant on the feed water side. An advantage is that the surface area is large compared to other configurations for a given volume or space.

13.7.1 Microfiltration & Ultrafiltration

Micro-filtration and ultra-filtration are forms of size exclusion filtration using semi-permeable membranes [22]. These membranes can be configured in either dead-end flow or cross-flow. These configurations were discussed in Section 13.3 (Tertiary Treatment Mechanisms and Equipment Types). They can be made of ceramics, etched polymers, foils and natural and synthetic compounds. The common materials used for ceramic membranes are zirconia, titania, and alumina. Common materials for non-ceramic membranes include cellulose acetate, polypropylene, polyamide, polysulfone, polyvinylidene fluoride or polytetrafluoroethylene. Each differ in its properties and resistance to heat, corrosion, bacterial attack, fouling resistance, resistance to aggressive chemical cleaning solutions, and abrasion. Manufacturers of these membranes will provide information on the properties and resistances of membrane materials. As discussed by Drewes [28], flux can range from 20 to 100 gpd/ft^2. Backwash is used periodically to clean the membranes in short duration (3 to 180 seconds), and relatively frequent intervals (5 minutes to several hours between backwash). Clean-in-Place systems can be used which provide periodic chemical cleaning. However, disposal of the spent cleaning solution can be a significant drawback. Membranes are a type of straining filtration analogous to screens. The separation performance does not depend on the build up a cake. Once the pores are plugged, head loss across the membrane climbs steeply.

Table 13.12 Membrane Separation Properties and Performance

Type of Membrane	Separation Mechanism	Pore Size (microns)	Molecular weight (amu or Da)	Operating Pressures (psi)
Reverse Osmosis	Screening and diffusion	<0.001	100-200	600-1,500
Nanofiltration	Screening and diffusion	0.001-0.01	300-1000	50-250
Ultrafiltration	Screening	0.01-0.1	1,000-100,000	3-80
Microfiltration	Screening	0.1-20	Over 100,000	1-30

One Dalton is one-twelfth the weight of a carbon atom, as defined by convention in 1960 and is approximately equal to $1.66053873 \times 10^{-24}$ g

The key to ensure long membrane life is crossflow cleaning coupled with chemical cleaning or frequent back flushing, depending on membrane types. If the contaminated liquid flows across the membrane and not normal to it, the membrane filtration run is improved because the *crossflow* removes the solid buildup which would otherwise plug the pores. With certain types of microfiltration membranes, the tubular hollow membrane is directly submerged in the aeration portion of the plant where it is in contact with bacteria, protozoans and viri in the wastewater as well as suspended materials and colloidal solids. In some tubular membrane applications, manufacturers have instituted frequent back flushing and an air-pump, which both shakes and scours the membrane and also promotes knocking off of the fouling layers. This pulsing and back flushing helps keep the pores open. In certain wastewater processes, the cleaning compounds may include hydrochloric acid, sodium hypochlorite, oxalic acid, ozone, hydrogen peroxide and citric acid. The primary use of the acid is to clean the carbonate build-up of the membrane pores. Cleaning cycles of back flushing and back pulsing can take up to a few minutes to an hour a day and may be required to be done every few hours per day, to once a day, depending on the severity of the clogging in the pores. The chemical cleaning of the membrane is performed weekly or monthly, depending on the manufacturer's instructions.

The table below gives a comparison of the various units of permeability or flux that are used. These units are used for deep bed filtration, dead-end filtration and cross-flow filtration. In deep bed filtration the most common units of flux are gallons per minute per square foot, and meters cubed per meter square per hour. As discussed previously, walnut shell filter systems can typically achieve 10 to 20 gpm/ft^2, which is equivalent to roughly 24 to 29 m^3/m^2 hour. In GFd units this is equivalent to 14,400 gal/sq ft day. In LMH units this is equivalent to 24,500 LMH. As will be discussed shortly, most cross-flow MF and UF systems can achieve a flux of about 100 to 1,000 LMH. As shown in the table, this is equivalent to about 0.4 gpm/sq ft. Thus, it is immediately apparent that cross-flow filtration requires greater size and weight than deep bed filtration for a given flux.

Table 13.13 Flux units and values

gpm/sq ft	GFD - gal/day sq ft	m^3/m^2 sec	m^3/m^2 hour	LMH - liter/m^2 hour
10	14,400	0.0068	24.4	24,446
0.41	590	2.7 x 10^{-4}	1.0	1000

Membrane microfilters are usually designed to be 30-45 L/H/M^2 or 16-25 gallons/day/ft^2 for straight microfilteration on immersed activated sludge systems, and usually between 50-60 L/H/M^2 for relatively clean wastewater after treatment, or pretreatment by a clarifier.

Seawater Filtration: In roughly the early 2000's, MF and UF started to become popular options in filtering seawater upstream of NF, and RO membranes [24, 25, 26]. Prior to this, there were only two filtration options available for water flooding, neither of which was satisfactory. One option was to use coarse strainers followed by multimedia filtration, which was followed by cartridge filtration. Multimedia filtration provides most of the removal of fine particulates and can be backwashed. Therefore, it is not consumed and no waste is created, because the backwash water containing the filtered solids is discharged directly overboard. The cartridge filters act as a guard bed, preventing the passage of solids during upsets and provide a final barrier to prevent fouling of the NF or RO membranes. MF and UF provides a cost effective practical alternative which has greatly encouraged the development of these technologies. However, it must be kept in mind that seawater filtration, where oil and grease are essentially not present, is significantly different than produced water filtration where oily solids are the main challenge.

Recirculation: MF and UF are dead-end filtration technologies in which the feed water passes through the membrane, and all particles larger than the pore sizes of the membrane are stopped at its surface. Periodic backwashing removes the filtered particles from the membrane surface. An advantage of MF and UF systems is that they withstand hypochlorite, making possible periodic deep cleaning of the membranes to ensure the prevention of biological fouling. Historically, MF and UF were considered as two distinct technologies. In some industries, this is still the case. However, nowadays, for seawater application, distinguishing between MF and UF has little meaning since there is overlap in the micron size. The typical MF and UF systems used for seawater filtration upstream of an NF or RO system use membranes with an average pore size between 0.02 and 0.06 micron.

Micro-filters are a positive barrier to contaminants. Bed and media filtration relies on particle trapping and on surface attractive forces, and chemical addition to achieve particulate removal. This is not the case the MF /UF or any of the straining or cross-flow filtration methods. MF / UF membranes do typically build up a filter cake of particles at their surface. This filter cake does enhance the separation efficiency and reduce the particle size of contaminants in the effluent, but this filter cake is not necessary to the filter mechanism. Initial performance after first installation, as well as initial performance after a backwash does not have the benefit of the filter cake and is dependent only on the surface properties of the material and its pore size. These are generally the parameters used for design, without taking into account the added separation benefit of the filter cake that develops. This makes MF / UF less sensitive to the surface, charge, and packing characteristics of the contaminants. The only parameter of importance is the size and shape; where:

- **Spikes** – Contaminants normally at low levels increase by one or more orders of magnitude for short periods (typically one-five days). They include events such as algae blooms and river first flush after rainfall and have short-term effects on backwashing and cleaning.

- **Constant low levels** – Contaminants continually present at low levels. They include reactive silica and manganese and have long-term effects on membrane resistance.

- **Seasonal** – Feed characteristics change markedly between seasons. These include temperature and algae levels and can have marked impact on operating conditions. High algae levels that persist through a season are often best addressed by some form of 'roughing' pre-treatment.

Nomenclature	Diameter (mm)
Tubular	>10.0
Capillary	0.5 – 10.0
Hollow Fiber	< 0.5

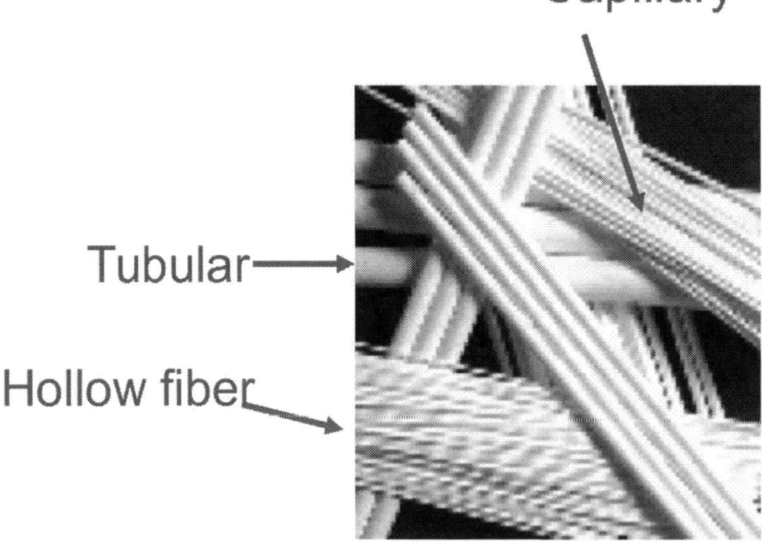

Figure 13.15 Capillary, Tubular and Hallow Fibers

<u>Fouling</u> Organic acids, resins, asphaltenes and polymers have a tendency to foul MF / UF membranes. Addition of organic polymer should be avoided upstream of an MF / UF system, if possible. If this is not possible, then a chemical backwash cleaning solution should be selected in order to preserve membrane performance in the presence of the polymer contaminant. Chlorination has proven to be effective in improving MF / UF performance and in extending the time interval between cleaning.

In most cases, a small concentration of coagulant (1 to 2 mg/L) is sufficient to improve the separation performance. It is not typically assumed in the selection or design of the MF / UF filter. However, it does aid in the buildup of a filter cake which in turn improves the separation of small particles. If a coagulant is intended to be used, then selection of the particular coagulant is an important step. Different coagulants have dramatically different properties depending of the chemical nature of the contaminants, and on the particular surface type of MF / UF membrane.

It is critical to recognize that MF / UF systems generate three types of waste stream:

1. Backwash
2. Spent cleaning chemicals
3. Rinse water (final stage of cleaning)

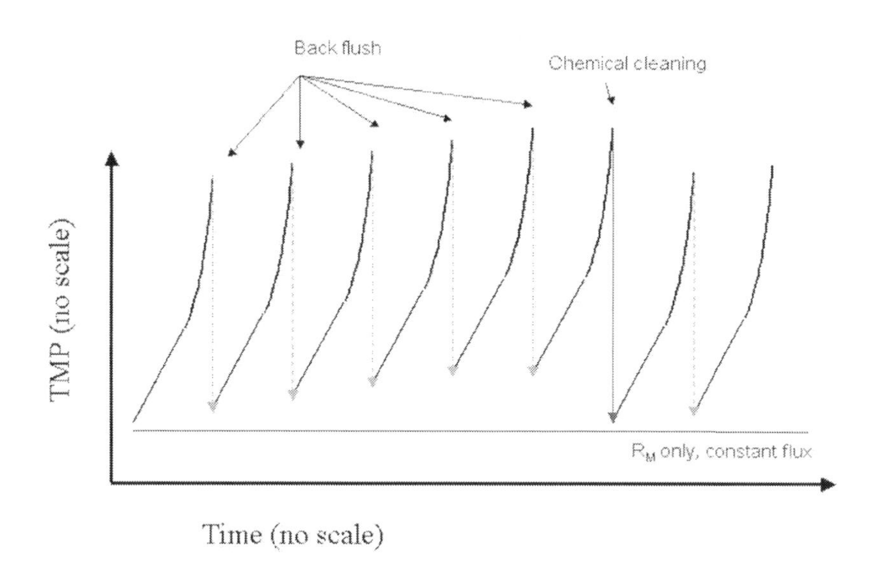

Figure 13.16 Simple fouling behavior in constant flux filtration for MF/UF operations

Performance for Oily Water Separation: There are several performance tests that have been carried out and reported. One example is given by the following [27]. Using a crude oil sample, having API gravity of 20, and a viscosity of 77 cST, from a producing well on the Marlim P-20, Campos Basin, an oily produced water mixture was prepared. No solids were added. An Ultra-Turrax homogenizer was used to generate small droplets on the order of a few microns diameter. Hollow fiber membranes were tested having an external diameter of about 1 mm, and an average (final outer layer) pore size of 0.3 to 0.5 micron). The MF membranes provided good separation efficiency, at a sustainable flux of 20 L/hm2 (LMH) for a short time period. Presumably, with a normal backwash cycle, such performance could be sustained for much longer. High flow, high pressure leads to (irreversible) fouling. Laboratory testing of permeate flux as a function of feed pressure and feed flow rate (Q).

Feed: 250 mg/L OiW
no solids

Effluent: 8 mg/L OiW
separation efficiency: 97 %

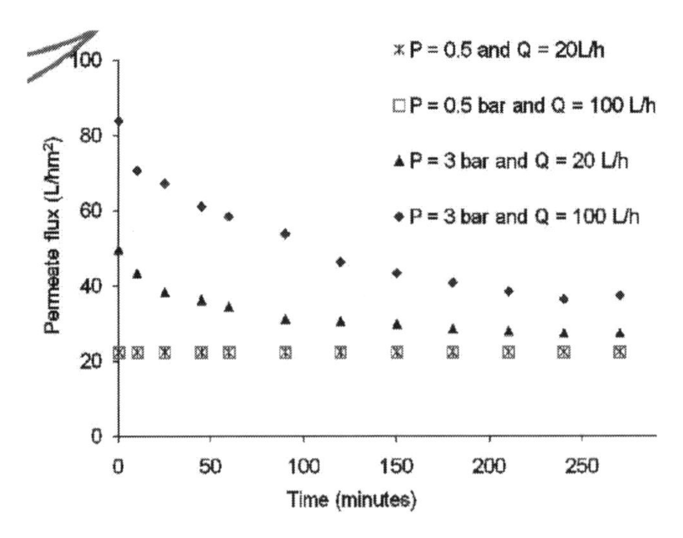

Figure 13.17 Performance test of MF membrane in removing oily water, without solids [27]. The permeate flux is given in units of L/m2 hr (LMH). Two different transmembrane pressures, and two different cross-flow flow rates were used.

The following table gives the estimated foot print for an MF skid operating at the capacity cited above. Note that in the "large" capacity skid, the modules are stacked horizontally.

Table 13.14 Foot print for an MF Skid

	Small Capacity	Large Capacity
Module surface area (m^2)	5	25
Modules per skid	10	50
Footprint (ft^2)	4W x 6L	8W x 12L
Flux (L/hm^2)	20	20
Capacity (BWPD)	150	3800
Capacity (m^3/hr)	1	25

<u>**Ceramic Membranes:**</u> As mentioned, ceramic membranes can be made of various metallic oxides including zirconia, titania, and alumina. They can be configured in either a dead-end or cross-flow configuration. The main advantage of ceramic materials is that they can withstand harsh cleaning treatments (caustic, acid, bleach) which help prevent long-term fouling in produced water applications. However, the spent cleaning fluids must be handled / disposed of in some economical way. Pedenaud and co-workers have carried out extensive laboratory and field pilot studies of ceramic membranes [25, 26], including evaluations of cleaning chemicals and field evaluation of Clean-in-Place (CIP) systems.

13.8 Ion Exchange

Ion exchange is a media filtration process. Produced water flows through an ion exchange bed (resin bed) for the removal of specific ions from the water [29, 30]. In a cation exchanger, multivalent metal cations are removed from the water by adsorption onto the resin material. When this occurs, the resin material desorbs simple group 1 cations (e.g. sodium) or protons. In the adsorption / desorption process electric neutrality is maintained. The net effect is that a multivalent target ion is removed from the water and replaced by replacement ions. In an anion exchanger, anionic groups (e.g. sulfate) are removed and replaced by simple halide ions (e.g. chloride). The resin material is typically composed of either naturally occurring inorganic zeolites or synthetically produced cross-linked polymeric organic material. When the replacement ions on the resin are exhausted, the resin is recharged with more replacement ions. In the process, the target removal ions are desorbed. The recharge process effectively reverses the adsorption sequence.

Ion exchange is used to remove arsenic, heavy metals, nitrates, radium, uranium and other elements from the produced water. It has several applications in water treatment processes such as low salinity desalination, and removal of radioactive waste, hardness, alkalinity, heavy metals (e.g. Hg^{+1}), ammonia, and boron, etc. Since divalent ions (Ca^{+2}, Mg^{+2} etc) are favored over monovalent (Na^{+1}, etc.) ions, secondary treatment for SAR (sodicity) reduction may be required, depending on the application. The nomenclature of the resin indicates which ion type is removed. A cation exchange resin removes cations. An anion exchange resin removes anions. The resins are further classified as follows.

Sodium Absorption Ratio (SAR): SAR is a somewhat specialized subject but one that is extremely important in produced water reuse in agricultural applications [30]. SAR is a function of the ratio of sodium to the sum of calcium and magnesium cations as shown in the following:

$$SAR = \frac{[Na+]}{\sqrt{\frac{[Ca2+] + [Mg2+]}{2}}} \qquad \text{Eqn (13.12)}$$

where the concentration are in meq/L. SAR values of greater than 12 are considered very high. The specific SAR value at which soil damage begins depends on the nature of the soil itself. The optimum conductivity and SAR must be determined on a site-by-site basis. Acceptable SAR values depend on the end use for the produced water; criteria for SAR values are also controlled to a high degree by each state. In many cases, SAR numbers of less than six are beneficial for the use in treatment systems. From the equation it can be seen that reducing the SAR from high values to acceptable levels can be accompanied through processes that either decrease sodium or increase magnesium or calcium.

Strong Acid Cation (SAC) Resins: Strong Acid Cation (SAC) Resins where the hydrogen or sodium forms of the cation resins are highly dissociated and H+ or Na+ ions are readily exchangeable over the entire pH range.

An example of salt removal with SAC:

$$2(R - SO_3H) + NiCl_2 \rightarrow (R - SO_4)Ni + 2HCl \qquad \text{Eqn (13.13)}$$

and, an example of Ca^+ softening with SAC

$$2(R - SO_3H) + Ca^+ \rightarrow (R - SO_3)^2\,Ca + 2H^+ \qquad \text{Eqn (13.14)}$$

In these examples, R refers to the bulk of the resin material. This resin uses sulfonic acid ($-SO_3H$) as its active group. These resins would be used in the hydrogen form for complete deionization (Ba, Na, Mg, Ca, etc. removal); they are used in the sodium form for water softening (Mg and Ca removal). After exhaustion, the resin is regenerated to the hydrogen form by contact with a strong acid solution, or to the sodium form with a sodium chloride solution.

Weak Acid Cation (WAC) Resins: Weak acid resin has carboxylic acid (COOH) group as opposed to the sulfonic acid group (SO_3H) used in strong acid resins. These resins behave similarly to weak organic acids that are weakly dissociated. WAC has high affinity for divalent salts. An example of Ca^+ softening with WAC where alkalinity present in bicarbonate form can also be removed by WAC.

$$2(R - COOH) + Ca^{+} \rightarrow (R - COO)_2\ Ca + 2H^{+} \qquad \text{Eqn (13.15)}$$

Free H+ ions can react with bicarbonate (present as hardness, Ca(HCO3)2) to form carbonic acid. The carbonic acid decomposes in carbon dioxide as shown below. Removal of carbon dioxide or decarbonation is necessary during the water treatment process.

$$HCO_3 + H^{+} \rightarrow H_2CO_3 \rightarrow CO_2 + H_2O \qquad \text{Eqn (13.16)}$$

Weak acid resins exhibit a much higher affinity for hydrogen ions compare to strong acid resins. This characteristic allows regeneration to the hydrogen form with significantly less acid than is required for strong acid resins. Almost complete regeneration can be accomplished with stoichiometric amounts of acid, and the degree of dissociation of a WAC is strongly influenced by the solution pH. Consequently, resin capacity depends in part on solution pH.

Strong Base Anion Resins: Strong base resins are highly ionized and can be used over the entire pH range. These resins are used in the hydroxide (OH-) form for water deionization. They react with anions in the solution and can convert an acid solution into nearly pure water. The equation below shows the reaction involved in an anion exchange step.

$$(R - NH_3OH) + HCl \rightarrow (R - NH_3Cl) + HOH \qquad \text{Eqn (13.17)}$$

Regeneration with concentrated sodium hydroxide (NaOH) converts the exhausted resin to the hydroxide form.

Weak Base Anion Resins: Weak base resins exhibit minimum exchange capacity above a pH of 7. The weak base anion resins sorbs anions associated with weak acid.

Powder River Gas, LLC: Powder River Gas, LLC proposed a project plan of development to drill and test for coal bed natural gas (CBNG) in eight federal and eight fee wells at eight locations in an area northeast of the Tongue River Reservoir, Big Horn County of southeastern Montana. Part of their 'No federal action' was to treat water produced from the wells using a Higgins Loop (which is a continuous counter-current ion exchange) treatment facility prior to discharging to the Tongue River. The proposed treatment facility will use 0.92 acres of private surface. The stationary Higgins Loop facility will be constructed along with a 0.5 acre-feet capacity impoundment and chemical storage tanks. All chemical storage tanks will be surrounded by a shallow spill containment berm to prevent any accidental chemical spills.

Produced water from CBNG wells are to be treated stepwise within the treatment facility. Settling of suspended sediments and releasing of residual gas will be within the impoundment. Barium, Na+, and other heavy metals from produced water will be removed using SAC resins in the Higgins Loop. Removal of CO_2 produced during the ion exchange process and adjustment of pH will be achieved by adding calcium hydroxide. CO_2 can be removed by air-stripping or membrane degasification. The physical law governing this process is the equilibrium between the concentration of the solute gas in the liquid phase and the gas phase.

The Higgins Loop is a vertical cylindrical loop containing a packed bed of strong acid ion exchange resin that is separated into four operating zones by butterfly (loop) valves. These operating zones (Absorption, Regeneration, Pulsing and Backwashing) function like four separate vessels.

Media: Typical media is in the form of beads. The size of the beads, the manufacturing process used, and the composition are typically proprietary details that distinguish one brand form or another, and which have a major impact on performance. The most common beads are 20 to 50 mesh (300 to 800 micron diameter).

Capacity: The ion exchange capacity is expressed in terms of gr-equivalents per liter of beads. This requires a bit of explanation but, as will be shown, it is a very convenient terminology. A chemist might express the ionic strength of a solution in terms of the normality. Normality is the number of moles of the ion of interest per liter of solution. For example, a 1.0 N solution of sodium chloride contains 1 mole of sodium ions and 1 mole of chloride ions. A typical cation exchange media has a normality of 2.0. This means that 1 liter of this media has 2 moles of exchangeable anions. A typical strong base media will have a normality of 1.3. This means that a liter of this material is capable of exchanging with 1.3 moles of base (anion).

Ion exchange capacity is also expressed in a number of different units. Almost all other units make reference to the number moles of ion exchange capacity per volume of material. An example is milli-equivalents per milliliter (meq/mL). This unit is identical to the Normality.

The convenience of this set of units is demonstrated as follows. The following equation allows calculation of the volume of media required to clean a certain volume of waste water with a certain concentration of contaminant:

$$V_m = \frac{N_c V_c}{N_m} \qquad \text{Eqn (13.18)}$$

Where

N_c	=	normality of contaminant in the waste water stream (mole/L)
V_c	=	volume of waste water stream (L)
N_m	=	normality of the ion exchange media (mole/L)
V_m	=	volume of ion exchange media required (L)

In the equation above, the volume of waste could be replaced by the volumetric flow rate (L/hour). If this is the case, then the volume of media required per hour would be calculated.

Cation Media Composition: Most ion exchange media are composed of copolymers of styrene and divinylbenzene. These co-polymers are often referred to as resin. Thus, media is also often referred to as ion exchange resin.

To produce a cation exchange resin, the resin material is reacted with sulfuric acid. During this reaction, sulfonic groups (SO_3H) become attached to the resin molecule. This is usually depicted as $R–SO_3H$, where the R refers to the rest of the resin molecule. The location of each of these sulfonic groups is referred to as a cation exchange site.

When waste water contacts this cation site, the reaction that occurs can be depicted as:

$$Na^{+1} + R - SO_3H ==> R - SO_3Na + H^{+1} \qquad \text{Eqn (13.19)}$$

As discussed, the resin structure is depicted as *R*. Note that the designation "cation" refers to the ion in the waste water that will be exchanged.

Selectivity of Cation Exchange Resins: Typical waste water contains a number of different ions, only some of which are required to be removed. Cation exchange resins are designed to remove ions with a certain preference. That preference is shown for one particular resin as follows:

$Ca^{+2} > Mg^{+2} > Fe^{+2} > Na^{+1} > NH_4^{+1}$

The selectivity shown above indicates that calcium will be exchanged preferentially over all other cations. Once the calcium is depleted, then magnesium will be exchanged, as so on. The selectivity indicated above is for a particular resin at a particular concentration. It is typical that at high concentrations, Na^{+1} is exchanged with greater selectivity than Ca^{+2}.

Anion Media Composition: Anion exchange resins can be made of Styrene

Typical Cycle:

1. Unit is operated to a predetermined leakage level
2. The regeneration cycle starts by upflow cleaning (backwash), then by down flow chemical elution
3. The resin bed is rinsed in down flow

Regeneration is almost never carried out to completion. It is generally uneconomical to carry out regeneration to completion. For example, a cation resin bed may have a capacity of 2.0 N. But only about half of this capacity is available for exchange at the end of a regeneration cycle.

Ion exchange beds are very effective filters. This is more of a disadvantage than it is an advantage. Most resin beads are roughly 20 to 50 US mesh (0.5 mm diameter). In waste water applications where there are non-adhesive suspended solids, the resin can act like a deep bed filtering media. Provided that the particles can be backwashed effectively, this is an advantage. The filtering tendency of the bed becomes a disadvantage when the suspended solids are sticky, or are a liquid with high viscosity. In either case, there will be a tendency for such solids to irreversibly foul the resin bed. Fouling causes mal-distribution of the flow through the bed. This reduces the capacity and it creates dead zones. The dead zones are then prone to microbial fouling which further reduces capacity, can lead to Microbial Induced Corrosion (MIC), and can lead to contamination from H_2S and biopolymer slime.

13.8.1 Boron Removal

Wastewater recycling units have been used since 2011 and have been very successful. Marathon, which owned approximately 211,000 acres of land in the Eagle Ford in 2013, spent USD 2.4 billion that year. They invested USD 10 million dollars on 5 miles of pipelines, water supply well upgrades, pond expansions, and brackish water supply wells to transport and manage water more effectively. The central element of this plan was a pilot water treatment unit called the HIPPO (Hydro Innovation Purification Platform for Oil and gas). This technology was created by Omni, a water treatment technology company based in Austin, Texas. Omni aimed to reduce boron levels from 90 ppm to less than 5 ppm in the freshwater stream, which makes it compatible with gel fracture formulations and able to be reused in future well operations without the need for more fresh water [31, 32]. The HIPPO unit has a real-time automated sense-and-respond system that allows operators to treat waste water at a lower cost and with more consistent results.

The HIPPO unit was used to treat more than 140,000 bbl of water in its first 3 months of operation. After a year, Marathon said it recovered approximately 130,000 bbl for reuse (Marathon 2013a). The following year, Marathon intended to use alternative sources for at least 50% of its water needs in the Eagle Ford. They estimated that these sources would allow them to reuse approximately 65% of the treated water stream. In May 2015, Nuverra, an environmental services company, purchased 180 acres in LaSalle County in south Texas,

13.9 Media Coalescence

Performax vessels are coalescer containing coalescing media designed to accelerate the separation of oil droplets. SP Pack Separators are considered free flow turbulent coalescers in the sense that agitation of the fluid increases by use of uneven walls so that an increase of collisions of the dispersed oil droplets occurs. This is done at a sufficient enough rate to cause coalescence but not too much that further emulsification results.

Coalescing Theory and Key Parameters: In produced water, common sources of oil based contaminations are liquid dispersions and emulsions formed during liquid-liquid extractions as well as free oil droplets, surfactants and both free oil and oil coated solids. The free oil droplets, dispersions and emulsions can often be treated with media that promotes coalescence. CETCO's coalescing technology [33] utilizes cartridge based coalescers that contain polymeric fibers that promote coalescence of the oil droplets and breaks the emulsions and dispersions with minimal retention time.

A liquid-liquid coalescing system usually consists of three operational stages: pre-filtration for separation of oily solids, coalescence of oil droplets, and separation of coalesced oil from water. The pre-filter stage is usually a separate vessel with filter cartridges, sock filters, or a back-washable filter system. The coalescing and separation stages are accomplished in a single vessel that can include several cartridges with each cartridge conducting both the coalescing and separation steps.

In order to prevent fouling of the coalescing media, it is recommended that solids be filtered immediately upstream of the coalescing media. Solids may impact the coalescing element in a variety of ways. Usually the removal of solids from produced water is achieved by a separate vessel containing pleated cartridge filters or by a regenerable backwash filter. The solids filtration stage is intended to increase the coalescing efficiency, protect the coalescing media, and increase the service life of the Hi-Flow canisters.

Following the solids filtration, the next step is the primary coalescence stage. The effectiveness of this stage depends on the porosity and hydrophobic/hydrophilic nature of the coalescing polymer and particles. The coalescence mechanism can be described by the following steps:

1. Influent oil droplet adsorption onto fiber.
2. Migration of oil droplet on surface of polymeric fibers by bulk flow of the produced water. The oil droplets will migrate and tend to accumulate at fiber intersections.
3. Coalescence of two or more droplets to form one larger droplet and repeated coalescence of small droplets into larger droplets at the fiber intersections.
4. Release of larger droplets from fiber intersections due to increased hydraulic drag on adsorbed droplets caused by bulk produced water flow.
5. Repeat of steps 1-4 with progressively larger droplet sizes.

Given the above mechanism, it is easy to see that for very clean streams with only oil droplets in the produced water (no surfactants or solids present); the coalescence will be very efficient and trouble free. It is also easy to predict that a number of factors will affect the coalescence performance.

Among the major areas that need to be investigated when evaluating coalescing media for site specific applications include:

1. **Effect of Surfactants** - Surfactants may contribute to the formation of stable emulsions. Stable emulsions usually require the presence of at least three components to form: the two bulk immiscible phases (oil and water) plus a small concentration of surfactant. Some common sources of surfactants include corrosion inhibitors, organic acids in the hydrocarbon feed, well-treating chemicals, sulfur compounds, and other chemical additives.

Surfactants hinder the coalescence efficiency because they tend to adsorb on and blind the coalescing fiber. By concentrating on the fibers, the surfactants shield the fibers from the passing oil droplets and prevent adsorption of the hydrocarbon droplet onto the fibers and particles. When the oil droplet is not adsorbed, the hydrocarbon droplet will continue to make its way through the tortuous path media and eventually end up in the effluent, resulting in poor oil separation efficiency and high effluent discharge concentration.

Another way that surfactants hinder the media efficiency is that they alter the interfacial tension (IFT) that is created at the interface of the two immiscible liquids (oil and water). When surfactants are present, they migrate to the interface and cause a reduction in the IFT which causes smaller sized droplets to stabilize and creates new emulsion particles in the effluent.

2. **Effect of Solids** - Higher concentrations of solids can be problematic and lead to excessive change out of disposable pre-filters. In cases where pre-filters are not used (like the Shell excursion simulation discussed later in the paper), the solids may blind the coalescing media and reduce the media lifetime. The solids can prevent oil droplet adsorption on the fibers and also inhibit the droplet migration on the polymeric fibers. Generally, the solids range at which liquid-liquid coalescers can economically operate with disposable filters is less than 50 ppm concentration. Above this concentration, further pretreatment may be required, such as backwash cartridge filters, mixed-media packed beds, or hydrocyclones.

Solids in the produced water influent can be detrimental to the coalescing of hydrocarbons from water for several reasons. First, the solids can increase the stability of an emulsion making the coalescence and separation more difficult. Second, solids can coat the polymer surfaces and impede the oil droplet initial adsorption and subsequent migration to coalescence by the polymeric media.

Removing the solids from the produced water influent can make coalescing easier, and lengthen coalescer service life. Solids can be removed by a separate disposable-cartridge or sock filter system, or by a regenerable backwash-filter system for high solids levels.

CETCO Hi-Flow™ System: The standard CETCO Hi-Flow system consists of a pre-filtration skid, vertical pressure vessels filled with canisters containing a coalescing media. The coalescing media canisters are installed over vertical centralizing rods supported by a radial header. Untreated fluids flow into the radial header, then up the inside of the column of stacked canisters, then radially outward through the media. Small oil and grease droplets coalesce into larger oil drops as they pass through the media. As the large droplets exit the outer surface of each canister, the oil droplets separate and rise to the upper portion of the vessel, above the stack of canisters. The coalesced oil accumulates at the top of each vessel where it is periodically removed for subsequent treatment. The oil typically has a low water content. The polished water flows downwards to an outlet on the lower head of the vessel. The unique design allows the vessel containing the media columns to actually perform as the separation vessel, bleeding the oil off the top and removing clean water off the bottom.

Each CETCO high flow canister is rated for 8.5 gpm. The standard 48 inch diameter vessel holds 70 canisters providing 20,000 bpd capacity and an equivalent flux of 46 gpm/ft^2. A 54 inch vessel has also been used commercially and accommodates 90 canisters providing a vessel flow rate of 25,000 bpd or an equivalent flux of 45 gpm/ft^2. Note that flux is used only to illustrate the loading of each vessel based on vessel cross-sectional area.

Coalescing media can be deployed in the produced water treatment line-up in one of two roles:

1. Continuous Tertiary Polishing process. In this role the media would typically be installed downstream of a flotation unit, where the Hi-Flow would remove residual oil and grease and provide discharge water in compliance with regulatory standards. This role is typically applied when equipment deployed in a secondary role is not able to produce quality discharge waters on a consistent basis. The installation of the Hi-Flow downstream of the typical polishing technology ensures compliance at all times. This role would also apply to offshore installations where Produced Water Reinjection (PWRI) is proposed. In this PWRI (Produced Water Reinjection) support role, the coalescing technology could be installed upstream of microfiltration and membranes (they are installed to remove sulphates and ions that contribute to scale formation) in order to protect the membranes from oil fouling. It is important to note that prefiltration of solids down to 20 micron [33] is required when coalescing media is deployed continuously to increase the life of the media. The pre-filtration requirement and application will be discussed further in subsequent sections.

2. The second potential deployment role for coalescing media technology is temporary "Excursion Intervention" polishing. In this role, the media is normally out of service and is only placed in service when needed. This can be done by automation or manually. In the automated mode, a continuous oil-in-water monitor device (e.g. Turner XD-4100, or Advanced Sensors) provides continuous indication of the oil concentration. When the oil concentration exceeds a set point value, produced water is diverted through the coalescing media before being discharged overboard.

A short backwash of no more than 5 minutes is recommended by CETCO for all continuous polishing applications. In an excursion role, a backwash can occur after each excursion event.

Coalescing media has been shown to reduce the so-called Water Soluble Organics, a key characteristic of produced water as measured by the EPA-1664 method. The mechanism of removal is rather complex and is discussed in detail in Chapter 21 (Applications – Dissolved & Water Soluble Organics). WSO removal has consistently been more successful than originally expected with the coalescing technology

For comparison, in-line coalescers such as PECT-F, Mare's Tail [56, 57], and G-Floc are in-line devices which enhance coalescence of oil droplets, but do not separate the oil from water directly. These devices require a down-stream separation step in order to capitalize on the larger oil droplets generated by coalescence. There are other coalescers with reported separation capacity, but they have not been tested and have a poor service reputation.

Mare's Tail: The Mare's Tail Coalescer was developed by Opus Plus through a joint industry program in the late 1990's. The device is configured as an in-line piping spool piece. The spool piece contains the Mare's Tail fiber element which is composed of oleophilic fibers that run in the same direction as the fluid flow. Fluid enters the spool piece via an axial inlet nozzle, and then flows through the spool piece in the same longitudinal direction as the fibrous coalescer bundle. As fluid travels along the fibers, small oil droplets are retained on the surface of the fibers. The droplets coalesce with other droplets on the fiber surface, and therefore grow as they migrate along bundle towards the outlet. Fluid drag increases as the droplet diameter grows, and eventually the larger droplets are released into the water flow at the end of the bundle. The coalescing action occurs within two seconds in the bundle, making a very compact device.

It is important to note that there is no phase separation in the coalescer itself. All the inlet fluid leaves through a common outlet, but the outlet mean droplet size is considerably enhanced, leading to easy gravity separation downstream. This is not a significant drawback per se. But it sometimes lead to installation of the device too far upstream where the produced water contains oily solids that have not been sufficiently removed. As a consequence, the Mare's tail has been installed in process locations where fouling is a problem. It is claimed that flow along the fibers, rather than across as conventional coalescers, together with relatively high fluid velocities, results in less fouling than other fiber based coalescing media. However, no data is provided in references [56, 57] to support this claim. The fiber media is typically replaced when the pressure drop across the Mare's Tail unit reaches 3 bar.

An unsuccessful Mare's Tail trial was carried out in Oman at Nimr. Basically the fibers became saturated with oil/sand before the device started falling apart. Before considering this type of device it is prudent to evaluate the properties of the crude, presence of solids, and the impact of any chemical injection upstream - corrosion inhibitor being the worst case. Corrosion inhibitor tends to reduce the attachment of oil droplets to the fiber element. The higher viscosity oil, together with solids, can cause rapid fouling.

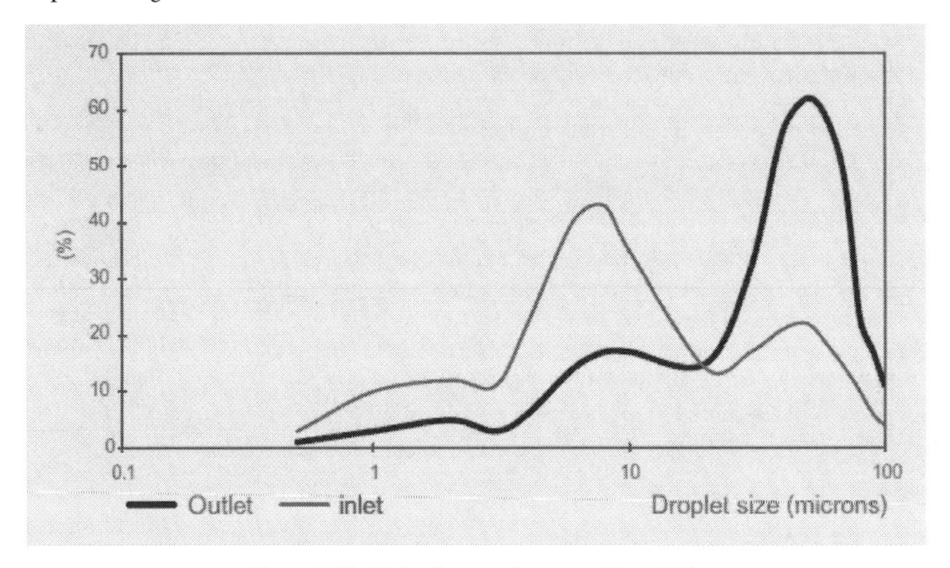

Figure 13.18 Oil droplet growth across a Mare's Tail.

PECT-F (Performance Enhancing Coalescence Technology): PECT-F [58, 59] was developed to improve the efficiency and extend the operating envelope and flexibility of deoiling hydrocyclone systems. The technology is aimed at existing systems which do not meet specification, and at new-build systems where the application of hydrocyclones is marginal due to difficult fluid characteristics (due to for example; cold fluids, chalk reservoirs, and heavy oil). This technology, apart from installation of a conventional four cell flotation unit downstream of hydrocyclones, is becoming the most proven method of improving hydrocyclone performance.

A small increase in the inlet drop size distribution by the PECT-F could significantly increases the separation performance - e.g. an increase in the drop size from 5 to 10 microns might increase the separation efficiency from 15% to over 90%. Many field tests of PECT-F have now been performed by Cyclotech, including trials at Shell Expro Dunlin and Norske Shell Draugen facilities. In order to install PECT-F, there are constraints on the internal geometry of the vessel and liner layout, e.g. a minimum triangular pitch of the liners.

PECT-F was installed on all the new Cyclotech hydrocyclones on Draugen during the produced water system upgrade project for 40,000 m3/d flow rate, sometime in 2000/2001. The system because blocked up. It was suspected that the problem was related to the cocktail of production chemicals that were being used. All the PECT-F units have now been removed (except for the test separator hydrocyclone) from the hydrocyclones on Draugen. Prior to the full scale installation, a one week trial was carried out for a small-scale test. The lesson learned here could be that a longer trial with a more representative sample of Produced Water could have helped. The major advantages of this technology is the low CAPEX & OPEX and easy retrofit. The major disadvantages of this and similar technologies is the difficulty in determining if fouling will be a problem before installation.

References to Chapter 13

1. G. Tchobanoglous, F.L. Burton, Wastewater Engineering: Treatment, Disposal, and Reuse, Third Ed., Metcalf & Eddy, Inc, Boston (1979).
2. R.M. Allen, K. Robinson, "Environmental aspects of produced water disposal," SPE – 25549, paper presented at the SPE Middle East Oil Technical Conference and Exhibition, Bahrain (1993).
3. Z. Khatib, "Produced water management: is it a future legacy or a business opportunity for field development," IPTC – 11624, paper presented at the IPTC, Dubai, UAE (2007).
4. P.J. Clifford, D.W. Mellor, T.J. Jones, "Water quality requirements for fractured injection wells," SPE – 21439, paper presented at the SPE Middle East Oil Show, Bahrain (1991).
5. T. Sparks, G. Chase, Filters and Filtration Handbook, 6th Ed., IChemE Series, Elsevier, New York (2016).
6. V. Jegatheesan, S. Vigneswaran, "Deep bed filtration: mathematical models and observations," Critical Reviews in Environmental Sceince and Engineering, v. 35, (2005).
7. E.D. Schroeder, Water and Waste Water Treatment, McGraw-Hill, Inc., Taiwan (1977).
8. S.J. Randtke, M.B. Horsley, Water Treatment Plant Design, American Water Works Association, 5th Edition, McGraw Hill, New York, (2012).
9. S. Kashaev, D.W. Lee, "Deep bed nutshell filter evolution," paper presented at the 20th annual Produced Water Seminar, Houston, TX (2010).
10. D. French, "Granular filter media: evaluating filter bed depth to grain size ratio," Filtration & Separation, December (2012).
11. T. Iwasaki, "Some notes on sand filtration," J. Am. Water Works Assoc., v. 29, p. 1591 -1602 (1937).
12. J.C. Crittenden, R.R. Trussell, D.W. Hand, K.J. Howe, G. Tchobanoglous, MWH's Water Treatment Principles and Design, 3rd Edition, John Wiley & Sons, Inc., Hoboken, NJ (2012).
13. G. Hirs, "Method of filtering oil from liquids," US patent 3993391, priority date 6-May 1975, Hydromation Filter Company (1975).
14. C.C. Patton, Applied Water Technology, Campbell Petroleum Series, Norman, OK (1974).
15. H. Rawlins, A. Erickson, C. Ly, "Characterization of deep bed filter media for oil removal from produced water," presentation to the Produced Water Society, Houston, TX (2010).
16. M. Stacy, Environmental Treatment Team, personal communication (2006)
17. F. DeSilva, Activated Carbon Filtration, Water Quality Products (Jan 2000).
18. ZIMPRO Wet Air Oxidation, Siemens Energy, Inc. - Water Solutions, 301 West Military Road, Rothschild, WI 54474

19. R. Dobbs, J.M Cohen, "Carbon adsorption isotherms for toxic organics," EPA Municipal Environmental research Lab, Water research Division (1980).

20. EPA-600/8-80-023, EPA Adsorption Isotherms for Toxic Organics

21. CETCO Carbon Vessel Design, 16350 Park Ten Place, Suite 217, Houston, Texas 77084

22. M. Jacob, N. Lesage, P. Pedenaud, "Seawater treatment for injection," paper presented at the SPE Global Integrated Workshop on Water Treatment, Rome (2012).

23. P. Pedenaud, S. Heng, W. Evans, D. Bigeonneau, "Ceramic membrane and core pilot results for produced water," OTC – 22371, paper presented at the Offshore Technology Conference, Houston (2011).

24. P. Pedenaud, C. Hurtevent, S. Baraka-Lokmane, "Industrial experience in seawater desulfation," SPE – 155123, paper presented at the SPE International Conference and Exhibition on Oilfield Scale, Aberdeen (2012).

25. P. Pedenaud, S. Heng, "Recent results on membrane filtration of produced water," paper presented at the TEKNA Produced Water Management meeting, Stavanger (2010).

26. P. Pedenaud, "Water injection – the FLEX filtration solution," SPE – 170992, paper presented at the SPE Oilfield Water Management Conference and Exhibition, Kuwait City (2014).

27. G.M. dos Ramos, "Membrane separation processes applied to oily water treatment," 3rd International Seminar Oilfield Water Man., Rio de Janeiro (2010).

28. J. Drewes, "An integrated framework for treatment and management of produced water. Technical assessment of produced water treatment technologies," 1st Edition, Report prepared by the Colorado School of Mines for RPSEA Project 07122-12 (2009).

29. J.D. Arthur, B.G. Langhus, C. Patel, "Technical summary of oil & gas produced water treatment technologies," ALL Consulting, paper produced in partial fulfillment of NETL project (2005).

30. T. Hayes, "Overview of emerging produced water treatment technologies," paper presented at the 11th Annual International Petroleum Environmental Conference, Albuquerque, NM, October (2004).

31. H.R. Goltz, A.M. Johnson, T. McCandless, "Boron removal from produced water," presentation given to the Produced Water Society, Houston, TX (2015)

32. S. Whitfield, "Unconventional resources: new facilities find solutions to limited water sources," paper published in the Water Treating Insights section of Oil and Gas Facilities (Dec 2014).

33. J. Darlington, "Evaluation of CETCO Hi-Flow™ Coalescing Media for Continuous Flow Filtration and Excursion Event Filtration of Produced Water," CETCO, Chicago (2012).

34. J.G.R. Eylander, "Suspended solids specifications for water injection from core flood tests," SPE – 16256, paper published in SPE Reservoir Engineering, November (1988).

35. A.J. Abrams, "Mud design to minimize rock impairment due to particle invasion," paper published in SPE Journal Petroleum Tech., May (1977).

36. J.H. Barkman, D.H. Davidson, "Measuring water quality and predicting well impairment," SPE – 3543 paper published in the SPE Journal of Petroleum Tech., July (1972).

37. E. van Oort, J.F.G. van Velzen, K. Leerlooijer, "Impairment by suspended solids invasion: testing and prediction," SPE – 23822, paper published in the SPE journal Production & Facilities, August (1993).

38. G.A. Gabriel, G.R. Inamdar, "An experimental investigation of fines migration in porous media," SPE – 12168, paper presented at the SPE Annual Technical Conference and Exhibition, San Francisco, October (1983).

39. C. Gruesbeck, R.E. Collins, "Entrainment and deposition of fine particles in porous media," SPE – 8430 paper published in the SPE Journal, December (1982).

40. T.W. Muecke, "Formation fines and factors controlling their movement in porous media," SPE – 7007 paper published in the Journal of Petroleum Tech., February (1979).

41. M.M. Sharma, S. Pang, K.E. Wennberg, L.N. Morgenthaler, "Injectivity decline in water-injection wells: an offshore Gulf of Mexico case study," paper published in SPE Prod. & Facilities, February (2000).

42. K.M. Bansal, D.D. Caudle, "A new approach for injection water quality," SPE – 24803, paper presented at the ATCE, Washington DC, October (1992).

43. P.K. Singh, R.G. Agarwal, L.D. Krase, "Systematic design and analysis of step rate tests to determine formation parting pressure," SPE – 16798, paper presented at the ATCE, Dallas (1987).

44. D.B. Bennion, D.W. Bennion, F.B. Thomas, R.F. Bietz, "Injection water quality – A key factor to successful waterflooding," Journal Can. Petroleum Tech., v. 37, p. 53 (1998).

45. C.L. Wu, Relationship between Particle Size, Aggregate Size, Pore Size, and Water Characteristics, Oregon State Univ. (1988).

46. P.J. Clifford, D.W. Mellor, T.J. Jones, "Water quality requirements for fractured injection wells," SPE – 21439, paper presented at the SPE Middle East Oil Show, Bahrain (1991).

47. J.F. Pautz, M.E. Crocker, "Relating water quality and formation permeability to loss of injectivity," SPE – 18888, paper presented at the Production Operations Symposium, Oklahoma City (1989)

48. New Mexico Oil Conservation Division. Underground Injection Control Program Manual. Published by the New Mexico Energy, Minerals and Natural Resources Department, Oil Conservation Division

49. http://www.rrc.state.tx.us/oil-gas/publications-and-notices/manuals/injectiondisposal-well-manual/summary-of-standards-and-procedures/technical-review/step-rate-test-guidelines/

50. C. Gao, "Factors affecting particle retention in porous media," Emirates J. Eng. Res., v. 12, p. 1 (2007).

51. J. Xiao, J. Wang, X. Sun, "Fines migration: problems and treatment," Oil & Gas Research, v. 3 (2017).

52. D.B. Bennion, F.B. Thomas, D.W. Bennion, R.F. Bietz, "Mechanisms of formation damage and permeability impairment associated with the drilling, completion and production of low API gravity oil reservoirs, SPE – 30320, paper presented at the SPE International Heavy Oil Symposium, Calgary (1995).

53. Railroad Commission of Texas, http://www.rrc.state.tx.us/oil-gas/publications-and-notices/manuals/injectiondisposal-well-manual/

54. J.O. Amaefule, A. Ajufo, E. Peterson, K. Durst, "Understanding formation damage processes: an essential ingredient for improved measurement and interpretation of relative permeability data," SPE-16232, paper presented at the SPE Production Operations Symposium, Oklahoma City (1987).

55. R.R. Matthews, "Evaluation of seawater filtration systems for North Sea application," J. Pet. Tech., p. 843 (1985).

56. M.B. Oyeneyin, G. McLellan, B. Vijayakumar, M. Hussain, R. Bichan, N. Weir, "The Mare's Tail – The answer to a cost effective produced water management in deepwater environment?" SPE – 128609, paper presented at the 33rd SPE ATCE, Abuja (2009).

57. B.L. Knudsen, M. Hjelsvold, T.K. Frost, M.B. Svarstad, P.G. Grini, C.F. Willumsen, H. Torvik, "Meeting the zero discharge challenge for produced water," SPE – 86671, paper presented at the Seventh SPE International Conference on HSE, Calgary (2004).

58. A.B. Sinker, M. Humphris, N. Wayth, "Enhanced deoiling hydrocyclone performance without resorting to chemicals," SPE – 56969, paper presented at the 1999 Offshore Europe Conference, Aberdeen (1999).

59. P.G. Grini, M. Hjelsvold, S. Johnsen, "Choosing produced water treatment technologies based on environmental impact reduction," SPE – 74002, paper presented at the SPE International Conference on Health, Safety and Environment, Kuala Lampur (2002).

CHAPTER FOURTEEN

Process Engineering

Chapter 14 Table of Contents

14.0 Introduction 285

14.1 Material and Flow Balance 286

14.2 Process Performance Diagram 294

14.3 Dynamics of Continuous Flow Systems 295

14.4 Process Configuration – Best Practices 299

References to Chapter 14 304

14.0 Introduction

To this point in the text, most of the discussion has been concerned with chemistry, fluid mechanics, emulsions and individual water treatment technologies. These subjects are of course important in design and troubleshooting. An equally important subject is the process configuration. The process configuration refers to the process flow diagram, i.e. the sequence of tanks and vessels, the connections and routing of process flows, and the all-important routing of reject streams. Many different process flow diagrams could be devised for a given project. Variations include use of degassing vessels, treating tanks, and the design of closed drain systems. The pumps, valves, and instrumentation that provide motive flow and control of levels and fluid flow are also an important part of the process configuration. The selection of separator pressures, valves and pumps can have an enormous impact on system performance. In the design stage of a project, the development of a process configuration is known as process integration. Generally no piece of equipment should be selected until the impact of that equipment on the overall process has been determined.

Included in this chapter are two important process engineering tools that are used for process integration. The first is the material balance. It accounts for solids and oil concentration throughout the facility and helps to identify proper versus problematic routing. It also helps to identify reject streams that should be further treated prior to recycle. The second is the Process Performance Diagram. It helps to identify the proper selection of equipment so that the capacity and performance of the selected equipment meets the requirements of the system as a whole. For example, some process systems require large multistage flotation units, whereas others do not. The Process Performance Diagram helps to distinguish between these two cases.

Valves, Instrumentation, Pumps: Many subjects within the broad subject of Process Engineering have already been discussed in this book. The subject of valves, for example, was discussed already in the chapter on shearing of oil droplets, Section 3.5.2 (Shearing Through a Valve). It was necessary to discuss this at an early stage of the book because valves cause shearing which breaks large drops into small drops. Drop size distribution is one of the most important characteristics of produced water. Much of the challenge of water treatment is a result of this shearing. Oil-in-water instrumentation, and drop size analyzers, were covered in Chapter 7 on sampling and analysis. Pumps were covered in Section 3.5.3 (Shearing Through a Pump) which emphasized the selection of low shear pumps in order to maximize oil droplet size. These subjects (valves, oil-in-water instrumentation, and pumps) will not be further discussed in this chapter.

Holistic Analysis: This chapter addresses, more or less for the first time in this text, the question of system integration – how individual pieces of equipment interact to give the performance of the whole system. Many people in the industry refer to this subject as holistic analysis. Often it is stated that the industry does not put enough emphasis on the holistic view of water treating either in design or in troubleshooting. By this it is meant that the industry tends to focus too much on individual pieces of equipment. A system can perform poorly even with excellent individual pieces of equipment, with optimal internals and well-designed capacities, due to a lack of holistic analysis in the process design.

Process Engineering Tools: In the oil and gas industry, process simulation tools are used to design new facilities for oil/water/gas separation, and to troubleshoot existing facilities. When applied competently, process simulation provides an optimized facility design for a given production flow rate, production profile, fluid composition, and given output specifications (gas dew point, pressure and hydrate formation temperature; oil BS&W, and vapor pressure). Process simulation provides the engineering link between fluid composition and facilities design. However most process simulation programs are focused on oil and gas flows, not produced water.

One of the reasons why the industry must deal with so many mistakes in produced water treatment is the lack of commercial software that would provide easy calculation of the material balance and Process Performance Diagram. If the industry had these tools then the performance of various process configurations could be easily compared. The scientific principles are known. But currently the only way to apply these scientific principles is for an engineer to do their own programming, usually in an Excel spreadsheet. This is rather straightforward but most process and project engineers do not have time to do this. Also, most oil and gas companies do not demand this type of analysis from the vendors and design firms because they are unaware of the performance impact of process integration.

Thus, the least expensive process configuration is often selected. The least expensive option will often involve mistakes such as mixing incompatible streams, lack of interface bleed lines, excessive shearing, excessive fluctuations in flow rate, and lack of flexibility to handle diverse fluid characteristics which typically emerge over time in any given facility, etc..

The subject of process configuration and integration is approached from two perspectives. The first perspective presented is a discussion of the engineering principles and tools that give insight and can be used for design. Such tools include the material and flow balance, and the Process Performance Diagram. (PPD).

The second perspective for the discussion of process configuration is that of Best Practices. These Best Practices are provided for the water specialists that do not have access to engineering design tools. They are based on observation, practical experience, and engineering modeling. The Best Practices are presented here for process integration. There is a more extensive set of Best Practices for a wide variety of subject in the chapters on Applications.

What makes one process flow diagram better than another can sometimes be subtle. There can be a significant difference in performance from one PFD to another, even if the two PFDs have the same equipment. A poor decision in the tie-in of a single recycle line, for example, can create a complex emulsion that renders a downstream separator ineffective. To those engineers with several years of field experience poor process flow diagrams can be obvious. Capturing this experience in a scientific, such that it can be applied systematically, is an emerging prospect for the industry.

14.1 Material and Flow Balance

The material and flow balance is one of the most important yet often overlooked aspects of process design and troubleshooting. A material and flow balance for a water treating system is a detailed accounting of the volumetric flow rates of all streams together with the oil content of the water and the water content of the oil. It is typically necessary to include the oil streams due to the fact that they will contain significant quantities of water which will be removed as the oil proceeds through the system. The "material" part of the "material and flow balance" refers to the oil content of the water, and the water content of the oil. These concentrations obviously change through the system by the action of the separation equipment and blending or mixing of various streams. When a material and flow balance is carried out for a system, it is often the case that internal flow streams such as recycle loops are discovered to have a significant impact on the oil concentration of the final effluent stream. This is one of the benefits of constructing a material and flow balance.

Essentially all water treating equipment has a single feed stream and two discharge streams, the effluent and the reject. In selecting equipment, it is often necessary to choose between the following two options: a piece of equipment with moderately effective water treating performance but a small reject flow rate versus a piece of equipment with very effective performance and a greater reject flow rate. All too often design engineers select the equipment with greater performance efficiency only to discover that in practice the performance of the system as a whole is poor due to difficulties in handling the higher reject flow rate. It is only through the analysis of the material and flow balance that the impact of the reject stream can be determined.

Equipment selection is just one example. Operability of a system is often tied to reject quality, flow rate, and process routing. System dynamics, surge volumes, shearing through valves, and shearing through pumps are all greatly influenced by the quality, flow and routing of reject streams. Understanding the impact of these variables on the system is only possible through construction of a material and flow balance.

The performance of the separation equipment depends critically on the drop size distribution however, the material and flow balance does not typically include this information. Such a comprehensive approach to modeling system behavior would be extremely useful but would require somewhat extensive software development. Instead, the performance of separation equipment is modeled externally to the material and flow balance. The results of that modeling are then inserted into the material and flow balance. The extent to which performance and material flows are coupled is not lost in this approach. They are captured by external iteration.

The art of generating a material and flow balance for a system hinges on a few key capabilities:

1. the ability to estimate flow rates based on system observations;
2. judicious approximation of certain variables;
3. general knowledge of equipment performance;
4. the ability to construct a spreadsheet which allows recycle loop calculations.

For many water treating engineers, the construction of a material and flow balance for a process is a difficult endeavor. The capabilities presented above are not common and are not easily developed. Nevertheless, the ability to construct a material and flow balance is one of the most important abilities for both troubleshooting and systems design.

As a water treating specialist progresses in his or her career they will be faced with troubleshooting challenges of greater and greater difficulty. In most cases, local staff will attempt to solve the problems. Whenever they are unable to do so, they call in the specialist. The specialist must have tools and knowledge that justifies their role as a specialist. The material and flow balance is one such tool. Problems that occur in a process as a result of inadequate reject, too high of a flow rate, inadequate retention time are often difficult to detect by simple system observation. The volume and characteristics of recycle loops often obscure what is happening in a system. Analytical tools are often required to decipher these features.

Constructing a material and flow balance is more difficult to accomplish for an actual system since flow rates are so seldom measured and reported in actual systems. This is where clever inference and approximation come into play.

There are two important and judicious approximations. In reality, there will be oil drops suspended in water. These drops will themselves contain small water drops within them. Likewise, water drops suspended in oil will contain small oil drops suspended within them. From experience it is found that minimal inaccuracy is obtained to approximate the oil in water as dry oil, and the water in the oil as not containing any oil. The oil content of the water is then expressed as mg/L or ppmv of dry oil in water. The water content of the oil is expressed as mg/L or ppmv of oil-free water in the oil. The oil-free water is referred to as pure water, although this terminology is a bit misleading. It just needs to be kept in mind that this refers to oil-free water drops.

Material and flow balance modeling can be carried out using software packages. While many of the commercially available packages will do an adequate job of modeling oily water systems, it is seldom worth the effort to learn the user interface, naming conventions and deciphering the results. All too often, after learning how to run a commercial program it turns out that the program does not have the flexibility or versatility to solve important aspects of the problem. It is far easier to construct a simple Excel spreadsheet. Further, a user-generated Excel spreadsheet is completely flexible and can be tailored to the user's needs. Finally, by constructing one's own material and flow balance, much is learned about the system. This section gives the methodology for developing a material and flow balance in Excel.

There are essentially three elements that must be modeled: separation equipment, mixing of two streams, and recycle loops. Mixing of two streams is trivial and will not be discussed. Calculating the material and flow balance for separation equipment requires some discussion.

Essentially all water treating equipment has a single feed and two discharge streams, i.e. an effluent and a reject. Dead-end filtration does not necessarily have a reject stream. However, in the discussion here, it too will be modeled as such. In reality it may have instead a buildup of waste which must be either backwashed or which requires the replacement of the media. In the case where backwash is required, the backwater is essentially the reject stream. The intermittent nature of that stream will be discussed below.

A gravity-based three phase separator will be discussed. The feed stream to the separator is a three-phase fluid (oil, water and gas). The presence of the gas phase will be ignored for this discussion. As discussed in other parts of the text, gas breakout, foam, and foam mitigation are important subjects. However, they are not essential subjects in the present discussion which focuses on the separation of oil from water.

The inlet or feed stream is assumed to be composed of two liquid phases, a wet oil phase and an oily water phase. The total feed flow rate (F) is the sum of flow rates of the dry oil and pure water:

$$F = F_O + F_W \qquad \text{Eqn (14.1)}$$

where:

F = total flow rate

F_O = dry oil feed flow rate

F_W = pure water feed flow rate.

If a sample could be collected that represented both liquid phases accurately, it could then be analyzed for total water and total oil. The volumetric ratio of water to total fluid is here referred to as the BS&W (Base Sediment and Water) of the feed, or $(BSW)_F$. The flow rate of pure water is the product of the feed flow rate times the BS&W. The flow rate of the pure (dry) oil is the difference between the feed flow rate and the pure water flow rate.

$$F_O = \left[1 - (BSW)_F\right]F \qquad \text{Eqn (14.2)}$$

$$F_W = (BSW)_F F \qquad \text{Eqn (14.3)}$$

where:

$(BSW)_F$ = BS&W of the feed

If the BS&W is measured accurately, it can be used to calculate the flow rate of dry oil and that of pure water. In reality though, each of these phases are not pure. They are contaminated. Each has dispersed droplets of the other phase suspended in it. Thus, the oil phase is composed of dry oil plus water droplets. Likewise, the water phase is composed of pure (oil-free) water plus oil droplets. The total flow rate of oil entering the separator is the sum of the oil in the wet oil stream plus the oil in the oily water stream. This is referred to as a component flow balance. The same logic is applied to water. The feed water flow rate is the sum of water in the two effluent streams, i.e. the water in the wet oil stream plus the water in the oily water stream. The objective of the separator is to minimize the oil in the discharged water stream, and to minimize the water in the discharged oil stream.

Applying the concept of flow balance, at steady state when there is no accumulation of fluid inside the separator, the feed flow rate into the separator must equal the sum of the wet oil flow rate out of the separator, plus the oily water flow rate out of the separator. These two streams are denoted as O and W respectively. Mathematically, these statements are expressed as follows.

$$O = O_O + O_W \qquad \text{Eqn (14.4)}$$

$$W = W_O + W_W \qquad \text{Eqn (14.5)}$$

Where,

W = flow rate of the oily water discharge (effluent)

W_O = flow rate of dry oil in the oily water stream

W_W = flow rate of pure water in the oily water stream.

O = flow rate of wet oil discharge (effluent)

O_O = flow rate of dry oil in the wet oil stream

O_W = flow rate of pure water in the wet oil stream

Having defined these quantities, it is now possible to distinguish between *Fo* and O, and likewise between *Fw* and *W*.

The amount of water suspended in the wet oil discharge is determined from the BS&W of the wet oil stream, as was discussed above for the feed. The amount of oil suspended in the water is determined from the OiW. Mathematically these statements are expressed as follows.

$$O_W = (BSW)_E O \qquad \text{Eqn (14.6)}$$

$$W_O = (OiW)_E W \qquad \text{Eqn (14.7)}$$

where:

$(BSW)_E$ = the water concentration of the wet oil in the effluent

$(OiW)_E$ = the oil concentration of the oily water in the effluent

In these equations it is understood that the BS&W refers to the wet oil stream and the OiW refers to the oily water stream. The subscript *E* refers to the effluent. Likewise, the following relationships exist:

$$O_O = [1 - (BSW)_E]O \qquad \text{Eqn (14.8)}$$

$$W_O = [1 - (OiW)_E]W \qquad \text{Eqn (14.9)}$$

At this point it is appropriate to discuss the selection of a set of units. Of course, the units are completely at the discretion of the engineer and the results will not be affected by the choice of units. In the derivation below, it is assumed that the flow rates are expressed as either BPM (barrel per minute), or LPM (liter per minute). The concentrations discussed above (BSW and OiW) are most conveniently expressed as fractions (e.g. mL/L – milliliter of oil per liter of water).

At this point, a number of substitutions can be made. Without going into detail, the following equations are derived.

$$F_O = [1 - (BSW)_E]O + (OiW)_E W \qquad \text{Eqn (14.10)}$$

$$F_W = (BSW)_E O + [1 - (OiW)_E]W \qquad \text{Eqn (14.11)}$$

These two equations are the basis of the solution. They represent two equations with two unknowns (*O* and *W*). These equations can be expressed using a different set of variables.

$$c_1 = c_3 a + c_4 b$$

$$c_2 = c_5 a + c_6 b$$

In matrix form, these equations written as follows.

$$\begin{bmatrix} c_1 \\ c_2 \end{bmatrix} = \begin{bmatrix} c_3 & c_4 \\ c_5 & c_6 \end{bmatrix} \begin{bmatrix} a \\ b \end{bmatrix}$$

$$a = \frac{c_1 c_6 - c_2 c_4}{c_3 c_6 - c_4 c_5}$$

$$b = \frac{c_2 c_3 - c_1 c_5}{c_3 c_6 - c_4 c_5}$$

Based on these matrix equations, the solution to the material balance equations can be written as follows.

$$O = \frac{F_O[1 - (OiW)] - F_W(OiW)}{[1 - (BSW)][1 - (OiW)] - (OiW)(BSW)} \qquad \text{Eqn (14.12)}$$

$$W = \frac{F_W[1-(BSW)] - F_O(BSW)}{[1-(BSW)][1-(OiW)] - (OiW)(BSW)} \qquad \text{Eqn (14.13)}$$

This provides a closed solution to the equations. In these equations, it is understood that OiW and BSW are measured quantities for the two effluent streams (oily water and wet oil).

The following example is taken from work done to troubleshoot a North Sea platform that had been operating for some twenty years and had been suffering poor water treating performance for roughly two years. The fundamental cause of poor performance was a dramatic increase in the use of corrosion inhibitor. Unfortunately, the use of this corrosion inhibitor was considered to be essential from an asset integrity standpoint. Therefore, the management of the asset had approved of a comprehensive upgrade of the water treating system in order to allow the use of the corrosion inhibitor and still meet the overboard discharge water quality requirements.

To support this comprehensive upgrade of water treating equipment, a material and flow balance was constructed, among other things. The following is a part of the Excel spreadsheet that was developed for the material and flow balance.

Table 14.1 Material and Flow Balance

Separator	Feed Flow	BS&W in (%)	BS&W Out (%)	Fo Dry oil in	Fo Dry oil out	Fw Pure water in	Fw Pure water out	OiW In (ppm)	OiW Sep Eff	OiW Eff (ppm)	O Wet Oil out	W Oily water out
V-1010	10,000	10.0	2.0	9,000.0	9,000.0	1,000.0	1,000.0	1,000	60	400.0	9183.3	816.7
V-1110	10,000	10.0	2.0	9,000.0	9,000.0	1,000.0	1,000.0	1,000	50	500.0	9183.3	816.7

In the cells shown above, input values are highlighted in darker background shade. Calculated values are in cells with white background. As shown, separator V-1010 had 10,000 BPD feed flow rate. The BS&W of the total feed (combined oil and water phases) was roughly 10 %. Thus, there were 9,000 bbl of dry oil per day and 1,000 bbl of oil-free water per day entering the separator. There are two other important characteristics of the feed that could be measured. One is the BS&W of the oil phase. The other is the OiW of the water phase. The former was not measured but the later was. Thus, there is a column for the OiW In (given as 1,000 ppmv). The column for BS&W In is for the total feed stream, as mentioned already. A model based on Stokes law was used to estimate the oil concentration in the water discharge. The model gave a value of 60 % separation efficiency, which when used in the material and flow balance, gives a value of 400 ppmv of oil in the water discharge. As shown, separator V-1110 had an oil separation efficiency of 50 % and therefore 500 ppmv of oil in the water discharge.

Also shown in the cells above is a total accounting of the oil and water. The equation for the wet oil flow rate (O) is Eqn (14.12). The equation for the oily water flow rate (W) is Eqn (14.13). Also shown in the cells above are the flow rate of dry oil out and pure water in. It is noted that these columns are designated as Fo and Fw respectively. These designations are actually for the inlet quantities. By material balance the inlet must equal the outlet. These columns are calculated using Eqn (14.10) and (14.11).

Other Separation Equipment: In the above discussion, a generic separator was discussed. The methodology can be easily extended to essentially any typical water treating equipment. As an example, the methodology is applied next to a hydrocyclone.

The feed to a hydrocyclone, as with most water treating equipment, does not have a bulk oil phase. This does not make much difference in the methodology of the material and flow balance. There is still a feed flow rate of dry oil and a feed flow rate of pure water. This is of course, not by accident. The methodology presented for the separator was developed for a general case that would equally apply to typical water treating equipment. However, there is one important difference between a typical separator and a typical piece of water treating equipment. To understand this difference, the particular case of a hydrocyclone is presented.

For a hydrocyclone, the mathematical formulation of the problem would be identical to that given above but for one significant difference. In the case of a hydrocyclone, as with most water treating equipment, the reject flow rate is typically known or can be calculated as a percentage of the feed flow rate. Because of this, Eqn (14.4) simplifies to:

$$O = F(R/100) \qquad \text{Eqn (14.14)}$$

Where:

F = total flow rate (same as above)

O = reject flow rate

R = reject rate (percent)

Likewise, Eqn (14.5) simplifies to:

$$W = F(1 - R/100) \qquad \text{Eqn (14.15)}$$

where

W = oily water flow rate (hydrocyclones treated water)

These are simple calculations. When applied to two hydrocyclones on the same platform discussed above the following results are obtained:

Table 14.2

Hydrocyclone	Reject (%)	OiW Eff (%)	Description	Feed Flow (BPD)	OiW	Dry Oil	Pure Water
			Feed	816.66	400	0.327	816.33
H-1010	**3.0**	**80.0**	**Effluent**	792	80	0.063	792.1
			Reject	24	10,776	0.264	24.2
			Feed	816.74	500	0.408	816.33
H-1110	**2.0**	**70.0**	**Effluent**	800	150	0.120	800.3
			Reject	16	17,650	0.288	16.0

The feed to hydrocyclone H-1010 is 816.66 BPD. The concentration of oil in the oily feed water is 400 ppmv. Thus, the dry oil content in the feed is 0.327 BPD.

The effluent flow rate is calculated using Eqn (14.17) which gives: W = 817 x 0.97 = 792 BPD. The separation efficiency for oil in water is given as 80 %, and the inlet OiW concentration is 400 ppmv.

This results in an OiW concentration in the effluent of 80 ppmv. The dry oil flow rate in the effluent is therefore 0.063 BPD.

The reject flow rate is calculated using Eqn (14.16) which gives: O = 817 x 0.03 = 24 BPD. The reject must contain the oil removed from the feed water which is given by difference between the feed and the effluent (0.327 – 0.63 = 0.264). Now that the dry oil content and flow rate of the reject are known, the oil concentration can be calculated by using:

$$O_O = F_O - W_O$$

This gives: 0.327 – 0.063 = 0.264 BPD. The concentration of oil in the reject is then calculated from:

$$OiW_O = O_O / O$$

This gives: 0.264 x 1,000,000 / 24 = 10,776 ppmv.

The flow rates of dry oil and pure water are used in subsequent calculations such that the spreadsheet provides an accounting of the flow rates of dry oil and pure water into and out of all separators and water treating equipment.

Recycle Loop: For the most part, the material and flow balance is relatively straightforward. The only complication arises when there is a recycle loop. On many platforms the reject stream from the flotation unit is recycled to the feed of a three phase separator. The complication arises due to the fact that the separator is upstream of the flotation unit. Thus, the feed to the separator has an impact on the feed to the flotation unit, which in turn affects the reject and hence the feed to the separator. This is the nature of a recycle loop.

In general, the material and flow balance equations involved in most recycle loops can be solved algebraically. This is typically not done however. The algebra is somewhat involved and hence prone to error. Also, there are variations in the details from one recycle loop to another. Hence, to develop a general set of equations that would apply to all recycle loops would be a difficult endeavor.

In general, recycle loops are calculated iteratively. In other words, an initial guess of the recycle stream flow rate and concentration is made. From this initial guess, the consequences of this guess are calculated. In other words, the material and flow rates are calculated for all downstream streams. The consequences invariably come back to allow calculation of the recycle stream itself. The spreadsheet is set up to allow easy comparison between the initial guess of concentration and rate versus the calculated concentration and rate. The difference between the guess and the calculated are then minimized iteratively.

In this case, the recycle stream is stream number 18. In this example, the initial guess of the water flow rate is 8 BPD and the initial guess of OiW concentration is 100,000 ppmv. These values are entered manually based on a general knowledge of the system. Based on these values, the material and flow for streams 37, 16 and all related streams are automatically calculated in the spreadsheet. By performing a material and flow balance around the hydrocyclone, stream 18 itself is also calculated automatically in the spreadsheet. In the example shown, the calculated flow rate is 8.748 BPD and the concentration is 92,987 ppmv. After a few iterations, the guess and calculated values agree and the final result is 8.748 BPD and 93,482 ppmv.

14.2 Process Performance Diagram

Process Performance Diagram: Two of the most important diagrams that characterize a water treating system are the Process Flow Diagram (PFD) and the Process Performance Diagram (PPD). Whereas the PFD gives the process schematic, the PPD gives the process performance. In other words, the PFD shows the equipment selected, the process line-up, the reject routing and all equipment and routing relevant to the system design. An example of a Process Performance Diagram is given in the figure below. The PPD gives the OiW concentration and oil droplet diameter (D_{v50}) for each important location within the process line-up. There are various measures of oil drop diameter that could be used in this figure. We have chosen here to use the Dv50 value (volume average 50 % value). The Dv50 value is typically used whenever the entire drop diameter distribution is calculated. In some cases however, only the maximum drop diameter is calculated and not the entire drop diameter distribution. In that case, the x-axis would be the Dv95 value or Dmax. In other cases, it is useful to use the Dv10 drop diameter. This is used in cases where a pronounced bi-modal distribution is observed. In that case, it is important to track the small drops through the system.

The separation efficiency of each piece of equipment can be seen directly on the figure. The first point in the system represents the condition of the fluids upon entering the inlet separator. It is in the upper right hand side. As the fluids proceed through the system both the D_{v50} oil drop diameter and the OiW concentration decrease. The final drop diameter and OiW concentration is seen on the lower left hand side. The length of the line from one point to another is a measure of the separation efficiency of each piece of equipment. The inlet separator efficiency is given by the distance between the first two points (upper right hand side, and next point proceeding to the lower left hand side). The distance between the second and third points is a measure of the hydrocyclone separation efficiency.

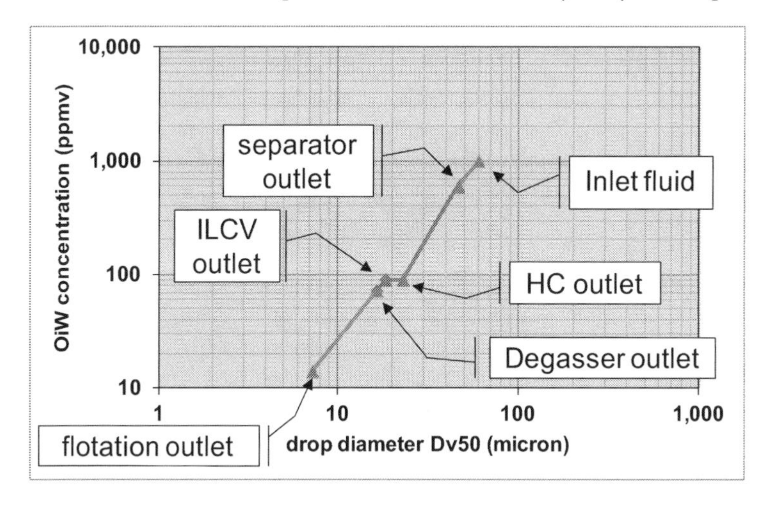

Figure 14.1 Process Performance Diagram (PPD). This diagram is for a typical platform and is for illustrative purposes. The PFD has a three phase separator, hydrocyclone, Interface Level Control Valve (ILCV) downstream of the hydrocyclone, degasser vessel and flotation unit

The data shown in the figure were selected from several surveys. Not all of the data from those surveys is shown. Much more data is available however when plotted, they appear scattered on such a figure. We have excluded those data that appeared to be outliers and which do not illustrate the differences between North Sea and deep water Gulf of Mexico systems. Unfortunately, high quality data that does not require such biased selection could not be found. Due to the fact that data was selectively chosen, this figure is only illustrative and not conclusive. This is part of the reason why modeling is also carried out.

As indicated in the figure caption, the points for the NS water treating systems include the primary separators and the hydrocyclones. This equipment, without flotation, is capable of treating water to the required quality. The points for the deep water GoM water treating systems include primary separation, hydrocyclones, and flotation. For the deep water GoM systems, large horizontal multistage flotation is used in order to meet the required water quality.

It is a bit counter-intuitive that the PPD for most systems should result in straight and parallel lines. However, with some degree of consideration, it stands to reason that such a correlation might prevail. Indeed, based on field measurements, a very good correlation is typically obtained.

Further, it is interesting to note that the slope of the lines is almost always found to be roughly 1.9 to 2.0. These lines are best-fit straight lines. In general, the slope on a log-log plot gives the exponential dependence of the dependent variable on the independent variable. According to Stokes Law, the exponent of drop diameter is 2.0. Rigorously, Stokes Law only applies to the motion of a single isolated drop in an otherwise stationary fluid. Perhaps the deviation from Stokes Law (exponent of 1.9 versus 2.0), is due to hindered settling in a fluid containing a large number of drops.

14.3 Dynamics of Continuous Flow Systems

In the upstream oil and gas industry, process upsets are far more common than in the downstream industry. The source of these upsets is the starting and ramping of wells, changes in well lineup, and slugging of multiphase fluids in wells and flow lines. The dynamics of upstream water treating systems are an important issue in the design and performance of a water treating system.

There are two important types of dynamics in water treating systems. The first is a sudden surge of flow. The second is a sudden surge in oil concentration. In many cases, the two types of surges occur together. In the material below, we analyze all three types of dynamics.

In general, predicting system dynamics is best done with the aid of differential equations. The basic equations are rather easy to write down. The difficulty usually arises in the complexity of the vessel internals. Such internals may include devices for slowing the momentum of the fluid, for more uniformly distributing the flow, or conical sections that establish reverse flow of an oil stream as in a hydrocyclone.

Relation between Dynamics and Flow Patterns: One of the means to understand the dynamics of a system is to first understand the flow pattern under steady flow. The flow pattern under steady flow can tell us much about how surges can be passed along to downstream vessels, or how they may be subdued in a given vessel.

Flow Pattern Modeling: In order to model system dynamics, it is necessary to have a good understanding of the flow patterns in the flow lines and vessels. Flow patterns in flow lines can be predicted based on a knowledge of flow rate, line diameter, and fluid properties. Both single phase and multiphase flow can be modeled reasonably well. The flow pattern in a vessel is much more difficult to predict. In general, there are two methods currently in use in the industry. Both methods are of relatively recent development and it is likely that further development and evolution will occur in the future.

Tracers & CFD: Vessel internals are critical in achieving the necessary separation performance of a vessel. However, they do add significant complexity to modeling the performance. To overcome the complexities of vessel internals, the industry has developed two techniques.

The first technology is empirical and based on the injection and detection of tracer compounds. A tracer is a compound that is added to the flow stream over a short period of time. It is added at a low concentration such that it does not have any effect on the flow pattern. It is added at a steady and constant rate. When added, the timing is recorded. Samples are taken downstream to determine the flow pattern in the line or vessel under study.

The second technology is Computational Fluid Dynamics. CFD uses well known equations of fluid mechanics to predict the flow patterns through the vessel. CFD is very useful in optimizing the type and location of vessel internals.

Both of these techniques provide detailed information about residence time, and mixing of fluids inside the vessel. Both techniques provide a detailed snapshot of performance for a given set of internals, and for a given set of flow conditions.

The advantage of tracers is that they require no assumptions about the vessel internals. In fact, tracers can be used to determine if the vessel internals are still in the original design condition, or whether the vessel internals have become fouled, plugged, or broken. The generalized flow model presented below is based on the concept of tracers.

Residence Time Distribution Functions: The most important property of a gravity based separation vessel is the residence time. This has been discussed to some extent in Chapter 8 – Primary Separation Equipment. In that chapter, the equipment available for smoothing out the flow pattern was discussed. It was pointed out that the residence time is not the same for all fluid elements that enter the vessel. Some fluid elements get sucked into a streamline that bypasses most of the volume of the vessel. For these fluid elements, residence time is short. As we discussed in the example, some other fluid elements are diverted away from the main streamline and end up getting caught in eddies. These elements may have longer residence time but because of the swirling motion of eddies, they do not necessarily contribute to good oil/water separation. The performance of a vessel depends on the sum total of all fluid elements. In this chapter we give more detail regarding residence time. We provide a general scientific description of residence time that is useful for not only steady state oil/water separation but also system dynamics. We define a property of the vessel referred to as the Residence Time Distribution Function, or distribution function for short. We give its properties, show how it can be measured using tracers or estimated using Computational Fluid Dynamics, and we demonstrate how it can be used to predict the dynamic performance of a vessel.

We start with a general description of two ideal types of residence time distribution functions, the perfect plug flow model, and the perfect mixing model. These are described below. They provide a general introduction to the concept although neither is ever found in practice.

Once the general subject has been introduced by way of these two ideal models, then the properties of residence time distribution functions are discussed together with some important details of how they are measured and calculated.

Perfect Plug Flow Mixing Model: In the plug flow mixing model it is assumed that fluids do not mix. Elements of fluid which enter the vessel at the same moment move through it with constant and equal velocity on parallel paths, and leave at the same moment. In this model, the retention time of fluid in a vessel is equal to the simple mathematical expression: V/Q, where V is the vessel volume occupied by the fluid, and Q is the volumetric flow rate.

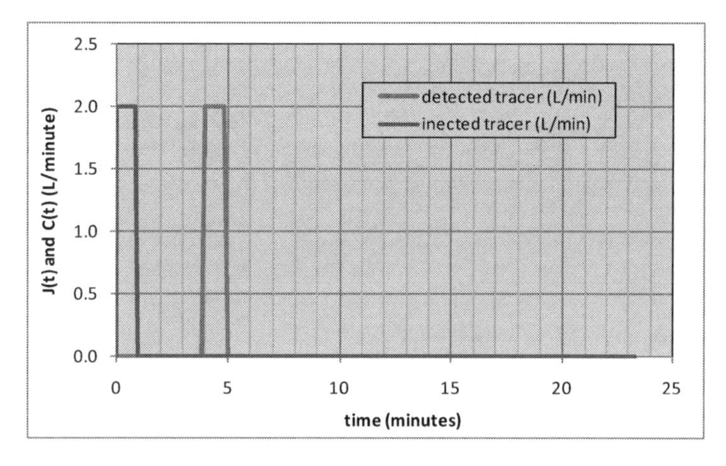

Figure 14.2. Ideal plug flow response to tracer injection. The flow rate of injected tracer is shown in red. Tracer injection started at time = 0, and continued for 1 minute, at a rate of 2 L/min. Thus, 2 liters of tracer were injected. The flow rate of detected tracer is shown in blue. The detection of tracer started at time = 4 minutes and lasted for a minute. The time difference between these two curves (4 minutes) is the vessel residence time

<u>**Perfect Mixing Model:**</u> When modeling fluid flow through a system of pipes and treating units, various assumptions are usually employed to account for the extent of mixing. In the perfect mixing model, the fluids are assumed to be uniform throughout the unit and the fluids are assumed to be completely mixed. In this model the outlet concentration is assumed to be the same as the concentration of material within the vessel or pipe. This assumption is known as the perfect mixing assumption. It is employed in the Continuous Stirred Tank Reactor model.

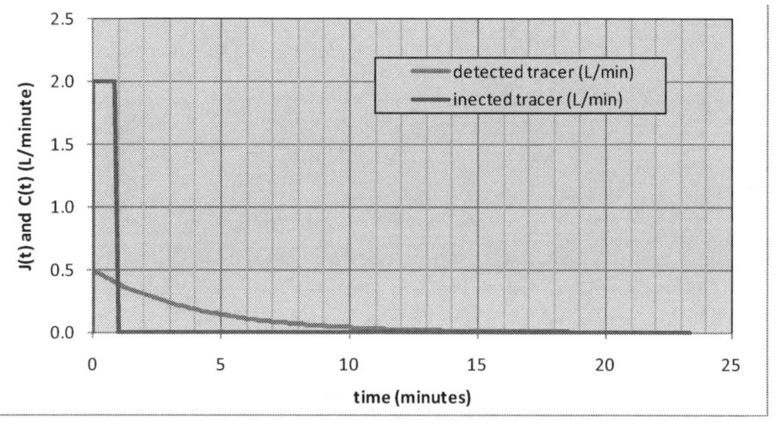

Figure 14.3 Perfect mixing flow response to tracer injection. The flow rate of injected tracer is shown in red. Tracer injection started at time = 0, and continued for 1 minute. Thus, 2 liters of tracer were injected. The flow rate of detected tracer is shown in blue. The detection of tracer started immediately and lasted for several minutes. This is typical of perfect mixing

<u>**Generalized Flow Models:**</u> In most cases, neither of the two idealized models above are accurate. In other words, few systems behave in practice as either a perfect mixed system, or a perfect plug flow system. We present here a general model that includes both of the idealized cases and all cases in between.

The generalized flow model is based on the concept of tracers. A tracer is a compound that is added to the flow stream over a short period of time. It is added at a steady and constant rate. When added, the timing is recorded. Samples are taken downstream to determine the flow pattern in the line or vessel under study.

If the vessel flow pattern is plug flow, then the tracer would be detected at some time later, at the same concentration as added. This time is known as the residence time. The detection of tracer would continue for the short period of time over which it was added.

If the flow pattern is perfect mixing, then the tracer compound would be detected at some concentration almost immediately. The concentration would die off over time.

Tracer Injection Function *J(t)*: The tracer is usually injected over a short period of time. It is usually assumed that the tracer injection is an idealized spike. But this idealization is not necessary. In a generalized model, we can treat any detail of the tracer injection.

The fundamental equation of tracer injection is [1]:

$$Q = \int_0^{t_1} J(t)dt \qquad \text{Eqn (14.16)}$$

where:

Q	total amount of tracer injected (mL)
t_1	total time over which the tracer is injected (minutes)
$J(t)$	tracer injection function (mL/minutes)

Referring to the two figures above, $J(t)$ is plotted in red. The tracer in those cases was injected at a steady rate of Q/t_1. As can be seen from these figures, the tracer injection rate was 2 L/minute, where Q was 2 liters, and t_1 was 1 minute. In these cases, the equation above can be evaluated easily.

The concentration of tracer injected can be calculated from the following relation:

$$q(t) = J(t)/v \qquad \text{Eqn (14.17)}$$

where:

n	flow rate of process fluids into vessel at the time that the tracer is injected (m^3/minute)
$q(t)$	tracer injection concentration as a function of time (ppmv: mL/m^3)

Tracer Response Function *C(t)*: Two examples of a tracer response function have already been shown, in the figures above. We now give the general case.

$$Q = \int_0^{\infty} C(t)dt \qquad \text{Eqn (14.18)}$$

where:

Q	total amount of tracer injected (mL)
$C(t)$	tracer response function (mL/minutes)

Note that the upper limit of time is now infinity, instead of t_1 as for the tracer injection function. In the case of the tracer response function, there is no way to know how long to carry out the integration therefore the upper limit is set at infinity. Obviously, once all of the tracer has been accounted for, then the integration can be stopped.

Residence Time Distribution Function $G(t)$ as a special case of the Tracer Response Function $C(t)$: In the examples already given, the tracer is injected over a short time interval. This is not always the case. In practice sometimes the tracer is injected over longer time intervals in order to study some aspect of the vessel performance. However, when the tracer is injected in a short sharp burst, the tracer injection function is referred to as a delta function.

$$G(t) = \int_0^\infty G(t-s)J'(s)ds \qquad \text{Eqn (14.19)}$$

Two examples of a tracer response function have already been shown, in the figures above. We now give the general case.

14.4 Process Configuration – Best Practices

It is difficult to provide general recommendations for process configuration since there are many possible options depending on the fluid characteristics, extraction techniques (primary, water flood, etc), disposal options for the water, etc.. Nevertheless, there are some generalizations that can be particularly related to handling rejects.

The science behind PFD design is quite complex. In the future, process design software packages will be available which allow the calculation of performance of a given PFD. When this software is available commercially, then various PFD can be evaluated and selection can then include not only cost, but also overall system performance. While the science is known, commercial software is not yet available. For now, Best Practices must serve the need for some simple approach to designing an effective PFD.

The term Best Practices, refers only to the performance of a system in removing oil from water. This is a very narrow definition which does not include capital cost, operating cost, schedule, operability, footprint, or weight. All of these factors are important in developing and executing a project. They typically, and in many cases justifiably account for deviation from Best Practices. To be most useful, the term Best Practice would include all of these factors. From a project perspective, all of these are important elements of a process lineup and they must all be taken into account. It goes without saying that a project must balance all of these factors to arrive at the best design for a given system. However, many of these elements are project specific, or depend on the local regulations, or the financial constraints imposed on a project due to the particular business cycle. It is not practical to take all of these elements into account in this report. In fact, it is not even apparent how one could do that in a general way in order to anticipate all regional and temporal possibilities.

Best practices for removing oil from water are given here in the case where there are no other constraints on the design of the system. The Best Practices are defined in terms of the following categories. These are described in detail below.

- minimize inlet shear
- apply heat upstream
- prevent solids production, separate and remove solids
- minimize the use of hydrate inhibitor
- minimize the use of corrosion inhibitor

- separate water early in the process
- provide an effective rejects handling system
- provide an effective chemical treating system
- provide an effective monitoring and control system

Minimize Inlet Shear: A fact which is often overlooked in water treating is that system performance is often determined by what happens to the fluids upstream of the inlet separators. The inlet drop size and the stability of oil drops are two important factors that determine the performance of the separation equipment. Inlet drop size is determined by energy dissipation (shear intensity) that the fluid experiences. The oil drop stability, or conversely oil drop coalesce tendency, is a function of drop size, oil drop concentration, and presence of production chemicals such as spent acid, and hydrate and corrosion inhibitor. These chemical factors are discussed further below.

The extent of inlet shear depends on the pressure drop of the fluids through the well head choke, and the platform boarding choke. The greater the pressure drop across these valves, the smaller the average drop size.

Apply Heat Upstream: Most fluid processing facilities have some form of heat input. Heat is typically added in order to meet the vapor pressure specification of the export crude oil. Since the vapor pressure requirement is applied to the export crude, heat can be added as far downstream as the dry oil tank. Typically however, heat is usually added upstream of this point in order to realize benefit from decreased viscosity and improved oil and water separation.

The point at which heat is added is a tradeoff between several factors. If heat is added upstream, then heating will be required for high pressure system with a greater volume of fluid and gas. When they are placed upstream, they must not only heat the oil but also the water. In addition to the added fluid volume, water has a heat capacity that is a least a factor of two higher than that of most typical crude oil. Thus, the heat generating capacity of the platform must be larger. This adds to the weight, space and capital expense of a project. However, the benefit of adding heat upstream is that more of the gas will break out at higher pressure, thus reducing the required capacity of the gas compressors. Also, if heat is added upstream then oil and water separation processes realize the benefit of lower viscosity fluids, hence achieving better water quality. If heat is added downstream, then less heat is required but there is no other process benefit.

Prevent Solids: The presence of solids has several detrimental effects on oil/water separation. Solids tend to become partially oil wet and migrate to the oil/water interface where they stabilize an oil/water emulsion. Most of the solids are reservoir related sand/clay production.

Handling of solids involves three phases. The first phase is potentially the most effective and that is to prevent solids from entering the production fluids. Depending on the source of solids (discussed below), this can be accomplished in various ways. Effective well bore sand screens are essential in high rate wells. The second phase of solids handling is to separate them from the production fluids at a point relatively upstream of the oil/water separation equipment. This typically involves the use of sand cyclones, screens or filters. The third phase of handling solids is to remove any accumulation of solids in the production separation and water treating equipment. This typically involves jetting with external nozzles, or the operation of an installed sand jetting system. Both solids separation and removal are somewhat difficult to carry out. By far the most effective means of dealing with solids is to stop them at their source, namely downhole in the wells.

Minimize Hydrate Inhibitor: Effective water treating starts upstream of the facilities. One of the most important decisions regarding water treating is the hydrate strategy and other flow assurance strategies, including Asphaltene and corrosion control of the subsea flowlines. Many of the chemicals that are needed for flow assurance applications have an impact in the oil water separation process. Some are worse than others. Many of the chemicals required will lower the interfacial tension thus making the oil/water mixture more likely to form small drops of oil as a consequence of valve shearing.

Select Corrosion Inhibitor on a Holistic Basis: Asset integrity for the lifecycle of a project can be a costly issue. Materials selection involves two kinds of information. The first is the CAPEX for the materials of construction. This is relatively straightforward in terms of making cost estimates on the basis of material selection for flow lines, risers, well's head and so on. Often there is pressure on a project to reduce CAPEX in order to meet project value requirements for investment.

The second is based on the net present value of operational costs over the lifecycle. The cost of corrosion monitoring, use of corrosion inhibitor, and the cost of shut downs and repairs are taken into account. This is a much more difficult calculation since it depends on a number of unknown factors. It is often difficult to assess accurately the corrosion rate prior to operation. For example, it is unlikely that the production fluid composition for 5 or 10 years will be known with any degree of certainty. Thus, it is sometimes the case that corrosion inhibitor is required later in field life.

Like so many other decisions in field development, it would be prohibitively expensive to design for the worst case scenario. Nevertheless, the use of robust and corrosion resistant materials which minimize the necessity for injection of corrosion inhibitor is a practice which will help ensure water treating effectiveness.

Separate Water Early in the Process: While the topic of oil/water separation is a broad topic, the definition of Best Practices for water treating can be easily stated: remove the large drops of oil before they become sheared either by valves or by pumps.

The required residence time for separating oil and water in a three-phase primary separator depends on the fluid characteristics such as oil density, water density, oil drop size, and water viscosity (related to temperature). The greater the driving force for separation, the more fluid can be processed in a given vessel volume per unit time. Conversely, if the driving force is diminished then the vessel must be larger to achieve the same level of separation. Stokes Factor, as described below is a good measure of driving force for separation.

Three phase separation and the use of a hydrocyclone on the water discharge is an effective means of separating oil from water before the detrimental effects of valve shear. The interface level control valve for the separator is downstream of the hydrocyclone.

Fluid characteristics can vary significantly from one reservoir to another. If for example, a light hydrocarbon is mixed with a hydrocarbon having unstable or marginally stable asphaltenes, the light hydrocarbon will act as an anti-solvent and cause the asphaltenes to precipitate. The precipitated asphaltenes will migrate to the oil/water interface and cause stable emulsions. Incompatible water streams are also problematic. Thus, having a number of primary separators rather than a single large separator is better form a water treating standpoint. Once the fluids are separated, then comingling has much lesser impact.

Provide an Effective Reject Handling / Recycle System: All water treating equipment generates a reject stream. In selecting equipment and designing an oil/water separation system, it is important to consider the flow rate and oil-in-water concentration of all reject streams. A material and flow balance of the overall process is required. If the primary separation equipment is not adequately sized to handle the reject flow, then the system will be bottlenecked. The operators will be forced to reduce either the oil retention time, the water retention time or the reject flow rate. Any one of these steps will result in poor oil and / or water quality.

While most water treating equipment provides better separation at higher reject flow rates, increasing the flow rate of the reject streams has serious negative consequences on the overall system design. Equipment selection should consider all relevant flows: the inlet flow and concentration, the effluent flow and concentration, and the reject flow and concentration. While this seems obvious, many equipment suppliers given only the inlet and effluent flows without specifying the separation efficiency as a function of the reject flow rate. This often leads to process design errors.

In many cases it is desirable to route the reject stream through the primary separation system. This greatly enhances the probability of coalescence and capture of the oil in the main oil stream destined for oil export. Realistic values of reject rate (percentage or flow rate) must be used in sizing the vessels which will handle both the reject and the primary production stream. Serious complications arise if vessel are not adequately sized for cross-sectional area, retention time, and discharge flow rate.

As discussed, it is necessary to minimize the volume of rejects while maintaining the performance of the water treating equipment. While it is imperative to do this in the design phase, it is also of great benefit to do this in real time during operations. This is achieved by measuring and controlling the reject flow rate in order to eliminate fluctuations. Control schemes for oily water reject streams are relatively simple but rarely used. Where they have been applied they result in a tremendous improvement of the overall operation of the oil and water separation system.

The design of facilities for offshore platforms is typically constrained in weight and space. Thus, it is often difficult to justify the addition of a slops or wet oil tank. Nevertheless, the presence of such a tank makes the handling and treating of reject streams much easier. Such a tank is used to provide settling time and another opportunity for chemical treating. Typically a solvent can be added which will prevent the formation of deposits due to recycling of water treating chemicals.

It is readily agreed that recycle loops should be minimized. Recycling of oily water generally results in smaller oil droplets due to pumping, and the conversion of water treatment chemicals into gunk. Most engineers would agree that recycling should be minimized. However, essentially all water treating equipment has a reject stream that must be handled in some way. On a deep water platform, there are typically only three discharge streams – dry oil pipeline, dry gas pipeline, and water for overboard discharge. Few deepwater or even shallow water platforms have an oily water waste disposal injection well. They are too expensive.

Since oily water reject streams cannot be discharged overboard, they must therefore be routed somewhere within the process. So if the reject is not routed to recycle, and if not to discharge, then where is it to be routed? This is a rhetorical question. The answer is that it must be routed to recycle.

Minimizing the impact of recycle streams is a key feature in a well-designed facility. In some cases this means minimizing the reject flow volume, but this is not always optimal. In many cases, water treating equipment provides improved separation performance at higher reject rates. Striking the right balance between reject volume and impact on the overall system can be difficult. Nevertheless, it is important.

Rejects from the hydrocyclone and flotation units are typically very well coalesced. In other words, the oil droplets are large and can be separated from water relatively easily. A small tank or vessel can be used to separate the oil and gently pump it using a low shear pump to an oil containing stream in the facility.

In [2] Frankiewicz and co-workers describe the addition of an "oil reject skid" to handle the hydrocyclone reject liquids, skimming from a surge drum, and the oily reject form an induced gas flotation unit. These streams contained oily solids and chemically stabilized emulsions which had been recycled to an upstream vessel and led to difficulty in meeting the overboard discharge limit. The oil reject skid was designed to separate oil, water, and solids. The vessel had a conical bottom for solids handling.

Select Effective Chemical Injection Points: Water treating systems are typically designed with the assumption that water treating and deoiling chemical will be used. Separation efficiency of separators and water treating equipment is almost always stated with the assumption that chemical will be used. In many cases, the performance of water treating equipment in the absence of chemical is actually not well known in the industry.

A well designed water treating system will include adequate chemical storage, premixing, mixing with process fluids, sufficient number and location of injection points, metering, and delivery.

Provide Effective Monitoring and Control Information: The successful operation of a water treating facility depends to a large extent on operator adjustments. Most separators and water treating equipment require adjustment of levels and interfaces, as well as the periodic operation of an interface bleed. Some separators and water treating equipment have the facility for online cleaning which must be periodically operated. Control systems must be periodically tuned to ensure smooth operation of the level control valves. Chemical injection rates must be adjusted to accommodate changing well lineups and production rates.

References to Chapter 14

1. P.V. Danckwerts, "Continuous Flow Systems. Distribution of Residence Times," Chem. Eng. Sci., v. 2, p. 1 (1958).

2. C. Yang, M. Galbrun, T. Frankiewicz, "Identification and resolution of water treatment performance issues on the 135 D platform," SPE – 90409, paper presented at the SPE ATCE, Houston (2004).

CHAPTER FIFTEEN

Troubleshooting

Chapter 15 Table of Contents

15.0 Introduction 309

15.1 Mechanical versus Chemical Causes of Problems: 310

15.2 The Dirty Dozen – 12 Common Causes of Poor Water Quality 311

15.3 Operations 318

- 15.3.1 Importance and Role of Platform / Facilities Operators 318
- 15.3.2 Knock-on Effect 319
- 15.3.3 Identifying Sweet Spots 319
- 15.3.4 Field Visit Checklist 319

15.4 Ishikawa Diagram - Five Categories of Causes: 321

15.5. General Methodology for Troubleshooting O/W Process Problems: 326

- Fact Based Problem Solving Process 326
- Step I: Capture: 327
- Step II: Problem Analysis and Definition: 327
- Step III: Root Cause Analysis and Definition: 328
- Step IV: Solution Development: 330

15.6 Summary of All Deliverables 331

15.7 Reporting 332

References to Chapter 15 334

15.0 Introduction

The main objectives of this chapter are to provide a systematic methodology to (1) determine the root cause of produced water treating problems, and (2) develop cost effective solutions to solve those problems. Much of the material in this chapter was developed based on three decades of personal experience in the deepwater Gulf of Mexico, North Sea, Brazil, and Nigeria, and onshore experience in the Permian Basin, and Oman. It was greatly advanced by discussion and collaboration with Ted Frankiewicz who has published relevant papers [1 – 3]. Also, the methodology borrows heavily from the business management literature on Fact Based Problem Solving, Root Cause Analysis, Failure Modes Effects Analysis, PDCA (Plan, Do, Correct, Act), Five Whys, and so on.

In this chapter, only a few examples of common problems are presented. More examples are available [4, 5]. Examples are not the main focus of the chapter. Instead, the focus is on the methods that can be used to identify the root cause(s) of a problem, to definitively exclude other potential causes, and to develop a stair-step approach to prove the identified cause and solve the problem.

This methodology is intended to be used for difficult water treatment problems. It is assumed that various simple solutions have already been attempted and failed. Such ad hoc methods, usually carried out by the local staff, are typically able to solve the majority of produced water problems.

The troubleshooting method described here is intended to be used for the difficult problems which have persisted for several months, or for those problems where the only known solution is expensive. In such cases it is necessary to follow the methodology as closely as possible. It is tempting to skip some steps but one thing to keep in mind is that it is difficult to accurately decide which tools, methods, and steps can be skipped until a detailed understanding of the problem has been reached.

Understand the Mechanism: The most important aspect of the troubleshooting method is to develop an understanding of the mechanism of the problem. The key to successfully designing and operating a water treatment system is to understand the fundamentals of the physics, fluid mechanics, and the chemistry behind the water quality problem. Once the mechanism is understood, a more robust solution can be identified. With this understanding, water treatment system designs can be made operationally more robust and problems are easier to resolve. This chapter presents the tools, methods, and steps to do that.

Multiple Stair-Step Solutions: By developing a strong understanding of the mechanism of the problem, a more precise set of cost effective solutions can be developed. It is typical that no single solution will be acceptable to the operating unit. Multiple solutions must be proposed which will range in cost, schedule, and likelihood of solving the problem. The operating unit will then select the solution that best fits with their objectives and constraints. This in-depth approach is justified particularly when large sums of capital expenditure are required to solve a problem. In order to develop multiple solutions, a strong understanding of the system is required. A complete characterization of fluids, equipment, process configuration, chemical injection system, and operating practices is required. In some cases, the verification of the mechanism will only be achieved after one or more stair step remedies have been implemented. As discussed below, stair step remedies should be designed both to verify the mechanism and incrementally fix the problem.

15.1 Mechanical versus Chemical Causes of Problems:

It is stated above that the main focus is not on examples. However, a few examples of specific problems are given in the next sections. Then the systematic methodology is described. Generally speaking, produced water troubleshooting is a complex process. There is a tendency to focus on either mechanical or chemical aspects. Successful troubleshooting generally involves a balance of both. Few, if any, problems are strictly chemical or mechanical. Chemical bottle testing should be carried out together with mechanical equipment optimization. The following aspects of a system should be considered together with equal emphasis:

- The properties and composition of the gas, oil and water
- The chemical treating system
- The stability (flow assurance) of the fluids
- The compatibility of the fluids
- The separation equipment
- Vessel design
- Equipment capacity
- Vessel internals
- Pumps
- Valves
- Process flow schematic

Chemical aspects include the oil and water properties, the chemical treating system employed, and the compatibility of fluids and chemicals. The mechanical aspects include the separation equipment, vessel design, vessel internals, pumps, valves, and process flow schematic. It is often the case that both chemical and mechanical solutions are required to improve water quality. This troubleshooting guide provides a systematic approach that will help differentiate between mechanical and chemical problems and will help provide integrated answers across all disciplines.

It must be pointed out that there is no single best or correct facilities design that will provide good results for all fluids. Water treating problems are often due to a miss-match between the chemical and mechanical aspects of a system. For example, a set of vessel internals may be appropriate for one type of produced fluid properties but not another. Also, the chemical treating program may not be appropriate for the particular location due to the particular fluid properties or the process configuration. Recycles that work well at one location may not work at another due to differences in the fluid properties, the process configuration, or the compatibility of the chemicals used and their gunking potential upon recycle. Usually, compromises in weight, space, cost, and construction time must be made in the system design in order to deliver a project within the required metrics. A typical system will have excess capacity in some sections, and bottlenecks in others. It will have limitations, and be sensitive to certain kinds of upsets. Equipment that works well at one facility may not work well at another due to differences in fluid properties.

One of the major challenges in such work is to differentiate between chemistry-related problems (fluid compatibilities, solids, WSO, etc.) and mechanical-related (process bottlenecks, vessel internals, etc.) problems. This Guide provides a systematic approach that will help the engineer make this differentiation.

It must be pointed out that there is no single best or correct design that will provide good results for all fluids. This may seem self-evident but we have often encountered design engineers who ask, "just tell me how to design a treating facility that will work in all cases." There may actually be such a design, but it would be prohibitively expensive. It would be costly, large, and would require a relatively long time to build. In other words, the only way to design and build a practical water treating system is to understand the fluid properties, the available design options, and to make appropriate compromises.

Water treatment problems are most often due to a miss-match between the fluid properties, the chemical treating system employed, and the process and vessel design compromises that have been made. For example, a set of vessel internals may be appropriate for one type of produced fluid properties but not another. Also, the chemical treating program may not be appropriate for the particular location due to the particular fluid properties or the process configuration. Recycles that work well at one location may not work at another due to differences in the fluid properties or the process configuration, or the compatibility of the chemical used and their gunking potential upon recycle.

15.2 The Dirty Dozen – 12 Common Causes of Poor Water Quality

Ted Frankiewicz published a list of common problems [1]. He referred to this list as "The Dirty Dozen."

1. Separators operating outside of their design range for oil, gas, and/or water.
2. Recycle streams feeding contaminant laden emulsions to separators.
3. Gas slugging below the water line in skimmers and separators.
4. Neutrally buoyant, oil coated solids.
5. Water soluble, film-forming corrosion inhibitors, especially in fresh waters.
6. High shear dispersion of oil droplets, especially as API gravity increases.
7. Excessive or incorrect chemical treatment.
8. Highly variable flow rates (e.g. snap acting valves).
9. Ultra-fine scale mineral solids (CaCO3, CaSO4, FeCO3).
10. High concentrations of organic acids in the water phase.
11. Iron sulfides.
12. Upstream separators filled with solids.

The more common causes of water treatment challenges can be grouped Info the following categories:

- Presence of solids (inorganic or organic)
- Excessive or highly varying fluid Flow rates
- Gas breakout or slugging in or into process equipment
- Improper chemical treatment programs

This section discusses several of the more common, often overlapping causes of produced water cleaning problems end suggest means to both identify and resolve those problems by either mechanical or chemical means. Examples are given where the solutions have been implemented a priori in process system designs and in field redo-fits. In addition, some general rules-of-thumb are presented for selecting an appropriate technology for addressing an identified water treatment challenge.

1. **Contributors to Poor Water Quality: Separators Operating Beyond Capacity**

Options for significantly increasing separator capacity and improving performance include:

- Install Vortex Tube inlet devices
- Install Perforated Plates for Flow Control
- Use a GLCC to by-pass gas around the separator
- Convert a bucket & weir configuration to oil-over-weir
- For oil-over-weir separators, install a distributed water discharge system

Gas-Liquid Cylindrical cyclone (GLCC)

- Tremendous weight and footprint savings
- Inclined inlet piping creates stratified flow
- Tangential entry creates centrifugal force and creates vortex inside separator
- Separated gas exists top of vessel and liquid exits bottom of vessel
- Ideal for test separator applications
- Also utilized for two-phase production separators

Improving Vessel Hydrodynamics Permits, Short Residence Time Separations

Porta-Test Revolution™ on inlet

- Contain inlet turbulence
- Reduce gas carry-under
- Pre-coalesce emulsions

Engineered Perforated Plate Baffling

- Controls liquid short circulating
- Contains slug-generated turbulence

Distributed Water Discharge for Separators with ILC

- Less oil/water interface disturbance
- Discharges cleaner water without coning

The Porta-Test Revolution

- eliminate foam
- Improve Separator efficiency
- Reduce the size of your separation equipment and its required footprint
- Used in horizontal or vertical separators
- Can be designed for any gas-liquid separation
- Always recommended for liquid to gas ratios of 500 (barrels per MMscf) and higher

Bucket & Weir Separators

- Advantages
  - Interface level control not required
  - cleaner water for a given residence time
- Disadvantages
  - Available separation time is lower for a given vessel size compared to spill-over weir type
  - Weir height adjustments require vessel entry
  - Oil pad depth is flow rate dependent
  - Oil pad cannot be adjusted as flow compositions change

2. **Contributors to Poor Water Quality: Recycle Streams**
   - Once separated, contaminants should be given a positive route for removal
     - Skim tanks
     - Accumulator vessels
     - Secondary hydrocyclone systems
   - Recycling solids-laden oil to an upstream vessel generally results in contaminants re-entering the water treatment system
     - The solids loading from recycle streams can easily double or triple the equilibrium solids loading of produced water.
3. **Contributors to Poor Water Quality: Gas Slugging into Skimmers**
   - Useful numbers for estimating the volume of dissolved gas which is released due to a pressure drop across a control valve or a hydrocyclone"
     - For every 15lbs of pressure drop in the partial pressure of gas, the following will break out:

- 0.15 SCF of CH_4 per bbl of water
- 3.0 SCF of CO_2 per bbl of water

Example – A gas with 90% CH_4 and 10% CO_2 is reduced in pressure from 300 PSIG to 0 PSIG.

The gas released is 18 (0.15) + 2 (3.0) = 8.7 SCF/bbl (1 bbl = 6.3 ft^3)

- Study piping runs to see if break-out gas can accumulate into slugs
  - Risers & down-comers
  - Long horizontal runs
  - Lack of highly turbulent flow
- Catch break-out gas to prevent skimmer upsets
  - Porta – Test Revolution vortex tubes
  - GLCC external to vessel

4. **Contributors to Poor Water Quality: Neutrally Buoyant, Oil-Coated Solids**

- Must be removed by flocculation and flotation
- High MW polymers are most effective at collecting solids:
  - Can be difficult to disperse into water
  - Simple systems are available to easily disperse high MW emulsion polymers into produced water
- Ferric Chloride or Dithiocarbamates are effective solids collectors
  - Both chemicals create heavy flocks that require paddle skimming

5. **Contributors to Poor Water Quality: Corrosion Inhibitors**

- Factors favoring dirty water:
  - Use of water soluble corrosion inhibitors
  - Fresh water (<10,000 PPM TDS)
  - High API gravity hydrocarbons (API >45)
  - High CO2 in gas phase (partial pressure > 5 psi)
  - Presence of fine solids, especially scale mineral particulate <10 microns
- Solutions:
  - Change CI's
  - Minimize use of CI's
  - Coordinate CI's with Mg, Zn, Fe, PEI or other suitable ion

6. **Contributors to Poor Water Quality: High Shear Dispersion of Oil Droplets**
   - Sources of Oil-Dispersing Shear:
     - High pressure drop across control valves
     - Pumps
     - High velocity flow through pipelines over extended distances
   - Solution:
     - Separate the oil from the water prior to the shearing device
     - Hydrocyclones are particularly effective for this service
7. **Contributors to Poor Water Quality: Excessive or Incorrect Chemicals**
   - Over-treating with demulsifiers
   - Use of water soluble corrosion inhibitors
   - Over-treating with certain water clarifiers
   - Failure to maintain an effective scale inhibitor program
   - Over-treating with or unnecessary use of biocides
   - Failure to adjust chemical treat rates to account for variable water flow rates
8. **Contributors to Poor Water Quality: Highly Variable Flow Rates**
   - Common when long flow lines are present, e.g. from remote platforms or headers
   - Deep-water riser slugging
   - Snap-acting valves in upstream equipment
     - Can result in +/- 50% changes in flow rates in the span of 5-15 minutes
     - Contribute to operator decision to over-treat with chemicals
9. **Contributors to Poor Water Quality: Ultra-Fine Scale Mineral Solids**
   - Precipitated mineral solids are typically in the 5-20 micron particle size range
   - Mineral precipitates tend to be oil-wetted
   - Most common precipitates include $CaSO_4$, $CaCO_3$, $FeCO_3$, $BaSO_4$
   - Identify precipitated minerals using XRD
   - Carbonate minerals are acid soluble
10. **Contributors to Poor Water Quality: High Concentrations of Organic Acids**
    - Common when production is form hot formations

- Naphthenic acids in crude have some water solubility
- Most active in the presence of low concentrations of divalent ions such as Ca^{++}, Mg^{++}
- Requires specialized analyses to find and identify organic acids
- Acetic acid, phosphorous acid, $AICI_3$, $FeCl_3$ will deactivate the surfactant character of organic acids

11. **Contributors to Poor Water Quality: Iron Sulfides**
    - Major sources of Iron Sulfides
        - Sour production
        - Sulfate Reducing Bacteria (SRB) activity
    - Iron Sulfides often form neutrally buoyant, oil-coated solids
    - Dried FeS can be pyrophoric
    - Best removed by froth flotation using chemicals specially selected from this service

12. **Contributors to Poor Water Quality: Upstream Separators filled with Solids**
    - Solids will occupy space and reduce the residence time for settling
    - Solids in the vessel provide a substrate for bacteria to thrive
    - Bacteria produce H_2S, iron sulfide, bio-polymer
    - Bacteria cause corrosion which risks asset integrity and which adds iron to the produced water

Particle Size Helps to Select the Equipment to Recommend

Technology	Removes Particles
API Gravity Separator	>150
CPI (Plate pack) Separator	>40
Induced Gas Flotation	>25 w/o chemicals
Induced Gas Flotation	>7 with chemicals
Hydrocyclone	>10-15
Mesh Coalescer	>5
Media Filter	>5
Centrifuge	>2
Membrane Filter	>0.01

Flotation vs. Hydrocyclones – Which should you recommend?

In the table above, it would appear that hydrocyclones and flotation remove oil droplets to roughly the same cut-off size. Flotation removes a bit smaller droplet but not much in actual practice. It is important to keep in mind that hydrocyclones and flotation remove oil and oily solids by two completely different mechanisms. Hydrocyclones remove contaminants according to their difference in specific gravity with respect to the water. Flotation removes contaminants by a gas bubble capture mechanism regardless of specific gravity. Thus flotation has a good chance of removing oily solids that are neutrally buoyant whereas a hydrocyclone cannot..

Factors favoring Flotation

- Highly variable water flow rates
- Low water source pressure
- Treating water from multiple sources at varying pressures
- Oil-coated solids are present
- Flocculent required to generate a coherent skim layer
- Chemically stabilized reverse emulsions present
- Feed water contains <300 PPM hydrocarbons

Factors favoring Hydrocyclones

- Feed water contains >300 PPM hydrocarbons
- High pressure drop across control valves are shearing oil droplets
- Testing shows that water can be cleaned without chemical treatment
- Space or weight constraints are important (offshore)
- Solids contribute to water quality problems

Note: A skimmer or 1-cell DGF/IGF downstream of hydrocyclone is generally recommended

Hydrocyclones: The Simple Rules for Running them Right

- Install an in-line screen ahead of Hydrocyclones to catch large solids
- Maintain the PDR (Pressure Drop Ratio) between 1.5 and 2.5
- Maintain a minimum pressure drop across the hydrocyclone (.25 PSI for de-oiling, >10 PSI for desanding)
- For each system, there is a maximum pressure drop, above which efficiency declines due to internal turbulence
- Vessel level control valves should be installed downstream of a hydrocyclone
- If possible, avoid pump feeding hydrocyclones
- A simple bench-top centrifuge test can predict the likely performance of a hydrocyclone system
- Install short residence time skimming downstream of the hydrocyclone to collect coalesced oil droplets
- Whenever possible, field test a hydrocyclone liner prior to selecting this technology

Summary and Conclusions

- Water Quality problems start in the first separator
- In one way or another, solids often contribute to water quality problems
- Flow slugs resulting from gas breakout must be contemplated in the design of a water treatment system
- Avoid recycle streams
- By understanding the cause of a water quality problems, selection of a mechanical and/or chemical means to resolve it, is reduced to common sense.

15.3 Operations

The one subject that has not been discussed in any great detail is that of Operations. Therefore a separate section is given here. Generally it is true that no one particular problem area alone will manifest itself as the sole source of the problems. Even if one particular area is deficient, more so than others, the impact of this one area will already have been minimized by the operators, as they find operating practices to compensate and keep the system running at the required capacity.

This compensation effect is common in continuously operated process systems. The good operators will utilize existing excess capacity and operating margins in one part of the process in order to overcome or compensate for a bottleneck somewhere else. One obvious example of this strategy is the raising of process temperatures in order to compensate for inadequate retention time in the oil dehydration system. In doing so, the operators may complain of limited heat exchange capacity. In this case, the complaints are symptoms of a more fundamental problem, that of poor demulsification. On the other hand however, as will be discussed later, the solution to the dehydration problem may actually lie in providing yet even more heat, depending on the situation.

In general, when problems persist in a process, it is not always straightforward to identify the root cause, or solution because operators have already compensated to some extent. It is however, an indication that there is not enough processing capacity or flexibility to handle problems or upsets, or that several problems exist in multiple areas. This is why the fact based methodology is important.

15.3.1 Importance and Role of Platform / Facilities Operators

Often much can be learned from understanding why the operators have chosen to run the system in the particular manner in which it is being run. Also, if different crews run the system in different ways, there is much to be learned. Sometimes this would indicate that a couple of sweet spots exist. On the other hand, this might indicate that there is not enough data or process information available to really know how best to run the system. Finally, it might also indicate that the system dynamics are so great that they obscure what is really happening.

In any case, it is important to pay particular attention to the observations of the operators, to experience the problems first-hand on-site, and to develop a deep understanding of why the operators operate the system the way they do. The explanations of operators cannot always be relied upon, but the observations are extremely important.

In the material above, we have emphasized the importance of talking to the operators and understanding the interaction of the operators with the system. Among the many tools available to a troubleshooting team, this is the most important. In the section below, we give a general methodology for troubleshooting a system that is experiencing oil / water processing problems. In addition to working with the operators, there are many technical tools that must be applied. In the rest of the paper, we give results of the application of these tools to a particular case study.

15.3.2 Knock-on Effect

Generally it is true that no one particular area alone will manifest itself as the sole source of the problems. Even if one particular area is deficient, more so than the others, the impact of this one area will already have been minimized by the operators, and other areas will also show problems. This "knock-on effect" is common in continuously operated process systems. It is due to the fact that most operators will compensate for, and overcome to a degree any relatively isolated sources of problems. In doing so, they search for the sweet spot that allows them to optimize the system. They utilize existing excess capacity and operating margins in one part of the process in order to overcome or compensate for a bottleneck somewhere else. One obvious example of this strategy is the raising of process temperatures in order to compensate for inadequate retention time in the oil dehydration system. In doing so, the operators may complain of limited heat medium capacity, or limited heat exchange capacity due to fouling of the heat exchangers. These complaints are symptoms of a more fundamental problem, that of poor demulsification. On the other hand however, as will be discussed later, the solution to the dehydration problem may actually lie in providing yet even more heat, depending on the situation.

15.3.3 Identifying Sweet Spots

In general, when problems persist in a process, it is usually an indication that either the sweet spot is unacceptable (low rates, poor quality, etc), or that there is not enough processing capacity or flexibility to handle problems or upsets, or that several problems exist in multiple areas. Often much can be learned from understanding why the operators have chosen to run the system in the particular manner in which it is being run. Also, if different crews run the system in different ways, there is much to be learned. Sometimes this would indicate that a couple of sweet spots exist. On the other hand, this might indicate that there is not enough data or process information available to really know how best to run the system. Finally, it might also indicate that the system dynamics are so great that they obscure what is really happening.

15.3.4 Field Visit Checklist

At some point in the troubleshooting process it will be appropriate to visit the field. In general, the main activities to be carried out in the field include: jar sampling and simple observation of fluids, sampling for analysis, process walk-through, and discussion with operators. The following list can be put into an Excel spreadsheet and carried out as a checklist. It should be reviewed prior to the field visit. Not all activities will be required for every visit. It is typical that only a portion of the activities listed here would be carried out. Also, it is important to review the list and decide on an order of execution.

- Walk through the system with the operators, one unit at a time, discuss and record the comments and observations made by the operators.
- Ask the operators what the problems are and record their comments.

- Make an assessment of operating practices – speak to the operators and ask them if operating procedures and practices are documented and carried out the same for all crews. If the crew-to-crew procedures are different, ask each crew what the other crews are doing different and record this.

- Examine the sampling points. Record which ones are working and which are not.

- Review the sampling and testing methods and determine which of the data coming from the facility can be trusted. Record which sampling and testing methods are, in your opinion, incorrect or inaccurate.

- For each bit of data ask, what is it telling us about the fluids or about the system?

- Look at the fluids yourself

- Gather samples and set them on a desk. Record the time to settle and the ratio of oil/water after various settling times. Make a rough estimate of drop diameters. Look for small black particles (iron sulfide). Look for solids settling to the bottom. Look for evidence of chemical stabilization (very little separation after 24 hours). Skim the oil off the top and check for solids.

- Take samples in glass jars and look for step changes in water quality through the system – where does the quality change?

- Look at the produced water under the optical microscope – look for emulsions, solids, oily solids.

- Verify chemical injection rates, chemistries, locations, look for over treating

- Once a walk-through has been made, and you are familiar with the overall layout and piping configuration, then take another walk-through and try to understand the system dynamics. Listen for gas surges. Listen for pumps and compressors starting up and shutting down. Observe control valve stem position and movement. How often do valves and pumps change operation and for how long? Record any observations about system dynamics.

- Obtain data that will allow the calculation of flow rate in and flow out of each piece of separation equipment. Record the interface levels. Calculate the residence times.

- Test BS&W & oil concentration in water into and out of each vessel. Calculate vessel separation efficiencies. Calculate a material balance around the equipment to verify that the readings make sense.

- Make a list of sources of shear – all valves and pumps. Record the pump rpm, whether variable speed drive or on/off. If on/off, then record the percentage of time on/off. impeller size, check for recessed impellors (cause of high shear). Record valve trim, percent open, percentage of time in open position.

- Check the flow rates against the vessel, equipment, pump and valve design parameters.

- Determine what causes the upsets and dynamic events – transients, slugs, surges, rates, snap acting valves, and consequences of surges in the system.

- Carry out drop size analysis if you have the equipment. This is not always necessary since simple visual observations and desk-top settling are usually sufficient.

- Do a pad analysis. If there is a pad suspected in the vessel, develop a plan with the operators to empty the vessel and capture the pad material as the vessel is emptied. Look at the pad material under an optical microscope. Analyze the pad material for oil in water, and water in oil concentration. Send samples for analysis for solids, composition. Determine the origin of the pad material.
- Take samples of both oil and water for solids analysis. Carry out filtration tests, if possible.
- Look for trouble streams:
  - condensate mixing with crude oil (asphaltene dropout)
  - synthetics entering system (lube oil, glycol, etc)
  - oxygenated water
  - corrosion inhibitors
  - hydrate inhibitors (methanol, glycol, anti-agglomerates)
  - sources of bacteria (seawater)
  - scale forming (incompatible) water
- Carry out chemical bottle tests – for both product selection & problem diagnosis. Often bottle testing gives a clue to the constituents within the oil or to the production chemicals that are stabilizing the oil in water emulsion.
- Set up a Trouble Journal for the operators to fill in that will capture relevant data and transients after you leave. This can be reviewed after a couple or few weeks.
- Generate a schematic of the chemical treating system including chemical types, injection rates, process fluid flow rates, and chemical dose rates.
- Record routing through heat exchangers, available heat sources, location of heat addition, temperatures throughout the process.
- treat solids and convert them to water-wet rather than oil-wet
- Review the PFD, P&ID, and perform a walk-through to identify piping where mixing of bulk phases occurs. For example, find locations where oil is mixed into water, or where water is mixed into oil.
- Review the PFD, P&ID, and perform a walk-through to identify all oily water recycle streams. Take desk-top samples upstream and downstream of the recycle tie-in points and identify whether or not this is a potential problematic recycle.

15.4 Ishikawa Diagram - Five Categories of Causes:

As discussed in Section 1.2 (Organization of the Book) the material in this book is organized into five subjects that correspond to a Troubleshooting Checklist. Essentially all aspects of water treatment can be categorized into one of five categories or subjects. Water treating problems can be complex and can involve a number of different causes. Troubleshooting is a search for an optimal solution. It is important to be thorough and to have a balanced viewpoint which covers all five categories in troubleshooting. If one category or subject area is left out or overlooked, a simple solution may then be missed.

Expert review of a system is guided by the use of a Fishbone or Ishikawa Diagram. Ishikawa diagrams are cause and effect diagrams created by Kaoru Ishikawa, a Japanese expert in quality control and process management. Common uses of the Ishikawa diagram are product design and quality defect prevention, to identify potential factors causing an overall effect. Companies such as Toyota, Mazda, and the Kawasaki shipyards have implemented the concept. These are highly successful companies that have come to dominate their business sector through the ability to control quality, reliability, and cost..

An Ishikawa Diagram for produced water is given below. It is based on the five main categories of potential causes:

1. Fluid properties;
2. Equipment performance;
3. Process configuration and control;
4. Chemical treating program;
5. Operating procedures.

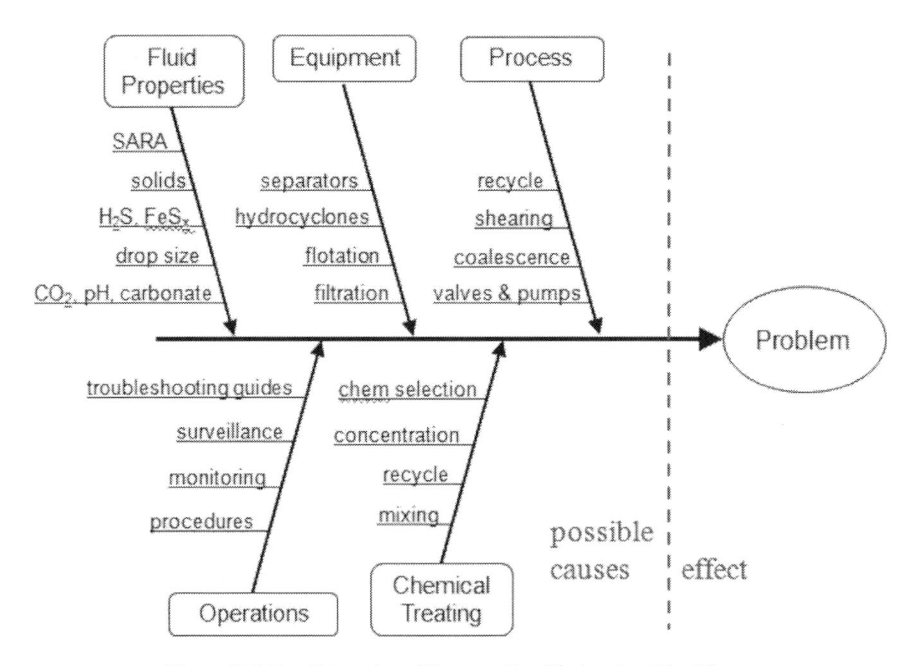

Figure 15.1 Five Categories of Causes – Troubleshooting Checklist

The fishbone or Ishikawa Diagram is used as a Troubleshooting Checklist. As discussed below, all potential causes of problems should be accompanied with the observed effect of the problem. This can be determined by starting with a statement of the potential problem and asking "why" for every answer that is given. This is typically done for five iterations of why question and answer. An example is given below under Step 3a of the Root Cause Procedure.

Cause-and-effect diagrams are intended to reveal key relationships among various variables, and the possible causes provide additional insight into process behavior. It is used here as a reminder of the 5 categories of water treating problems.

Causes can be derived from brainstorming sessions with the operators and engineers. The following discussion gives some detail on the 5 categories of water treating problems.

1. **Gas, Oil & Water Characterization:** The first activity is to characterize the oil and water and determine the source of problematic characteristics (e.g. small drops, stabilized emulsions, solids, conglomerate, schmoo, gunk, etc.).

   Relevant Questions:

   - Is the oil biodegraded and does it have high TAN, or high organic acids?
   - Is there H2S, iron sulfides, Schmoo?
   - What is the solids content, are they oil-wet vs water-wet, what is the composition, source, size?
   - Are naturally occurring surfactants, acids, emulsion stabilizing constituents present?
   - How stable is the oil stability in terms of asphaltene appearance pressure, asphaltene content, asphaltene / resin ratio?
   - What is the water chemistry, bicarbonate, pH and how stable are the mineral solids?

   A proper oil and water characterization is vital for the next steps, which aim to determine if the equipment, process and chemical treating program are appropriate for the fluids being processed. For example, heavy fluids require different handling than light fluids and problematic fluids containing iron sulfide require a different processing approach than fluids that do not have iron sulfide. Note that it may be necessary to review fluid rates, BS&W, temperature ranges and other fluid characteristics for various individual wells, especially when production is from different fields. In some cases, a single well can be responsible for the majority of problems.

   If a proper characterization is carried out, then the type of equipment, the process configuration, the chemical program and the operating procedures required to treat the water can be determined fairly quickly. In the initial phase of troubleshooting it is important to determine if the equipment, process, chemical treating program, and operation are suitable for the particular fluids. For example, heavy fluids require different handling than light fluids. Problematic fluids containing iron sulfide for example require a different processing approach than fluids that do not have iron sulfide.

2. **Equipment Capacity and Technical Limit:** The second activity is to determine the technical limit of the existing equipment. The technical limit of each piece of equipment must be determined as well as the technical limit of the system as an integrated whole. The technical limit is defined as the separation efficiency or performance for a range of feed flow rates, reject and discharge flow rates. Flow rates are systematically varied and limit diagrams are constructed and operating points plotted on the diagram. The technical limit of the equipment is compared to the actual flow rates being used. If the actual flow rates are close to, or above, the technical limit, then the particular piece of equipment is identified as a bottleneck.

Relevant Questions:

- Is the reject flow rate suitable for the equipment design?
- What is happening inside the vessels (pads, bypassing, surging, skateboarding)?
- Are the internals in good shape or are they fouled, plugged, busted apart, etc?
- Are the individual vessels operating at good efficiency? Measure BS&W and oil in water into and out of each vessel.

3. **Process Configuration:** The piping connections, and process routing of equipment is nearly as important as the equipment itself. This is particularly true of reject and recycle streams. While the process flow diagram will be different for each facility, there are some generalizations that can be made. High shear and flow variations may upset the process and particular attention must be paid to recycle streams. Some examples are:

   Relevant Questions:

   - How severely do the fluids get sheared?
   - What is the effect of the process recycles on the fluid properties?
   - Are all sources of heat exchange being effectively utilized?
   - Where are the bottlenecks in the process? Do a mass balance to determine the flow rate versus rated vessel capacity.
   - What are the flow rate and pressure dynamics of the system? Can they be smoothed out or eliminated?

   The presence of water clarifier and flocculant in fluids which are recycled from a flotation unit to an upstream separator will often result in the formation of stable interface emulsions which can upset the treatment process. Since it is impossible to know the character of the water clarifier beforehand, it is particularly important to avoid the recycle of contaminants recovered from a flotation unit into any upstream separator or tank.

   Clarified water from the top of the slop or wet oil tank should be returned to the inlet of the water treatment system and not to an upstream separator where residual chemicals and solids can contribute to the formation of stable interface emulsions. Solids recovered from the slop or wet oil tank should be sent for disposal.

   The slop oil or wet oil tank should have facilities to send oil to the dry oil tank. Typically the oily emulsion recovered from the produced water will contribute to the BS&W of the sales oil, so this must be done within the limits of the oil dehydration specification.

4. **Chemical Treating System:** The type of chemicals dosed, their injection locations, and amounts injected may all contribute to treatment problems. All upstream chemicals (e.g. methanol, corrosion inhibitors, scale inhibitors, oil demulsifiers, etc.) can negatively impact the water treatment system. It may have been a long time since the chemical dosage program has been evaluated. It is helpful to know when it was last optimized, what were the results of the optimization, what are the flow assurance challenges of the field.

Relevant Questions:

- Ensure that selected chemicals, injection locations, and injection rates are suitable for current production conditions. When was the last time that the entire chemical injection program was optimized? Was optimization done under supervision of company surveillance engineer?
- Are the chemicals the best for the location, or merely the most readily available in the region?
- If solids are a problem, does the chemical treating program address the presence of solids?
- If the process has extensive internal recycles, are the chemicals compatible?
- Is there a program of optimization and continuous improvement of the chemical treating program?

5. **Operating Procedures:** The operating procedures of the facility in general and water treatment equipment in particular are an important troubleshooting area although it is difficult to make a systematic review of this area. One way to assess the robustness of the operating procedures is to assess the level of detail of the procedures and the record keep od facility operations. If the documentation is poor, then there will be a lack of consistent understanding across all operators staff.

   Relevant Questions:

   - Are operating procedures available, and are they being followed?
   - Do the operators have a good understanding of how to operate the equipment?
   - Do the operators have a good understanding of the oil, water, and solids properties?
   - Are there good handoff notes and do the different crews agree on how to operate the platform?
   - Is there adequate collection of data to allow good analysis and troubleshooting by the operators and engineers? Do the data backup what the operators say is happening on the platform?

   The availability of procedures and their correct implementation should be evaluated for each treatment unit. For example, for hydrocyclones operating procedures may include daily backwash, routine cleaning of the liners, adjustment of the number of active liners, and maintaining the DPR in the recommended range. Not following these procedures will ultimately lead to deterioration of equipment performance.

In troubleshooting, it is important to periodically review the progress being made by the team and consider if enough balance is being given to each of these areas. It is easy for a team of troubleshooting experts to gravitate toward those areas where data are readily available, or where other groups have concluded that the source of the problem lies. It is necessary to avoid this trap and to periodically consider if enough troubleshooting work is being applied to all four areas.

15.5. General Methodology for Troubleshooting O/W Process Problems:

The following is a step by step approach to solving produced water problems. We make use of Root Cause Analysis and Fact Based Problem Solving Process. The methodology consists of the following steps:

Step I:	Capture
Step II:	Problem Analysis and Definition
Step III:	Root Cause Analysis
Step IV:	Solution Development

It is not necessary that all steps be carried out in all cases. Sometimes the entire troubleshooting process can be carried out in a few days. In other cases, particularly those where problems have been persistent for a few months or years, it is necessary to follow the methodology as closely as possible.

Fact Based Problem Solving Process

Before we present the details of troubleshooting a produced water system, we provide a short review of the fact based problem solving process.

The fact based problem solving process is commonly referred to as Root Cause Analysis. It is a structured process that provides a common set of terms and tools that can be used to solve problems through cause elimination. There are essentially four phases in the process as given in the table below.

Phase	Description	Steps
1. Incident Capture	The recording of an incident or problem with relevant information and the decision as to whether an RCA must be carried out and if so, at what level the investigation will be conducted.	1. Incident reporting 2. Incident ranking
2. Problem Analysis	The breaking apart of a complex situation into manageable pieces. Providing answers to the question "what is/are the problem(s)?"	3. Problem identification 4. Problem definition
3. Root Cause Analysis	The systematic search for causes of a problem. Providing answers to the question "why?"	5. Possible causes 6. Data verification 7. Cause verification
4. Solution Development	A systematic technique to select best balanced choice of solutions (select a solution that eliminates causes without creating new/worse problems)	8. Decision identification 9. Criteria selection 10. Alternative solutions 11. Decision analysis

The following material is intended to (1) characterize the fluid, (2) characterize the process, (3) determine the technical limit of the system, and (4) help in the development of solutions. These steps will provide the data necessary for several of the steps above in the RCA process.

Step I:	Capture
Step IIa:	Problem Identification – Gather Background Information
Step IIb:	Problem Identification – Field Visit
Step IIc:	Problem Definition
Step IIIa:	Root Cause Analysis – Possible Causes
Step IIIb:	Root Cause Analysis – Data Verification
Step IIIc:	Root Cause Analysis – Cause Verification
Step IVa:	Solution Development – Identification & Selection
Step IVb:	Solution Development – Presentation for Decision Makers

Step I: Capture:

Step I of the Fact Based Problem Solving Process is to discuss and agree that a significant problem exists. Step I is complete when a plan for Step II is agreed.

Step II: Problem Analysis and Definition:

Step IIa: Problem Identification – Gather Background Information:

Make a comprehensive checklist list of information required based on an initial understanding of the problems. Keep in mind that some information may be difficult to obtain so assign priority to each bit of information required.

As a minimum collect:

Basis of Design, Control Narratives,

P&ID, PFD, PEFS, PFS

Oil in water monitoring results

PI trends

Hysys (Unisys) models

Equipment vendor data sheets

Chemical vendor reports and data sheets

Based on this initial gathering of information, identify which items in the checklist have been obtained and which items must be gathered in the near future. This checklist then becomes part of the plan for a field visit. At this stage, it is useful to have a small brainstorming workshop with the surveillance engineers. It is encouraged to generate as many hypotheses of the cause of problems as possible and to discuss these ideas with the surveillance engineers.

Deliverables: Information Checklist

Step IIb: Problem Identification – Field Visit:

Develop a plan for a field visit. Try to be self-sufficient in gathering samples and minimize resources required from field staff since they have other obligations. Keep in mind that you may not have time to carry out all of the intended activities so your plan must be prioritized and easy to follow. Include a sampling plan, and questions for operators. Keep in mind that the primary observations of the operators are very important. During the field visit, try to document as much information as you can before you leave the facility. Have a close-out meeting with the operations staff to verify observations and as a courtesy since additional field visits may be required.

Visit the platform. Review the troubleshooting questions given above. Carry out the Field Visit Checklist.

Deliverables: Field Visit report including all data and observations

Trouble Journal

Step IIc: Problem Definition:

Once this information is compiled, then the problem statement can be written. It should be precise. It should break the problem down to manageable pieces. It should provide answers to the questions, what are the problems, where and when they occur. It should also suggest the possible causes of the problems, but the root cause will only be developed in the next steps.

It is often helpful to map the problems onto a flow schematic together with relevant information. PowerPoint may provide a suitable format for a problem statement since supporting data can be included.

Deliverables: Problem Statement in the form of a PowerPoint slide pack

Step III: Root Cause Analysis and Definition:

Step IIIa: Root Cause Analysis – Possible Causes

Compile all of the information obtained to this stage. By now you should have as a minimum:

- Characterization of the oil and water (see Chapter 2), with an understanding of the challenges of the particular fluids, the source of solids, the source of shear (sources of small drops), and the source of complex emulsions.
- Technical limit of each piece of equipment, including a comparison with actual performance, with an understanding of which equipment is and is not performing to specification. This is best done using schematics of each piece of equipment and a process flow schematic of the system with data overlain on these schematics.
- Material balance of oil and water flow rates throughout the system, with an understanding of where the bottlenecks occur, and whether sufficient reject flow rates are being used.
- Process trends with an understanding of the system dynamics.

The search for root causes is facilitated by asking the question “why” in an iterative manner. An example of this is the following:

The vehicle will not start. (the problem)

1.Why? - The battery is dead. (first why)

2.Why? - The alternator is not functioning. (second why)

3.Why? - The alternator belt has broken. (third why)

4.Why? - The alternator belt was well beyond its useful service life and not replaced. (fourth why)

5.Why? - The vehicle was not maintained according to the recommended service schedule. (fifth why, a root cause)

6.Why? - Replacement parts are not available because of the extreme age of the vehicle. (sixth why, optional footnote)

Start maintaining the vehicle according to the recommended service schedule. (possible 5th Why solution)

Adapt a similar car part to the car. (possible 6th Why solution)

At this point in the troubleshooting process you should have a good characterization of the fluids, the equipment performance, the process performance and the operational approach. Now it is time to explain why these elements are as you have characterized them. Initially it is helpful to brainstorm all possible causes. Software such as MindMap may be useful. An Excel spreadsheet of possible causes, indicators supporting, indicators refuting, and data required to verify may be useful.

Deliverables: bulleted items above (Characterization, Technical Limit, Material Balance, Process Trends)

Spreadsheet of Possible Causes

Step IIIb: Root Cause Analysis – Data Verification

At this point it is useful to have a workshop or teleconference with the surveillance engineers to review the deliverables that have been generated. The primary objective at this stage is to identify the information that is missing and to develop a plan to obtain that information.

The Deliverables to this point include:

Problem Statement

Characterization of Fluids

Technical Limit Schematics

Material Balance Spreadsheet

Process Trends

Spreadsheet of Possible Causes

New Deliverables: Information Checklist - updated

Final Sampling / Analysis / Data Gathering / Field Trial Plan

Step IIIc: Root Cause Analysis – Cause Verification

Once the final data gathering has been completed, the Spreadsheet of Possible Causes should be updated with verifying or refuting information.

At this point, it is not necessary to separate the search for causes from the search for solutions. In fact, they become synonymous at this stage. Possible solutions should be discussed with operators and surveillance engineers, particularly if it is anticipated that field trials will be required to evaluate a particular solution.

Deliverables: Spreadsheet of Possible Causes - updated

Step IV: Solution Development:

Step IVa: Solution Development – Identification & Selection

At this point, possible solutions have inevitably been discussed with operators and engineers. A set of preferred solutions has already begun to emerge. Nevertheless, engineering analysis must be carried out on a wide range of possible solutions in order to provide the facts upon which a final solution shall be agreed. An analysis of potential solutions should be generated using the Spreadsheet of Possible Causes as a starting point. For each solution on the list, a SWOT analysis should be carried out, together with a rough cost estimate and a rough estimate of schedule to implement.

A SWOT Analysis consists of the following considerations for each possible solution:

Internal Factors (these are usually technical factors):

Strengths

Weaknesses

External Factors (these are usually organizational or business related factors):

Opportunities

Threats

Once the SWOT Analysis has been carried out for the various potential solutions, the SMART Recommendations are then generated.

SMART Recommendations are:

- Specific
- Measureable (Quantifiable)
- Actionable
- Recordable (Trackable)
- Time-Relative

Each of these **SMART** characteristics are described below.

Specific recommendations are well defined in terms of what equipment or procedure, who should be involved, where in the process, when it should occur, and specifically what is being recommended. The reasoning behind the recommendation should also be included.

Measureable (quantifiable) recommendations state how much of a benefit is expected from the recommendation.

Actionable recommendations are those that point out what can be done to alleviate a problem or fix a process. It is not enough to simply point out that something is not working properly or that an aspect of the system is sub-optimal. Any such comments should be accompanied with a recommended action, and benefit from taking such action.

Recordable recommendations are those that can be measured or tracked. If not mechanisms is in place for monitoring or surveillance, then such mechanism should be recommended to be put in place before the recommended action takes place. It is important to include control, benchmarks, or other means of establishing the effect of the recommended action.

Time-Relative recommendations give an indication of how soon an expected improvement will be seen.

Once this analysis has been carried out for the possible solutions, individual ideas should be bundled together into a few different strategies. The benefits and drawbacks of the different strategies should be clearly and concisely articulated for presentation to management.

Deliverables: Spreadsheet of Possible Causes & Solutions
SWOT of solutions
Cost & Schedule
Solution Strategies

Step IVb: Solution Development – Presentation for Decision Makers

This is the final step and potentially the most important step in the process of troubleshooting. The solution strategies must be clearly and concisely articulated to the decision makers. It is worthwhile to review the troubleshooting process in order to give management a sense that a thorough and systematic process was followed. It should be kept in mind that certain strategies may be unacceptable to management due to various constraints. One or more steps in the troubleshooting process may need to be repeated depending on the preferred strategy of the decision makers.

Most often, decision makers need a very robust assurance that any money spent will give the promised results. Typical E&P managers may not understand the technical details, but they usually appreciate how complex the technical details can be. Most often they prefer a Stair-Step approach involving investment / verification of results / further investment / further verification.

Deliverables: Recommended Solutions
Stair-Step Implementation Plan
Final Presentation
Final Technical Report
All other deliverables in concise presentation format

15.6 Summary of All Deliverables

A complete troubleshooting study could involve all of the following deliverables. Not all studies require this complete list. A scope of work and list of deliverables shall be agreed with the customer at an early stage of the work.

Complete list of Deliverables:

- Problem Statement
- Field Visit report(s)
- Summary of information:
    - Characterization of fluids
    - Technical Limit for each piece of equipment
    - Process Performance Diagram, material and flow balance
    - Chemical optimization results
    - Operations procedures, troubleshooting guide, monitoring and surveillance recommendations
    - Possible Causes & Effects (Ishikawa diagram & 5 Whys)
- Final statement of Root Causes
- Potential Solutions, SWOT, cost & schedule
- SMART Recommendations
- Stair-Step Solution Strategies
- Final Presentation
- Final Technical Report

All deliverables in concise presentation format

15.7 Reporting

The final technical report should be structured according to the deliverables that are listed above. It should also include the following:

- timeline of discussions and meetings with the client
- agreed scope of work
- timeline of field visits
- itemized third party costs

The following check list can be used to rate a troubleshooting study. This check list is intended to guide the final report toward higher quality.

Problem statement	
	Is the problem clearly identified?
	Was the problem statement verified with the customer?
Field Reports	
	Were field reports gathered?
	Are they clear?
	Are they balanced and do they include the 5 areas of water treatment?
Summary of Information	
	Did the customer hand over information on the facility?
	Is the information clear?
	Is the information balanced and does it include the 5 areas?
Possible Causes & Effects	
	Are the causes accompanied by effects?
	Has a Five-Whys analysis been carried out?
Final Statement of Root Causes	
	Are they clear?
	Are they missing any potential causes?
Potential Solutions	
	Was a SWOT analysis carried out?
	Are the options given with cost estimates?
	Are the options given with a cost / benefit analysis?
SMART Recommendations	
	Are the recommendations rated in terms of SMART characteristics?
Stair-Step Solutions	
	Has a Stair-Step analysis been carried out?
Final presentation	
	Is it clear, concise, complete?
Final Technical Report	
	Is it clear?
	Is it complete?

References to Chapter 15

1. T. Frankiewicz, "Understanding the fundamentals of water treatment. The dirty dozen. Twelve common causes of poor water quality," paper presented at the Produced Water Society, Houston (2001).

2. T. Frankiewicz, J. Lee, C.-M. Lee, "Diagnosing and resolving chemical and mechanical problems with produced water treating systems," paper presented at the Produced Water Society, Houston (2000).

3. T. Frankiewicz, J. Clemens, "Diagnosing chemical and mechanical limitations in produced water handling and cleaning processes," paper presented at the Produced Water Society, Houston (1999).

4. J. Walsh, J. Fanta, W. Bryson, C. Toschi, J. Petty, J. Lee, T. Frankiewicz, M. Stacy, "Part 1 - Troubleshooting produced water – methods and lessons learned," World Oil, p. 111 (March 2007).

5. J. Walsh, J. Fanta, W. Bryson, C. Toschi, J. Petty, J. Lee, T. Frankiewicz, M. Stacy, "Part 2 - Troubleshooting produced water – methods and lessons learned," World Oil, p. 151 (April 2007).

CHAPTER SIXTEEN

Applications: Characterization

Chapter 16 Table of Contents

16.0 Introduction 339

16.1 When to Characterize Produced Water 342

16.2 How to Characterize Produced Water 342

16.3 Sources of Data (Data Mining) 345

16.3.1 Geological Data 346

16.3.2 Reservoir Engineering Data 346

16.3.3 Process Engineering Data 346

16.3.4 Flow Assurance Data 347

16.3.5 Water Analysis Data 348

16.3.6 Field Data 349

16.4 What to do with the Analytical Results 349

16.5 Cook Book 350

16.5.1 Stokes Factor 354

16.5.2 Oil Flow Assurance Factors 355

16.5.4 Mineral Precipitation Tendency 356

16.5.5 Dissolved Organic Factors 357

16.5.6 Solids 357

16.5.7 Iron Sulfide 361

16.5.8 Surface Active Production Chemicals 361

16.5.9 Small Drops 361

16.6 Example – Deepwater Gulf of Mexico 362

References for Chapter 16 369

16.0 Introduction

This chapter brings together material from several of the previous chapters and addresses the critical question: once the analytical data have been gathered, what should the engineer do with the data? Or stated in a different way, this chapter answers the question: what does the analytical data tell me about how to design a Green Field facility, or how to fix a problem with an existing water treatment system?

Subject of this chapter: The subject of this chapter is data interpretation. Providing guidance on how to interpret analytical data is not always intuitively obvious. The author had the good fortune to be the designated specialist for water treating in a large multinational oil and gas company. Due to the large number of projects, he could not be involved in every one. He was compelled provide a systematic methodology that could be handed to others. The objective was to develop a methodology that could be used by the typical process engineer and which relates analytical results to suitable design and operation of water treating systems. Section 16.5 (Cook Book) is the result of this management mandate. Over several years the cook book was used by engineers to develop new designs and to fix existing facilities. The most important outcome of this experience was the direct link that the engineers were able to make between produced fluid characteristics and proper design.

The main concepts presented in this chapter are:

1. you need much more than just a water ion analysis;
2. make use of data already available from other disciplines such as reservoir engineering (crude oil properties, and geological setting), process engineering (gas composition), flow assurance (stability of asphaltenes, wax, scale components, and need for production chemicals);
3. comparison is the key;
4. compare your fluid properties to benchmark fluid properties;
5. compare your fluid properties to analog fields;
6. compare your fluid properties to other fluids for which you know what equipment, process, chemicals, and operating procedures will work.

Laboratory versus Field Characterization: Laboratory studies of water chemistry and equipment have the advantage that most variables can be controlled and systematically varied, such as contaminant concentration, flow rate, temperature, flux, etc.. This allows a direct cause and effect relationship to be established and quantified. The disadvantage of laboratory studies is that the properties of the water cannot, in general, be controlled to match those of a particular facility. In general, oil and water samples that are used in the laboratory have been equilibrated with their surroundings (typically ambient), typically exposed to air, weathered, and aged. If bacteria are present, then hydrocarbon and acids will have been consumed, and other short-chain acids will be generated. The significance of these effects is explained in Chapters 1 (Chemistry) and 3 (Emulsions). Such samples can be significantly different from the pressurized, anaerobic, and dynamic state of freshly formed oil/water interfaces and emulsions. Thus, at least some design and testing work must be carried out in the field in pilot studies. Understanding how to interpret laboratory results and how to translate laboratory studies to the field is an important skill. It is a skill that must be mastered since it is impractical to conduct all necessary design, scale-up and demonstration studies in the field.

Field data, where the equipment is tested online or as a side-stream can eliminate fluid change. However, this does not mean that field data is always superior to lab data. Field data often suffers from a lack of water characterization and often the process configuration is not reported. These are common and serious drawbacks. Most equipment will not work effectively on all types of produced water. Therefore it is necessary to analyze and report the water characteristics and the process configuration within which the new equipment was tested. If the field trial is carried out without an understanding of these elements, the equipment performance data are essentially useless. Even if samples are collected in the field for later analytical laboratory characterization, there is still the problem that the sample may change in important ways by the time it is analyzed. One way to minimize these effects is to carry out as much analysis in the field as possible. However, practically speaking, this can only go so far.

However obvious it may seem that fluid characteristics (influent and effluent) should be used to design the facilities, there are people in the industry who think otherwise. The problem that they express is that sampling and laboratory analysis of produced fluids will give inherently false information. According to them, even sampling and analysis carried out in the field is not accurate. The process fluid is a "living" and dynamic system. Preserving the produced fluid is inherently not possible. The fluids oxidize, emulsions age, surface active agents migrate to the interface. They are susceptible to temperature, pH and oxidation effects and otherwise change, and interpretation is difficult and uncertain. It is said that such analyses do not provide any relevant information on the behavior of crude oil / produced water systems [1, 2].

This thinking led to the development of the Multi Test Unit (MTU) which allows direct evaluation of detailed fluid properties and characteristics as well as providing a means of pilot testing equipment on site. An operating company, such as Statoil that has perhaps a half dozen very large facilities can afford to take such an approach. Testing can be carried out onsite using complex expensive testing apparatus. But a company with hundreds of facilities, or an equipment design firm trying to sell equipment across an entire industry cannot possibly take the time or spend the money needed to do so much onsite pilot work for every facility and project.

The industry has proven many times over decades, that techniques are available to solve problems without resort to such elaborate and expensive methods. There is no doubt that samples must be carefully taken and analyzed. For some characteristics, it is best if measurements are made in the field. However, not all analyses can or should be done in the field. Some pilot testing on simulated fluids has to be carried out both in the field and in the laboratory. A comparison between lab results and field results must be made in order to understand and control the variables. Predictions of how fluids and equipment will behave in full scale in the field must then be made based on small scale and synthetic fluids in the lab. A good understanding must be obtained of the fluids in the field and the fluids in the lab. The design of equipment and processes must take into account the changes that occur to produced fluid in the sampling, transportation and storage processes. This is the only practical approach to design and troubleshooting and is the approach used by essentially all chemical engineering (fluid based) industries. Making correlations between the analytical data, laboratory tests, pilot tests and full-scale field results is essential. It is part of the reason why the first several chapters were presented with so much mechanistic detail.

System Designs for Particular Water Types: Correct characterization of a produced water stream is essential to ensure the appropriate selection, design and operation of equipment and process system configurations. One of the goals of this chapter is to provide a bridge between the chemistry and analytical discussion of previous chapters to the discussion of equipment and process configuration that is to follow in subsequent chapters. The bridge between these subjects is based on a firm understanding and recognition of the fact that there are distinct types of oily water and that certain pieces of equipment work well for certain types of oily water, and that there is no equipment that works well for all types. There are no magic bullets.

There is a tendency in the oil and gas industry to expect water treating technology to work well for all types of produced water. The thinking assumes that a "good" piece of water treating equipment should be capable of performing well with all different kinds of produced water. If it cannot, then it is a "bad" technology. As will be demonstrated in this and subsequent chapters, this is seldom, if ever , the case in practice. Some equipment works well for some produced water types, but not others. It is important to recognize that all equipment has limitations and to know what those limitations are. As with any process equipment, water treating equipment has limitations in temperature, pressure, flow rate. Also, fluid composition will be an important limitation which includes oil/water emulsions, suspended solids, oil-wet solids, pH, salinity, etc..

The Concept of an Operating Envelope: Process and facilities engineers often refer to the "operating envelope" for separators, vessels, pumps, compressors, and various other pieces of process equipment. For example, a three phase separator has a range of inlet flow rates and pressures for which it will provide adequate oil / water / and gas separation performance. If the inlet conditions are outside of this range, poor performance will result. The concept of operating envelope is deeply embedded in the thinking and practice of process and facilities engineering.

The concept of operating envelope should be deeply embedded in the thinking and practice of water treating. In water treating, the operating envelope is defined not only in terms of flow rate, temperature and pressure, but also in terms of the chemistry and characteristics of the feed water. The definition of an operating envelope for equipment must include consideration of all these features. It cannot be expected that the same equipment will perform well for a chemically stabilized emulsion as for a light oil. Similarly, equipment that performs well for a moderate API oil cannot be expected to perform well for a heavy oil and water system. The limits of equipment performance are therefore very much limited to an operating envelope. This envelope is defined by characteristics such as asphaltene content and stability, solids content, dissolved organics, etc. Once this important concept is understood, the statement "such and such equipment works well," seems unworldly because it leaves out the most important information, i.e. the conditions under which the equipment was applied.

This chapter provides a road map to defining those conditions, which includes the oil, water and gas characteristics. Thus, the operating envelope from a water treating perspective includes the process and facilities engineering parameters (temperature, pressure, flow rate, process configuration) but also the oil characteristics, SARA analysis, ionic composition, dissolved organics, pH, alkalinity, production chemicals, dissolved gases, etc, etc etc.. No one said that water treatment is easy.

Characteristics of Produced Water: This chapter discusses a number of the elements which define the characteristics of a waste water stream. Our primary goals in this chapter are to provide practical advice on how to characterize produced water, when to characterize it, and what to do with the information that is obtained, in order to design a new facility or improve the performance of an existing facility.

For a checklist of what properties to include in a characterization, see Table 1 below.

Much of this chapter focuses on how to interpret the results and what to do with the information in terms of equipment selection and design. These subjects are also discussed in later chapters.

Difficulty of Obtaining Good Data: The importance of proper characterization of produced water cannot be overstated. Nevertheless it is a complicated task and most attempts to characterize produced water are deficient in one way or another. While the technology available has improved dramatically over the past several years, there are still many difficulties involved in the proper sampling, handling, analysis, and interpretation of results.

Each of the major steps involved in characterization have associated problems:

1. Sampling - poor sampling technique often changes the drop size distribution;
2. Transportation – the time between sampling and analysis often allows drops to coalesce;
3. Handling - sample handling often allows oxygen to intrude or CO2 or H2S to escape;
4. Analysis - limitations of the analytical methods can lead to incorrect conclusions;
5. Interpretation – insufficient knowledge or experience which leads to erroneous conclusions;
6. Implementation – mistakes in detailed design which compromise overall design intent.

These problems are described further and practical advice about how to overcome them is given.

Sampling: Sampling is not discussed in this section in detail since it was covered in Chapter 7 (Sampling and Analysis). Instead, in this section the emphasis is on what should be analyzed, and what should be done with the analysis results. Guidelines and manuals for how to perform various analyses are readily available [3 – 5].

16.1 When to Characterize Produced Water

The question of when to characterize produced water becomes a significant issue for a new project. During the design phase of a new facility, there is usually little knowledge of the produced fluid properties. While the water treating system may not be required until sometime after startup, space and weight must be allocated to water treatment and all other operations. Even in the best case where water and oil samples have been taken at an early stage of the process design, and have been carefully analyzed, such samples will not represent the drop size distribution of oil in water which can only be assessed upon actual production. In such cases, water characterization should still be carried out using the methods given here for characterizing the chemistry of the water and hydrocarbons. For an estimate of drop size, theoretical considerations can be used, and analog fields and facilities should be investigated for drop size.

Therefore, the use of analogs becomes of great value. Analogs are nearby facilities already in operation where the fluid properties are well known and are likely to be similar to those expected for the new facility. This chapter discussed the use of analogs and the gathering of data for a geological region.

Also, what is needed, and the guidance provided in various parts of this book, is practical advice on how to make judicious design decisions that build in flexibility at a reasonable cost and with minimal space and weight requirements. In other words, the original system design must have sufficient built-in flexibility to allow debottlenecking and capacity increases later, when the produced water characteristics are better known, without major facility modifications. In fact, the timing of when to make design decisions is one of the most important aspects of designing or debottlenecking a produced water system. That timing is driven by the availability of information related to proper produced water characterization.

16.2 How to Characterize Produced Water

Produced water can contain a wide variety of contaminants. When considering produced water in E&P operations, the concentration of the produced hydrocarbons (oil in water concentration) is normally the focus of both system design and environmental monitoring. However in addition to the dispersed and dissolved hydrocarbons, produced water streams typically contain a wide variety of contaminants, many of which have an impact on the effectiveness of the separation equipment, as well as a potential impact on the environment. In order to minimize these effects, and in line with best practices in environmental management, all the constituents of produced water should be identified and accounted for. This includes the solids, the injected production chemicals, and the full ionic composition including the heavy metals, and of course, any priority pollutants that may be present.

Location defines the Important Contaminants: As discussed in Chapter 1 (Introduction), the relative importance of contaminants in produced water depends on the regulations regarding reuse, discharge, disposal, corrosivity, as well as the environmental impact, the potential for sustained biological activity in the case of waterflooding, and the effluent quality required for reuse options., discharge water or disposal streams is dependent upon the receiving process downstream or the environment into which the water will be released. Environmental regulations vary depending on location. An example of this is water salinity. Saline produced water may be relatively benign when discharged into the sea. But it may have a significant environmental impact when discharged into a fresh water or agricultural environment. Another example is in subsurface disposal (Produced Water Reinjection – PWRI) where produced water characteristics must meet specified limits for oil and solids content in order to insure that disposal zones are not fouled by contaminants. A third example is where H_2S scavenging is carried out using amine-based chemistries. In that case, carbonate mineral scaling may occur as a result of pH changes to the water unless prudent design and treatment decisions are made.

Formation Water Characteristics: To fully understand the characteristics of produced water, one must ultimately consider its source in the formation. In the formation, water is in equilibrium with the fluids contained in the pore space, and the formation rock material. As the water is produced, the pressure and temperature decrease, causing a shift in the equilibria. This shift may cause various components to precipitate as solid particles. Other components may migrate to the oil/water interface. A complete characterization of produced water must take into account the components of the co-produced gas and oil phases as well as the composition of the reservoir rock. In some cases phase equilibrium modeling will be necessary to identify phase instability and to provide any missing information about the fluid components.

Important Components of Produced Water

Before discussing the characterization of produced water, we give a list of possible components of produced water:

- Dispersed oil
- Dissolved oil (hydrocarbons, BETX, phenols, PAH, etc)
- Dissolved organic acids and other organics
- Dissolved or precipitated minerals (NaCl, $CaCO_3$, $FeCO_3$, FeS_x, $BaSO_4$, $SrSO_4$, etc)
- Dissolved metals (Fe, Zn, Cr, Mn, etc)
- Process & Production chemicals (corrosion inhibitor, water clarifier, methanol, glycols)
- Produced formation solids (clay, sand, carbonate)
- Dissolved and precipitated corrosion products (dissolved metals, solid metal hydroxides and oxides)

- Dissolved gases (O_2, H_2S, CO_2)
- Combinations of the above (e.g. Schmoo)
- Various bacteria and by-products (e.g. Sulfate Reducing Bacteria, General Heterotrophic Bacteria)

The above list is given from the perspective of chemical entities, rather than types of analyses. Many of these entities are determined in typical or standard analyses. The table below gives the recommended checklist for water characteristics required for design and troubleshooting.

Table 1. Summary Checklist of Water Characteristics required for design and troubleshooting

1. Geological Data
    a. Geological setting (depth, T, P, mineralogy)
    b. Geological history of oil, gas and water
    c. Likelihood of oil biodegradation
2. Reservoir Engineering Data
    a. Water Cut
    b. API Gravity
    c. Oil yield
    d. TAN
    e. Biomarkers
3. Process Engineering Data
    a. H_2S
    b. CO_2
    c. BTEX
    d. Flow rates and ranges
4. Flow Assurance Data
    a. SARA
    b. Wax stability
    c. Scaling Tendency
    d. Flow Assurance Chemical Requirements (for hydrates, scales, asphaltenes, etc)
5. Water Analysis Data
    a. Composite Characteristics (COD, TOC, BOD)

   b. Individual Component Concentrations (Geochemical analysis for anions, cations, heavy metals)

   c. Alkalinity and pH (hardness, dissolved gases, etc)

6. Field Data

   a. Desktop settling

   b. Visual observations of the produced water

   c. Optical microscopy

   d. Oil droplet size distribution (on-line and off-line)

   e. Oil in water concentration

   f. Solids particle size analysis

   g. Solids particle composition

In the material below, we discuss the typical ranges of measured values, and indications of problems, and how to interpret the results of such a characterization.

16.3 Sources of Data (Data Mining)

In order to design or troubleshoot a treating system it is necessary to know the characteristics of the water. Essential questions must be addressed such as:

- What is the size of the oil drops and solid particles?
- What is the fouling tendency of the contaminants?
- What is the plugging tendency (for water injection or media filtration)?
- What is the tendency for additional solids to precipitate?
- To what extent are the oil drops and solid particles stabilized by other constituents?
- What toxic, heavy metals, and priority pollutants must be removed from the water?

This is a short list of many questions that must be answered in order to competently design a system or provide effective troubleshooting. Answering these questions requires the appropriate analytical data. Not all of the analyses must be carried out by the water treating engineer. In most operating companies there is a wealth of information available as a result of the work of other disciplines.

In the rest of this section, sources of data are discussed. In developing a hydrocarbon field, it is typical to employ many specialized disciplines who gather information that can be used by the water treating engineer. This section gives a summary of the types of information that is relevant to water treating, and it explains why.

16.3.1 Geological Data

As discussed in Section 2.2.1 (Origin of Crude Oil), hydrocarbon reservoirs have geological histories that span several million, if not tens or hundreds of millions of years. During that time, many processes have taken place that change the composition of the oil, gas, and water. The geologist will have some knowledge of these processes and can help identify suitable analogs in the region, or elsewhere in the world, and can help make rough estimates of the composition of the oil and water.

16.3.2 Reservoir Engineering Data

One of the main concerns of the reservoir engineer is the most economic means to develop the field. This typically requires knowledge of the quality and composition of the crude oil and gas that will be produced and whether it will suffer price penalties for characteristics such as low refinery yield, high residuum content, corrosive components, catalyst poisoning components such as nickel and vanadium, and so on. These characteristics are useful to the water treating engineer as well.

The simplest oil characterization parameter is the density or API gravity. Light gravity oils have a greater density difference with water which thus helps drive gravity settling. They also have lower surface elasticity which promotes drop coalescence. But their lower viscosity promotes drop break up in a shear zone.

Oil characterization is important for produced water because there are many components in the hydrocarbon phase that contribute to the stability of drops of oil in water. Organic acids, asphaltenes, and resins, are just some of the components that are initially found in the oil phase but which can transfer to the water phase or migrate to the oil/water interface and thus have an impact on produced water quality.

Oil analysis should include the Total Acid Number (TAN) which provides a simple first indication of naphenic acid content. The recommended procedure is ASTM D-664. Besides TAN, organic acid content and composition should also be measured. The definitive test for biodegradation is done by gas chromatographic hydrocarbon fingerprinting which reveals the presence of naphthene biomarkers [6]. Biodegradation does not generally occur in reservoirs above 80 °C. However, hydrocarbon fluids often have complex histories over geological time scales. Thus, current reservoir temperature is not always a valid indicator of biodegradation since the hydrocarbons may have migrated. Although both anaerobic and aerobic bacteria can be active in a reservoir, it is the aerobic bacteria in the presence of a source of migrating or flowing water with dissolved oxygen that gives rise to the most profound microbial oxidation of crude oil.

Progression of Biodegradation: Mild biodegradation of crude oil will reduce the fraction of saturated hydrocarbons, and increase the organic acidic and natural surfactant fraction of the residual oil. As biodegradation proceeds, the progressive loss of n-alkanes is followed by loss of isoprenoids which leaves an oxidized naphthenic residue. Aromatic functional groups, resins, and asphaltenes are not attacked and will therefore increase in mole fraction as the other components are consumed. As discussed below, biodegradation often results in a high concentration of polar organics, organic acids, natural surfactants, and surface active components which stabilize oil in water emulsions and therefore make water treating difficult.

16.3.3 Process Engineering Data

One of the main concerns of the process engineer is gas conditioning. By that it is meant, gas compression, gas dehydration, gas dew pointing, and separation of particular gas components, if necessary. All of these concerns require a fairly detailed knowledge of the gas composition. Thus the process engineer must know how much of the following constituents are contained in the gas: H2S, mercury, carbon dioxide, or other contaminants such as nitrogen. Such information is useful in calculating the pH through the system and the response of the produced water to the addition of acidic and alkaline production chemicals.

Operating Conditions: The operating conditions for a produced water treatment system obviously play a critical role in the design of a water treatment system and its constituent equipment. In addition to nominal operating pressures, temperatures, and flow rates, the range of flow rates which the system needs to handle should be included when sizing and designing equipment. Slug flow can be especially challenging for a water treatment system to handle, but slug flow can be handled if equipment and the control philosophy for the equipment are properly specified.

In addition, expected contaminant levels along with the character of the expected contaminants should be specified as part of the design basis for equipment or a treatment system. A Geochemical Analysis of the water is also required along with the produced gas analysis or a gas analysis from an appropriate vessel or tank in the water treatment system. Remember, if there is CO_2 and/or H_2S in the well head gas, the composition of the gas with which the water is saturated (and thus its pH) will change with pressure.

16.3.4 Flow Assurance Data

The subject of flow assurance is typically staffed by specialists. These engineers typically address the following subjects:

- Scaling
- Hydrates
- Wax
- Asphaltenes

Although it may not be obvious at first, each of these flow assurance factors can have a significant impact on water treating. The connection is described below.

Scale versus Mineral Precipitation: The flow assurance engineer is concerned with the formation of scales on piping and vessel walls. Obviously, if scale is to form, then mineral salts must be precipitating from solution. The flow assurance engineer must obtain a complete mapping of temperature and pressure throughout the facility. This is needed for the determination of dissolved carbonate and sulfate salt stability. Carbonate stability is coupled to the concentration of carbon dioxide in the gas, the pH, alkalinity, temperature, pressure, flow rate and pressure drop across valves. For sulfate scaling, concentrations and temperature are the main variables. In general, the Flow Assurance Engineer is an excellent source of information about many aspects of the facility.

In some cases, precipitation and sticking to the surface occur essentially simultaneously. This happens when a tiny seed crystal deposits on the surface of a pipe, valve, or vessel and provides a location for dissolved carbonate or sulfate salts to precipitate in the form of a growing crystal that coats the surface of the pipe, vessel, etc. The process of scale growth on equipment walls and in pipes is generally insidiously slow and sometimes be difficult to recognize prior to there being a significant flow problem.

The flow assurance engineer tends to be most concerned with this crystal growth form of scale mineral precipitation. The important result being an assessment of whether or not scale will build up on the pipe or equipment wall. Precipitation, in the absence of scale formation can and does occur under some (hard to predict) flow conditions or when a scale inhibitor chemical treatment is utilized. The precipitation of scale mineral solids which remain dispersed in the produced water and/or the crude oil is a particular challenge for water treatment. It should be noted that although high velocities or turbulence can reduce scale formation under some circumstances, in other circumstances, scale formation is enhanced by high fluid velocities.

The water treating engineer must also be concerned with scaling processes. However, the water treating engineer must be even more concerned with mineral precipitation, even in the absence of scaling. Water laden with suspended solid particles is one of the most difficult contaminants to deal with.

Dissolved Hydrocarbons: All produced water contains dissolved organic compounds. In the past, until roughly ten years ago, produced water specialists have tended to downplay the importance of these compounds because they are difficult to characterize. Also, most legislative bodies do not emphasize the importance of dissolved hydrocarbons as much as they do dispersed hydrocarbons. But it is now recognized that dissolved hydrocarbons in particular, and organic carbon in general play a major role in flow assurance and in the performance of separation equipment and that they have a major impact on produced water quality.

Hydrates: In E&P operations, oil, gas and water are often transported and processed together. Whenever this is the case, there is a potential for hydrate formation. Hydrate prevention is typically carried out using chemical inhibitors, by insulation, or by heating. Chemical inhibitors include monoethylene glycol (MEG), tri-ethylene glycol (TEG), methanol, anti-agglomerates, and kinetic hydrate inhibitors. All of these chemicals affect the stability of oil drops in water. They also affect the stability of mineral scaling components, and some of the polar hydrocarbon components. It is therefore necessary to have a good understanding of the hydrate forming tendency and the options being considered or being used to prevent hydrate formation. Methanol, for example, reduces the solubility and increases the precipitation of divalent salts such as $CaSO_4$ and $CaCO_3$.

Wax and Asphaltenes: These two components are placed in the same category for convenience. The stability criteria and the processes that cause their precipitation are significantly different. However, both can lead to suspended particles, and both can contribute to the stability of oil and solids in water.

The modified IP 143 / 57 for SARA (Saturates, Aromatics, Resins, Asphaltenes) analysis method is available from many oilfield labs. The method consists of an initial topping step to allow volatile components to escape, and generate a dead crude sample. This eliminates variability due to light ends. This is followed by asphaltene precipitation and gravimetric analysis, and liquid chromatography to determine the concentration of the remaining components (saturates, aromatics, and resins). In order to determine the stability of asphaltenes, and determine their tendency to help stabilize oil in water emulsions, the results of a SARA analysis must be compared with analog data and field experience. Generally, the ratio of asphaltene to resin and aromatics to saturates gives an indication of asphaltene stability. This was discussed in Section 2.6.4 (Resins and Asphaltenes). Further details are available elsewhere [7].

16.3.5 Water Analysis Data

As discussed above, water analysis data is initially collected by the geologist who correlates the water composition to other formations in order to determine various aspects about the geological history of the hydrocarbons. In some cases, quite detailed information may be available. It is important to recognize that samples taken during drilling and completion of the well may be contaminated by drilling and completion fluids. Analysis of the data, together with a knowledge of the composition of the drilling and completion fluids used can generally be used to assess the degree to which contamination has occurred.

A complete water analysis is difficult to obtain in any phase of a project or during troubleshooting. Part of the reason for this is that what constitutes a complete analysis depends on the composition of the water, which obviously is not known until the analysis is carried out. Guidelines are given below to help determine what analyses are needed. In general, it is best to know as much as possible about the composition of the water. Many times water treatment performance problems can be avoided by knowing the composition of the water and designing a facility accordingly.

16.3.6 Field Data

If the task is to design a new facility, there will probably be limited, if any, field data. However, if the task is to troubleshoot an existing facility, then gathering of field data is one of the most important tasks. In designing a new facility, field data should be gathered from analog facilities. Thus, the subject of field data is relevant to both troubleshooting and Greenfield design.

Dissolved Oxygen: Produced water does not normally contain dissolved oxygen. In fact, most produced water has never been in contact with oxygen. Thus, the dissolved and suspended components in produced water are typically in a reduced state, in the sense of oxidation/reduction potential. Once produced water is exposed to oxygen its nature changes dramatically and for the worse in several regards. This fact cannot be overstated.

Introduction of oxygen laden water into the process stream is extremely detrimental because it (1) oxidizes the iron in solution in the water, creating iron oxide solids and (2) creates very aggressive, damaging corrosion in the form of pitting caused by the formation of anodic sites on internal metal surfaces. Iron in the fluids, when oxidized, will precipitate and contribute strongly to the stabilization of the emulsions.

Sources of Oxygen Contamination: A few common sources of oxygenated water entering into a separation system include seawater sump systems, holding tanks for off-spec produced water, or an open drain system. Such water systems should always be segregated from the production streams. Also, rigorous biological control should be practiced in such systems.

Another common source of oxygenation of produced water is by way of an open API separator system. Such systems are very cost effective and have been used successfully in the industry. However, they are open to the atmosphere and as such must have a means of mechanically collecting the scum that will form on the surface and the sludge that will accumulate on the bottom of the separator. These accumulations may contain reservoir and precipitated mineral solids along with sand and dirt that is blown into the separator. *Most importantly, any of the reject streams from an API separator (scum, sludge and oily reject) should absolutely never be routed back into any upstream section of the oil/water separation train or water treating system.*

16.4 What to do with the Analytical Results

As discussed in the material above, one of the major challenges in designing or troubleshooting a water treating system is sampling and analysis. As difficult as this may be, it is only the first step. Once an analysis has been obtained, the next challenge is to use the results properly. Knowing what to do with the results, is an important skill. The analysis only gives you the opportunity to make the right decisions. It does not make the decisions for you. This section is about the crucial step of determining the correct course of action on the basis of the analytical results.

Comparison is the Key: To borrow an amusing statement from the real estate business; there are three critical steps in interpreting analytical data, they are: compare, compare, compare. The point of this statement is that there is really one important step in interpreting analytical data. The idea is to compare, whenever possible, the data for your fluid with that from fields where you have both analytical data and field experience. Compare the analytical results from other fields. Compare your design or system with the systems used for treating the other, similar fluids. If the systems are adequate, copy the design. If the systems are inadequate, determine the gaps and improve the design. This process requires knowledge of facilities and fluid properties for several fields. This can be thought of as developing a database of fields that includes both the fluid properties and the facilities used to treat the fluids. The material in this chapter provides guidance on how to construct a database containing the necessary information.

It is often the case that such a database is not available. In the material below, it is possible to carry out the comparisons recommended without the need to develop a database. Two tools are used. The first is a numerical tool that helps quantify the areas that are likely to be the greatest challenge. This is referred to as the Difficulty Index. This index is then used to identify similar systems which are described here. Based on those systems, a first-pass design can be developed.

Comparison via Difficulty Index: One of the most effective ways to understand the character of a set of fluids is to apply a numerical ranking or rating to the various aspects. The procedure for doing this is given in the next section. This has several advantages. The most important advantage is that it forces a certain level of discipline in the work process.

16.5 Cook Book

In this section, the so-called "Cook Book" for water treating is presented. It is a systematic procedure for identifying the likely problems that will be encountered, identifying suitable analog fields, and for recommending an appropriate water treatment system. The strength of this methodology is that it is systematic, relatively straightforward, and can be applied in a short amount of time. The weakness is that it is perhaps over-simplified, and can be easily misused by inexperienced engineers. To help overcome this weakness, the material in the preceding sections should at some point be read and understood. That material provides the background and understanding for the methodology presented here. Also, once the methodology has been applied, the input data and the results should be reviewed with a specialist in water treating.

Cook Book for Water Treating: The methodology was originally referred to as the Cook Book for Water Treating since it provided a step-by-step method or recipe for designing or troubleshooting a water treating system. Eventually, the method became part of a Deoiling Manual.

The Cook Book is comprised of two parts. The first part is given in Section 16.5 (Cook Book) where the Difficulty Index is introduced. In this section, a method is given for making a numerical assessment of the character of the fluids. Nine categories are identified as being significant for making the assessment. Analytical results in each of these categories is given a numerical score. By applying the methodology in this section, a quantitative, multidimensional score can be assigned to the fluid properties.

The second part is given in Section 16.6 (Example – Deepwater Gulf of Mexico) where data for the deepwater Gulf of Mexico are analyzed. Together, Sections 16.5 and 16.6 provide a methodical means to proceed from analytical data to a rough understanding of the equipment, process configuration, chemical treatment program and operations procedures that are likely to result in satisfactory water quality.

To an extent, each produced water stream is unique, with characteristics defined by a wide range of variables such as the water source, the characteristics of the oil, the mineralogy of the reservoir, the upstream processing operations such as well head type, flow lines, and directly or indirectly added chemicals. The material in this chapter suggests that there is only a certain number of fluid and system types. To help overcome this weakness, the material in the rest of this chapter (above) should at some point be read and understood. That material provides the background and understanding for the methodology presented. Also, once the methodology has been applied, the input data and the results should be reviewed with an expert in water treating.

Advantages and Disadvantages of the Cook Book: The Cook Book for water treating is presented as a systematic procedure for identifying the likely problems that will be encountered, identifying suitable analog fields, and for recommending an appropriate water treating system. The strength of this methodology is that it is systematic, straightforward, and can be readily applied by non-specialists. The weakness is that it is perhaps over-simplified, and can be easily misused. To an extent, each produced water stream is unique, with characteristics defined by a wide range of variables such as the water source, the characteristics of the oil, the mineralogy of the reservoir, the upstream processing operations such as well head type, flow lines, and directly or indirectly added chemicals. The material in this chapter suggests that there is only a certain number of fluid and system types. In reality, this is not the case. To help overcome this weakness, the detailed material in the rest of this chapter and other chapters should at some point be read and understood. That material provides the background and understanding for the methodology presented. Also, once the methodology has been applied, the input data and the results should be reviewed with an expert in water treating.

In-Depth Material: While this chapter gives a cook-book instruction for interpreting analytical data, it hopefully does much more than that. Most of the material in this chapter is aimed at educating the interested engineer in the details. Explanation is given that shows why certain analytical results require certain water treating strategies. Using the material here, an engineer could construct their own guidelines, and assemble their own database from which to select analogs.

How the Cook Book was Developed: The methodology was developed in collaboration with process engineers who were responsible for new projects and for supporting existing assets. These process engineers were experienced but typically had a limited knowledge of water treating issues. Some had a knowledge level. None had a high level of skill or a mastery of the subject. Yet, they were being asked to provide a high skill-level of water treating expertise. These engineers were the users, so to speak, of the methodology presented here. They provided the valuable feedback as to what makes sense and what does not. Based on their feedback, the methodology was refined. Over several years some of these process engineers became skilled in water treating. At one point, the author had the interesting experience to attend an industry-sponsored workshop on water treating where the method was presented. The presenter was completely unknown to the author. The presenter did an adequate job of presenting the material and the presentation was well received by the audience. Setting aside for the moment the lack of due credit, the experience demonstrated that the material presented in this chapter is sufficiently well described that it can be understood and applied by most process engineers.

The following table summarizes the characteristics that will likely lead to water treating problems.

The following characteristics of Produced Water will likely lead to problems:

- **Solids (in general) and Iron Sulfides (in particular):**
    - Oil wet solids (also known as conglomerate)
    - Neutrally buoyant conglomerate
    - Small particle size and/or high concentration of solids

- **Biodegradation of Crude Oil:**
  - High TAN or high concentration of organic acids
  - Presence of calcium naphthenate
- **Unstable asphaltenes:**
  - High ratio of asphaltene/resin concentration together with a high saturates / aromatics ratio
- **Production Chemicals:**
  - Presence of methanol, Anti-Agglomerate Hydrate Inhibitors, Corrosion Inhibitors
  - Over-dosing of these chemicals

In this section a checklist is presented of the main parameters that should be measured in a characterization of produced water. A point system is used so that an assessment can be made of the difficulty of oil / water separation. For any water characterization, the total ranking points should be added. The table below then provides a ranking which can further be used to determine the difficulty of the separation problem. Equipment selection and system integration are discussed in subsequent sections of this book. Reference is made in those sections to the characterization types given here.

Table 16.1 Difficulty Index – Quantitative Assessment of Water Characteristics

Characteristic	Source	Design Detail	Possible Points	Characteristic	Ranking Points
Stokes Factor	API, temperature, water density	Longer residence time in primary separators	10	< 200,000	10
				400,000 to 800,000	5
				> 800,000	0
Oil flow assurance factors	Wax, paraffin, asphaltene stability, incompatible hydrocarbons	Inhibitors, heating for wax and paraffin	5	No inhibitors or heating required	2
				Inhibitor or heating required and none used	5
				Inhibitor or heating required and is used	0
Biodegradation	TAN, oil fingerprinting, biomarkers	Secondary separation equipment, optimized chemical treatment	15	0 wells	0
				1 to 2 wells Delta API > 2	10
				> 2 wells Delta API > 2	15
Scaling tendency	Mineral scales e.g. carbonates, sulfates; incompatible water	Water wetting chemicals, filtration / tertiary separation equipment	5	No inhibitors required, or inhibitor required and is used	0
				Inhibitor required and none used	5

Characteristic	Source	Design Detail	Possible Points	Characteristic	Ranking Points
Dissolved organics	Acids, naphthenates	Secondary separation equipment, optimized chemical treatment	5	pH > 6; acid < 100 mg/L	0
				4 < pH < 6 100 < acid < 500	3
				pH < 5 acid > 500 mg/L	5
Solids	Formation fines, scale particles,	Water wetting chemicals, filtration / tertiary separation equipment	15	Solids < 100 lb/MBbl	0
				100 < solids < 400	10
				Solids > 400 lb/MBbl	15
Iron sulfide	H2S, iron	Secondary separation equipment, optimized chemical treatment	20	FeSx < 10 mg/L	15
				10 < FeSx < 50	18
				FeSx > 50 mg/L	20
Surface active or shear enhancing chemicals	Corrosion inhibitor, methanol	Secondary separation equipment, optimized chemical treatment	10	No change upon turning off chemical	0
				Moderate water deterioration w/chemical	5
				Severe water quality deterioration w/ chemical	10
Small droplets	High shear	Reduce shearing	15	D < 10 micron	15
				10 < D < 50	10
				D > 50 micron	0

Severity of treating challenge:

Type 1 System:

Total points less than 35

No iron sulfide.

No biodegradation.

No sources of high shear.

For this level of challenge, a typical system may consist of primary separation followed by hydrocyclones. Flotation may be required, depending on the Stokes Factor. See the example below.

Type 2a System:

Total points between 35 and 55

No iron sulfide.

No biodegradation.

No sources of high shear.

For this level of challenge, a typical system may consist of primary separation, hydrocyclones and flotation. Care should be given to the handling of reject streams from the water treating equipment in order to ensure that a stabilized emulsion is not generated.

Type 2b System:

Total points between 35 and 55

Presence of iron sulfide or biodegradation.

No sources of high shear.

For this level of challenge, a typical system may consist of primary separation, hydrocyclones, multi-stage flotation and some means of treating the reject from the water treating equipment. Chemical application will be critical both in terms of demulsifier and water clarifier selection and optimization, but also in terms of minimizing the use of methanol and corrosion inhibitor.

Type 3 System:

Total points above 55

Presence of iron sulfide and / or biodegradation.

Source or sources of high shear.

Excessive use of corrosion or hydrate inhibitor.

For this level of challenge, a typical system may consist of three-phase primary separation, use of low shear valves and pumps, hydrocyclones installed on the three-phase primary separators, multi-stage flotation, some means of treating the reject from the water treating equipment, and some form of filtration / coalescence. Reject treating systems include a Slops Tank, or Wet Oil Tank with the ability to inject chemical or solvent. Chemical application will be critical both in terms of demulsifier and water clarifier selection and optimization, but also in terms of minimizing the use of methanol and corrosion inhibitor. Some form of tertiary water treating equipment may be required such as filtration, coalescence, or centrifugation.

16.5.1 Stokes Factor

A typical produced water stream from an E&P operation will contain a wide variety of organic compounds derived from contact with the produced oil. These organic compounds can be broadly grouped as dispersed oil, dissolved hydrocarbons, dissolved polar organics, and dissolved gases. The relative importance of the components depends on the regulations regarding discharge and disposal, as well as the environmental impact, the potential for sustained biological activity in the case of waterflooding, and the effluent quality required for reuse options. Hydrocarbons in water are considered to be dissolved if they pass through a 0.45 micron filter. Solids or dispersed oil in water is that which is retained on the filter.

Oil characterization is important for proper characterization of produced water because there are many components in the hydrocarbon phase that contribute to the stability of oil in water drops. Organic acids, asphaltenes, resins, are just some of the components that are found in the oil phase which can transfer to water phase or to the oil/water interface and thus have an impact on produced water quality.

The simplest oil characterization parameter is the density or API gravity. Light gravity oils have a greater density difference with water which thus helps drive gravity settling. They also have lower surface elasticity which promotes drop coalescence. But their lower viscosity promotes drop break up in a shear zone.

Stokes Factor: One of the most useful characterizations of produced fluids is to calculate the Stokes factor on the basis of oil density, water density, and water viscosity. An example of this is given in the Examples section. The Stokes Factor is a measure of the settling tendency of oil in water, without taking into account the drop size. The higher the Stokes Factor, the greater the speed with which oil drops rise to the surface of the water phase in the separator vessel. It is one of the characterizations that can be carried out in the design phase before drop size is known.

$$S_F = \frac{(\rho_w - \rho_d)}{\mu} \qquad \text{Eqn (16.1)}$$

Where:

S_F = Stokes Factor (sec/m^2)

r_w = density of water phase (kg/m^3)

r_o = density of oil phase (kg/m^3)

μ = viscosity of water phase (Pa sec)

When SI units are used, no unit conversion factors are required. An example calculation is given by the following:

$$S_F = \frac{(1100\ kg/m^3 - 800\ kg/m^3)}{(1x10^{-3}\ kg/(ms))} \qquad \text{Eqn (16.2)}$$

$$S_F = 300{,}000\ \text{sec}/m^2$$

16.5.2 Oil Flow Assurance Factors

As discussed in Section 16.3.4 (Flow Assurance Data), SARA analysis provides an indication of the asphaltene content and with proper analysis, it can also provide an indication of whether the asphaltenes have a tendency to precipitate. This was discussed in detail in Section 1.6.2. The SARA analysis will also indicate the fraction of crude oil that is composed of the resins. This is a very broad class of compounds and it is not possible to correlate water treatment challenges with the resin content or its characteristics, except to say that resin concentration above 15 wt % are characteristic of crude oils that contain a high concentration of acids, dissolved organics, and water soluble organics.

Naphthenates: The analysis of naphthenic acids, and the ARN acids that are responsible for naphthenate scale is becoming more routine since it can indicate a potential scale problem. By way of example, a class of compounds of major importance are the tetra-acid naphthenates. These compounds were first identified in the oilfield in the 1960's. Their importance was demonstrated with the shutdown of Chevron's Kuito field in the 1990's due to calcium naphthante scaling. The industry responded with significant activity directed at analysis and understanding of this important class of compounds. Analytical tests were developed to detect naphthenates, and phase equilibrium models were developed to estimate their stability. It has since been recognized that naphthenates are not only responsible for well publicized shutdowns, but they are also responsible for a much more pervasive occurrence of water treating difficulties. One expert in naphthenates scaling believes that naphthenates solids precipitation occurs in roughly 10 % of North Sea crude oils, 20 % of west African crude oils, and 30 % of south east Asian crude oils [8]. But naphthenates are only one of many organic compounds in produced water.

Biodegradation: All produced water contains dissolved organic compounds. In the past, until roughly ten years ago, produced water specialists have tended to downplay the importance of these compounds because they are difficult to characterize. Also, most legislative bodies do not emphasize the importance of dissolved hydrocarbons as much as they do dispersed hydrocarbons. But it is now recognized that dissolved hydrocarbons in particular, and organic carbon in general play a major role in flow assurance and in the performance of separation equipment and that they have a major impact on produced water quality.

Oil analysis should include the Total Acid Number (TAN) which provides a simple first indication of biodegradation. The recommended procedure is ASTM D-664. Besides TAN, organic acid content and composition should also be measured. The definitive test for biodegradation is done by gas chromatographic hydrocarbon fingerprinting which reveals the presence of naphthene biomarkers [6]. Biodegradation does not generally occur in reservoirs above 80 °C. However, hydrocarbon fluids often have complex histories over geological time scales. Thus, current reservoir temperature is not always a valid indicator of biodegradation since the hydrocarbons may have migrated. Although both anaerobic and aerobic bacteria can be active in a reservoir, it is the aerobic bacteria in the presence of a source of migrating or flowing water with dissolved oxygen that gives rise to the most significant microbial oxidation of crude oil.

Progression of Biodegradation: Mild biodegradation of crude oil will reduce the fraction of saturated hydrocarbons, and increase the organic acidic and natural surfactant content. As biodegradation proceeds, progressive loss of n-alkanes is followed by loss of isoprenoids which leaves an oxidized naphthenic residue. Aromatic functional groups, resins, and asphaltenes are not attacked and will therefore increase in mole fraction as the other components are consumed. As discussed below, biodegradation often results in a high concentration of polar organics, organic acids, natural surfactants, and surface active components which stabilize oil in water emulsions and therefore make water treating more difficult.

Dissolved Organics – Definition: Organic material is conventionally considered to be dissolved if it passes through a 0.45 micron filter. Particulate or dispersed oil in water is that which is retained on the filter. Obviously this is an arbitrary criterion. Nevertheless, it is universally adopted in the E&P industry. The 0.45 micron filter is commonly used for many other purposes and is readily available. One of the problems with this criterion is that drops of light viscosity oil tend to pass the filter whereas drops of heavier oil do not. Nevertheless, the 0.45 micron filter provides a simple and readily available test and has been accepted as the industry convention.

16.5.4 Mineral Precipitation Tendency

The treatment of waste water streams from many E & P operations is focused on the removal of dispersed hydrocarbons and oily solids. The two most critical characteristics of these contaminants are the size distribution and the density. The performance of most deoiling equipment will be ultimately limited by the smallest hydrocarbon droplet size that can be efficiently removed from the water stream. Droplets of dispersed hydrocarbons smaller than the minimum cut off size will not be removed. Droplet size distributions are discussed in more detail in Section 3.3 (Particle and Droplet

Size Distributions). The majority of deoiling equipment currently installed in E & P operations is only capable of removing dispersed hydrocarbons from waste waters.

16.5.5 Dissolved Organic Factors

Once the sample has been filtered, then a simple and commonly used test to measure the concentration of organic compounds is the Dissolved Organic Carbon (DOC) test. DOC is measured by converting all of the organic material in solution to CO_2 and then measuring the CO_2 that is produced. Total Organic Carbon or TOC is the organic content measured using the same test as DOC but run on the unfiltered sample. ASTM D 2579-78 discusses test methods for measuring Total and Dissolved organic carbon concentration. It references information relating the TOC to other measures of water quality such as biological oxygen content (BOD), and Chemical Oxygen Demand (COD), which are both used to characterize produced water effluent into rivers lakes, streams, and other biologically sensitive areas.

DOC is often used to assess the biological nutrient content of injection water for waterflooding. High DOC content water will likely result in significant reservoir souring (conversion of sulfates to H2S) if sulfate is present. Seawater, collected at least a few hundred meters offshore, typically contains about 0.5 to 5 mg/L of dissolved organic matter, as measured by DOC. Produced water can contain a few mg/L to many hundreds of mg/L.

Identification of dissolved organic compounds in produced water is usually made from two perspectives. One perspective is to determine the concentration of specific relatively low molecular weight species such as benzene, toluene, the xylene isomers, ethyl benzene, phenol, benzoic acid, and the Short Chain Fatty Acids (SCFAs) such as formic acid, acetic acid, propanoic acid, and butanoic acid. In fact, SCFA is a misnomer since most of the compounds usually included in this list have no or almost no aliphatic chain. Another name for these compounds is Volatile Organic Acids (VOCs) but the former name is more common. Below we discuss the importance of all of these compounds.

<u>Polycyclic Aromatic Hydrocarbons:</u> Environmental risk assessment studies have shown that the aromatic components in produced water constitute a major contribution to the Environmental Impact Factor (EIF). Recent studies have shown that alkylated phenols and Polycyclic Aromatic Hydrocarbons (PAH) have finite partitioning into both the oil and water phases, as shown in the table below.

Table 16.2 Partitioning of PAH components for two different oil in water concentrations. Partitioning percentages are given in mass percent [9]

	10 mg/L		100 mg/L	
EIF Group	oil	water	oil	water
naphthalenes	57	43	81	19
2 - 3 ring PAH	67	33	92	8
4 - 6 ring PAH	61	39	94	6
C0 - C3 phenols	0.05	99.95	0.5	99.5
C4 - C5 phenols	4	96	29	71
C6 - C9 phenols	64	36	95	5

16.5.6 Solids

Solids in hydrocarbon and produced water streams can originate from the formation, or by precipitation of inorganic scale forming minerals, or precipitation of organic materials. Typical solids in produced water systems include formation fines (calcite, sandstone), mineral scales such as carbonate (calcium carbonate, magnesium carbonate, iron carbonate), sulfate (calcium sulfate, barium sulfate, strontium sulfate), or sulfide (iron sulfide), corrosion products (hematite and magnetite), salt (halite), mineral/organic combinations (sodium and calcium naphthenates), organics (asphaltenes, waxes), and production chemicals (carbamates, polymer flocs), etc.. As far as oil in water emulsions are concerned, the most important properties of produced water solids are the size distribution, the overall quantity of solids, and the surface wetting characteristics. The treatment and discharge of drilling cuttings and drilling mud is not discussed in this manual.

NACE Test Method 0173 – 2005: A good starting point for the analysis of solids in produced water is the NACE Standard Test Method TM 0173 – 2005. Without going into details, solids are typically sampled using filtration through a 0.45 micron filter at a constant pressure drop across the filter. Several sets of sample are collected so that the samples can be subjected to various analyses which include gravimetric determination of the fraction of organic material, inorganic material, and the composition of the organic and inorganic constituents. A typical result is given here using example data for solids which were collected on the Bullwinkle platform in the Gulf of Mexico in October 2006. These solids were collected from the produced water discharge of the wet oil tank. This discharge line was fed into a centrifugal pump and the stream was then recycled into the Bulk Oil Treater. A solids size analysis was not carried out but based on settling tests, the solids were very small diameter (less than a few microns) and had oil attached making them roughly neutrally buoyant in the produced water. At the time, the platform has several subsea production systems from significantly different reservoirs and the TDS of the combined produced waters was high (above 200,000 mg/L).

Table 16.3 Solids Analysis for solids collected on the Bullwinkle Platform (Oct 2006)

Gravimetric Wash Test	weight % of dry sample	Comments
Deionized water wash	19.7	Includes substances soluble in water such as salts
Xylene wash	9.6	Includes substances soluble in xylene such as paraffin, oil, and organics
Acetic acid wash	27.7	Includes substances soluble in weak acetic acid such as carbonate mineral scales
Hydrochloric acid wash	27.4	Includes substances soluble in 15 % HCl acid such as iron sulfide, and iron oxide.
Acid Insolubles	15.6	Includes substances insoluble in 15 % HCl acid such as sulfate scale, sand, silica fines
Total	100	

A portion of the sample was washed using deionized water followed by xylene and then subjected to inorganic solids composition analysis using EDAX, XRF and XRD. The results are given as:

Positive for calcium carbonate scale

Positive for barium sulfate scale

Positive for silica fines

Positive for iron compounds

As given in the table above, we can conclude that roughly 70 wt % of the sample is composed of these inorganic solids. Another portion of the sample was washed with deionized water, followed by xylene, followed by HCl. This leaves the Acid Insolubles fraction and this sample was analyzed using EDAX and XRD which gave a positive indication of sulfate scale and silica fines. The acid wash did not evolve noticeable levels of sulfur which rules out the presence of iron sulfide compounds. Optical microscopy verified the presence of cubic crystals which are typical of halide precipitates.

Organic solids analysis was performed using DSC/TGA, GC, and H NMR. The results are positive for waxes and asphaltenes. Results such as these must be interpreted by the analytical laboratory since they involve a combination of qualitative and quantitative techniques. However, the final result in this case turned out to be close to that given in the table of gravimetric wash results.

Table 16.4 Final results of solids sample analysis

Substance	Weight %
Organic material	10
Halide salt from high salinity	20
Calcium carbonate precipitate	28
Iron oxide solids	27
Barium sulfate precipitate and silica fines	15
Total	100

The conclusions from these analyses and observations were:

- mineral solids were being formed from a combination of processes (incompatibility leading to barium sulfate precipitation, pressure drop leading to carbonate precipitation);
- the solids were a conglomerate of inorganic precipitates and oil making them roughly neutrally buoyant in the high salinity produced water;
- high iron oxide solids content indicated corrosion processes which was verified by the frequent requirement to repair and replace section of the produced water piping;
- high solids content was a major contributing factor in poor water quality, and was a complicating factor in implementing clean-up technologies which suffer performance degradation in the presence of solids.

Over a period of several months, the problems were solved by a combination of chemical treatment, better corrosion prevention, segregation of fluids, re-routing of recycle streams.

From an environmental standpoint, suspended solids may have a number of potential impacts, including:

- Some solids may be toxic themselves, contain toxic elements, or have radioactive constituents.
- Solids may trap of collect other contaminants (i.e. oil).
- Discharged solids may accumulate as mud or silt in the local environment.
- Discharged solids may result in turbidity in receiving environments with poor dispersion characteristics.

Suspended solids may have a significant impact on the performance of deoiling equipment. The turbidity caused by discharge of solids may have an environmental impact as well as being undesirable visual pollution.

Many of the solids found in produced water systems, particularly the mineral precipitates and corrosion products are initially water-wet. In some cases, the solid particles remain water wet and are discharged in the overboard water stream. Some regions of the world have a limit on the amount of solids that can be discharged. In other regions there is no limit, however the turbidity caused by discharge of solids may have an environmental impact as well as being undesirable visual pollution.

The solids may settle in the bottom of separators where they occupy volume and thus reduce the residence time for process fluids. Such solids can contribute to under deposit corrosion or become suitable habitat for bacteria. Solids can also cause clogging of the piping and precipitate on the walls of vessels and piping to form scale which is a major flow assurance threat.

Oily Solids – Surface Activity: Often a fraction of the particles will become at least partially oil wet by attachment of sticky crude oil components such as asphaltenes, waxes, naphthenates and naphthenic acids, or by attachment of various production chemicals including corrosion inhibitors and water clarifiers. A solid particle that is initially water wet and which becomes partially oil wet will then be surface active. Thus, any amphoteric component that is itself moderately surface active has the potential for wetting the solid surface and enhancing its emulsion stabilizing properties. In produced hydrocarbon fluids, potential wetting agents include organic acids, naphthenic acids, asphaltenes, resins, waxes, and other polar compounds.

Small Solids – Small Oil Drops & Tight Emulsions: In general, the smaller the solid particles, the tighter the emulsion (smaller emulsion drops) which can be stabilized by the solids. The greater the amount of solids, the greater will be the concentration of oil in water. As a general rule, particles that are predominantly water-wet will stabilize an oil in water emulsion. Likewise, particles that are predominantly oil-wet will stabilize a water in oil emulsion.

Conglomerate: We refer to partially oil wet solids as conglomerate. Their properties are different from those of oil drops or solid particles. The density of a conglomerate particle depends on the mass ratio of oil to solids in each particle, which can vary significantly. The combination of a light component (oil), plus a heavy component (solid) will result in a specific gravity decrease for the solid particle. When this occurs, separators and hydrocyclones will be less effective because there is less density difference to drive the relative movement of the conglomerate from the water. In addition, a fraction of the conglomerate will be neutrally buoyant in the produced water which will render all gravity separation devices ineffective. Induced gas flotation can be and often is used to remove neutrally buoyant solids from produced water.

From a water treating standpoint, the presence of oily solids (conglomerate) can be devastating. Once a solid stabilized emulsion forms, it is particularly difficult to separate the components from each other and to separate the conglomerate from water, due to the shift in density. The more successful technologies for removing such conglomerate include flotation, chemical treatment, and filtration. Chemical treatment is discussed below. Filtration carries huge risks in terms of media usage and cost. The capacity of a filtration system must be tested prior to full scale implementation, due to the potential for high cost of frequent media replacement.

Flotation for Treating Solids: Flotation is somewhat effective in separating neutrally buoyant solids stabilized emulsions, through the attachment of bubbles which then create the density difference needed for separation. The gas bubbles stick to the solids themselves and to the oil-wet surface of the solids and carry them to the top of the water where they are floated to the water's surface and recovered via a spill over weir (See Chapter xx where the specific mechanisms by which induced gas flotation works are discussed in some detail.). This is an effective means to perform the separation of oil and water. It should be noted that while flotation can be highly efficient, it's application is generally limited to relatively low levels of conglomerate (e.g., less than 100 to 300 ppm). However, the use of chemically assisted flotation can lead to problems in a process if the rejects (oil-wet solids with attached polymer) are recycled through the system.

16.5.7 Iron Sulfide

Iron sulfide is one of the most problematic solids in produced water treating. It was discussed in some detail in Section 2.7.4 (Iron Compounds). Iron sulfide, like iron oxide is very insoluble. According to the Gas Machinery Research Council, Schmoo is the least understood and most prominent contamination problem in pipelines and gas compression equipment.

Schmoo: There have been many reports on the composition of Schmoo, and its precise definition is the source of ongoing debate. The composition varies considerably depending on water composition and the use of various production chemicals such as corrosion inhibitors. For our purposes, we classify Schmoo as an oily solid conglomerate formed fundamentally from iron solids such as iron sulfide and to lesser extent iron oxides.

Unique Properties of Iron Solids: It is generally true that when a solution changes conditions rapidly, such that a relatively insoluble species reaches supersaturation rapidly, precipitation of the species will form many small particles rather than fewer large particles. This is particularly true of the iron compounds. Solids formed from iron tend to have size distributions that average less than a micron in diameter, and their numbers are relatively enormous.

16.5.8 Surface Active Production Chemicals

It is beyond the scope of this section to provide a complete discussion of how to address solids problems. However, some general guidance is proffered.

The effects of injected chemicals to oil/gas production streams can be a significant factor in the selection and design of deoiling equipment. Many treatment chemicals are surface active and may stabilize small hydrocarbon drops in the water phase. The corrosion inhibitors used in gas production operations and the demulsifiers used to assist oil dehydration can sometimes result in such stabilization. The small drops may be very difficult to separate with conventional deoiling equipment.

Some treatment chemicals may be incompatible with other chemicals. Examples are demulsifier chemicals interfering with subsequent deoiling chemicals, or deoiling chemicals reacting with dilute polymers present in the water as a result of enhanced recovery schemes.

Efforts should be made to reduce the use of treatment chemicals whenever possible on the basis of minimizing both operating costs and the ingress of additional chemicals into the environment.

16.5.9 Small Drops

The fluids in the deepwater GoM are particularly suitable to forming conglomerate stabilized emulsions. They have relatively high asphaltene content, ranging from a few to several percent. Resin content is also somewhat high, varying from a few percent to as much as 15 %. Due to the presence of a relatively high aromatics and resin content, the asphaltenes in the deepwater GoM oils tend to be stable to marginally stable, with none of the fluids in the unstable region. While this means that asphaltene precipitation is not a problem in the reservoir, near well bore or tubing locations, asphaltenes do nevertheless precipitate to some extent in the topsides facilities and contribute to the stability of both water in oil and oil in water emulsions. They also adsorb onto solids particles, causing solids stabilized emulsions. This, together with the relatively short residence times found in the deepwater process equipment, leads to a relatively challenging situation regarding oil and water separation. As already discussed, sophisticated electrostatic treater designs and well managed chemical programs are required to separate drops of oil that are stabilized by asphaltene particles. Thus, it is not surprising that induced gas l flotation would be of such great use.

As a specific example of processes that can lead to water treating problems due to solids, we consider an example from Shell Brazil. When a reservoir is waterflooded with any water other than source water, there is the potential for scale formation as the two water chemistries intermingle. In a typical seawater breakthrough scenario, barium sulfate particles will form in the produced fluids. The particles can be very small (<1 micron to 2 micron), particularly in the presence of a scale inhibitor. As the barium sulfate particles form, the asphaltenes may also be precipitating. As asphaltenes precipitate, they may adsorb onto available surfaces including the barium sulfate particles. Such fine particles, with an asphaltenic coating are excellent water-in-oil emulsifying agents.

16.6 Example – Deepwater Gulf of Mexico

In this section we give examples where the results of produced water characterization were used in the design of a new field or to solve a water treating problem. In the examples given here, only the highlights of the characterization are given, rather than the complete characterization. Characterization results can be quite lengthy and in the interest of brevity, only the important highlights are given.

As discussed above, one of the simplest characterizations that can be carried out for fluids is to calculate the Stokes Factor. A Consultancy Services Benchmarking report gives Stokes Factors for BP and Shell platforms in the North Sea. Below we provide data from which the Stokes Factor can be calculated for the Mars platform in the deepwater Gulf of Mexico. The Stokes Factor is a measure of the settling tendency of an oil/water fluid, without taking into account the drop size. The higher the Stokes Factor, the greater the speed with which oil drops rise to the surface of the water phase in the separator vessel. The Stokes Factor does not take into account the drop size.

Mars fluids:

API 17.3 to 31.5

Water density: 1.14

Water temperature: 120 F

Water viscosity: 0.56 cP

Stokes factor 337,605 to 485,539

Also provided in the Consultancy Services report is the volumetric flux of liquids through the main oil/water separator. The volumetric flux is a measure of the flow related duty of the separator. It is calculated as the ratio of the total liquid volumetric flux and the interfacial area between the oil and water in the separator. As the flux increases, separation becomes less efficient.

Mars FWKO:

Total liquid volumetric flow: 200,000 BFPD = 0.368 m3/sec

FWKO: S/S length: 40 feet; Inside diameter: 120 inch; NWL from bottom: 3.3 feet

Flux rate: 0.0090 m3/sec/m2

Data for both Shell and BP in the North Sea, together with the Mars value for the deepwater Gulf of Mexico are given in the figure below. As shown, the flux number is significantly higher for Mars which means that vessel residence time is lower. It is noted that in typical North Sea platforms, water treating includes hydrocyclones but not flotation. In the deepwater GoM, both hydrocyclones and flotation units are required to meet very similar specifications.

In this example we have given a very simple characterization of the produced fluids properties. For the purposes of oil/water separation we have demonstrated that both oil and water properties are required. We have also shown the tight link between separator sizing and fluid properties.

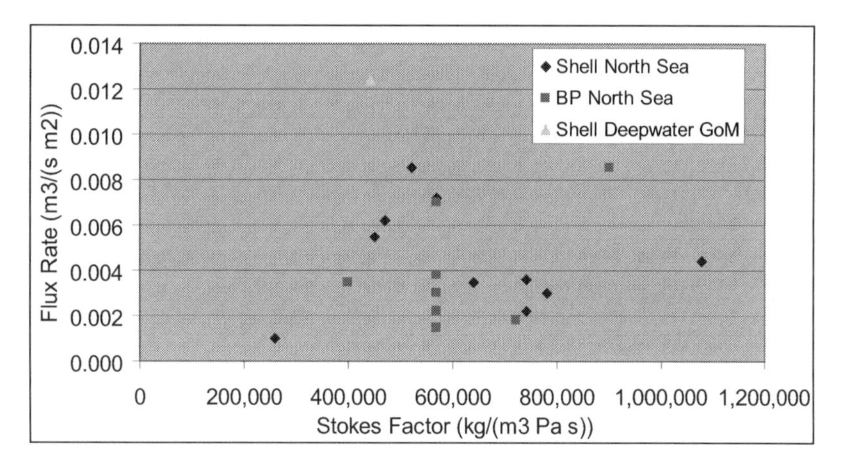

Figure 16.1 Flux in the main oil/water separator versus the Stokes Factor which provides a simple means of characterizing the fluid properties. Note that the flux for the deep water floating platform (yellow) is much higher than that used in the North Sea where the platforms are predominantly supported on the sea floor

At the time of this writing, Shell operated six deepwater platforms in the Gulf of Mexico (Bullwinkle, Auger, Mars, Ram Powell, Ursa, Brutus). The platform with the most problems at the time was Bullwinkle. Rather than present the fluid characterization of just that platform, we present characterization of fluids including all of the platforms. The other platforms were included in a search for analogs. It was determined that there were fluid similarities between Mars and Bullwinkle, which suggested that water treatment techniques used on Mars would have a good chance of working on Bullwinkle as well.

The other feature of this example that we wish to highlight is the fact that it includes not just water characterization but also characterization of the reservoir sands, and of the hydrocarbons. As has been discussed, the properties of produced water are strongly affected by both.

As shown in the figure below, the fluids range in API from 24 to 32, with most wells producing in the range of 27 degrees API. The Ram-Powell field, and the early Auger field are exceptions. Both had relatively high gas production with gas condensate liquids. The gravity of the condensate was in the range of 32 to 34 API. Over time, Auger reservoir targets moved from gas to oil.

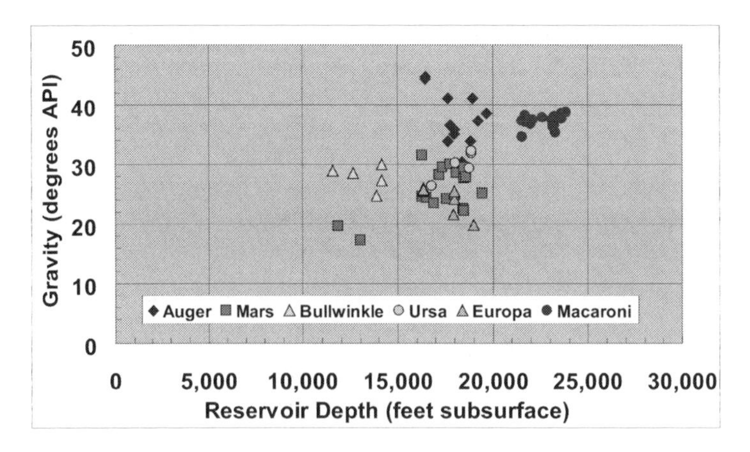

Figure 16.2 API gravity as a function of the reservoir depth for reservoirs in the deepwater Gulf of Mexico. As discussed in the text, the shallower reservoirs were subjected to biodegradation

Most of the hydrocarbon fluids produced at the Shell deepwater locations are of the Miocene age. The reservoirs are typically deep (deeper than 15,000 feet below subsurface). Reservoir temperatures are in the range of 165 to 180 F. These factors suggest that biodegradation is possible but not highly probable, and that biodegradation is likely determined by geologically historic access to percolating, low saline and nutrient rich water. As previously mentioned, given the low occurrence of aquifer volumes, and the fact that there is a high occurrence of massive salt bodies, there is little evidence that access to such percolating low saline water has occurred. The fluid properties do indeed reflect this analysis.

While biodegradation is relatively rare across the Shell deepwater GoM portfolio, it was seen in two notable exceptions. The Pink reservoir on Mars and Bullwinkle are both considered to be biodegraded. For Mars Pink, the acid number is high (see table below), the relative fraction of saturates is lower and the aromatic fraction is higher. GC fingerprint analysis (not given in this report) also showed characteristics of biodegradation. Platform personnel noticed that when the well containing biodegraded fluids (A-1) were flowing, oil dehydration (< 1 % BS&W) and water deoiling (< 29 ppm) targets were much more difficult to achieve. Biodegradation was also detected in the Bullwinkle Pink reservoir which also had high Total Acid Number and caused water treating problems when online. Also, naphthenic acid content for the platform as a whole was rather high and this was thought to be due to the wells flowing from the Pink reservoir. Across the other Shell deepwater platforms, it is thought that the biodegradation ceiling is below the Pink reservoir but above all the other reservoirs.

Biodegraded fluids are problematic because they contain a relatively high concentration of surface active species such as short chain fatty acids, acidic resins and acidic polynuclear compounds. Mars not only had such fluids to contend with but it also had short separator and water treatment system residence times as well. The water treating system was typical of GoM deepwater consisting of a FWKO, hydrocyclones, a Wemco flotation unit with recycle to a slop tank and ultimately to the Bulk Oil Treater. To handle the biodegraded Pink fluids on Mars, a rapid acting high molecular weight polyol type demulsifier was selected based upon bottle testing. This demulsifier was somewhat expensive but it helped maintain good control of BS&W without degrading produced water quality. Vessels were kept clean of pads. A carbamate water clarifier was used. More details can be found in [10].

While wax is not a general problem at Shell deepwater locations, there are a few wells that require wax inhibitor or some form of thermal or mechanical treatment to remove wax buildup. For the most part, wax problems are isolated to a few wells.

Most of the Shell deepwater GoM production is sweet with no H2S. Some reservoirs that were water flooded have low concentrations of H2S. There were no signs of iron sulfide or Schmoo. The CO2 content generally falls in the range of 0.1 to 0.4 mole % by volume for gas recovered from a multi-phase sample at stock tank conditions. This is relatively low and helped to maintain relatively low bicarbonate concentrations. Although the produced water was highly saline, and contained high concentrations of calcium and magnesium, calcium carbonate scaling was not a major problem, due to relatively cool temperatures and low bicarbonate concentrations.

The atomic sulfur in the Mars crude is mostly incorporated in aromatic species such as thiophenes, benzothiophenes, dibenzothiophenes, and higher molecular weight species.

As a detailed example of fluid property variation within a field, fluid properties for the Mars reservoirs are given in the table below. The asphaltene and resin content are shown in the figure below. The impact of the asphaltenes on topsides processing is discussed below.

Table 16.5 Fluid Properties for the Mars Reservoirs

Reservoir	Depth	API	Saturates	Aromatics	Resins	Asphaltenes	Atomic S (wt %)	ACID No.
Pink	13,036	17.3	21.0	61.6	14.0	3.4	2.7	4.35
Lower Green	16,287	31.5	40.9	47.6	9.5	2.1	1.8	0.34
Ultra Blue	16,301	24.6	41.3	37.5	13.9	7.3	2.6	
Orange	16,550	24.4	27.4	54.5	14.5	3.6	2.4	1.00
Upper Green	16,910	23.5	24.3	55.6	14.9	5.2	2.6	
Magenta	17,610	24.2	28.1	54.1	13.5	4.3	2.6	0.61
Violet Ic	18,419	22.5	25.0	59.3	12.6	3.1	2.7	1.20
Lower Yellow	18,476	27.6	30.8	51.8	15.1	2.2	2.2	
Terra Cotta	18,476	22.1	25.2	50.2	13.4	11.2	2.8	0.95

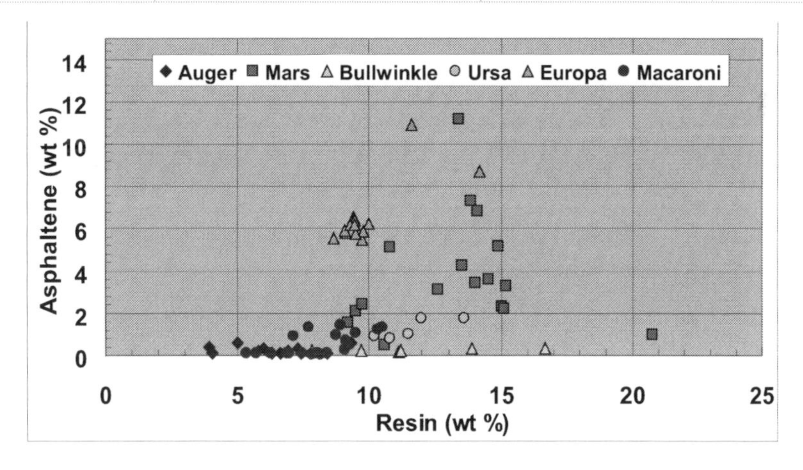

Figure 16.3 Resin and Asphaltene content of hydrocarbon fluids in the Gulf of Mexico

In the deepwater Gulf of Mexico, asphaltene precipitation in the wells and flow lines was anticipated to be a problem for at least a few subsea systems. Europa and Macaroni were started up and operated for a few years with continuous injection of asphaltene inhibitor down hole. Data for the Europa system are given below as an example. Heavy asphaltene deposits have been observed (Macaroni A-2) on subsea choke bodies however these were determined to have been from sludge formed by contact of the hydrocarbon fluids with acidic completion fluids. In general, asphaltene problems and deposits have not been experienced down hole or in the well bore despite the fact that asphaltene inhibitor injection has been discontinued.

Most of the Shell deepwater reservoirs are bounded by large salt bodies. What little aquifer water there is in these basins is typically in communication with these salt bodies along their outer edges. As the typical analysis shows in the table below, dissolved mineral content of the produced water tends to be very high.

Table 16.6 A typical Shell deepwater GoM water analysis – Mars Terra Cotta A-4 well

Specific Gravity	1.13
pH	6.5
Cations	**mg/L**
Sodium	61,300
Calcium	5400
Magnesium	1630
Barium	160
Iron	20
Anions	**mg/L**
Chloride	109,000
Bicarbonate	120
Sulfate	1
Total Dissolved Solids (TDS)	177,640

Solids in Crude Oil: The incidence of solids in crude oil in the GoM has only been sparsely studied. Some results are given in the figure below. The solids found in the GoM are small diameter (3 micron and smaller) sand stone fines. On Mars (GoM-1 in the figure) and Bullwinkle (GoM-2 in the figure) there was a tendency for these fines to be coated with asphaltenes which contributed to produced water problems.

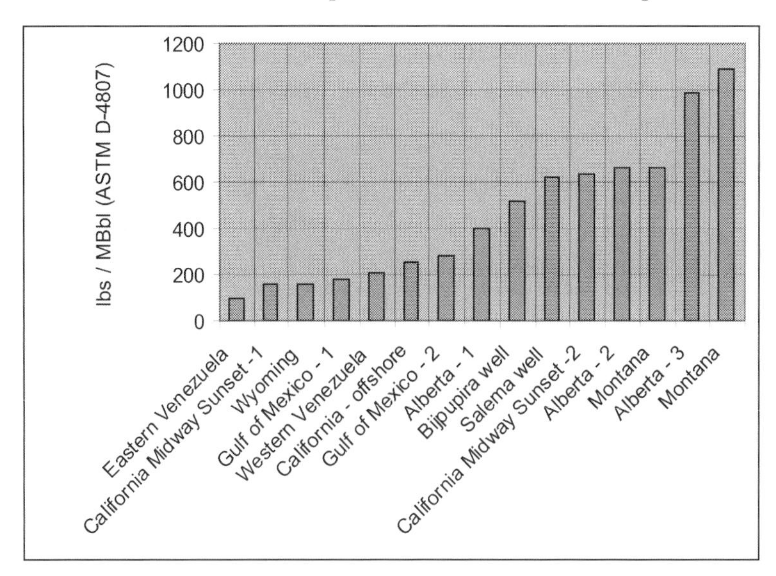

Figure 16.4 Solids content of produced fluids from various regions of the world. It is important to compare one region to another in order to gain a perspective on the relative magnitude of solids problems. California Midway Sunset-2, Alberta, and Montana have some of the most challenging solids problems in the world. Fields such as California Midway Sunset-1 do not have significant solids problems

It is helpful to express the solids concentration in units of mg/L. The conversion factor is:

1 lb/M bbl = 2.86 mg/L

It must be emphasized that solids in the crude oil are not necessarily indicative of oily solids in the produced water. Solids in the crude oil typically have their origin in the oil coated part of the reservoir. The solids get swept out of the reservoir by the produced fluids. In sandstone reservoirs, the solids are a combination of silica (sand) and clay, depending on the rock composition. As discussed in Section 4.4.8 (Surface Wetting), both silica and the flat surface of clay particles are negatively charged. Crude oil typically contains a predominance of negative functional groups such as carboxylic acids. Thus, in order for crude oil to stick to sandstone solids a positively charged group must first adsorb onto the surface. The solids are oil coated due to a fairly complex surface chemistry.

Solids Stabilized Emulsions: Once a solid stabilized emulsion forms, it is particularly difficult to separate the components. A fraction of the emulsion drops, if dispersed into the water phase, will be neutrally buoyant in the produced water. That is, depending on the relative quantities of oil and solids, the combination of a lighter than water component (oil), plus a heavier than water component (solid) can result in a specific gravity close to that of water. When this occurs, separators and hydrocyclones will not provide effective separation because there is insufficient density difference to drive the relative movement of the oil and water. There is some initial evidence that solids related oil and water treating problems are not being recognized as such.

In 2001, drop size distribution was measured on the Mars platform. The volume average drop diameter is given for various locations in the figure below. Unfortunately, not all locations could be sampled during this study. Particular emphasis was placed on locations that would elucidate the effect of pump shear. Recycle systems (shown in blue) often require pumping in order to push the fluid into an upstream location. As shown, the discharge from the two produced water centrifugal pumps did in fact have very small average drop diameters. A sample of fluid was taken upstream of the Bulk Oil Heat Exchanger. As observed under the microscope, the fluid was composed of a continuous oil phase in which drops of water were dispersed. In addition, the drops of water had very small drops of oil dispersed in them. This is typically referred to as a complex emulsion (see the discussion in Chapter 3, "Emulsions"). The cause of the complex emulsion was the recycling loop shown in the process schematic. Another feature of the data given in the figure is the relatively small drop diameters that were observed on the platform. In particular, the overboard discharge stream, from the Wemco Induced Gas Flotation Unit had an average drop diameter of 6 microns. This is considerably smaller than would be expected without the use of chemicals. But in fact, a fast acting and highly active carbamate chemical was being used. This could account for the higher than expected performance of the IGFU.

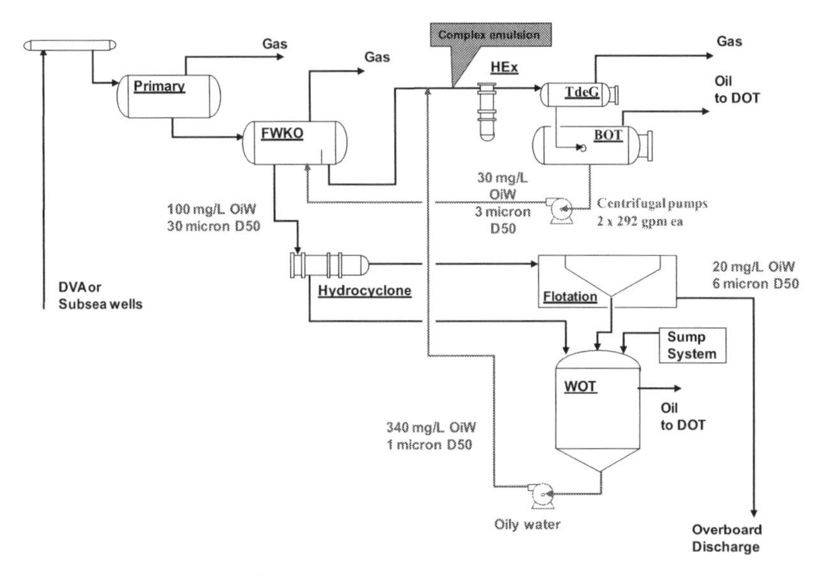

Figure 16.5 Process Schematic of the Mars TLP oil and water treating system. Volume average drop size is shown as measured by isobaric sampling and optical microscopy determination of drop size distribution

A summary of the water treating experience for the deepwater platforms in the time period from 2001 to 2006 is given in the table below. This table identifies the characterization area and gives the characterization result together with the observations of the water treating performance and strategies used to overcome the fluid properties that lead to water treating challenges.

Table 16.7 Correlation between produced fluids characterization and the performance of water treating equipment.

Characterization	Learning from Equipment Performance
Subsurface setting (depth, T, geochemistry)	High salinity seen across the region which gave high density differences between oil and produced water; some scale problems addressed with scale inhibitor; separate trains required in some cases to segregate incompatible fluids.
Possibility of biodegradation	Hand full of wells caused significant oil dehydration and water deoiling problems when they were producing; extensive vessel internals modifications were required for oil dehydration; process re-routing and chemical treatment work required for water deoiling in order to stay on specification when those wells were producing.
Gas constituents (CO2, H2S) & process conditions (T, P)	Sweet production thus no iron sulfide problems; carbonate was high in some cases but with low CO2 (< 1 mole %), moderate bicarbonate, and no need for alkaline H2S scavengers, calcium or magnesium carbonate was kept under control with scale inhibitors, and scale monitoring.
Oil characterization	Resins and asphaltenes were high; asphaltenes were moderately stable due to high resins and aromatics, with a few exceptions; caused foaming, emulsion, and water treating problems. Had to apply special chemistries, keep water treating equipment clean by flushing with solvents, developed special acid flow back procedures to prevent acid precipitation of asphaltenes, changed process routing to reduce complex emulsions, minimize recycles, avoid condensate mixing w/ oily water, minimize shearing, improve treatment of recycle streams.
Brine analysis by asset & by region	High iron was used to advantage w / DTC chemistry.
Possible solids from reservoir	Cleaned vessels frequently; practiced rigorous sand surveillance; Auger applied an acid treatment to strip oil off of solids. Solids wetting demulsifiers applied at Mars.

References for Chapter 16:

1. P.E. Gramme, "Fundamentals of separation. Stability of emulsions. Stability of dispersions," presentation given to NTNU (2011).

2. P.E. Gramme, B. Dybdahl, O. Holt, J.D. Friedemann, B. Riise, "MTU - The Multi Test Unit for investigating offshore separation problems and optimizing the gas / oil water separation process," SPE - 56847, paper presented at the SPE ATCE, Houston (1999).

3. "Standard Methods for the Examination of Water & Wastewater," 21st Ed., Managing Editor Mary Ann H. Franson (2005)

4. C.C. Patton, "Oilfield Water Systems," Campbell Petroleum Series, Norman, OK (1974).

5. A.G. Ostroff, Introduction to Oilfield Water Technology, National Association of Corrosion Engineers, Houston, TX (1979).

6. L.M. Wenger, C.L. Davis, G.H. Isaksen, "Multiple Controls on Petroleum Biodegradation and Impact on Oil Quality," SPE 71450 (2001).

7. E. Tegelaar, "New Guidelines for the Modified IP 143 / 57 Method," Baseline Resolution, Inc. (1999).

8. "Naphthenate Deposits, Emulsions Highlighted in Technology Workshop," Journal of Petroleum Technology (July 2008).

9. M.A. Reinsel, J.J. Borkowski, J.T. Sears, "Partition Coefficients for Acetic, Propionic, and Butyric Acids in a Crude Oil / Water System" J. Chem. Eng. Data v. 39, p. 513 - 516 (1994).

10. J. Walsh, T. Frankiewicz, "Treating Produced Water on Deepwater Platforms: Developing Effective Practices Based Upon Lessons Learned" SPE 134505 (2010).

CHAPTER SEVENTEEN

Shale Produced Water Treatment

Chapter 17 Table of Contents

17.0 Introduction 375

17.1 Produced Water Characterization from Select Fields 376

17.2 Water Quality Specifications for Reuse 387

17.3 Water Quality Specifications for Disposal Injection 388

17.4 Water Treatment Strategies and Options 390

17.4.1 Chemical Treatment 391

17.4.2 High Rate Clarifiers 396

17.4.3 Field Experience 396

17.5 Electrocoagulation Practical Applications 397

17.5.1 Mechanism of Separation & Performance Variables 401

17.5.2 Design Differences and Characteristics 407

17.5.3 Process Configuration, Waste Handling, Control 408

17.5.4 Representative Performance Data 409

17.5.5 Boron Removal 412

17.5.6 Softening and Silica Removal 415

17.5.7 Fine Solids Removal: 418

17.5.8 Benefits and Drawbacks of Electrocoagulation: 419

References to Chapter 17 421

17.0 Introduction

This chapter focuses on produced water in the Permian basin (west Texas and south east New Mexico). As discussed by Patton [1] the shale wells in the Permian produce a large volume of water compared to other shale plays. In the Permian there is both conventional developments and shale. Just within the shale fields, the ratio of water production to oil production ranges from a factor of 3 to a factor of 5. This is significantly greater than in the Bakken and Eagle Ford. Including conventional production this value goes up to 20 in the Permian region. A relatively small percentage of the total water volume is composed of flow back fluids. Thus, the produced water contains only a small percentage of polymer. The net result is that there is sufficient produced water that could be treated and used for fracturing. McLin et al. [2] discuss the case for ConocoPhillips which has three main production fields in the Permian: Maverick and China Draw in Texas, and Red Hills in New Mexico. ConocoPhillips has piloted large scale produced water reuse.

Sourcing water for hydraulic fracturing and disposing of produced water are well-known constraints and significant cost items in the development of shale formations particularly in the Permian Basin. Utilizing a water life-cycle approach, some of the produced water can be treated and reused. However, there is usually more produced water than needed and some must be disposed, typically by injection into a disposal well. Whether the water is to be reused or disposed, it must be treated to some extent. Given the volumes of water involved, treatment technology must be robust and inexpensive. This suggests that the selected technology should be tailored to the characteristics of the water and the quality requirements of the final purpose (reuse or disposal).

The starting point for developing a water treatment system for any application is to understand the water quality and characteristics of the water to be treated. Once this is done, then the target quality or specifications must be understood. The specifications come about form a combination of regulations, practical considerations such as disposal well injectivity, reuse quality requirements, and compatibility with other water sources. The difference between the produced water characteristics and the specifications of the treated water provides the basis for the design of the water treatment system.

It is sometimes the case that a water treatment system must be designed for a green field facility. When this is the case, there is limited sample or process data available. Nearby analog fields must be used to provide data. For the moderate to large oil and gas companies suitable analogs can be found within the company. For the smaller companies, suitable analogs may not be available and an educated guess must be used to fill in the missing data. The risk of over or under design becomes greater. One of the objectives of this chapter is to provide generalizations that can be used when minimal field data is available for a particular location.

The water quality required for reuse is driven by the fracturing chemistry and process. Salt tolerant polymers are available such that desalination is not required. The main water quality requirements are the removal of iron, H2S, and most of the suspended solids [3, 4]. This is further discussed below.

The water quality required for disposal depends on federal and state regulations, and on the economics of well stimulation which will be required when plugging has occurred. There are several indirect costs associated with shutting down and stimulating an injection well not the least of which is trucking the deferred produced water to alternative disposal sites.

Once the produced water characteristics are known, and the quality requirements for the disposition (disposal or reuse) are known, then appropriate water treatment facilities can be designed. As will be discussed in greater detail below, this approach is a data driven approach and is more typically used in industrial water treatment than in oil and gas.

17.1 Produced Water Characterization from Select Fields

A competent sampling and analysis program must anticipate a wide range of contaminant size and composition. It is usually not possible to predict these accurately ahead of the site visit. The equipment brought to site must cover a wide range of parameter values. Measurements made in the field have significant scatter, even with the most highly trained staff. Therefore multiple measurements must be made for each sample point and for each characteristic. Also, for the important characteristics such as particle size and contaminant composition, it is prudent to use multiple measurement techniques in order to ensure accuracy. These sampling and analysis issues are discussed below.

The types of sampling and analysis that are carried out must always be directly correlated with the anticipated final disposition. There are many different characteristics of a water sample that could be measured. The analyses required for municipal, chemical, food and beverage water treatment involve measurements such as salinity, BOD, COD, TOC, aquatic toxicity and several others. Most of these characteristics are not needed for shale applications. Instead, water characterization in the Permian focuses on iron, H2S, and suspended solids.

One of the objectives of this discussion is to point out what information is required to design a produced water treatment scheme for a shale field in the Permian given the two possible final dispositions, reuse and injection well disposal. While these two dispositions are significantly different from each other in many regards, the water characteristics that must be measured are qualitatively similar.

Characteristics of Shale Produced Water: Shale produced water from various locations in the Permian/Delaware region has been analyzed and results are reported here. Individual wells, and flow lines containing produced water from a number of wells, and flowlines entering gathering stations have been sampled and analyzed. The selection of locations has not been systematic nor comprehensive. Three fields were studied in detail. A number of other fields were studied through limited sampling and analysis. Also, data from the literature has been included below.

Field 1: Bass Facility, West of Midland

Field 2: Red Hills, South East New Mexico

Field 3: Midstream water gathering facility in Lea County, New Mexico

The following is a list of sampling and analysis for shale produced water.

- Suspended Solids:
    - The objective of this testing is to (1) calibrate the particle size analyzer, and (2) turbidity meter, (3) to obtain solids samples that can be analyzed subsequently to determine the composition of solids, and (4) obtain a survey of solids composition throughout the facility to understand how the solids change through the facility.
    - This testing involves Jorin particle size analysis, tandem multi-filter analysis, turbidity, Total Suspended Solids analysis, Barkman-Davidson testing, chemical spot testing (NACE TM0173 test), and a range of analytical laboratory analyses (XRF, XRD, SEM, EDAX, etc.) performed on filter residue samples.

- Dissolved Solids:
  - The objective of this sampling and analysis is to determine the dissolved composition of the water including the ionic composition, and dissolved organic matter.
  - On-site pH and H2S measurement at the time of sampling
  - Most of this analysis is carried out in an offsite analytical laboratory using samples from the field. These samples must be stabilized.
- On-line oil-in-water measurements:
  - The objective of this analysis is (1) to determine the average concentration of oil-in-water throughout the facility, (2) to determine the frequency and magnitude of upsets caused by oil and oily solids.
- Chemical bottle testing:
  - Various chemistries will be tested in order to determine the optimal chemical treatment.
  - Include pH scan for Zeta Potential estimation.
- Biological survey:
  - Determine and record all biocide practices and chemical injection
  - Carry out a survey of bacteria using the ATP test kit

Samples should be taken throughout the facility and observed on site over at least a couple of days to determine the settling rate of contaminants, color changes of the liquids and solids, and whether colloidal materials precipitate and form flocs. These observations, if carefully made, can provide important clues about the overall make-up of the water and how it changes through the facility. The contaminants often found in produced water, particularly from shale reservoirs, cannot be described accurately in simple terms of dissolved or suspended solids. Often the dissolved and suspended solids chemistry is a changing quantity that depends on many factors. The sampling and analysis program must take this into account. One way to do this is to maximize the portion of this work that is done onsite.

In addition to sampling and analysis of the fluids, it is helpful to gather process system data as well. Process field data includes temperature, pressure, flow rate, residence time in the vessels, sources of oxygen, and thermal imagining of process vessels to determine pad thickness.

The following table gives a summary of the main characteristics found in the Permian shale fields that were studied. Samples were taken from the discharge of various gun barrel facilities, prior to water treatment. There is of course variation from one field to another in the Permian as there is in all shale and conventional produced water fields. There are also variations within any one facility over time. The quality of water flowing through a gun barrel facility can experience wide variation depending on the manner in which the facility is operated. Excursions in oil and solids concentration are common. However, certain generalizations can be made. As the results show, there is significant variation in TDS, iron as well as other parameters.

Table 17.1. Main characteristics found for produced water from three Permian shales.

Parameter	Value	Units
Oil API gravity	> 38	degrees
Temperature inlet injection pumps	80 to 100	F
Total iron	60 to 200	mg/L
Oil-in-water concentration at facilities discharge	10 to 10,000	mg/L
Total Dissolved Solids (TDS)	100 to 200	k mg/L
pH	6.5 to 7.5	
Turbidity	160 to 360	NTU
Total Suspended Solids at facility discharge	100 to 2,000	mg/L
Number-Average solids particle size at facility discharge	1 to 10	micron

In the next few sections the analysis of suspended solids is presented. This includes Total Suspended Solids, solids particle size distribution and composition of the solids.

Total Suspended Solids Measurement: As discussed above, solids were characterized in a number of ways. An online Millipore filter (0.45 micron) apparatus was used to measure the solids concentration using the method described in NACE TM0173 [5]. Representative results of TSS measurements are given in Table 2. It must be emphasized that these results were obtained at the discharge of the gun barrel facility just upstream of the SWD (Salt Water Disposal) pump. This location was selected because this would be the location to install a water treatment facility. We did find that in general, the total solids content decreases as expected from the feed side to the discharge side of the gun barrel facility. Strictly speaking, the water cannot be referred to as produced water. Instead, this is processed produced water having been processed according to primary separation (settling in gun barrel tanks). Lord and co-workers [6, 7] report TSS of produced water from a Permian shale sampled upstream of a gun barrel facility and measured as 10,000 mg/L TSS. The sample location (upstream of primary separation) would be a contributing factor for the very high value of TSS. Other factors are discussed below. The TSS should be less for water emerging from the gun barrel facility. Nevertheless, the values measured by us are high compared to conventional onshore or offshore produced water.

Table 17.2. Total Suspended Solids measured using 0.45 micron filter. Samples taken at the discharge of an oil/water separation and storage (gun barrel) facility.

	TSS (mg/L)
Field 1	300
Field 2	400
Field 3	2,000

Particle Size by Sieve Analysis: Solids were further characterized for particle size using the sieve filter test. As with TSS, particle size was measured throughout the facility but only those results at the end of the gun barrel facility just prior to the feed to the injection pump are given. We did find that the concentration of large particles (larger than 60 micron) decreases significantly as expected. But the concentration of very small particles (1.2 micron to 8 micron) actually increases through the facility. This suggests mineral precipitation, oxygen intrusion, or iron precipitation, or perhaps some combination of mechanisms occurs in the facility. In every case studied, the facility allowed air intrusion mostly through the use of air for tank blanketing instead of natural gas. In the next two figures below, the solids concentration, in terms of number of particles per mL, is plotted versus the particle size.

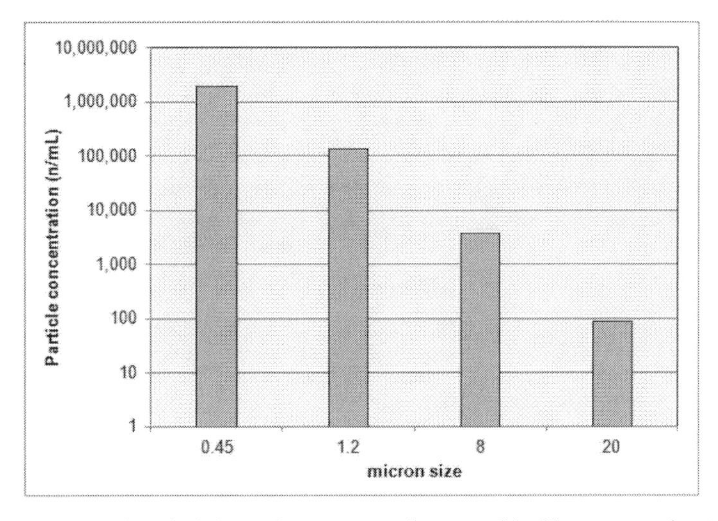

Figure 1. The number of solid particles per mL as a function of the filter pore size for Field 1.

The above graph shows the number of particles per mL of water sample captured by the four-filter sieve measurement.

The above graph shows the number of particles per mL captured by the five-filter sieve measurement. Figures 1 and 2 show a very high number of particles present in the water for these two fields. Both figures are a composit of several water samples. Sieve particle analysis was not carried out for Field 3.

Particle Size by Jorin ViPA Analysis: Solids were also characterized using the Jorin Visual Particle Analyzer (ViPA). Particle size distributions are shown in the following figures as measured by the Jorin Visual Particle Size Analyzer (ViPA). As shown, for Fields 1 and 3, most of the particles are in the range of five micron and smaller. This data was taken on line. The sample locations were downstream of the gun barrel system such that the larger particles would have been removed. Particle size was measured over several hours. It must be kept in mind however, that the Jorin is an optical instrument and as such it has a limit as to the size of object it can detect. This is true of all optical based particle size analyzers. That limit is about 2 to 3 micron. Below 2 to 3 micron visual particle analyzers are not accurate. This can be better understood by comparing the sieve analysis shown above to the optical analysis shown below. The sieve analysis shows very high particle concentrations in the range of a few microns or less whereas the optical analysis does not show any particles in that range.

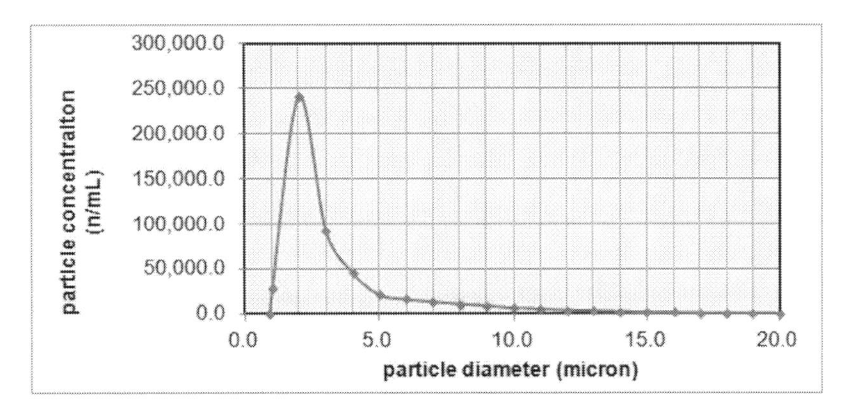

Figure 4. Particle size distribution for Field 1 as measured by an optical particle size analyzer. The number of particles per mL of produced water is shown as a function of particle size (micron)

As was the case for the solids sieve analysis, the Jorin data verifies the presence of very small particles in the produced water. As discussed by Walsh [8], particle size distributions for produced water in the deepwater Gulf of Mexico and in the North Sea are in the range of 15 to 150 micron. As shown here, the particles found in these Permian produced waters, at the discharge end of the gun barrel facility, are generally in the range of less than 10 micron. From a water treatment standpoint this difference is significant. Small particles are more difficult to separate than large particles. Stokes settling times are long. Small particles have exponentially greater surface area which has an impact on the effectiveness of threshold chemical treatment such as coagulant and flocculant. As discussed below, reuse requires the elimination of particles in the 5 to 10 micron range because these small particles are incompatible with fracturing polymer and they block the pores in the proppant pack. Also, in some cases, small particles can contribute to injection well impairment by adsorption and bridging mechanisms.

It is not known at this time how representative of other locations within the Permian these TSS and particle size data are. Lord and co-workers [7] have published a particle size distribution for produced water from a Permian shale. It appears that the water was sampled at the front end of the gun barrel system since it contains a high concentration of larger particles that would likely settle out within the system. However, there is a significant quantity of small particles. The main peak in the distribution is at 15 microns, which is relatively small. In addition, almost half of the particles are in the 1 to 10 micron range. This provides at least some confirmation of our findings. In addition to the data presented here, some operators have confirmed, through personal communication [9], the presence of small particles for other Permian shale fields. In those cases small particles were indicated by desktop settling, short run times between well re-stimulation, and through testing of various filtration technologies. It is not accurate to extrapolate the present results to an entire region. But those fields that do have a high concentration of small particles will require a specific type of water treatment system as described below.

Barkman-Davidson Injectivity Test: The final method of solids characterization that was applied is the Barkman-Davidson test [10]. The values of B-D Slope measured for the three test Fields are given in the table below. They are all significantly less than the desired value of 300.

Table 17.3. Barkman-Davidson Results.

Location	B-D slope
Field 1	8 to 44
Field 2	26 to 58
Field 3	0.6 to 0.8

In the discussion of water treatment technologies given below the Barkman-Davidson results are reported before and after water treatment.

Composition of Solids: The composition of the solids was determined. Our interest in the composition was three-fold. First, we wanted to know the likely composition of the contaminant surface since that would help in selecting coalescing and flocculating chemistries. The second reason was to understand source of the solids, the chemical and physical mechanisms that were generating them and whether they could be eliminated by scale inhibitor or some other treatment. The third reason was to compare the measured composition of the solids to expected compositions based on the geological processes of Permian shale deposition.

The solid residue from several filter papers was analyzed using a variety of techniques in order to determine the composition of the solids. Chemical spot testing using NACE TM0173 [5] was carried out on a large number of samples. It is a simple and easy to execute and it provides qualitative results. The composite results from dozens of samples taken at Field 1 are given in the table below.

Table 17.4. Chemical Spot Test (NACE TM0173 [5]) results for filter residue from Field 1.

chemical	indicative of	percent weight
xylene wash	hydrocarbons	21
acetic acid wash	carbonates	47
HCl	iron sulfide, oxides	19
acid insolubles	shale fines, sand	13

Based on the above table, and observations made during the testing, the following conclusions were made regarding the composition of the filter residue solids:

- Carbonates are likely iron carbonate.
- HCl solubles are likely iron oxide compounds.
- No indication of sulfide smell.
- Acid insolubles are likely kerogen, shale, silica, clay, and wax particles.

Similar results were found for Field 3. Field 2 on the other hand had iron sulfide present which was indicated by hydrogen sulfide smell when the residue was contacted with acid. Hydrogen sulfide was also observed in the field.

Chemical spot testing was also carried out for the residues collected on individual filters of the multi-filter sieve test. Each of the individual filters in the multi-filter sieve test captures particles of a certain micron size range. As described in Appendix I, the filters are stacked one above the other such that each filter represents a range of particle sizes which depends on its pore size and that of the filter above it. Composition was estimated on the basis of the chemical spot testing and on electron microscopy. The following table gives the size and composition of particles in Field 3.

Table 17.5. Chemical Spot Test (NACE TM0173 [5]) results for filter residue from Field 1.

Chemical	0.45 to 1.2 micron percent weight	1.2 to 8 micron percent weight
hydrocarbons	14	28
carbonates	12	24
iron sulfides, oxides	37	16
shale fines, sand	36	32

The results in the table above suggest that the smallest particles are composed of iron oxides (no sulfide was present), shale fines and crushed sand. The carbonates are slightly larger. The composition of the hydrocarbon was not determined. It is possibly wax particles.

In addition to the chemical spot testing, a variety of analytical tests were carried out on a relatively smaller number of samples (XRF, XRD, FESEM, FTIR, XEDS, TGA-DSC). Filter autopsies were carried out on a number of sock filters. In a filter autopsy test a number of squares of filter material are cut from a used sock or cartridge filter. The squares are subjected to chemical spot testing as in the NACE TM0173 [5] test. Also a number of analytical tests are performed (XRF, XRD, FESEM, FTIR, XEDS, TGA-DSC).

Table 17.6. XRD analysis of filter material from Field 3.

Mineral Phase	Formula	Weight percent normalized to exclude salt
quartz	SiO2	83
coesite	O2Si	5.4
magnetite, iron oxide, diiron (III)	Fe3O4	10
hematite	Fe2O3	0.2
misc silica compounds	SiOx	1.4

The final step in chemical spot testing is to immerse the filter square in strong acid. The acid extract is then subjected to ICP analysis. When HCl is used as the strong acid it must be kept in mind that the analysis result will include the chloride from the acid. Results of a representative sample are given in the table below. As shown, there is a significant amount of iron in the acid extract. This indicates that solid iron compounds were captured in the filter. These likely include iron sulfides, iron carbonate, and iron oxides.

Table 17.7. Analysis of acid extract from Field 3.
Results are given as mg/L of each element dissolved in the acid.

Al	B	Ba	Ca	Fe	K	Mg	Mn	Na	P	S	Si	Sr	Zn
27	6	5	67	179	19	5	0	364	5	50	12	11	2

As shown in the table above, the acid extract contains a high percentage of iron. There is essentially no manganese. This suggests that the origin of the iron is not pipe corrosion and is more likely from the formation itself.

The composition data was reasonably consistent across all three fields with the exception that Field 2 has hydrogen sulfide and iron sulfide and the other fields do not. The presence of shale fines was verified by TGA-DSC (Thermal Gravimetric Analysis – Differential Scanning Calorimetry). This is not surprising given the tremendous pressures exerted on the shale during the fracturing operation. Likewise, crushed sand seems to be the result of overburden pressure which causes the proppant sand to be crushed. All fields had significant quantities of iron compounds. As pointed out, this is consistent with other shale plays besides the Permian. The iron in the produced water appears to originate predominantly in the shale itself as dissolved iron which contacts air in the gun barrel facility and becomes oxidized. It then forms a hydrated iron oxide floc. This floc is seen throughout the Permian and probably in other shale fields. There are differences of opinion in the literature regarding the geologic mechanism of iron deposition in the original shale deposition processes. But there is general agreement that the iron comes predominantly from the shale formation itself rather than pipeline corrosion. The fact that the iron is predominantly in the form of iron oxides (ferric hydroxides) with relatively little in the form of dissolved iron (ferrous) is consistent with the Oxidation / Reduction Potential (ORP) of the fluid containing oxygen at thermodynamic saturation.

Lord and co-workers [7] have detected small solid particles in the Permian produced water "generated from a wide range of sources, such as formation fines and clays, sand, and corrosion products from the injection / production lines." This is in fairly good agreement with our findings. What they report as fines is probably what we have identified as small clay and shale particles. They explicitly identify sand as do we. They identify corrosion products from the injection / production lines. This is potentially a bit different from our suggestion that the iron is produced from the shale and that the oxygen is from the air blanketing of the gun barrel facility. Otherwise, there is fairly good agreement both in terms of composition and particle size.

All things considered, the origin and mechanism of formation of the small particles has been reasonably explained and is consistent with the particle sizes reported. There is some variation in particle size depending on composition. For example, it appears that the carbonate particles are slightly larger than the iron particles. However, all of the particle types that were detected were smaller than 10 to 15 micron. The small size together with the high concentration of suspended solids makes shale produced water relatively unique.

Analysis of Dissolved Contaminants: Inductively Coupled Plasma is an elemental analysis method typically used for dissolved cations in water. This testing was carried out by Tomson Technologies. The results are given in the table below.

Table 17.8. Main characteristics found for produced water from a Permian shale.

Parameter	Value		Units
parameter	average value this study	values from Lord and co-workers [6]	units
temperature	72		F
pH	7.0	4.8	
specific gravity	1.18	1.2	
conductivity		257	mS/cm
dissolved oxygen	8.0	8.2	mg/L
alkalinity as HCO3	2,000		mg/L
hardness as CaCO3	42,000		mg/L
Si	13.		mg/L
Ba	1.6	5.7	mg/L
Sr	570		mg/L
Ca	12,500	29,222	mg/L
Fe (total)	180	35	mg/L
Mg	3,130	4,347	mg/L
K	2,200	1,660	mg/L
B	50	20.3	mg/L
Mn	8.0		mg/L
Na	68,000	70,342	mg/L
Cl	140,000	163,637	mg/L
Br	1,149		mg/L
Al		1.4	mg/L
SO4	1,040	40	mg/L
Sulfide	0.2		mg/L
Total Dissolved Solids		267,588	mg/L
TSS	2,000	10,623	mg/L

Total iron was reported which is a combination of dissolved iron plus iron that had precipitated and was returned to solution by the process of sample digestion which generally involved reducing the pH and agitating. Total iron from other shale plays is reported in the table below. It appears that high iron concentration is typical of shale produced water.

Table 17.9. Summary of total iron content of several shale fields. The total iron is a combination of dissolved and suspended iron.

Shale Play	Iron (mg/L)
Woodford	30
Wolfcamp	100
Haynesville	150
Eagle Ford	70

Bacteria Analysis: Two bacteria analyses were carried out: ATP test, BART test. The ATP test is a quick test to confirm the presence or absence of bacteria. Results for field 3 indicated the presence of high concentration of bacteria throughout the facility (sample points 1 through 4). Values above 1E+05 are considered high. Values above 1E+06 are cause for concern over a time frame such as a few years or so.

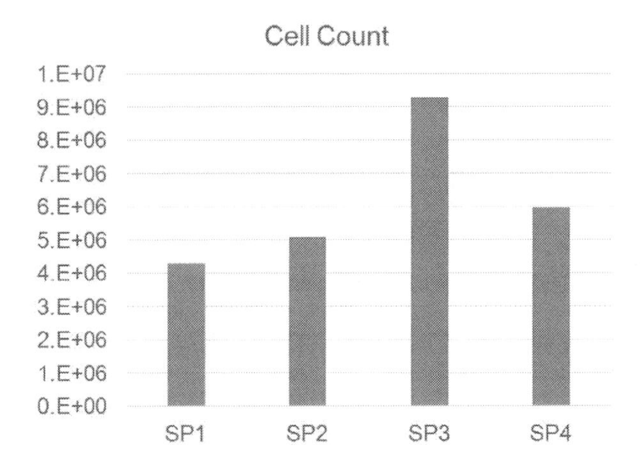

The ATP test provides a quick on-site test for bacteria. It does not indicate the type of bacteria. For that analysis the BART test is carried out. Iron reducing bacteria are particularly detrimental because they lead directly to corrosion and they contribute to high concentrations of iron in the produced water. Also, sulfate reducing bacteria are detrimental since they lead to souring of the facility and of the injection well. Once souring has become high enough to cause noticeable levels of H2S in the gas venting system, removing the bacteria to eliminate the H2S is somewhat costly since it requires several aggressive chemical treatments.

- SRB: sulfate reducing bacteria
- IRB: iron related bacteria
- APB: acid producing bacteria

	SRB	IRB	APB
SP1	Anaerobic SRB is present	Present	Present
SP2	Absent	Present	Absent
SP3	Absent	Present	Absent
SP4	Aerobic SRB is present	Present	Absent

Chemical Jar Testing: The overall objective of chemical jar testing is to determine the optimal treatment chemistry. Given the many different chemical products available on the market, this optimization is best done systematically in stages. The objective of the first stage is to determine the surface charge of the contaminants. This is done by running a pH screen.

Several identical samples are taken. Jar testing is done to determine the pH at which the contaminants coagulate and settle most quickly and the water has the highest clarity. In the initial screening test, acid (HCl) is used to lower the pH and base (NaOH) is used to raise the pH. In effect, this test determines the pH at which existing surface charge on the contaminant particles become neutralized. Once neutralized the particles are able to stick together by natural van der Waals, hydrogen bonding, and electrostatic forces. This is a reasonable first step in searching for coagulants and polymeric electrolytes that can coagulate and flocculate the contaminants.

Figure 6. Sample bottles with HCl (from left to right, added HCl concentration: 0, 7, 15, 19, 24, and 29 mmol/L)

Figure 7. Sample bottles with NaOH (from left to right, added NaOH concentration: 0, 2.5, 5, 10, and 20 mmol/L)

These figures show that both HCl and NaOH can clarify water samples to some extent. However, the greatest clarification in the shortest time was achieved by raising the pH. This suggests that the contaminant surface is positively charged.

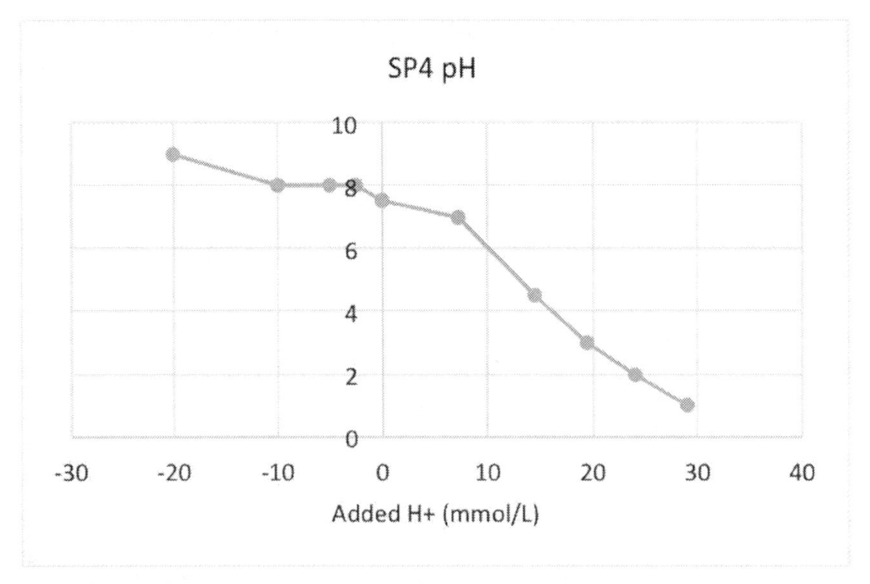

Figure 8. pH results of acid and base chemical bottle test for SP4. Positive x (added H+) means HCl was added, and negative x (added H+) means NaOH was added.

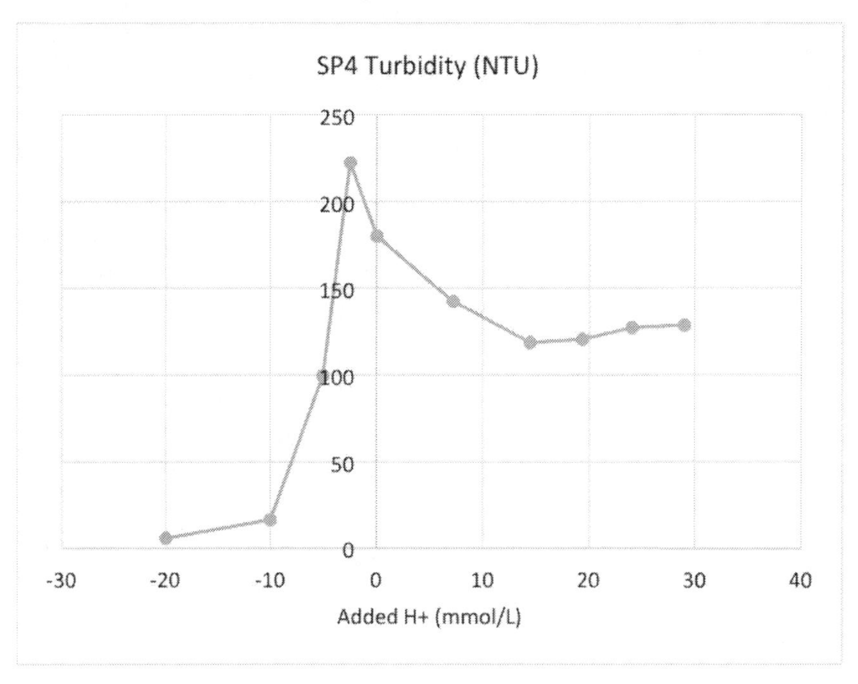

Figure 9. Turbidity results of acid and base chemical bottle test for SP4. Positive x (added H+) means HCl was added, and negative x (added H+) means NaOH was added.

Following the simple acid / base screening, a relatively small number of anionic coagulants and flocculants were tested for their capability to clarify the water. Visual observation and turbidity measurement were carried out to check the clarity of the water samples. Among all the tested chemicals, an anionic chemical (coagulant or flocculant) was shown to be more effective than cationic chemicals. This is in agreement with the observation that NaOH is effective and that the surface charge of the suspended solids is cationic. When the two treatments were used together (NaOH and anionic coagulant) the effective concentration of each could be reduced significantly. Chemical treatment of shale produced water is discussed below in the section on water treatment. This is just one case study of one field. Only one of the fields was tested due to the fact that the method required some development. It was only ready for use on Field 3.

17.2 Water Quality Specifications for Reuse

Iron and Suspended Solids: If the produced water is to be used for hydraulic fracturing, then water quality must be compatible with the hydraulic fracturing fluid and it must promote the properties required of a successful fracturing job. Over the past few years the industry has learned that monovalent salinity is not a problem. However, as discussed by Sharma and Bjornenm [3] some multivalent cations such as iron must be minimized or eliminated in order to prevent compatibility problems with HPAM (Slickwater) polymer and other additives. HPAM is used as a hydraulic fracturing polymer but it is also a well known flocculating agent used extensively in industrial water treatment. As such, it has a strong tendency to flocculate when iron and suspended solids are present in the water. By removing iron and suspended solids, the performance of the polymer (friction reduction, proppant transport and placement, and conductivity) are improved. When a high-pH cross-linked gel is used, this is particularly true. When HPAM-AMPS is used, sensitivity to multivalent cations is less dramatic. The presence of suspended solids results in consumption of other chemical treatments as well. Additives such as organic and oxidative biocides adsorb on the high surface area of the suspended solids which then requires a higher dosage. Exactly how much additional chemical to use becomes a bit of a guessing game.

Fracture Conductivity: It is necessary that the make-up water have low suspended solids. When a proppant is used to support the fracture, it may seem counter intuitive to require removal of suspended solids in the recycled produced water. After all, if hundreds of thousands of pounds of proppant are being injected, why must suspended solids be removed at all? It turns out that when fine particles are present they can clog the small pores of the rock restricting flow of hydrocarbons to the well bore. In addition, they can clog the small pores of the proppant pack. Those pores are the conduits for hydrocarbon to flow out of the shale and into the well. If that pore space is clogged, then the well will have poor productivity [7]. High fracture conductivity is key to high productivity and is greatly influenced by the presence of fine particles.

Quality Control of Fracturing Fluids: Furthermore, removing suspended solids from the recycle water has the added advantage that it helps to return the fluid to baseline characteristics. Drilling and completions engineers prefer to use makeup water that is as clean as possible. Despite the fact that recycling has been carried out successfully with a wide range of contaminated makeup water, there will always be a preference for clean water. This baseline may actually be higher quality than necessary but establishing a consistent and reproducible baseline is helpful to quality control in making up the fracturing fluids in the field. Often the success of a fracturing job is analyzed in terms of the variables that differ from one job to the next. If water quality is poorly characterized then it becomes a variable that changes from one job to another. This makes the job-to-job quality control more difficult. This is true for both the dissolved and suspended solids content of the makeup water.

Many volumes of produced water have been treated and used for hydraulic fracturing in the Permian. In a few cases the produced water quality used in the jobs is now documented. In one of the nicely documented cases, XTO Energy together with Halliburton used 100 % produced water with CMHPG (carboxymethyl hydroxypropyl guar) as polymer with zirconia as cross linking agent [6]. Electrocoagulation was used to treat the produced water. Scale inhibitor was injected. The use of EC as a water treatment technology for Permian shale is discussed below.

The important point to make is that the total suspended solids of the produced water was in the range of 10,000 mg/L. The TSS of the water after treatment and as used for fracturing was less than 100 mg/L. Further testing confirmed:

- achieved target viscosity
- low pipe friction
- low gel residue
- good proppant transport
- good dynamic break with sodium chlorite

The hydraulic fractures were applied to laterals of an existing well. Good production from the laterals was obtained. As demonstrated by this example, successful reuse was achieved by removing the small suspended particles from the produced water. Operators have begun to recognize this fact and have started specifying particle size rather than total suspended solids in their bids. The specification in these bids require complete removal of suspended solids larger than 5 or 10 micron.

17.3 Water Quality Specifications for Disposal Injection

For some locations, injectivity of produced water into an SWD is becoming a significant economic issue. The term, "injectivity" suggests that there is minimal resistance to injection. In other words, water is injected at a high rate with low pressure drop across the reservoir. There are a number of factors that impact the injectivity including the water quality or cleanliness. The cleaner the water the less frequently the well must be shut down for stimulation.

Regulations for Disposal Injection: The Underground Injection Control (UIC) program of the US EPA defines different types of wells depending on the use and fluid types to be injected. The overall goal of the program is to protect ground water from contamination. A Class II well is used exclusively to inject fluids associated with oil and gas production. States can, and many states do provide guidelines and regulations for the permitting of Class II disposal wells.

A critical issue is whether or not the water is injected above or below the parting pressure of the reservoir rock. If the water is injected above parting pressure, the reservoir rock will fracture. Parting pressure depends on the integrity of the rock itself and can be measured by performing a step rate test. This test involves measurement of injection flow rate, topsides and downhole pressure at progressively greater rates of injection [11, 12]. The test is designed to measure the pressure at which reservoir fracturing will occur. Fracturing opens new conduits for injection water to penetrate through and beyond the reservoir. This lowers the Injectivity Index but it also reduces the control of where the water goes. This increases the risk that the water will find its way into a drinking water aquifer. When water is injected at pressures below the parting pressure the location of the injected water is compartmentalized and there is much less risk of contamination of drinking water. This is referred to as matrix injection. Determining the maximum injection pressure and controlling the pressure in order to prevent contamination of drinking water sources is a key objective of the UIC program.

In New Mexico the UIC Program Manual [13] states that all injection wells are subject to a surface injection pressure maximum. In general, the permitted injection pressure is limited to 0.2 psi / ft, to the uppermost perforation. This means that the surface injection pressure, for an injection target at a depth of 1,000 ft, could be as high as 200 psi. Under this guideline higher surface pressure is allowed provided that a step-rate test or other test is documented in order to prove that injection will not initiate fractures in the confining rock and potentially contaminate nearby aquifers.

In Texas, in order to operate a Class II injection well, the operator must obtain a permit from the Texas Railroad Commission [14]. The intent of the permitting process is to ensure that sources of fresh water are protected from contamination. The maximum surface injection pressure is limited to 0.5 psi / foot of depth to the top of the disposal interval. Higher pressure is allowed provided that a step-rate test or other test is provided to prove that injection will not initiate fractures in the confining rock. Guidelines for step rate testing are available [15].

Cost of Injection Well Stimulation: Stimulation has direct costs in terms of chemical, manpower, fuel, and equipment. A less direct but significant cost is the handling of produced water while the injection well is shut down for the remediation job. If trucking is required to move the water to another site, then costs escalate quickly. Another cost that is beginning to be recognized is the cost associated with capacity limitation. When an injection zone has been injected with poor quality water over a few or several years, the pore space becomes blocked and the available capacity of the zone can be dramatically reduced. Thus, the closely related issues of produced water cleanliness and injectivity are beginning to be recognized for some of the shale developments in the Permian.

Injectivity Impairment: An extensive discussion of injection well impairment is given in Section 13.2 (When is Tertiary Treatment Required?). For the most part, injection wells become impaired when the disposal water contains large particles that plug the pores in the injection reservoir. This mechanism is referred to as geometrical matching. The usual criterion that is applied is to remove essentially all particles that are 1/3 of the pore diameter or larger. This is usually sufficient to prevent plugging in most applications around the world.

However, in some locations somewhat smaller particles can form bridges and accumulate across the pore throat. This behavior typically occurs when the concentration of fine particles is relatively high (50 to 100 mg/L and higher) and the surface chemistry is conducive. When the surface chemistry of the solids and the surface chemistry of the reservoir rock are attractive to each other particle bridging across the pore throat is greatly favored. Evaluating this surface chemistry is typically done using a small core flood test apparatus on site. On the other hand, when the concentration of fines is moderate and the surface chemistry is not attractive, then small particles will readily pass through the pores and not cause plugging.

Given the very small size of particles observed in the field data, and the relatively high concentrations of solids, injectivity impairment should be evaluated. While none of the operators were able to provide quantitative information regarding pore size or permeability, generally speaking permeability for injection reservoirs in the Permian basin tend to be in the range of hundreds of millidarcies to several Darcies. That implies pore sizes of 30 to 40 micron and larger. This is several times greater than the particle sizes measured. Yet, some degree of injection well impairment was reported at all three locations. This apparent discrepancy caused us to revisit the literature on injection well impairment discussed in Section 13.2.

The general conclusion is that other mechanisms besides simple geometrical matching must be present in those wells that have a tendency to plug. Schmoo formation and precipitated asphaltene particles are obvious candidate mechanisms in conventional oilfields. However, shale oil is so light that sticky oil components are not likely and were not detected in this study. Precipitated wax particles are a candidate mechanism for shale fields. Field 2 did have some wax precipitation. However, the literature study summarized in Appendix II also suggests that a combination of high solids concentration (as observed) plus colloidal surface forces could cause particle buildup and bridging across pore throats and lead to plugging over time, even for very small particles. Some evidence for colloidal surface forces is seen in figure 3 where various contaminant particles were stuck together. This is the most likely mechanism in Permian shale fields that have small contaminant particles and injection well impairment. The remedy for such fields is to reduce the concentration of suspended solids through water treatment.

17.4 Water Treatment Strategies and Options

Several technologies have been successfully applied in the Permian to treat produced water to the quality needed for either disposal well injection or reuse. Some of these technologies have already been mentioned. Obviously the produced water quality requirements for these two applications differ. The main difference between injection water quality and reuse is the extent to which suspended solids must be removed. Disposal injection can tolerate higher concentrations of solids and larger particles than reuse. However, in some cases the contaminant can stick to the injection reservoir sand grains and cause bridging which leads to the development of an impermeable filter cake. In that case, more extensive water treatment is required than would normally be the case for an injection well. For those fields where impairment is not a problem, then simple water treatment strategies can be employed. Often this means filtration using sock or cartridge filters. While such media are consumed, and therefore must be changed out periodically, the cartridge or sock filter provides a convenient package for disposal. The contaminant is neatly packaged and requires only to be disposed and not further treated.

ConocoPhillips has carried out a survey of water treatment technologies [3, 4]. Many technologies have been evaluated in the field in pilot and full scale studies. In addition, a competent assessment of water treatment technologies [16] is available which takes into account the economics of the situation following the collapse of oil prices in 2015.

Oil Removal: In general, the removal of oil from produced water in shale fields is not a major challenge. The oil typically has high API and therefore low density which makes it separate from the water after a short settling time. The concentration of oil at the discharge of a typical gun barrel facility is relatively low, on average. Light oil has a tendency to pass through a TSS filter and not contribute to injectivity problems. However, there are exceptions to this. For example, if paraffin compounds are present and wax precipitates, then the wax particles will attach to other solids particles creating a suspension of sticky particles which can impair injection. Also, if valves are suddenly opened or closed, solids suddenly entrained from the bottom of tanks or flow lines, pumps cycled on and off abruptly, then oil and oily solids can fluctuate dramatically and cause wide excursions in the quality of the discharge water. Bansal [11] points out that for some fields it is not possible to find a correlation between water quality and injectivity because injectivity damage occurs as a result of episodic upsets or excursions. In that case, high concentration of oil in the excursions can cause injectivity problems. Installation of an online turbidity meter can provide an inexpensive means to monitor and diagnose such excursions and prevent their occurrence.

Difference with Respect to Conventional Fields: In conventional fields with moderate to low API gravity oil, the presence of oily solids can be a major challenge. In those cases, the oil can form a relatively thick coating on the solids and can become a paste that binds other oily solids into larger contaminant particles. The oil/solids mixture has high viscosity and tends to stick to piping and vessel surfaces. In some cases the contaminant becomes neutrally buoyant due to the fact that the solids are heavier than water and the oil is lighter, which further complicates the removal. The solids material that we found in Permian produced water does appear to have some oil attached but it is difficult to say at this time whether the oil that is present with the solids has any role in the rapid formation of a tight filter cake and injection well impairment. Small solids at high concentration are confirmed but the impact of oily solids is still an open question. In any case, the water treatment technologies that have been shown to work well (see below) do not specifically target the oil associated with the solids and do not appear to be negatively impacted.

A conventional offshore water treatment system comprising of primary separation / hydrocyclone / flotation would not be effective, at least with the Permian produced waters that we tested. The Permian water contains too high of a solids loading and the particles are too small for those technologies.

Also, iron removal is not typically practiced for offshore platforms. On the other hand, the fluids involved in offshore systems are typically not exposed to air which means that any iron present would be in the soluble ferrous form and would not need to be removed.

Similarity with Municipal Water Treatment: Given the low density of the oil and the ease with which it separates, the water treatment technologies for shale produced water have much more in common with municipal water treatment than they do with conventional produced water. River water can in many cases have a high concentration of small suspended solids. A well known example would be the Mississippi river which carries tons of suspended solids downstream for thousands of miles. The gentle turbulence of the flowing river, and the small size of the solids is sufficient to keep the particles suspended. Municipal water treatment systems along the river must contend with high concentration of small particles which must be removed in order to provide high quality drinking water. Coagulation and flocculation (also known as "floc-n-drop") are classic water treatment processes for this purpose. The common treatment strategy is coagulation, flocculation, and either settling with some form of filtration or flotation or perhaps a combination. In many cases the solids are dewatered using a filter press and hauled to a disposal site.

17.4.1 Chemical Treatment

In this section, iron and suspended solids removal for shale produced water is discussed. These are just a couple of technologies among many that are suitable for this application. However, these technologies are well developed having been applied in the industrial and municipal water treatment industries for decades. Customization for the oilfield is of course required. But the basic chemistry and technical features are well developed and understood.

Regardless of what technology occurs downstream, it is almost always advantageous to apply course filtration as an initial treatment step. This pre-treatment step is usually carried out with course screens or sock filters. This removes the large particles and reduces the load on downstream units. Filter micron size is generally in the range of 75 to 100 micron. Multi-ply sock filters are available that can remove particles down to 10 micron or so. However, they would require frequent replacement which is time consuming and labor intensive. The key to this step is to select filters that have a high dirt holding capacity and a moderate particle size removal. This will reduce the frequency of change-out. The downstream water treatment equipment can then be designed to focus on the much smaller particles without labor intensive consumable media.

Iron Removal: As discussed above, most shale produced water contains a high concentration of iron. While there is debate over the details, most geological studies of the origin of the iron in shale agree that it was deposited concurrently with the shale itself. This indicates that it is not predominantly from pipeline corrosion. Field studies where both iron and manganese have been measured suggest that a small percentage does come from pipeline corrosion. In any case, iron is found in produced water throughout most shale basins and must be removed in order for the produced water to be used for hydraulic fracturing.

Iron Removal for Reuse: As discussed above in Section 17.2 (Quality Specifications for Reuse), the Slickwater class of water soluble polymers are sensitive to the presence of iron. In order for the polymer to behave as required, most if not all of the iron must be removed.

Iron Removal in Industrial and Municipal Water Systems: Iron removal from water is a common problem in industrial and municipal water treatment industries that has been around for many decades. Thus, there is vast knowledge of the chemistry and technologies for removing it. In industrial and municipal water systems iron can precipitate as carbonate, hydroxide, or oxide. In drinking water systems, the presence of iron oxides discolor the water and give it a foul smell and taste. Clothes washed in water that contains iron will discolor. Municipal water that contains iron will also often contain iron bacteria (crenothrix) which cause a foul smell. All of these negative features are unacceptable to the general public. Many municipalities have adopted an upper limit (SMCL – Secondary Maximum Contaminant Level) of 0.3 mg/L for iron.

In the pulp and paper industry iron will cause similar problems of discoloration. This significantly degrades the value of printing and office paper. Industrial cooling water is a very large sector of the water industry. Iron containing cooling water will cause deposits to form on heat exchange surfaces which will reduce heat exchange capacity and eventually lead to overheating of the surface, corrosion, leaks and failure. Such systems will also become plugged with deposits. All of these problems have been recognized by the municipal and industrial water industries for several decades. Various technologies have been developed to remove iron. Selection of the appropriate technology depends on the price point (value of iron-free water), the concentration of the iron, pH, temperature, carbon dioxide content, alkalinity, and available residence time.

Selecting the most appropriate iron removal technology depends on various details of the produced water chemistry. Shale produced water generally contains iron in the range of 30 to 200 mg/L. This is a relatively high concentration which immediately excludes technologies such as ion exchange, and various complexing, chelating, and sequestering strategies.

Oxidation of Iron in Shale Produced Water: Oxidation followed by settling, flotation or filtration are viable and cost effective strategies for iron removal. Oxidation will generate iron oxide solids which can be used as a flocculating agent and can then be removed. Iron oxide solids are very effective as a flocculating agent to remove other solids from produced water. There are several options for oxidation:

1. Aeration
2. Chlorine
3. Chlorine dioxide
4. Hydrogen peroxide
5. Ozone
6. Fentons' reagent

Aeration has the lowest operating cost. However, it is the slowest of the oxidation technologies, is sensitive to pH, and may require a two-stage process. Adding retention time and a second stage can contribute to higher relative capital cost. It will oxidize ferrous iron (soluble form) to ferric iron (usually an hydroxide). Aeration works best at pH 7 and above. Aeration can be quite slow at pH 5 and below. Theoretically 1 mg/L of oxygen will oxidize 7 mg/L of ferrous iron to ferric iron. Reaction rate data are given tin the table below. The data are based on a starting concentration of 100 mg/L of ferrous iron and 15 minutes of reaction with air.

pH	ferrous iron (mg/L)
5.0	9.0
7.0	0.1

Aeration will degas carbon dioxide which will raise the pH. In many cases of shale produced water, degassing of carbon dioxide is sufficient to raise the pH to 7 and above. However, aeration could also degas compounds such as aromatics, light hydrocarbons, and hydrogen sulfide. This can is some cases be problematic.

Oxidation of iron using chlorine is faster and more complete than aeration. At ph of 5, chlorine will oxidize 100 mg/L of ferrous iron to ferric in less than 5 minutes with essentially no residual iron. There is no need to raise the pH. Theoretically 1 mg/L of chlorine will oxidize 1.6 mg/L of ferrous iron. Based on this theoretical requirement, a system operating at 50,000 BWPD with a feed concentration of 100 mg/L of ferrous iron will require roughly 500 kg of chlorine per day. That is equivalent to about 1.3 IBC (Intermediate Bulk Container) tote tanks of 30 % chlorine.

Like chlorine, ozone, chlorine dioxide, hydrogen peroxide, and potassium permanganate all provide essentially instantaneous oxidation of iron. The use of hydrogen peroxide has the added benefit of initiating a free radical chain reaction with ferrous iron that results in good yield. The combination of added hydrogen peroxide together with the ferrous iron in the produced water results in a combination known as Fenton's reagent. Between pH of 3 to 5 Fenton's reagent results in rapid oxidation at low dosage.

All of these oxidizing agents, including air, can potentially suffer from one major problem. Iron has a strong tendency to form a complex with carboxylic acid groups. Any dissolved or dispersed organic acid material will bind the iron and make it difficult to oxidize. Carboxylic acid groups are prevalent in dissolved and naphthenic acid compounds. The so-called dissolved and water soluble organics have a high concentration of acid groups. Not all shale produced water have these compounds. Those that do, will resist oxidation of the iron. This problem has been experienced in some municipalities depending on the water source. In municipal water industry these acids are referred to as fulvic and humic acids. Together they constitute a class of contaminants known as Natural Organic Matter (NOM). The higher the molecular weight of the NOM and the greater the acid content, the stronger the tendency to bind iron. There is an enormous body of literature related to these compounds and their tendency to complex iron. The strategy for dealing with NOM in the municipal water industry is to remove the NOM by coagulation prior to the oxidation of the iron. Alum can be used as a coagulant in the pH of 5 to 8.5, which is a fairly broad range. Polymeric coagulating agents can also be used effectively. NOM is not difficult to remove. In doing so, a significant quantity of the iron is removed as well. This can be followed by oxidation to remove the residual iron.

Suspended Solids Removal: Once course filtration has been carried out, the removal of mineral and organic suspended solids is carried out. As discussed above, these solids are composed of crushed sand, carbonate solids, iron solids, clay particles, and fine particles of the shale. The objective of this treatment is to remove suspended matter such that the remaining solids do not impair an injection well or plug a proppant pack. As discussed, the extent of TSS removal required will vary depending on the application. The treatment strategy will be similar and only vary in intensity of applications.

The most important characteristics of the suspended solids are their particle size and the concentration. A small concentration of fine particles will have little detrimental effect on injectivity or proppant pack permeability. On the other hand, a significant quantity of fine particles will cause impairment. "Significant" is defined by the surface chemistry of the particles, the surface chemistry of the shale or proppant, and the particle size.

Fine particles are a challenge to remove. Being small, they do not readily settle on their own. Primary separation therefore has distinct limitations. Particles smaller than 10 micron are not readily removed by flotation either unless an effective flocculating agent is used with dissolved gas and multiple stages. Two important technologies for the removal of fine suspended solids are coagulation/flocculation and electrocoagulation. As discussed below, these technologies are related to each other. It is helpful to have an understanding of coagulation/flocculation because it is being applied successfully now. Electrocoagulation is also being applied with some remaining challenges as discussed below. However, there is a good chance that the challenges will be worked out and it too will become an important technology.

There is good scientific and engineering justification for their application. In the discussion below, coagulation and flocculation are described first, before the discussion of electrocoagulation. The level of detail is intended to provide an understanding of why these technologies are suitable for produced water recycling in shale fields.

The main mechanism of coagulation/flocculation is the aggregation of suspended solids. Once the solids have formed aggregates, separation can be carried out by settling, filtration or by flotation. In the case where flotation is used, flocculation of the suspended solids improves separation by creating a larger target for flotation bubbles to collide with.

The term, "coagulation" can be a bit misleading since it is a common and broadly used term outside of water treating. In biology it describes blood clotting. In cheese manufacture it describes the process of making curd from milk. Both of these processes imply aggregation of suspended material. In water treating, coagulation does indeed include aggregation. But, the important aspect is the prior step of chemically modifying the surface of the suspended particles to reduce the inherent electrostatic repulsion between particles. Without this step, flocculation and aggregation in water treatment would be limited.

In produced water treatment, the surfaces of most suspended solids have a negative charge. This is true whether the solids are composed of mineral, clay or organic material. There are various reasons for this. In part, it is due to the natural abundance of anions in minerals such as clays, carbonates, silicates, and metal oxides; the effect of lattice imperfections; and the abundance of negatively charged functional groups in organic chemistry.

These negatively charged suspended solids repel each other. This makes separation more difficult because the particles are so small that they tend to stay suspended. According to DLVO theory, the charge-charge repulsion can be reduced by increasing the salinity, which is not practical, or by adding a relatively small amount of trivalent cations. The Schultz-Hardy Rule indicates that trivalent cations are over 700 times more effective than monovalent cations, on a molar basis, at reducing the electrostatic repulsion of suspended solids. The simplest such coagulants are salts of iron (iron chloride) and aluminum (alum). By adding a coagulant, the electrical double-layer becomes compressed and charge-charge repulsion is reduced to the point where the short-range van der Waals attractive forces can bind particles together.

Coagulation is enhanced by intense mixing, applied immediately after chemical addition. Intense mixing promotes the migration of the coagulant chemical to the surfaces of the suspended particles. However, it must not be so intense to shear the particles and make them smaller since that would defeat the overall purpose. Through the process of coagulation, aggregates of particles begin to form since the repulsive barrier has been reduced in size and intensity.

Flocculation, takes the aggregation process a step further. Flocculation is the process whereby a flocculating agent, usually a polyelectrolyte polymer, is added to promote the formation of macroscopic aggregates or flocs. The polymer binds suspended particles together by forming bridges, strands or by forming a mesh. Once aggregated, the suspended material will migrate either up or down, depending on its density relative to water. The larger the flocs and the more tightly aggregated the suspended solids, the faster the migration rate.

Gentle mixing is important for uniform distribution of the chemical. But flocs are delicate objects and too much mixing will break the flocs apart. Flocs are typically settled in a quiescent tank in order for separation to occur. As mentioned, filtration or flotation can also be used.

Coagulation and flocculation are considered to be separate processes in large part because they involve separate chemical treatment, and separate unit operations. As described, the two processes differ in the intensity of mixing and in processing time. They differ somewhat in the chemicals used as well but there is also quite a bit of overlap.

In conventional coagulation / flocculation chemical selection, dosage and treatment time are important. Overall, there are three types of chemicals used in coagulation/flocculation. The trivalent salts (iron chloride, alum) were historically developed first. They are mostly used for coagulation. In industrial and municipal water treatment where there is a steady water quality and flow, they can be optimized. They are effective at breaking down the electrical double layer but are a bit tricky to use in the field. They require a precise knowledge of the water chemistry, and pH adjustment in some cases in order to avoid side reactions, unwanted precipitates, and to assure their effectiveness. Under optimum conditions they form polymer-like poly-hydroxide structures which reduce the electrical double layer and help bind suspended solids particles together.

The polyelectrolytes are effective for both coagulation and flocculation but tend to be used primarily for flocculation. As a class of chemicals, they include synthetic polymers, natural compounds, and derivatized natural polymers. When used as coagulating agents, they are more expensive than the simple salts but are far more robust in terms of compatibility with other ions, and are more effective over a wider pH range. They too can have limitations in terms of compatibility with the waste water. But pre-mixing is often effective at overcoming compatibility problems. Given the large number of products on the market, on-site bottle testing is often necessary. When a proper coagulating step is included upstream, the effectiveness of a polyelectrolyte flocculating agent can be exceptional. This reduces the cost of the polyelectrolyte significantly. Due to their wide applicability and robust performance, they are widely used.

The third class of chemicals are the pre-hydrolyzed alum salts. Generically they are known as PACl (polyaluminum chloride, pronounced packle). But there are many varieties. They are pH stabilized such that there is essentially no pH shift when applied. They are formulated with an optimum hydroxide, sulfate, and alumina ratio to promote the formation of large cationic aluminum hydroxide polymer chains which are effective as a flocculating agent. They are more expensive than the simple metal salts but they have essentially none of the compatibility and performance problems.

Both coagulation and flocculation are more complex than described above. Only the dominant mechanisms, and most common chemicals have been described. An excellent in-depth discussion of all aspects is given in the MWH Water Treatment book [17].

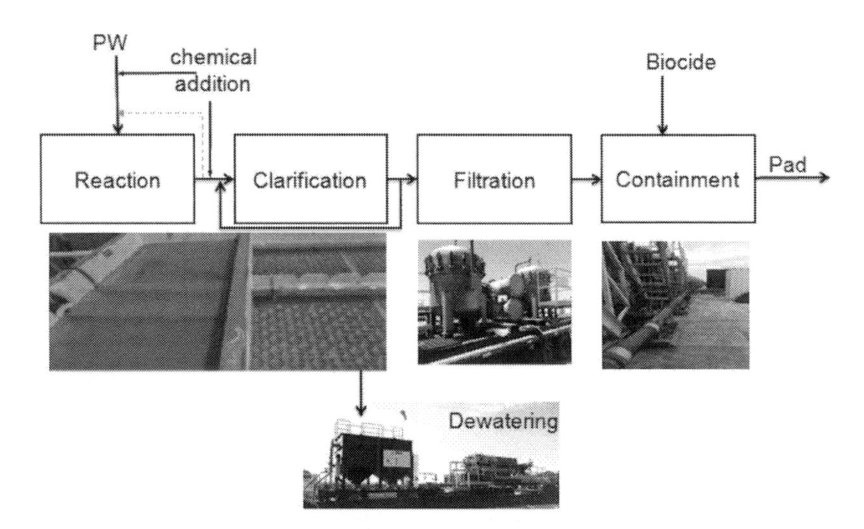

Figure 17.1 Block diagram of floc-n-drop system [3].

All things considered, coagulation and flocculation are well understood and there is precedent and good reason why they should be successful in treatment of HF flow back water. The significant challenges that have been overcome by the service companies are in the areas of logistics, cost, universality, robustness, and all of the challenges associated with bringing such technology to remote locations and applying it to unknown fluids with highly variable characteristics.

17.4.2 High Rate Clarifiers

Sedimentation can be accelerated by increasing the particle size or by decreasing the capture distance – i.e. the distance that a particle must fall (or rise) in order to be separated. Particle size can be increased by coagulation/flocculation. See table below which gives average settling times for various floc sizes [17]. Reducing the settling distance can be achieved in much the same way that a conventional Corrugated Plate Interceptor achieves reduced capture distance. In a CPI a plate pack provides narrow channels such that particles are captured after traveling only the relatively short distance within the channel. In a high rate clarifier, parallel plates or parallel tubes are used in the sedimentation basin. As in the design of a CPI, the clarifier plate pack must be inclined in order for the particles to slide down the bottom of the channel to the catch basin. The fluid velocity must be adjusted such that there is sufficient residence time within the plate pack to achieve separation of particles and flocs of a certain size.

Floc Type	Settling speed (cm/minute)
small alum floc	3.3 to 7.5
medium size alum floc	5.0 to 8.4
large alum floc	6.7 to 9.2
iron hydroxide floc	3.3 to 6.7
PACL floc	3.3 to 6.7

17.4.3 Field Experience

<u>Fountain Quail Rover:</u> As described in several papers and presentations, Fountain Quail Energy Services has been successful in applying its Rover technology to flow back water recycling. The process flow diagram for the Rover is described in [18] and involves pre-filtration, clarification, flocculation, settling and sludge handling. The details of the clarification and flocculation steps, and the chemicals used, are proprietary but are most likely based on well-proven technology. In terms of process design and modularization, the sludge handling steps would be the main challenge. However, this appears to be the area where the extensive oilfield experience of FQES becomes an advantage.

<u>ConocoPhillips / Aquatech:</u> In a large-scale pilot, ConocoPhillips has achieved success using specific oxidation for iron and hydrogen sulfide removal, followed by chemical coagulation and flocculation for the removal of total suspended solids [3, 4]. The water treatment pilot system was provided by Aquatech Energy Services (AES) and consisted of: three mobile treatment trailers each having a 3-chamber reaction tank for specific oxidation of the iron using hydrogen peroxide, flash mixer, coagulation tank, high rate clarifier, gravity thickener, and filter press. The most effective coagulant was anionic which is consistent with the produced water characterization given above. It is likely that iron was the dominant cationic contaminant. Gravity settling in the gun barrel system was generally found to be adequate to remove free oil. Occasional upsets involving high oil concentrations resulted in oily skimmings from the clarifiers. Relatively large suspended solids were removed by filtration with sock filters. Dissolved iron and fine suspended particles were removed. The high rate clarifiers required on

the order of 30 minutes for separation. The treatment process generated 0.35 lbs solids / bbl produced water treated. The treated water iron concentration was consistently below 5 mg/L and the turbidity was below 10 NTU. Low turbidity suggests low concentration of solids. The Barkman-Davidson slope for the feed water was about 28 to 56. Downstream of the water treatment facility it increased to over 100. This is a significant improvement in filter cake permeability.

Dissolved iron and fine suspended particles were removed. The high rate clarifiers required on the order of 30 minutes for separation. The treatment process generated 0.35 lbs solids / bbl produced water treated. The treated water iron concentration was consistently below 5 mg/L and the turbidity was below 10 NTU. Low turbidity suggests low concentration of solids. The Barkman-Davidson slope for the feed water was about 28 to 56. Downstream of the water treatment facility it increased to over 100. This is a significant improvement in filter cake permeability.

Table 17.10. Influent and effluent water quality for ConocoPhillips [21].

Parameter	Produced Water	Treated Water
pH	6.0 – 6.5	7.5 – 7.8
total iron	65 – 80	0.6
NTU (turbidity units)	> 500	1.1 - 1.7

The treated water was applied to hydraulic fracturing jobs with success [3, 4]. This treated water was compatible with salt-tolerant friction reducers. The friction reduction using this recycled saline brine was comparable with that found for fracturing solutions made up with fresh water. Also, compatibility with formation water using this hydraulic fracturing fluid is assured since the fluid is made up of produced water.

The final step of coagulation / flocculation is the settling step. When high rate clarifiers (slanted plates) are used, the settling time can be reduced compared to simple frac tanks. When small particles need to be removed extensive flocculation must be achieved with gentle mixing and relatively longer reaction times. In order to shorten the time and hence volume of the treatment equipment a number of technologies can be used such as back washable deep bed filtration, back washable screen filtration, dissolved air flotation (or induced air flotation).

For reuse, back washable filters can also be used in some cases. The limiting factor is the presence of sticky gunk forming material. The most reliable means of testing this is through field trail. As discussed, the operators have started specifying solids removal in terms of an absolute limit of particle size. It is now common to see specifications that require complete removal of particles larger than 5 or 10 micron. There are several technologies that can achieve this level of filtration on a continuous and reliable basis. The backwash frequency and hence volume varies from one technology to another. It is not necessary to use a disposable cartridge or bag filter. Selection of the most appropriate technology should be made on the basis of a verifiable operating envelope. For solids filtration in a high API field depends mostly on total suspended solids and particle size to be removed.

17.5 Electrocoagulation Practical Applications

In industrial water treatment, electrocoagulation is used to remove metal and metalloid (e.g. boron, arsenic) ions from water. It is also effective at removing suspended solids and in this application has been used in pulp and paper mills, tanneries, food and beverage, slaughter houses, and canning factories, as well as in processing produced water for reuse in shale development. The mechanism involved in EC treatment are similar to those described in Section 5.3 (The Mechanisms of Coagulation, Flocculation and Coalescence) for coagulation and flocculation, but with less chemical required for

EC. The electric field and the dissolution of the sacrificial anode are usually sufficient to collapse the electrical double layer. This is an advantage in treatment of HF flow back where any waste generated must be transported and disposed which adds to the cost of the operation.

An electrocoagulation application has been well documented [6, 7]. EC operates on the basis of a coagulation / flocculation mechanism provided by an electrical anode / cathode system. It does require a small amount of chemical in order to speed the settling and separation of the floc. However chemical use compared to conventional coagulation / flocculation is greatly reduced. This can be an important factor when removing small particles that have very high surface area and hence high chemical demand.

As discussed in [6, 7], using EC in a Permian field was effective to significantly reduce suspended solids concentration from roughly 10,000 to less than 100 mg/L. There was a slight increase in dissolved aluminum due to the use of an aluminum sacrificial anode. Iron was removed completely. The pH was adjusted to 8. There was a very slight decrease in alkalinity which included a drop in boron of about 20 %. All things considered, EC was effective in the elimination of suspended solids. While EC has proven to be technically effective in the Permian and other shale developments, it requires specialized staff to operate and the equipment is capital intensive. Given the scale of shale developments in the Permian, local operating staff are scarce. There are simply not enough people living in the area to support the scale of Permian operations. Operators must be brought into the region from other locations and their transportation and accommodation can be expensive.

In general, the oil and gas industry has migrated away from the use of EC. This is more or less the experience of EC in the industrial and municipal water markets as well. In those markets, EC is not competitive against conventional chemical based coagulation / flocculation / settling. It is primarily used in niche applications to remove heavy metal ions.

However, the technology is not quite as simple as suggested. The chemistry of the sacrificial anode is similar to that of the simple (metallic salt) coagulants discussed in the section on coagulation / flocculation. Thus, compatibility and performance problems associated with the use of simple coagulants (metallic salts) can occur. "overdosing" of the metal coagulant can occur if the voltage is set too high. When this occurs, residual levels of iron or aluminum can exceed the effluent specifications. These problems can usually be overcome but depending on the complexity and variability of the flow back water chemistry, a certain level of expertise is required to make the adjustments. This may involve selecting the right anode material, adjusting the flow rate, tuning the voltage, adjusting the pH, and an ability to determine when to clean the electrodes. All of this suggests the need for a relatively high level of on-site expertise.

As mentioned, the suspended solids migrate to the counter-ion electrode. This has benefits and drawbacks. The concentrated solids that have migrated to the electrode form what is referred to as a sludge or floc blanket. This blanket promotes the capture of smaller flocs and suspended solids. This improves the separation performance. It is a design feature of many industrial clarification separators. However, in EC, it leads directly to fouling of the electrode. The guar-based HF flow back fluids have tremendous fouling potential. This necessitates proper sparing of electrodes so that some of the electrodes can be taken offline for cleaning. It also necessitates a somewhat larger staff count onsite to cater for the labor intensive job of cleaning the electrodes.

The technology is inherently modular in the sense that it is composed of small units that can be packaged conveniently. For many technologies, modularization is a difficult design challenge. The electrode units of EC lend themselves to modularization.

However, as with most suspended solids treatment technologies, sludge thickening and handling is not so easily modularized. In fact, the design of the sludge handling unit has been a drawback of the initial EC units deployed in the field. The same could be said of the coagulation / flocculation technologies except for the fact that

EC is an inherently modular technology. This is a significant advantage. As mentioned, this allows individual units to be taken offline for cleaning. It also allows lends itself easily to a flexible and mobile process design. Flexibility and mobility are key design advantages in being able to deliver technology to remote locations.

Electrocoagulation is an intensified form of coagulation. It employs sacrificial anodes (aluminum, iron, sometimes both) and involves electrolysis which generates flotation bubbles. When operated properly it can provide effective removal of suspended solids and some of the dissolved divalent cations.

In an EC unit, two electrodes immersed in the produced water are used to apply an electrical field to the water. The anode is composed of either iron, aluminum or an inert metal. In the case of iron or aluminum, the anode is referred to as a sacrificial anode. It dissolves in the water thus providing an in-situ source of coagulation ions (Fe^{+3}, Al^{+3}). These ions hydrolyse to form $Al(OH)_3$ and $Fe(OH)_3$ which are effective flocculating agents. The chemical reactions are discussed below.

The Figure below shows a schematic of the electrochemical cell, the electrodes, and the half-reactions that occur. It shows that the flocculated contaminants can either rise to the top or settle to the bottom. This depends on the composition of the contaminants and the concentration of flocculent that is generated. Most HF flow back fluids when flocculated would rise to the top. However, with sufficient coagulant addition, they have been known to settle.

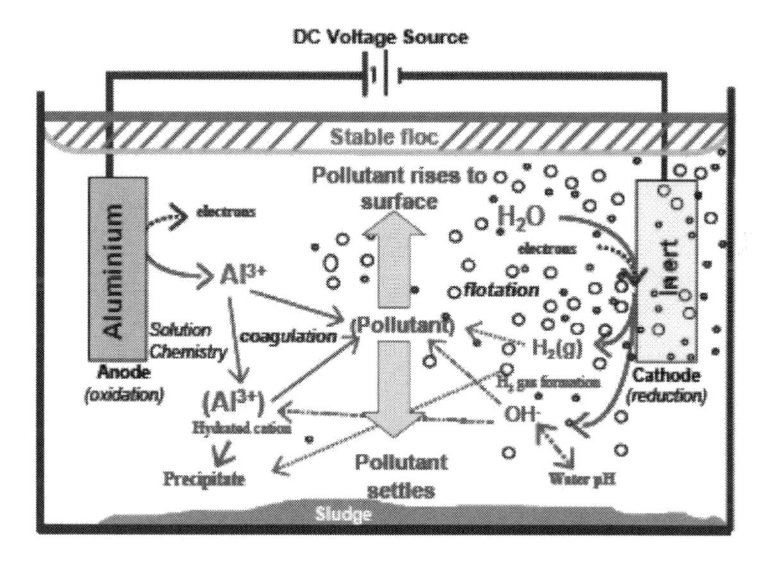

Figure 17.1 Schematic diagram of electrocoagulation cell

Design Correlations: One of the challenges with electrocoagulation is that, for a given application, it is difficult to achieve optimal design without extensive experience in design and operation [19]. In other words, the mechanisms of decontamination are sufficiently complex that accurate and easy to use design correlations are not readily available [20]. This makes it difficult to optimally specify a number of design features such as amperage, voltage, flow rate, electrode surface area, gap width between electrodes, chemical selection and dosage required, etc.. Without design correlations, design and application depends on direct experience. This is more of a problem in industrial applications where the costs of competing technologies, such as coagulation/flocculation, have been driven down and optimized over decades of use. Simple coagulation / flocculation is the main competing technology which is inherently simple, well understood over a broad range of applications, scalable, and design is straightforward and not likely to require costly retrofit. For electrocoagulation, the chemistry and mechanisms of separation are somewhat complex and interrelated. This makes the technology more difficult to operate and more difficult to compete economically.

Batch and Manual Operation: Another challenge with electrocoagulation is that for variable feed characteristics, it requires adjustment of operating parameters such as current (voltage), chemical (if being used), backwash frequency and duration, and residence time in each stage. This manual adjustment is in some cases facilitated by bottle testing. All of this activity is costly in an industrial setting where the number of operators per ton of waste water treated is usually a small number. In the setting of a hydraulic flow back such activity is not necessarily a problem. In contrast to an industrial setting where the process is intended to operate continuously, in a flow back the equipment will operate in a batch mode. Continuously changing feed characteristics is normal. The number of operators is already somewhat high compared to an industrial setting since field staff are required to mobilize and demobilize the equipment anyway. This demonstrates that the economics of flow back treatment are quite different from the economics of industrial water treatment.

Another advantage that makes electrocoagulation suitable for flow back is that the technology is inherently modular. It is (or can be) composed of small units that can be packaged conveniently and brought online as needed. For many technologies, modularization is a difficult design challenge. The electrode units of EC lend themselves to modularization. Capacity, if needed, is achieved by adding more modules. This makes it particularly suitable for mobile application.

Overtreatment: An advantage that is often overlooked is the consequence of overtreatment. EC is relatively forgiving in an overtreatment situation. If the electrodes are run too hard (defined below), the most detrimental consequence is that the electrodes require more frequent backwash or cleaning. Typically EC machines for flow back operation must be designed for easy backwash and cleaning because the guar containing HF fluids are prone to fouling. But the effluent water quality does not usually deteriorate significantly on overtreatment. This is an important advantage in an application where feed characteristics and flow rate vary dramatically in just a few minutes time. In competitive technology such as floc-n-drop (coagulation, flocculation, settling) a chemical overtreatment situation is as detrimental to effluent water quality as under treatment and in some cases the range of chemical is somewhat narrow.

pH Insensitivity: Another advantage that is sometimes overlooked is the range of pH over which EC is effective. This is in contrast to conventional chemical-only treatments (floc and drop) where the coagulants are most effective over a relatively narrow pH range. Often, EC does not require pH adjustment of the incoming fluid. This reduces the amount of chemical that must be transported to site.

Fouling of Electrodes: The technology is prone to electrode fouling. This is due to the migration of suspended materials to the electrode where they passivate the electrode. This migration is known as electrophoretic migration. As such, it is well understood and is an inherent problem. However, the extent to which the electrodes foul depends to some extent on the design and operation of the electrodes. As in heat exchange designs, plates tend to foul less in general than tubes and cylinders. A uniform distribution of concentration across a flat plane is more easily attained than across a cylinder. Also, by maintaining a high Reynolds number between the plates, the deposition and build up of particles on the plate can be reduced. Periodically reversing polarity also helps to reduce fouling. Backwashing at regular intervals helps. All things considered, the most important design feature is facilitation of electrode cleaning. The electrode housing should be designed such that it can be drained and opened quickly, and the electrodes cleaned without difficulty. It is best if the design allows the electrodes to be cleaned in place. Mild acid may be required to remove carbonate buildup.

Depending on the application, and the difficulty of waste disposal, solutions of cleaning agents may be used. If electrodes must be removed, in order to be cleaned or replaced after their useful life, the design should facilitate this to occur easily.

17.5.1 Mechanism of Separation & Performance Variables

An electrical field is applied to the contaminated water. The electric field can be applied by adjusting the voltage across the electrodes or by adjusting the current. In the case where voltage is applied, the electric field strength depends on the voltage, electrode separation, and the conductivity of the water. The current which then flows from the electrode through the water and to the other electrode is a function of the applied voltage, and the electrical resistance of the water. A current density can be calculated as amps per unit area of electrode (amps / cm^2). Many HF flow back fluids have high conductivity which reduces the resistance of the water and hence the voltage required to achieve a target current density. This also results in lower power consumption of EC for HF flow back fluids.

There are three main independent variables in design and operation of an EC unit:

- current density (amps / cm2 electrode surface)
- initial pH, and chemical addition upstream of the EC unit
- electrolysis time (residence time in the electrolytic cell).

These variables are discussed in this section and in the section below on applications. Before a complete discussion of EC can occur, certain electrical engineering concepts need to be defined and reviewed.

Review of Electrical Engineering Concepts: It is necessary to review certain definitions and relationships in order to fully describe how electrocoagulation works.

$$P = IV \quad \text{Eqn (17.1)}$$

and

$$V = IR \quad \text{Eqn (17.2)}$$

where:

P = power (Watt = Joule / second)

I = current (amp = coulomb / second))

V = electrical potential (volt = Joule / coulomb)

R = electrical resistance (ohm = volt / amp = Watt / $coulomb^2$)

Power (Watts): Power is the most logical place to start a discussion of electrical engineering since it is the measure of energy per unit time. If an electrical circuit is designed to do a certain amount of work or provide a certain amount of energy, then power is the measure of how much work per unit time will be done. It is expressed as energy per unit time (Watts = Joule/second). It is calculated as volts times amps.

Current (Amps): electrical charge per unit time (i.e. a measure of the number of electrons that flow per unit time). An amp is equal to one coulomb of electricity flowing per second. A coulomb is the charge equivalent to a certain number of electrons.

Potential (Volts): electrical potential. It is a measure of the energy that would be released (or put to useful work) per unit time if an electrical current was allowed to flow. Thus, power is often calculated as $P = I\ V$, where power is the rate of energy release as a result of current (amps) flowing across an electrical potential (volts). One volt is equal to one Joule / coulomb.

Resistance (Ohms): electrical resistance. It is a measure of the relationship between volts and amps. The unit of electrical resistance is the Ohm. An Ohm is defined as the electrical resistance that results in 1 amp of current flowing in a circuit that has 1 voltage of electrical potential. One ohm is equal to one Watt per coulomb squared, as suggested by the relationship $P = I^2R$.

Specific Amperage (Current Density): The amperage applied to the electrode, as well as the power consumed in treating the water, should both be expressed in specific units. Specific units are those units that are expressed in terms of surface area and flow rate. Specific units allow scale-up from a test unit to a full scale unit. They also allow comparison of one application to another. According to Degremot between 2 to 4 kWhr of energy are required per m3 of water.

In the case of amperage, specific units are amps / cm2 or amps / m2 of electrode surface area. This is also known as current density. Current density determines the coagulant dosage rate. It is a measure of how hard the electrodes are being driven. The harder they are driven the more of the anode dissolves and the greater the oxidation rate at the cathode. However, the specific amperage units must also include the volumetric flow rate in order to provide a complete description on the system. Specific amperage should include the amps / unit of water volume. The amps per unit flow of water is then expressed as amps / m3 of water. These two quantities (amps / m2 electrode area and amps / m3 water), provide a complete specification of the electrode design and can be used to calculate the coagulant dosage rate. These values are also necessary for scale-up. The equation which describes these calculations is given as:

$$\chi = I / Q = q / L \qquad \text{Eqn (17.3)}$$

where:

c	= electric charge per unit volume of sample (coulombs / liter)
I	= current (amps = coulombs / sec)
Q	= volumetric flow rate of water (L / sec)
q	= amount of electric charge (coulombs)
L	= volume of sample (L)

Specific Energy Units: energy units are usually reported in terms of kWhr / m^3 water. A kWhr (kilo Watt hour) is a unit of energy. It is equivalent to 3.6 million Joule. Specific energy consumption is commonly reported as the power being consumed at steady state operation divided by the volumetric flow rate during that period. Power is expressed as kW, and the volumetric flow rate is expressed as m3 / hr. Thus, the energy consumption is just kW / (m^3/hr).

$$E = P / Q \qquad \text{Eqn (17.4)}$$

where:

E	= energy per unit volume of water (usually as kWh / m^3 water; J/m^3)
P	= power consumed (Watts = J / sec)
Q	= volumetric flow rate of water (m^3 / sec)

As discussed above, both power (P) and current (I) must be measured in order to completely specify the electrical energy situation.

Conversion Rate at the Electrodes (Faraday's Law): The following formula allows the calculation of electric consumption. It also provides a calculation of the conversion of anode to dissolved ions, or the generation of oxidants at the cathode:

$$\chi = I / Q = q / L = F \frac{C}{M} \qquad \text{Eqn (17.5)}$$

where:

c	= electric charge per unit volume of sample (coulombs / liter)
I	= current (amps = coulombs / sec)
Q	= volumetric flow rate of water (L / sec)
q	= amount of electric charge (coulombs)
L	= volume of sample (L)
C	= Concentration (g/L)
F	= Faraday constant (96,500 coulomb / mole)
M	= molecular weight of substance (g/mol or mg/m-mole)

When this equation is applied to the calculation of the amount of hypochlorite generated, the following values can be inserted:

M	=	52.5 mg / m-mol
C	=	150 mg/L HOCl (0.15 g/L)

Calculated values are:

c	=	274 coulombs/L

If the flow rate of sample is 5 BPM (13.3 L/sec), then the current setting must be:

Q	=	13.3 L/sec

$$I = \chi Q = (274 \text{ coulombs/L}) \times (13.3 \text{ L/sec}) = 3{,}685 \text{ coulomb/sec} = 3{,}685 \text{ amps} \qquad \text{Eqn (17.6)}$$

Thus, according to this calculation, 3,700 amps would be required to generate 150 mg/L of hypochlorite at the cathode for a volumetric flow rate of water of 13.3 L/sec. it is not necessary in this calculation to specify the electrode surface area. Thus, specific amperage or current density is not calculated. Also, neither the voltage nor the resistance of the solution are required in the calculation. Therefore, power and specific energy cannot be calculated.

Anode Function and Composition: The anode is composed of either iron, aluminum or an inert metal. In the case of iron or aluminum, the anode dissolves in the water thus providing an in-situ source of coagulation ions (Fe^{+3}, Al^{+3}). These ions hydrolyse to form $Al(OH)_3$ and $Fe(OH)_3$ which are effective flocculating agents.

The electrodes also promote electrophoretic migration of suspended particles which results in a higher concentration around the anode. These suspended particles become flocculated. Overall the effect is to promote the formation of a floc blanket. A floc blanket is a high density of flocculated material. The floc blanket, together with the electrophoretic motion of suspended particles is effective at capturing suspended materials. This mechanism accounts for most of the performance of electrocoagulation units. However, as discussed, the buildup of suspended material also runs the risk of fouling. Thus, the operation of an EC unit requires adjustment in order to optimize the treatment between separation effectiveness on the one hand and fouling tendency on the other. A certain level of oxidation of dissolved components takes place at the anode, which also aids in the breakdown of hydrogen sulfide, organics, and bacteria. Hydrogen sulfide is converted by oxidation to elemental sulfur.

Cathode Function and Composition: The cathode is composed of either iron, aluminum or an inert metal. Hydrogen (H_2) and chlorine (Cl_2) gas bubbles are generated at the cathode which provide an in-situ gas flotation effect. This flotation effect increases the rise velocity of the flocs.

pH: The use of sacrificial anodes composed of iron or aluminum results in an inherent shift in pH. Both the iron and aluminum cations consume hydroxyls when the hydroxide compounds are formed. In doing so, they lower the pH of the solution. Protons are consumed at the cathode where they are converted into hydrogen molecules. This raises the pH. The overall reaction of anode material dissolving, the cations forming hydroxide, and the generation of hydrogen gas is pH neutral. In other words, there is no net gain of protons or hydroxyls. This would suggest that the pH remains unchanged in the EC reactions. But this is not the case. pH does change. The pH changes because the overall reaction, which involves as many protons and hydroxide ions, does not typically occur in practice as explained presently.

Elementary Reactions: The anode and cathode reactions are given in the formulas below. In addition to the reactions that occur at the electrodes, the hydrolysis reactions are given as well. As discussed below, the rate of the electrode reactions is not necessarily related to the rate of hydrolysis of the metals to form the hydroxides. Thus, the pH can shift during the operation depending on the relative rate of electrode reactions versus hydrolysis.

Anaerobic Reactions:

Anode – Coagulant Generation:

$$Fe(s) \rightarrow Fe^{+3} + 3e^{-1}$$

$$Fe(s) \rightarrow Fe^{+2} + 2e^{-1}$$

$$Al(s) \rightarrow Al^{+3} + 3e^{-1}$$

Cathode – Hydroysis:

$$H_2O \rightarrow H^{+1} + OH^{-1}$$

$$2H^{+1} + 2e^{-1} \rightarrow H_2(g)$$

Reaction of Cations & Hydroxides to form Flocculating Agents:

$Fe^{+3} + 3OH^{-1} \rightarrow Fe(OH)_3(p)$

$Fe^{+2} + 2OH^{-1} \rightarrow Fe(OH)_2(p)$

$Al^{+3} + 3OH^{-1} \rightarrow Al(OH)_3(p)$

Overall Reaction:

$Fe(s) + 2H_2O \rightarrow Fe(OH)_2(p) + H_2(g)$

The metal hydroxide compounds (e.g. $Fe(OH)_3$) are written with the designation (p). This indicates that a precipitate is formed. The precipitate is not a solid in the sense of a crystalline or ordered structure. The metal hydroxide compounds form small particles and amorphous structures which are ideal as flocculating agents since they have very high surface area for a given weight of material.

The overall reaction assumes that the elementary reactions balance each other. This is usually not the case. Electrical neutrality must, of course, be satisfied, but the rate at which the electrodes react does not necessarily equal the rate at which the metal hydroxides are formed. When these two reaction rates are different, the pH goes up due to the accumulation of hydroxide anions. The solubility of the hydroxide compounds will influence the extent of that reaction. Thus, not all of the metal cations will form hydroxides. The more cations that are left in solution, the less hydroxide that are formed and the less the hydroxyls are consumed. This has the net effect of raising the pH.

Metal Hydroxide Flocculating Agents: With adequate stirring, the metal hydroxides will form. They hydrolyze to form polymeric hydroxide complexes that have a gel-like character. This polymeric metal hydroxide gel is very effective at capturing suspended contaminants in the water. Also, the gel material has a net negative charge which induces movement toward the anode. This movement is referred to as electrophoretic mobility as discussed in Section 4.3.3 (Electrophoresis and Measurement of Zeta Potential). Movement toward the anode further concentrates the gel material and creates a flocculation blanket. This is completely analogous to the process of flocculation clarification wherein a floc blanket is formed. The more material that is captured by the floc blanket, the more effective the blanket is at capturing suspended material. The blanket becomes thicker and tighter with more added material. The formation of the floc blanket is an added effect of electrocoagulation that is not seen in standard floc-n-drop (coagulation / flocculation / settling).

Bacteria Removal: EC has been demonstrated to be effective at killing bacteria. The mechanism is based on the generation of chlorine / hypochlorite at the anode. The chlorine immediately forms hypochlorite by the typical hydrolysis reaction. Hypochlorite ruptures the bacteria cell wall which kills the bacteria. The hypochlorite also oxidizes some of the the organic carbon left behind by the lysed bacteria. Not all of the organic carbon material will be oxidized. The part that does not get oxidized may be lifted by gas flotation or captured by flocs formed from the EC mechanism.

Electrode Design: The shape of the electrodes and the spacing between them are critical design features from the standpoint of performance as well as fouling and cleaning. Flat plates have been found to work best in hydraulic fracture flow back. It is easier to attain a uniform flow across the plates compared to cylinders (rods). Also, the space between electrodes is uniform in a flat plate configuration. Electrode spacing is an important parameter. Generally speaking, the closer the spacing, the greater the separation efficiency for a given residence time. However, there is often an optimum spacing. In those cases where there is an optimum, smaller and larger spacing, at a given residence time, results in less separation efficiency. Close spacing means greater number of electrodes and higher capital cost. As discussed below, capital cost is less important in batch operation where throughput is critical both in terms of keeping up with demand and reducing the amount of time that staff are required to be at site.

Operation: The technology is not quite as simple as suggested in much of the literature. The chemistry of the sacrificial anode is similar to that of the simple (metallic salt) coagulants discussed in the section on coagulation / flocculation. Thus, compatibility and performance problems associated with the use of simple coagulants (metallic salts) can occur. Overdosing of the metal coagulant can occur if the voltage (current) is set too high. When this occurs, residual levels of iron or aluminum can exceed the effluent specifications. These problems can usually be overcome but depending on the complexity and variability of the flow back water chemistry, a certain level of expertise is required to make the adjustments. This may involve selecting the right anode material, adjusting the flow rate, tuning the voltage (current), adjusting the pH, selecting and adding other chemicals, and an ability to determine when to clean the electrodes. All of this suggests the need for a relatively high level of on-site expertise.

Fouling and Electrode Cleaning Requirement: As mentioned, the suspended solids migrate to the counter-ion electrode. This has benefits and drawbacks. The concentrated solids that have migrated to the electrode form what is referred to as a sludge or floc blanket. This blanket promotes the capture of smaller flocs and suspended solids. This improves the separation performance. It is a design feature of many industrial clarification separators. However, in EC, it can lead to fouling of the electrode. The guar-based HF flow back fluids have tremendous fouling potential. This necessitates proper sparing of electrodes so that some of the electrodes can be taken offline for cleaning. It also necessitates a somewhat greater number of staff onsite to cater for the labor intensive job of cleaning the electrodes.

Sludge Handling: At the end of the separation train, two products are formed. The first is the treated effluent water. The second is the waste sludge. As with most suspended solids removal technologies, sludge thickening and handling is not so easily modularized. In fact, the design of the sludge handling unit has been a drawback of the initial EC units deployed in the field. The same could be said of the coagulation / flocculation technologies except for the fact that Fountain Quail has developed a robust design based on their experience in the heavy oil industry in Alberta. In a process application, the electrodes are accompanied by a settling tank, filter system, and sludge de-watering unit. Sludge handling is critical to reducing manpower requirements.

Modularization: EC is an inherently modular technology. This is a significant advantage. As mentioned, this allows individual units to be taken offline for cleaning. It also lends itself easily to a flexible and mobile process design. Flexibility and mobility are key design advantages in being able to deliver technology to remote locations.

Process System Design: The process is usually designed for three distinct stages:

1. pre-treatment with chemical (not always required)
2. electrocoagulation
3. additional chemical (if needed), settling of flocs, separation of flocs from effluent water
4. sludge dewatering and handling

Current density has a large effect on the kinetics of the process [21]. The higher the current density, the shorter the latency period, and the shorter the residence time can be for electrocoagulation to occur. The effect of current density is linked to production rate of the coagulant from the sacrificial electrode.

17.5.2 Design Differences and Characteristics

A study was carried out to understand the effect of various design parameters on the separation performance of EC with municipal waste water as feed. The objective of the study was to achieve treated effluent water quality suitable for use in landscape irrigation and plantation. By doing so, less water would be drawn from the municipal water system.

Summary findings are given in the table below. For these results, the electrode current density was 25 mA/cm2, and the spacing between electrode plates was 5 cm. The residence time for the water in the system was 30 minutes. The feed water chloride concentration was 72 mg/L, which is much lower than typical produced or hydraulic flow back water.

Table 17.11

Water Quality Parameter	Feed Water	% Reduction
Turbidity (NTU)	415	92
COD (mg/L)	420	77
TSS (mg/L)	236	69

More detailed results are given in the Figures below.

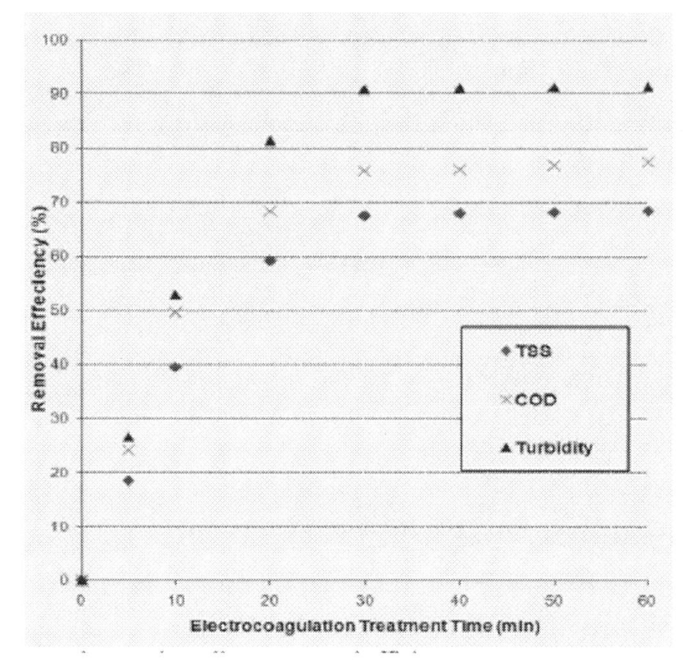

Figure 17.2 This figure is from reference [22]. It shows the effect of treatment time on the removal efficiency of TSS, COD, and turbidity.

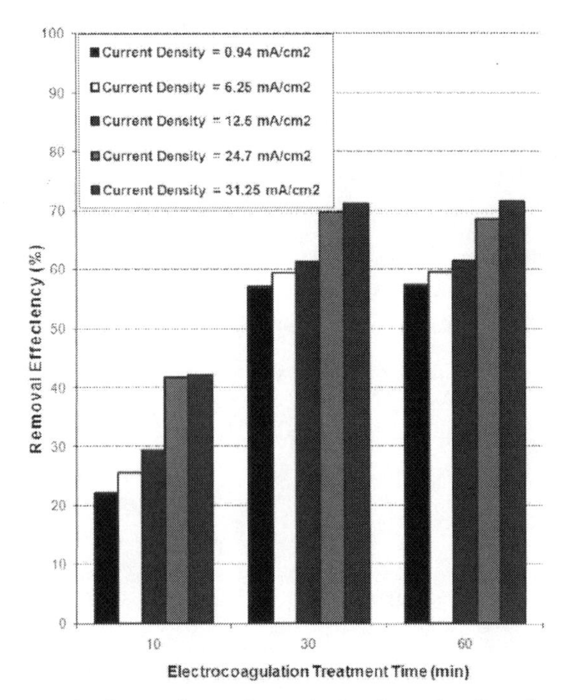

Figure 17.3 This figure is from reference [22]. It shows the effect of treatment time and current density on the TSS. The feed water has low TDS (450 mg/L).

Anode Choices: Besides the shape and spacing of the anodes, there are several chemical varieties that can be selected.

- DSA (Dimensionally Stable Anodes) – do not provide soluble cations (DSA do not dissolve)
- Anodes are typically composed of a base metal (Titanium) with catalyst coating
- Catalyst is applied using Chemical Vapor Deposition (CVD)
- Catalyst coatings include: Ir, Os, Ru, Rh, Pd, Pt
- Planer catalytic surface

17.5.3 Process Configuration, Waste Handling, Control

Typical Configuration – Single Step: Almost all EC applications in HF flow back start with a small volume that can be used for chemical addition, including pH adjustment, and utilize a single step EC unit followed by settling / separation, which is then followed by sludge handling. This is a typical configuration. It is referred to as single step since the generation of coagulant and the reaction of the waste water occur in the same compartment.

Two-Step Configuration: As will be discussed below, in the separation of boron, other configurations have been evaluated such as the separation of coagulant generation and reaction with the contaminated waste water. In this configuration, the EC unit is used solely to generate the coagulant. An electrolyte solution is prepared and flowed across the electrodes. The electrolyte is formulated to assist the generation of coagulant. The EC solution, containing the coagulant is then flowed out of the EC reactor and into the waste water reactor. This configuration has the advantage that the electrodes will not be fouled by the contaminated waste water. Since the waste water never comes in contact with the electrodes, fouling is only a function of the electrolyte used to generate the coagulant. The disadvantage is that the EC process benefits discussed above are not realized. The enhanced filtration caused by the formation of a floc blanket (electrophoretic effect), the flotation effect of hydrogen bubble generation, and the benefit of oxidation are not realized.

17.5.4 Representative Performance Data

Mickley [23] gives the following overall assessment of EC performance:

Table 17.12

Removal Focus	% Removal
BOD	90% +
TSS (clay, silt, coal, etc)	99% +
Fats, oils and grease in water	93 to 99% +
Water in sludge	50 to 80% +
Heavy metal	95 to 99% +
Phosphate	93% +
Bacteria, viruses, cysts	99.99% +
TDS	15 to 30% +

While it is helpful to have such a summary of results, the actual performance of an EC unit will depend on a number of factors which are discussed below. For our purposes here, the Mickley summary should not be taken too literally. Instead, what it seems to indicate is the following; Suspended contaminants such as TSS, fats, oil, grease, bacteria can be separated with high efficiency, under certain design and operating conditions. Heavy metals, such as mercury and zinc, can also be removed with strong dependency on the chemistry of the water and the pH. These compounds can occur in produced water in a range of complexes and compounds. It appears that EC can remove the colloidal and suspended forms as well as a significant portion of the elemental, ionic and dissolved forms. This is somewhat expected given that flocculation chemicals have similar capability, though at high dosage. The group II alkali earth ions (Mg^{+2}, Ca^{+2}, etc) can be removed to a limited extent. Other dissolved components that contribute to TDS cannot be removed to any great extent. These observations agree with the known mechanisms involved in EC technology, and are therefore expected.

As part of an evaluation of salt tolerant polymers for hydraulic fracturing, produced water was treated using electrocoagulation [6]. A parallel bench study was carried out, under controlled conditions and in order to obtain more detailed data than available under field conditions. Produced water from a well near Carlsbad, New Mexico was used as feed to a bench scale EC unit.

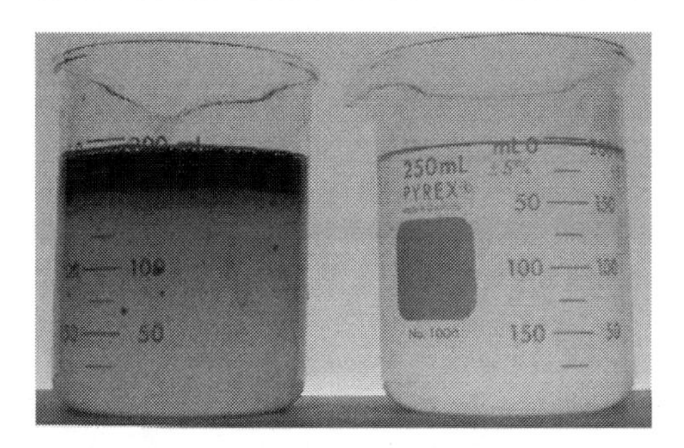

Figure 17.4 Produced water, untreated (left) and treated with electrocoagulation (right). Produced water obtained from a shale well in Carlsbad New Mexico. This figure from reference [6].

Few details are given on the operation of the EC unit. However, the detailed compositional analysis given in the table below is excellent. The EC treatment brought about a significant change in the suspended solids content of the water as can be seen visually in the Figure above, as well as in the values of Turbidity, TSS and TPH. Since this was designated as a produced water sample, and not a flow back sample, it is difficult to say how much spent guar is in the sample. However from the appearance (Figure) the sample does appear to have some polymer. The TPH of the feed sample was much greater than 20 mg/L (see table). The EC treatment reduced the TPH significantly.

Table 17.13 Produced water analysis before and after treatment with electrocoagulation. Ref [6]

Source	Produced Water – Influent Sample	Effluent EC – Treated Produced Water
Specific Gravity	1.200	1.203
pH	4.83	8.00
Conductivity (μS/cm)	257	258
Turbidity (NTU)	182	15.4
Dissolved Oxygen	8.24	8.45
Chloride	163,637	164,951
Sulfate	40	38
Aluminum	1.42	2.28
Boron	20.3	16.6
Barium	5.69	6.03
Calcium	29,222	28,845
Iron	34.6	0.264
Potassium	1,660	1,689
Magnesium	4,347	3,148
Sodium	70,342	75,517
Strontium	2,204	2,020
TDS	267,588	273,552
TSS (mg/L)	10,623	92
TPH (ppm)	>20	Between 5 and 20

**Measurements are expressed in ppm unless otherwise noted*

Iron was significantly removed. Aluminum increased as a result of the treatment, indicating the aluminum anodes were used. Roughly 20 % of boron was removed. The pH increased significantly. This is likely due in part to the EC process itself, and to the addition of caustic which is somewhat suggested by the increase in the Na / Cl due to the treatment. There is essentially no decrease in TDS. There is an increase in Al (anode material); Slight decrease in B; Significant decrease in iron; very slight softening (Ca, Mg).

The actual current density used is not reported. However the intensity of the treatment can be inferred from various observations. When sacrificial anodes are used, as in this case, the mildness or severity of the treatment can be judged, to a rough approximation, by the concentration of anode material that is dissolved in the treated effluent. This is not always a good indication of the intensity of EC treatment since anode material is consumed in the floc material. In this case though, a small increase in Al concentration together with the limited extent of softening, indicates a relatively mild treatment was applied. But apparently this treatment achieved the goals which were to eliminate the suspended solids.

Table 17.14

Pollutant	Electrical requirement	Electrodes	Electrode type	Pollutant removal	Reactor type
Bentonite	0.2 - 1A	A=SS C=SS	Plates	Settling (implied)	Batch (41)
Carbon	120 - 170 Am^{-2}	A=Al C=Fe	Plates	Flotation	Continuous
Silica, Alumina	2.5 - 10.0V cm^{-1}	A=SS C=SS	Mesh	Settling	Batch (31)
Clay	3.4 – 27 Am^{-2}	A=Al C=SS	Plates	Floatation and settling	Batch (71)
Suspended solids	5 - 50 Am^{-2}	A=Al C=Fe	Plates	Settling	Continuous
Kaolinite	0.01 Am^{-2}	A=SS C=SS	Mesh	Settling	Stirred batch
Suspended solids	50 – 70Am^{-2}	A=Al C=SS	Plates	Flotation (inseparate chamber)	Continuous

Textile Waste Water: In the evaluation of technology it is often useful to look outside of the oil and gas industry for relevant data. Cotton and woolen textile processing requires large volumes of water and generates large volume of contaminated waste water. The contaminants include aromatic and heterocyclic dye compounds, bleaching agents, and spent finishing solutions containing acidic and alkaline organic compounds. Finishing involves a wide range of treatment solutions that may include softening or strengthening of fabric, sizing with starch, cross-linking and waterproofing with polymeric resins. Textile waste water tends to be high in COD, BOD, and turbidity. Relative to HF flow back waste water, the TSS tends to be moderate. The conventional water treatment technology is coagulation flocculation sedimentation. Oxidation is also used to remove the turbidity particularly due to dye compounds.

An evaluation of electrocoagulation was carried out [19] on textile waste water with the contaminant concentrations as given in the Table below. Current density varied from 6 to 14 mA/cm2, and the EC reaction time varied from 30 to 90 minutes. No chemical was added except to adjust pH. The initial pH was 7. DC current was applied which was kept constant during the evaluation. For the mid-range of these values (10 mA/cm2, 60 minutes reaction time, pH 7) the following results were obtained.

Parameter	Initial Value	% Reduction
COD (mg/L)	1,260	60
TSS (mg/L)	1,750	55
Turbidity (NTU)	1,310	99

It is interesting that the separation performance of EC for this kind of waste was somewhat less than would be obtained with EC application at these conditions on HF flow back water. COD and TSS reduction in the range of 55 to 60 % is oaky but not great. Even at the upper end of the current density and reaction time, the separation performance was not dramatically better. It is possible that in this type of waste water much of the contaminant is dissolved. That would explain why relatively high values of current density and reaction time only result in 60 % COD removal.

SAGD Pilot Test: Devon evaluated the use of EC in produced water recycling for SAGD (Steam Assisted gravity Drainage) operation [24] in the Jackfish project, Alberta. In that field, OTSG (Once Through Steam generators) are used.

17.5.5 Boron Removal

In 1993 the World Health Organization (WHO) provided the following recommended guideline for boron concentration in drinking water: 0.3 mg/L [25]. This was based on the No-Observed-Adverse-Effect Level (NOEL) which was studied and reported by Jansen et al. [26]. Most surface and ground waters contain less than this level. Nevertheless, this is a low level. Seawater contains anywhere from 26 to 72 mg/L of borate (BO_3^{-3}) depending on which part of the world. These values correspond to 5 to 13 mg/L of boron (ratio of molecular weights: 10.8 B / 58.8 BO3).

Many produced waters and HF flow back waters have relatively high borate concentrations. For agricultural reuse, and surface discharge, boron must be removed. Also, in order to achieve good control of the gel cross-linking reaction, it is much better to start with water that has a low borate content. Thus, if produced water is being used, there is an incentive to remove the borate before the cross-linking components are added.

For onshore applications in the oil and gas industry, the main treatment method for boron removal is precipitation / coagulation / flocculation / settling. This is a proven process. The main drawback is the rather high chemical consumption and the rather high volume of sludge generated.

Other processes for removal of boron include pH adjustment to form the borate anion, followed by reverse osmosis, and ion exchange. As shown in Figure 17.5 at high pH the borate anion is dominant species which makes RO a viable technology. However, it should be noted that RO will foul quickly in HF flow back service unless there is extensive pre-treatment.

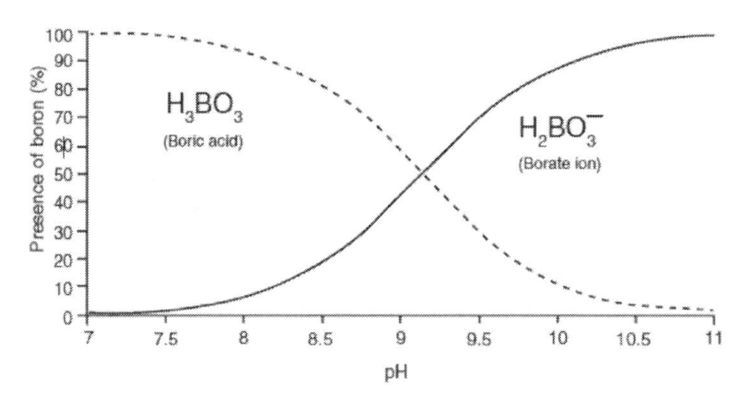

Fig. 1—The behavior and chemical composition of boron varies in an aqueous environment. Borate ions dominate at higher pH, and boric acid dominates at lower pH.

Figure 17.5 Boron species as a function of pH. At high pH, the borate anion is the dominant species. The borate anion has high rejection across an RO membrane which makes RO a viable technology

Boron removal using electrocoagulation has been studied in the lab and has also been applied in the field. In one study reported by Sayiner et al. [25], a two-step EC process was used as shown in the Figure below. The removal efficiencies for this process were encouraging as shown in Figure x and y below.

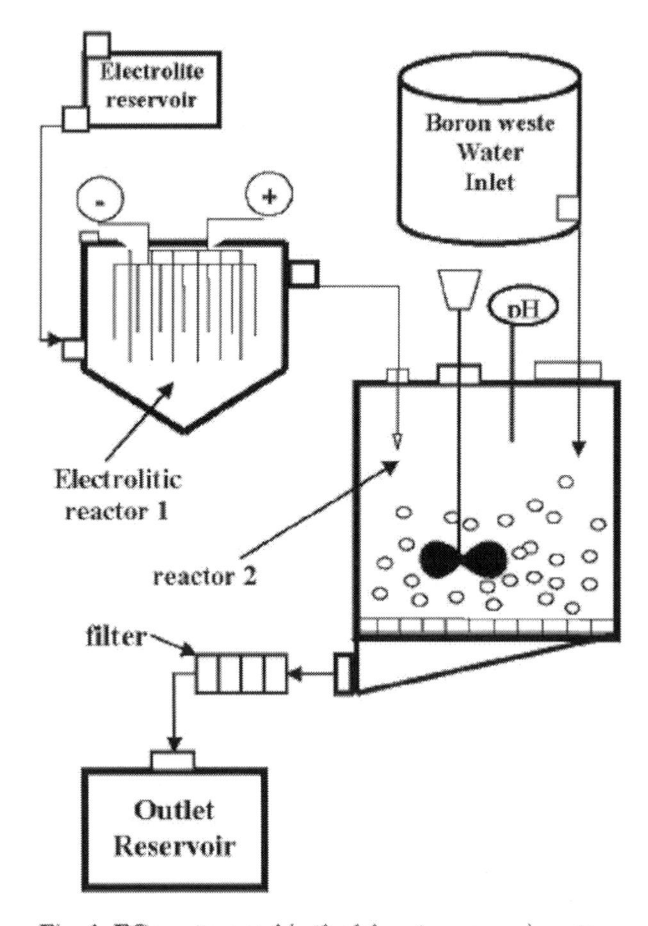

Fig. 1. EC reactor used in the laboratory experiments.

Figure 17.5 A two-step EC reactor used in [25]. In this process, the electrodes do not contact the contaminated waste water. This essentially eliminates the problem of electrode fouling. However, as discussed in the text, the full benefit of EC mechanisms is not realized. Results are shown in Figure 17.5

The specific electrical energy consumption was 0.8 to 1.5 kWh / m^3 water treated. The current density of the electrodes was in the range of 10 to 30 mA/cm^2. According to Degremot [27] between 2 to 4 kWhr / m3 would be considered relatively high energy consumption for electrocoagulation. So in this application the specific energy requirement is not particularly high.

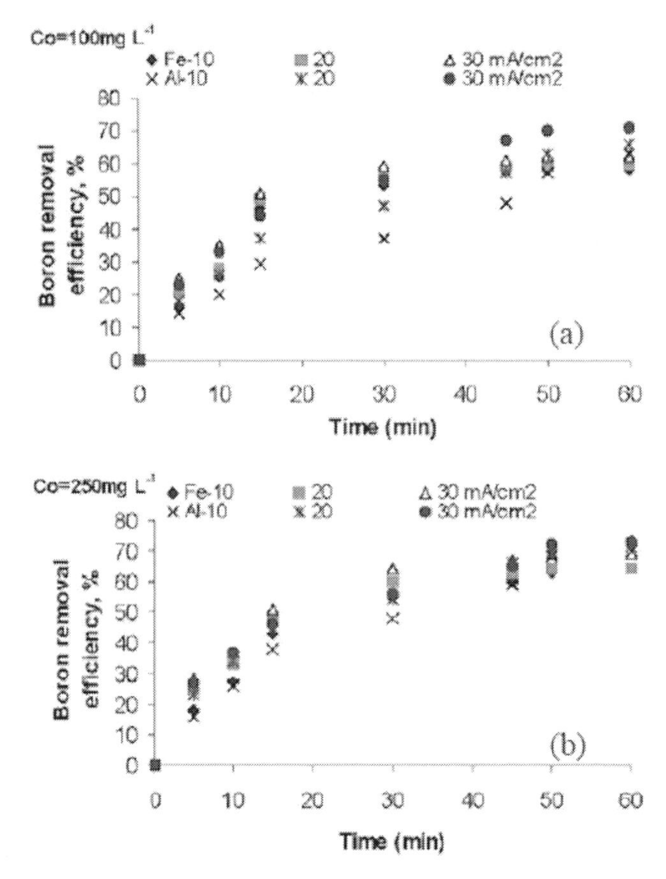

Fig. 3. Boron removal % versus time for initial concentrations 100 ppm (a) and 250 ppm (b) under different CD.

Figure 17.5 Separation efficiencies obtained from the two-step EC process shown in Figure 17.5. Data are from [25]. Boron removal efficiency is shown as a function of two inlet concentrations (100 and 250 mg/L), and as a function of the anode type, anode current density (20 to 30 A/cm^2), and residence time (5 to 60 minutes)

A two-step EC reactor used in [25]. In this process, the electrodes do not contact the contaminated waste water. This essentially eliminates the problem of electrode fouling. However, as discussed in the text, the full benefit of EC mechanisms is not realized. Results are shown in Figure x.x.x. The salinity of the treated water was in effect 10,000 mg/L, composed of NaCl.

Yilmaz et al. [28] also studied the removal of boron using electrocoagulation. They used a single-step design, in which the contaminated feed water contacts the electrodes. This is by far the more common arrangement. High concentrations of boron in the feed water were studies, from 100 to 1,000 mg/L.

The specific energy consumption was in the range of 6 to 141 kWh/m3. The current density applied was in the range of 1 to 10 mA/cm^2. These values suggest that a rather long residence time was used and indeed, residence times are reported as 30 to 120 minutes. As shown in the figure below, removal efficiencies were not impressive for the 30 minute residence time. But removal efficiency is quite good for 60 minutes and higher.

It is interesting to see that at higher values of pH, where the borate anion is the dominant species, removal efficiency is lower.

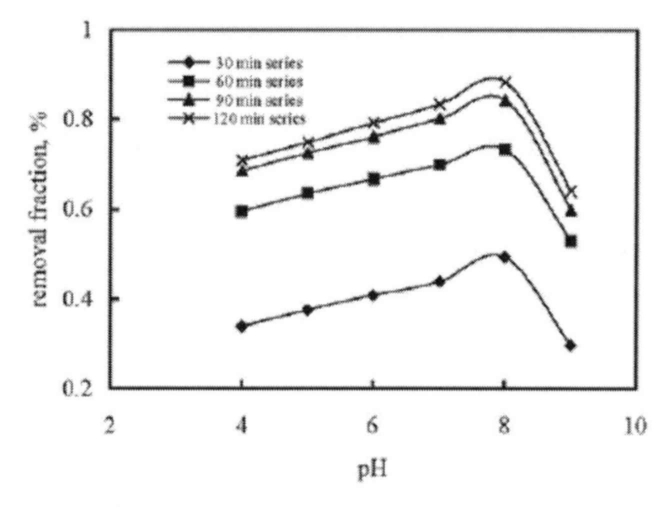

Fig. 2. Effect of pH on boron removal.

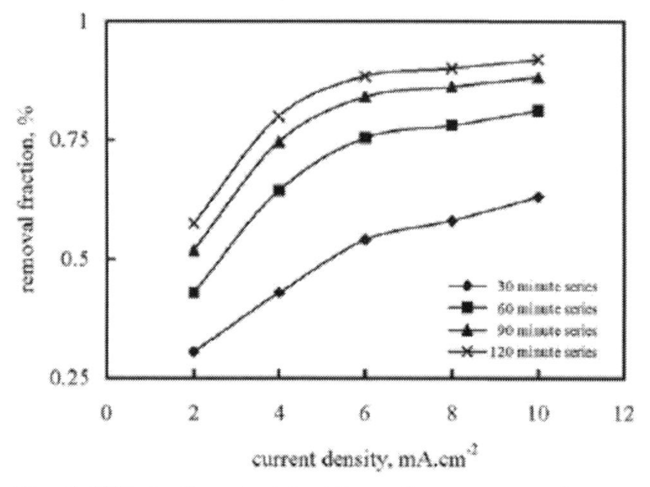

Fig. 3. Effect of current density on boron removal.

Figure 17.8 Separation efficiencies for removal of boron as a function of the pH, current density, and reaction time. Data are from [28]

17.5.6 Softening and Silica Removal

Lime Softening: The lime softening process is used extensively in pre-treatment of produced water for steam generation in steam flood applications. The objective of the process is to remove hardness and silica. Both of which can cause scaling in the steam generation units (Once Through Steam Generators or Mechanical Vapor Compression). In steam flooding applications, lime and magnesium hydroxide are added which remove the carbonate in the form of $CaCO_3$ and $MgCO_3$. Silica removal is also a major objective of softening. When magnesium hydroxide precipitates, it forms a large floc, with flocculation properties analogous to that of iron and aluminum hydroxides. The $Mg(OH)_2$ floc adsorbs and entraps suspended silica particles. For each pound of silica to be removed from the water, one pound of magnesium hydroxide is required in the water. If sufficient magnesium is not present in the contaminated feed water, then additional magnesium must be fed. This is typically done in the form of $MgCl_2$. Removal of suspended material such as oil, organic compounds, and bacteria is also accomplished. Some dissolved organics are also removed.

Composition of Lime: Lime is a general term which can be applied to a number of different minerals which have calcium oxide, calcium hydroxide and some calcium carbonate, in varying concentrations depending on the particular mine from which it was recovered. Limestone or chalk is composed primarily of calcium carbonate. As a chemical product, if purchased from a chemical supplier, lime is composed of calcium oxide (CaO) and calcium hydroxide ($Ca(OH)_2$). Calcium oxide is also referred to as Quicklime. It is highly caustic when added to water which makes it the preferred form of lime for softening.

Lime Softening Reactions: The chemical reactions involved in softening depend on the particular form of lime that is used. In terms of the bicarbonates and hydroxides, the reactions are:

$$Ca(HCO_3)_2 + Ca(OH)_2 \rightarrow 2CaCO_3 + 2H_2O$$

$$Mg(HCO_3)_2 + Ca(OH)_2 \rightarrow 2CaCO_3 + MgCO_3 + 2H_2O$$

Lime Softening Benefits and Drawbacks: In order to understand the opportunity for EC in this application, it is worthwhile to list the benefits and drawbacks of conventional lime softening:

Pros:

- well established technology, well known design methods
- capable of removing a broad range of contaminants
- not subject to damage as a result of poor operation or maintenance neglect
- all things considered, it is a robust technology

Cons:

- requires large amounts of chemicals
- generates large amounts of waste sludge
- the sludge blanket requires a couple days to re-establish once it is lost
- manpower intensive

EC applied to Softening:

Electrical consumption: 1.0 kWh/m3 (4 kWh/1000 gal)

Consumption of anode: 24 g/m3 (0.2 lbs / 1000 gal)

Representative design information is given in the Table below. This information was gathered from actual operating data from the lime softening process and estimated performance information for the EC unit [29].

Table 17.15 Representative Design Information [29]:

	Raw Water (mg/L)	After Lime Softening (mg/L)	After Lime & pH Adjust (mg/L)	After EC (mg/L)
Cations				
Na	2,586	2,586	2,586	2,586
Mg	239	30	30	16
Ca	42	2	2	30
Anions				
Cl	3,656	3,656	3,656	3,024
HCO3	466	544	0	466
CO3	672	60	0	672
SO4	139	139	665	90
Alkalinity (as CaCO3)	1,503	546	0	1,503
Silica as SiO2	219	10	10	4
pH	7.8	11.3	7.0	7.0
TDS	8,088	7,095	7,017	6,953

EC Softening Benefits and Drawbacks: In order to compare EC to conventional lime softening, the following benefits and drawbacks for EC are given:

Pros:

- requires no or only a small amount of chemicals
- generates only a small amount of waste sludge
- effective at removing hardness and silica
- relatively insensitive to wide variation in feed water quality
- low power consumption

Cons:

- design of new installations still not well understood
- anodes must be replaced periodically and electrodes must be cleaned periodically
- requires specific operator training (most operators are not familiar with the technology)

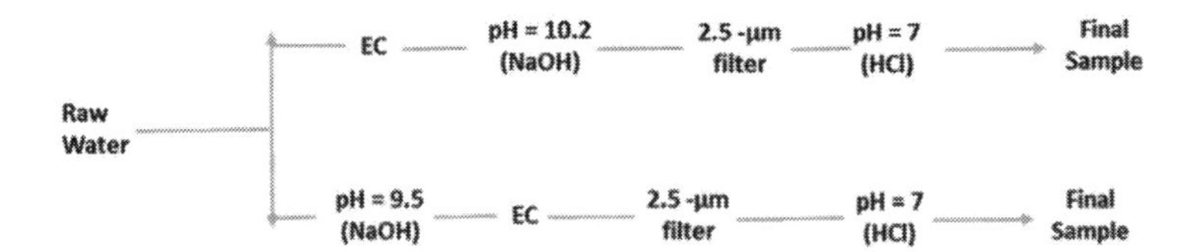

Figure 17.9 Process configuration options when EC is used together with a softening step [30]

This represents a mild application of electrocoagulation [30]. Under such conditions, the data indicate that chemical softening followed by EC is better than EC/softening.

Table 17.16 Electrocoagulation applications

cation	feed (mg/L)	EC/softening effluent (mg/L)	softening/EC effluent (mg/L)
Ca^{+2}	280	200	40
Mg^{+2}	48	36	24

The lime softening option was based on the conventional lime softening process. Although the design water analysis is high in total dissolved solids, it has very low hardness (calcium and magnesium) and significant magnesium must be added to the water to precipitate the desired amount of silica. Magnesium was added in the form of magnesium chloride but other forms can be used depending on availability at the specific site.

17.5.7 Fine Solids Removal:

As discussed already in Section 17.1.4 (Representative Performance Data), in 2013 Halliburton, working together with XTO published a paper on the success of using produced water for hydraulic fracturing [6]. The HF wells were in Eddy County New Mexico. The produced water salinity had a TDS of 270,000 mg/L. The gels used were formulated with carboxymethyl hydroxypropyl guar gum (CMHPG), a zirconium-based cross linker, sodium chlorite breakers, and other ingredients. A parallel bench study was carried out, under controlled and measured conditions, in order to obtain more detailed information than could be collected in the field. In the bench study, the inlet produced water had a Total Suspended Solids content of 10,623 mg/L. The effluent form EC treatment had a TSS of 92 mg/L. This represents an outstanding separation performance for the suspended solids.

The XTO/Halliburton paper does not provide a characterization of the solids. The particle size, composition and source are not reported. But as Gruber [9] points out, produced water from shale formations often contains large concentrations of colloidal solids. Solid particles are often in the size range of 0.001 to 1 micron. Removing such solids often requires chemicals to coagulate and flocculate the solids, followed by settling, flotation, filtration or some combination of these technologies. The paper by Lord and co-workers also shows that EC is capable of such colloidal particle removal.

Lord and co-workers published a subsequent study on the importance of removing suspended solids from produced water when that water is recycled for HF stimulation [7]. "Removal of TSS is critical. The solids are generated from a wide range of sources, such as formation fines, clays, sand, and corrosion products from the injection / production lines. Clogging or bridging of the pore space in the formation by these suspended solids can be rapid and severe if these particles are not removed before use." They discuss the various options for fine particle removal and provide test results for the use of EC. As shown in the Table below, the results using EC are excellent. The particle size distribution for the feed water (produced water) was in the range of 10 micron. However, no details are given on the means of measuring the PSD and it is possible that the method did not detect particles smaller than 1 micron.

	TSS feed water (mg/L)	TSS effluent – after EC (mg/L)	Solids composition
Bakken	301	1	75 % CaCO3; 5 % clay; 20 iron oxides
Permain	100	3	17 % CaCO3; 2 % clay; 66 % organics; 15 iron oxides
Marcellus	375	10	15 % CaCO3; 50 % clay; 25 % barite; 10 iron oxides

The benefit of using EC for recycling produced water for removal of small particles which would otherwise cause injectivity and production problems was demonstrated.

17.5.8 Benefits and Drawbacks of Electrocoagulation:

Initial Oilfield Introduction: When electrocoagulation was first introduced as a flow back treatment technology it was met by the oil and gas industry with skepticism. It was unlike anything that the industry had encountered previously. Although it was invented over a hundred years ago for industrial water application, and has been applied ever since, it has only occupied a small niche in the industrial water segment due to the low cost of the competitive chemical-only approach.

EC was initially promoted to the oil and gas industry as a silver bullet and at the same time, very poor explanations about how it works left people feeling that it was based on mysterious electrical phenomena. Initial performance seemed to justify the skepticism. Stories circulated of equipment problems, fouling and overall poor on-site service. This was a pity since there is actually good reason why electrocoagulation makes sense as a HF flow back treatment technology. It is also a story that has been repeated countless times with new technology in the oilfield. No technology, no matter how good, is worth anything if it is not operated and maintained competently in the field.

Recent Experience: More recently, on-site service, reliability, and performance of EC have all improved. This is in part due to the oilfield service companies getting involved. They have recently added EC to their portfolio of technologies. These companies have good service capability, they understand the industry and the challenges of on-site services, and they know how to manage the costs of those services. As a result, many operating companies are reevaluating EC. Personally, I feel that the technology is a bit tricky to apply and will require an expert onsite to make it work. But overall it has inherent advantages and it is inherently scalable and can be easily modularized.

Does not compete well (cost-wise) with floc-N-drop in an industrial (centralized, continuous) setting. This is not a serious drawback in HF operations where other benefits can be overriding.

Pros:

- effective at treating complex waste waters
- suitable for highly variable feeds, with experienced operators
- removes oil, TSS, divalent cations
- uses less chemicals than floc-n-drop
- chemicals (anodes) easier to handle than liquids
- generates less waste sludge (due to less chemical)
- suitable for modulatization – a number of small units can be put together easily in parallel

Cons:

- requires significant power generation
- oil is separated from water but usually not recovered
- the sludge volume is smaller than for floc & drop but still significant
- requires operator attention, and experience and skill to operate
- is prone to fouling and passivation of the electrodes
- difficult to control voltage as it is sensitive to water chemistry
- difficult to train operators adequately

Compared to floc-and-drop, EC has the following advantages and disadvantages.

Advantages of EC compared to chemical coagulation:

- No process chemicals required (but acid is required for cleaning the electrodes)
- Reduces waste disposal and operating expense (Opex)
- Minimal manpower requirements (but operators must be well trained)
- Reduced waste disposal
- Tendency for sludge volume to be less, composed mostly of oxides and hydroxides and simpler to dewater. EC Floc's tend to be larger than chemical floc and typically contain less bound water.
- Higher removal percentages for critical dissolved species
- Generally insensitive to changes in inlet conditions (flow, quality) unlike chemical coagulation which can be upset · Ability to remove a wide variety of different contaminants in a single operation
- Operates over a wider range of temperatures and pH

Disadvantages of EC compared to chemical coagulation:

- Limited exposure as it is an emerging technology within the oil and gas industry
- Replacement of electrodes periodically required
- Passivation of electrodes must be addressed and can impact EC performance
- Electricity dependent process and implementation dictated by energy cost
- High conductivity of waste stream necessary for efficient EC process

References to Chapter 17

1. M. Patton, "Produced water management opportunities in a challenging environment," Shale Play Water Management, p. 12 (March-April 2016).

2. K.S. McLin, J.M. Freeman, S. Aarekol, E.S. McKean, R.R. Sharma, J. Rincon, J. Speaker, "Permian Unconventional Water Management: Collaboration to Develop Full Cycle Solutions," SPE-174538-MS, paper presented at the SPE Produced Water Handling and Management Symposium, Galveston (2015).

3. R.R. Sharma, K. Bjornen, "Systematic approach for developing a "fit-for-purpose" treatment for produced water reuse in hydraulic fracturing," URTeC-2902544, paper presented at the Uncon. Resources Tech. Conf. (URTeC), Houston (2018).

4. R. Sharma, K. McLin, K. Bjornen, A. Shields, Z. Hirani, S. Adham, "Fit-for-purpose treatment of produced water for hydraulic fracturing – a Permian basin experience," paper presented at the IPTC, Doha (2015).

5. NACE TM0173, Standard Test Method – Methods for Determining Quality of Subsurface Injection Water Using Membrane Filters,

6. R. LeBas, P. Lord, D. Luna, T. Shahan, "Development and use of high-TDS recycled produced water for cross-linked-gel-based hydraulic fracturing," SPE – 163824, paper presented at the SPE Hydraulic Fracturing Technology Conference, The Woodlands, TX (2013).

7. X. Ye, N. Tonmukayakul, P. Lord, R. LeBas, "Effects of total suspended solids on permeability of proppant pack," SPE – 165085, paper presented at the SPE European Formation Damage Conf and Exhibition (2013).

8. J.M. Walsh, "Produced water treating systems: Comparison of North Sea and deepwater Gulf of Mexico," SPE-159713-PA, Oil and Gas Facilities (2015).

9. E. Gruber, "Recycling produced and flowback wastewater for fracking," published by Ecologix Environmental Systems, LLC, (2014).

10. J.R. Coleman, W.G. McLelland, "Produced water re-injection; How clean is clean," SPE – 27394, paper presented at the SPE Int. Sym. Formatin Damage, Lafayette, LA (1994). W.G. McLelland, "What do we mean by water quality," paper presented at the SPE Produced Water Handling and Management Symposium, Galveston, TX (2015).

11. K.M. Bansal, D.D. Caudle, "A new approach for injection water quality," SPE – 24803, paper presented at the ATCE, Washington DC, October (1992).

12. P.K. Singh, R.G. Agarwal, L.D. Krase, "Systematic design and analysis of step rate tests to determine formation parting pressure," SPE – 16798, paper presented at the ATCE, Dallas (1987).

13. New Mexico Oil Conservation Division. Underground Injection Control Program Manual. Published by the New Mexico Energy, Minerals and Natural Resources Department, Oil Conservation Division

14. Railroad Commission of Texas, http://www.rrc.state.tx.us/oil-gas/publications-and-notices/manuals/injectiondisposal-well-manual/

15. http://www.rrc.state.tx.us/oil-gas/publications-and-notices/manuals/injectiondisposal-well-manual/summary-of-standards-and-procedures/technical-review/step-rate-test-guidelines/

16. P. Xu, G. Ma, Z. Stoll, Assessment of treatment technologies for produced water to improve water supply sustainability in southeastern New Mexico," a publication of the New Mexico Water Resources Research Institute and the Dept. Civil Eng., New Mexico State University (2016).

17. J.C. Crittenden, R.R. Trussell, D.W. Hand, K.J. Howe, G. Tchobanoglous, "Water Treatment, Principles and Design," MWH, John Wiley and Sons, Third Ed. (2012). For a discussion of coagulation and flocculation, see chapter 9.

18. Brent Halldorson, "Successful Oilfield Water Management," American Association of Drilling Engineers (AADE), paper number AADE-13-FTCE-14 (Feb 2013).

19. S. Zodi, O. Potier, F. Lapicque, J.-P. Leclerc, "Treatment of the industrial wastewaters by electrocoagulation: optimization of electrochemical and sedimentation processes," Desal., v. 261, p. 186 (2010).

20. P.K. Holt, G.W. Barton, C.A. Mitchell, "The future for electrocoagulation as a localized water treatment technology," Chemosphere, v. 59, p. 355 (2005).

21. M. Khemis, J.-P. Leclerc, G. Tanguy, G. Valentin, F. Lapicque, "Treatment of industrial liquid wastes by electrocoagulation: experimental investigations and an overall interpretation model," Chem. Eng. Sci., v. 61, p. 3602 (2006)

22. M. Saleem, A.A. Bukhan, M.N. Akram, "Electrocoagulation for the treatment of wastewater for reuse in irrigation and plantation," J. Basic Appl. Sci., v. 7, p. 11 (2011)

23. M. Mickley, "Pretreatment capabilities and benefits of electrocoagulation," report prepared by Mickley & Associates for the Office of Naval Research under contract No.: N00014-04-C-0027 (2004).

24. T. Erickson, J. Paddock, "Evaluation of opportunities in SAGD water processing: electrocoagulation, filter press," report submitted to ENCH 665, Devon.

25. G. Sayiner, F. Kandemirli, A. Dimoglo, "Evaluation of boron removal by electrocoagulation using iron and aluminum electrodes," Desalination, v. 230, p. 205 (2008).

26. J.A. Jansen, J.S. Schou, B. Aggerbeck, "Gastrointestinal absorption and in vitro release of boric acid from water-emulsifying ointments," Food Chem. Toxic., v. 22, p. 49 (1984).

27. Degremont, Water Treatment Handbook, Lavoisier, Paris (2005).

28. A.E. Yilmaz, R. Boncukcuoglu, M.M. Kocakerim, E. Kocadagistan, "An empirical model for kinetics of boron removal from boron-containing wastewaters by the electrocoagulation method in a batch reactor," Desalination, v. 230, p. 288 (2008).

29. Hamilton Engineering Report, "Evaluation of lime softening versus electrocoagulation for treatment of produced water," report prepared for Chevron Energy Technology Company, Denver, CO (2009).

30. N. Esmaeilirad, K. Carlson, P.O. Ozbek, "Influence of softening sequence on electrocoagulation treatment of produced water," J. Haz. Mat., v. 283, p. 721 (2015).

CHAPTER EIGHTEEN

Applications: Solids

Chapter 18 Table of Contents

18.0 Solids in Produced Water 427

18.1 Origin and Types of Solids 428

18.1.1 Formation and Reservoir Solids (Sand, Clay, Proppant, Organics) 429

18.1.2 Iron Oxides 432

18.1.3 Iron Sulfide 433

18.1.4 Inorganic Scale Forming Minerals 434

18.1.5 Organic Solids 435

18.1.6 Bacterial Contamination 435

18.1.7 Naturally Occurring Radioactive Material (NORM) 435

18.2 Sampling and Analysis of Solids 435

18.2.1 Total Suspended Solids (TSS) 436

18.2.2 NACE Chemical Solvent Test 437

18.2.3 Particle Size Distribution (PSD) 439

18.3 Solids Management 442

18.4 Solids Separation from Fluids 444

18.4.1 Flotation Solids Removal 444

18.4.2 Conical Separator Solids Removal 445

18.4.3 Cyclonic Desander 447

18.5 Solids Removal from Separators 448

18.6 Solids Handling 450

18.6.1 Disposal in a Landfill Site 450

References to Chapter 18 452

18.0 Solids in Produced Water

Much of the material in previous chapters is fundamental in nature and deals with the chemistry, fluid mechanics, and characterization of oil and solids in produced water. In this Chapter, practical aspects of solids separation, handling and management are discussed. The scope of discussion here is intended to cover all types of solids encountered in the oil and gas industry. The emphasis is however on sand.

Prevent Solids in the First Place: Generally the most effective strategy to deal with solids problems in water treating systems is to eliminate the source of the solids. Prevention of sand production, formation of organic solids, precipitation of mineral scale, and prevention of corrosion products are all prudent practice from the standpoint of well bore integrity, asset integrity, flow assurance, and water treatment. The design of wells with effective sand screens is an enabling technology that allows production from reservoirs that would otherwise produce excessive amounts of sand [1]. Formation of deposits can lead to under deposit corrosion and rapid loss of integrity.

Producing with Solids: When prudent operating practices have been applied, and solids production is still present, then a solids management system must be put in place in order to continue hydrocarbon production. Co-production of sand and other solids is a fact of life for some facilities [2, 3]. In fact, some degree of sand production is inevitable for many fields particularly later in life and when water cut increases. When that is the case, facilities sand management program must be implemented with the objective to sustain hydrocarbon production in the presence of sand, while minimizing its impact [4]. This is the main subject of this chapter.

Solids Separation, Solids Handling and Solids Management: It is helpful to discuss solids separation, handling and management as three distinct activities. When dealing with a solids problem, it is important to address all three activities at an early stage of the project. If any one of these activities is neglected then all of the work on the other two can become irrelevant due to an unforeseen bottleneck in the third neglected activity.

There is a variety of technologies that can be considered in order to separate solids from fluid streams. Some examples are flotation, settling in surge vessels, cyclonic desanders, and/or vessel jetting systems. The oil and gas industry must deal with a variety of fluids including multiphase (oil, water, gas) fluids, oily water or wet oil. In selecting the technology the properties of the fluid, solids, and the flow rates and concentrations must be taken into account.

Once the solids have been separated, they typically are in slurry form and must be moved from one location of the facility to another, and usually must be further concentrated and moved from the facility location to the point of disposal. This movement, possible cleaning, and ultimate disposal of solids is referred to as solids handling. The engineering involved in solids handling is relatively straightforward but most oil and gas engineers do not have experience in this area.

Solids management is the more general activity which encompasses both solids separation and solids handling. It encompasses the entire lifecycle of dealing with solids starting with separation, and including handling, transportation, treatment, regulatory testing and analysis, application for a disposal permit and the final step of actually disposing of the solids. Simply stated, solids separation and handling are important but they are only the beginning of the solids management journey.

18.1 Origin and Types of Solids

Solids in hydrocarbon and produced water streams can originate from the formation as sand, clay or carbonate fines, or it can originate by precipitation of inorganic scale forming minerals, or precipitation of organic materials. Examples of solids in produced water include:

- formation and reservoir materials (calcite, sand, silica, clay, shale particles, propant);
- mineral scales (halite, calcium carbonate, magnesium carbonate, iron carbonate, calcium sulfate, barium sulfate, strontium sulfate, radium sulfate, iron sulfide);
- corrosion products (hematite and magnetite);
- mineral/organic combinations (sodium and calcium naphthenates) [5];
- organic solids from the oil (asphaltenes, waxes);
- organic solids from the production chemicals (water clarifier, demulsifiers, polymer flocculating agents);

Types of Solids in Production Systems: An important distinction needs to be made regarding various types of solids encountered in hydrocarbon production. From a water treatment perspective, there are four basic types of solids:

1. oil-wet solids that primary come from the oil bearing part of the reservoir;
2. water-wet solids that can either come from the water bearing part of the reservoir, or have been formed during the production process; these solids include minerals from the reservoir, minerals that precipitate during production, and corrosion products;
3. partially oil-wet solids; these solids are referred to as partially oil-wet but they are also therefore partially water wet; due to the presence of both oil and water wetted surfaces these particles tend to be surface active and can be a major problem in the water treatment system;
4. organic solids which include asphaltene, wax, or production chemicals; in hydraulic fracturing, the dominant form of solids are proppant and friction reducing polymer (HPAM, HPAM-AMPS, guar, gelled guar).

In general, oil wet solids originate in the oil-bearing part of the rock. These solids include sand, clay and carbonates and have been wet by the oil over a very long time scale. The water-wet solids originate in the lower, water bearing part of the reservoir and are composed of similar, if not very similar rock types as found in the oil-bearing part of the reservoir.

Problems Caused by Oily Solids: Oil-wet solids, or oily solids, can cause significant water treatment problems. The combination of a lighter-than-water oil together with a heavier-than-water solid particle can create an oily solid particle that has a density close to that of water. When that occurs, the oily solid particle is difficult to separate using settling, hydrocyclone or centrifuge since all of those methods are based on a density difference between water and the contaminant. Flotation is often effective at separation of oily solids. Oily solids when discharged overboard or to other surface water will contribute to sheening and a measurable oil-in-water concentration. On the other hand, if the disposal option involves a disposal well, then oily solids can significantly contribute to injectivity problems, fracture control problems, and overall well impairment which can result in significant cost for stimulation.

Produced Solids in Separators: Oil and gas wells often produce sand or solids with the well fluids. Despite well completions design [1], solids can be detected on surface facilities through equipment failure, completion limitations, or a step change in production profile, such as water breakthrough.

Produced solids may erode chokes and flowlines, but the first place of accumulation in large quantities is the production separator [6 - 8]. Solids settle in production vessels and piping when the transport velocity drops below the limiting deposition velocity. The preferred location for facilities sand removal is before the choke with a wellhead desander [4]; however, in cases where this technology is not currently used, the produced solids will collect in the primary separating vessel. Accumulated solids in gravity=-settling vessels result in loss of residence time, corrosion-enhancement zones, increased sediment in oil, increased oil content in produced water, and degraded injectivity of produced water [3]. Sand and solids are removed from production separators either off line (shut down for physical removal) or on line by use of jetting system.

18.1.1 Formation and Reservoir Solids (Sand, Clay, Proppant, Organics)

The dominant forms of solid to emerge from the formation are sand and clay. Proppant used in hydraulic fracturing can be thought of as being produced from the reservoir although it does not originate there. Particles of asphaltene and wax can also emerge from the reservoir.

Sand is composed predominantly of silica (SiO_2). Feldspars and clays may also be present as minor constituents. Clay components include chlorite, iolite, kaolinite and smectite. These occur in the fine size fractions of a distribution, the size of a dehydrated clay particle being about 4μm. Sands containing less than 5% clay are classed as clean and those with more than 5% wt as dirty. Clay generally reduces formation permeability. If present in sand, deposits in process equipment, it could tend to make those deposits more difficult to remove.

Proppant compositions include sand, bauxite, alumina and zirconia. Proppants may also be resin coated. Shapes of sand particles can range from angular to rounded particles depending on their origin. Sand from igneous or metamorphic rocks are very angular whereas sand from previously deposited sand stones are rounded. At present there is little understanding of a relationship between shape and erosive nature. It is proposed that a research project will commence at the University of Tulsa, to address this, supervised by SGSi & SEPTAR. Clay occurs as thin plate like particles which can cluster together in stacks. Scale particles can occur in a wide variety of shapes including flat flakes (Forcados Terminal, Nigeria) and near spheroidal (El Fateh, Dubai).

The hardness of particulate materials relative to that of the equipment influences particle erosiveness. Silica, especially in the form of quartz, is much harder than most metals. See the following table.

The hardness of particulate materials relative to that of the equipment influences particle erosiveness. Silica, especially in the form of quartz, is much harder than most metals (Table 3.4).

Table 3.4: Hardness Of Sand Constituents and Metals

Composition	Mohr scale	Vickers, kg/mm^2
Quartz	7	1200-1300
Silica sand	6.5	800
Smectite clay	1	-
316 stainless steel	3	175
Carbon steel		145
Cast Iron		240

Wetting State of Reservoir Solids: Most solids that originate in the reservoir tend to be completely oil wet or completely water wet. But this is not always true. Of course, a significant portion of the reservoir may contain both oil and water in close proximity in the pore space. In that case, the solids can be partially oil wet and partially water wet. Very fine solids of this nature are attracted to the oil / water interface. Significant study has been devoted to developing flooding chemistries that desorb the oil from the rock surface when the flood sweeps through the reservoir. Low salinity water flooding has been developed with this mechanism in mind.

Whether sand in the wellhead fluid is oil-wet, water-wet, or mixed (both oil and water wet), will have a significant effect on the way it behaves throughout the fluid production system and can have a significant effect on its settling behavior. It can impact the ease of removal from vessels, and the effort needed to clean it to acceptably low levels of oil prior to overboard discharge. Oil wet solids are more difficult to remove from separators and obviously more difficult to clean. Solids from oil reservoirs will usually be coated in oil [9]. Examples from the Gulfaks and other Norwegian fields are described though the emphasis is on solids in produced water.

In studies of sand jetting, samples from the primary separators on Hutton, Merchison and Ninian and from test wells on Alba (all independent UK offshore platforms) were definitely well wetted by oil and for conservative design of jetting systems, these were assumed to have been oil-wetted before the separators. Sands from separators on Forties (BP), at least in earlier years, and on Magnus(BP) appear to have been sufficiently oil free for discharge to sea with no or little pre-cleaning, implying, that they were water wet at the wellhead.

Fundamental studies on wetting states of reservoir rocks have primarily been concerned with the wetting of the rigid porous rock rather than released individual grain. Until relatively recently, there was the general assumption that all petroleum reservoirs were originally water-wet and that oil migration only occurred later [8]. However, it has since been shown that strongly water-wet reservoirs are the exception rather than the rule [8]. Intermediate states occur, fractional wettability due to variations in mineralogy and mixed wettability due to oil entering larger pores and water entering the smaller pores.

Wetting of Various Minerals: Regarding mineralogy, silica is more likely to be water-wet whereas kaolinite clay and siderite are more often oil-wet. Given equal exposure, the wetting fluid is that for which the molecular forces of attraction between solid and fluid are greatest. On Brent, and intermediate wet sandstone, the kaolinite and feldspar were oil-wet whereas the quartz (silica) was water-wet [10]. In the processing of oil sands [11] some inorganic mineral surfaces are oil-wet rather than the preferred water-wet condition due to a tightly bound surface layer of organic humic matter which is insoluble in toluene.

Crude Oil Influence on Wetting: The type of wetting is also influenced by the composition of the crude oil, pH, and brine composition and by additives such as surfactants. High asphaltene in crude favors oil-wetting. Base numbers, acid numbers and resin fraction can also influence wetting. At low pH, positive charges in the oil are attracted to negative silica, thus, promoting oil-wetting. At higher pH the oil becomes negatively charged and is repelled giving a water-wet surface. Hence, base and acid contents are useful indicators of the wetting properties of an oil.

Production Chemicals Influence on Wetting: It is possible that the wetting state can become altered as solids make their way through the production system and are separated from oil and water and recycled through the reject system. For water-wet sand, clay or shale, cationic surfactants which carry positive charge promote transition to the unwanted oil-wet state. Such surfactants may be present in drilling mud filtrates, workover and well simulation fluids. Corrosion inhibitors and bactericides are often cationic surfactants as are some emulsion breakers. Oil base muds containing blown asphalt and crudes containing high percentages of asphaltenes also promote oil-wetting. Low API gravity crudes usually contain high percentages of asphaltenes.

For dolomitic and limestone formations anionic surfactants promote the oil-wet state up to a pH of 8 but the water-wet state at pH ≥9.5. Conversely, cationic surfactants promote the oil-wet state at those higher pH values and water-wetting at pH up to 8. A strong water-wetting surfactant may convert some oil-wetting surfaces to water-wet surfaces. However, if oil wetting has been caused by cationic surfactants the cationic are very difficult to remove. The overall recommendations is to avoid treating sandstone wells with cationic surfactants. A strongly oil-wet rock makes water-flooding less effective and for present purposes would promote production of oi-wet sand.

Whether the sand or other solids from a given well is oil-wet or water-wet, depends on initial mineralogy. Oil and water properties and multiple of operational factors also contribute. The combined feed stream to a separator might therefore contain sand with different degrees of water or oil wetness, and the extent of oil or water-wetness could change with time over the life of the field.

Whether a given sample of sand is oil-wet or is water-wet with an outer layer of oil, could be assessed from the relative ease of reduction of oil content to low levels by agitation in water under some standard conditions. Prediction of the extent of oil-wetness on a new field is more difficult due to the many factors noted above, which can be influences.

Information on mineralogy, oil composition and other characteristics, gathered during reservoir evaluation, should provide some guidance through the potential effects of surfactants; and other additives can complicate the assessment.

Risks Associated with Sand Production: Sand is just one type of solids that are encountered in a hydrocarbon producing facility. It is typically larger than other forms of solids, although this is not always true. There are two main risks in operating sand producing wells. The first and most common is the risk of erosion of valves, pipe lines, pipe bends and process equipment and instrumentation such as flow meters. Erosion is usually associated with high particle velocities and high gas to liquid ratio. Also, if oxygen is present in a stream containing sand, then erosion-corrosion can occur. This is an extremely detrimental situation that can result in loss of asset integrity within weeks. The second risk is that sand deposition in pipes and vessels can lead to under deposit corrosion. A third risk is buildup in pipelines and flow lines such that pigging is no longer possible. These two risks are referred to as Asset Integrity Risks, for obvious reasons. In addition to these risks, sand production causes a number of problems that can force the operator to reduce production rates. Maintaining the quality of export oil and of overboard water are the two main problems that can cause production curtailment.

Inevitable Production of Sand: For many platforms, operation of sand producing wells is the only economic alternative to drilling new wells. In the Gulf of Mexico, where most hydrocarbon reservoirs are composed of unconsolidated sandstone, sand production is common. Unconsolidated sandstone is composed of sand particles that are only loosely cemented together. In some cases, when a core sample of unconsolidated sandstone is placed on a benchtop, it will crumble and form a mound of sand. Obviously, such material has a tendency to get pulled along with the fluids being produced from the reservoir.

Often sand production occurs later in the field life when water cut has begun to increase and when overall hydrocarbon flow rates and pressures have begun to decline. Higher water cut implies that greater fluid volumes must be handled in order to maintain hydrocarbon production. This translates into lower residence times for oil/ water/ sand separation throughout the facility. Since higher flow rates also implies greater risk of erosion, production may have to be curtailed to reduce flow velocities and prevent erosion. In addition, lower reservoir pressure implies less driving force for cyclonic separation. In any case, an assessment must be made of the bottlenecks and locations where sand may accumulate within a facility.

Of course, when sand production occurs, risk mitigation, asset integrity and flow assurance measures must be implemented. There is no doubt that sand production adds a higher dimension of risk. Nevertheless, far too few production facilities are designed to handle the inevitable production of sand. It is understandable that sand management equipment is not installed in early field life. But catering for later addition of equipment is simply a matter of providing tie-in points, sample ports, and laydown areas. This lack of acknowledgement that sand production will occur is a real problem since it adds even more risk and cost to an already challenging situation.

18.1.2 Iron Oxides

Sources of Oxygen Contamination: Produced water does not normally contain any dissolved oxygen. In fact, below a certain depth, which varies from location to location but is generally quite shallow, most underground water has zero oxygen content. Thus, the dissolved and suspended components in produced water are typically in an electrochemically reduced state. Once produced water is exposed to oxygen its nature changes dramatically and for the worse in several regards. This fact cannot be overstated.

Introduction of oxygen is extremely detrimental because it (1) oxidizes the iron in solution in the water, creating iron oxide solids and (2) creates very aggressive, damaging corrosion in the form of pitting caused by the formation anodic sites on internal metal surfaces. Iron in the fluids, when oxidized, will precipitate and contribute strongly to the stabilization of the emulsions.

A few common sources of oxygenated water entering into a separation system are by way of a seawater sump system, holding tanks for off-spec produced water, or an open drain system. Such water systems should always be segregated from the production streams. Also, rigorous biological control should be practiced in such systems.

Another common source of oxygenation of produced water is by way of an open API separator system. Such systems are very cost effective and have been used successfully in the industry. However, they are open to the atmosphere and as such must have a means of mechanically collecting the scum that will form on the surface and the sludge that will form on the bottom of the separator. These accumulations will contain solids and sometimes sand and dirt that is blow into the separator, and a separate waste treatment facility must be employed for ultimate treating. Most importantly, any of the reject streams from an API separator (scum, sludge and oily reject) should absolutely never be routed back into any upstream section of the oil/water separation train or water treating system.

18.1.3 Iron Sulfide

Iron sulfide is one of the most problematic solids in produced water treating. Iron sulfide, like iron oxide is very insoluble, at typical produced water pH values. Again, like iron oxide, it precipitates as very fine solid particles having a diameter in the sub-micron range. It has an affinity for oil, particularly the resin and asphaltene fraction of oil, as well as certain production chemicals such as corrosion inhibitor. These affinities often result in the particles becoming partially oil wet and partially water wet. Their small size and wetting state makes them surface active and one of the more aggressive stabilizers of emulsions.

Unique Properties of Iron Solids: It is generally true that when a solution changes conditions rapidly, such that a relatively insoluble species reaches supersaturation rapidly, precipitation of the species will form many small particles rather than fewer large particles. This is particularly true of the iron compounds. Solids formed from iron tend to have size distributions that average less than a micron in diameter, and their numbers are enormous.

When the presence of iron solids are suspected, it is particularly important to practice good sampling technique. If the produced fluids have relatively high CO_2 content (> 1 mole %), then loss of CO_2 upon sampling must be avoided. Loss of CO_2 will raise the pH and cause an initially clear sample to turn black due to precipitation of iron sulfide at the higher pH. In that case, iron sulfide may be stable at the lower pH of the in-situ fluid. Further, if oxygen is allowed to enter the sample then precipitates of iron oxides will form, again where they may not have been present in the process fluids under process conditions. Release of CO2 or intrusion of oxygen will also shift the dissolved CO2/bicarbonate equilibria which will affect subsequent analysis of carbonate stability.

In addition, solids formed from iron are easily oil wet thus making them amphoteric (oil and water loving) and therefore driving them to the oil/water interface. Further, various surface active compounds (acids, asphaltenes, corrosion inhibitors) also tend to bind to the surface of iron solids thus making them attract oil to an even greater degree. These properties of iron solids have been exploited in the cleanup of produced water systems through the use of carbamate chemicals which are described further below. However, from what has been said thus far regarding iron solids, one can imagine that the use of carbamates must be undertaken with care. Given their small size, large numbers, and wetting properties, solids formed from iron are extremely effective at stabilizing oil in water emulsions.

Schmoo: As is typical of the E&P industry, problematic systems are often given descriptive if not colorful names. In the case of solids formed from iron, the designated name is "Schmoo." In gas fields, Schmoo causes severe operating problems in compressors, pipelines, stabilizer columns, gathering systems, and storage facilities. According to the Gas Machinery Research Council, Schmoo is the least understood and most prominent contamination problem in pipelines and gas compression equipment. There have been many reports on the composition of Schmoo, and its precise definition is the source of ongoing debate. The composition varies considerably depending on water composition and the use of various production chemicals such as corrosion inhibitors. For our purposes, we classify Schmoo as an oily solid conglomerate formed fundamentally from iron solids such as iron sulfide and to lesser extent iron oxides.

Pyrophoric Iron Sulfide: Some iron sulfides can spontaneously combust. This is not the case with all iron sulfides. It is true of iron sulfide that is typically generated in a gas well where iron sulfide and condensate combine without the presence of oxygen. If that iron sulfide is then exposed to air, the conversion of iron sulfide to iron oxides can provide enough exothermic reaction energy to combust the condensate. This reaction does not generally occur with iron sulfide that is water-wet. Generally the presence of water suppresses the combustion of the condensate. However, in the oilfield it is typical practice to assume that any iron sulfide generated in the facilities has the potential to combust. This is particularly true of pig trap debris. Safe handling methods are implemented which generally include preventing the iron sulfide from exposure to air and keeping the iron sulfide water wet.

18.1.4 Inorganic Scale Forming Minerals

Suspended solids in produced water span the range from the very small, < 1 micron, to relatively large, several hundred microns. Because of this size range and the corresponding variation in the surface properties of these solids, their impact on water treatment, emulsions, oil/water interfaces is complex and requires careful consideration before a water treatment technology is selected.

Oil Wet Solids: The properties of oil-wet solids are different from those of oil droplets or water-wet solid particles. The density of an oil-wet particle depends on the mass ratio of oil to solids in each particle, which can vary significantly. The combination of a light component (oil), plus a heavy component (solid) will result in a specific gravity decrease for the solid particle. When this occurs, separators and hydrocyclones will be less effective because there is less density difference to drive the relative movement of the particle from the water. Small neutrally buoyant particles can stay suspended for months in a desk top sample.

One of the most common wetting mechanisms is that of asphaltene and wax adsorption onto sand, clay, carbonate, iron oxides, iron sulfides, and mineral scale particles. Experiments carried out using whole crude oil, and various crude oil fractions dissolved in aromatic solvents, have shown that asphaltenes can indeed adsorb on produced solids. Before discussing these wetting studies, it is useful to review the broad subject of solids contaminants in production and water treatment systems.

From a water treating standpoint, the presence of oily solids can be devastating. Once an oily solids emulsion forms, it is particularly difficult to separate the components from each other and to separate the oil-wet solid from water, due to the shift in density. The more successful technologies for removing such oil-wet solid include flotation, chemical treatment, and deep bed filtration.

The fluids in some geographical areas are particularly suitable to forming emulsions which are stabilized by small, partially oil-wetted solids. The Organic component of oil-wet solid tends to have a relatively high asphaltene content, from a few to several percent. Resins are also somewhat high in the oil-wet solid, varying in the range of a few percent to over 15 %. Due to their relatively high aromaticity and associated resin content the asphaltenes tend to be marginally stable. While this means that while asphaltene precipitation may not be a problem in the reservoir, near well bore or tubing locations, asphaltenes can and do nevertheless precipitate and contribute to the stability of both water in oil and oil in water emulsions. This, together with the relatively short residence times found in offshore and other remote facilities can lead to a relatively challenging situation regarding oil and water separation.

Chemical treatment of oil-wet solid has been effective in many cases but chemicals for this purpose must be applied with care. The injection of an acid, for example, is intended to displace the oil from solid particle surfaces and return the solid particles to a water-wet state which then facilitates gravity separation and eventual discharge. However, acidic produced water will almost certainly cause corrosion problems. Also, the presence of bicarbonate in the water increases the amount of acid required to reduce the pH of the produced water – which results in both logistic and cost issues for this strategy. The injection of so-called wetting agents can have a positive effect to separate the oil from the solid surface. But this too must be done in a limited manner otherwise emulsions will be stabilized due to the surfactant nature of the wetting agent.

18.1.5 Organic Solids

There are two main types of organic solids. One is formed from wax or asphaltene precipitation. The second is due to chemical over-treatment and recycling of rejects from chemically treated streams. Both types of organic solids tend to be sticky. They have a strong tendency to adhere to other solids in suspension. They also have a strong tendency to stick to vessel and piping walls. As with most solids encountered in a facility, the most effective means of dealing with organic solids is to determine their source (cause) and eliminate it. As discussed below, analytical methods are available for determining if organic solids are present.

18.1.6 Bacterial Contamination

Bacterial contamination becomes a greater problem in the presence of oxygen since aerobic bacteria are between 10 to 1,000 times more active than anaerobic bacteria. Aerobic bacteria grow significantly faster, and multiply significantly more frequently than anaerobic bacteria. Bacterial activity results in corrosion, and generation of sticky biopolymers which contribute to the stability of oil-in-water emulsions and which can contribute to pad formation in separator vessels.

18.1.7 Naturally Occurring Radioactive Material (NORM)

Solids extracted with oil, gas and water from reservoirs may contain naturally occurring radioactive materials (NORM). These can accumulate in vessels or deposits, as scale, on internal surfaces. Deposition of NORM increases if seawater is injected due to formation of barium sulphate and other solids by reactions between dissolved components in sea and formation waters. In the past, such radioactive deposits were termed Low Specific Activity (LSA) materials, based on materials on account of the much lower levels of radioactivity than those for man-made radioactive materials. The UK National Radiological Protection Board published an article describing where and how NORM occurs. Describing its detection, measurement, precautions and procedures for handling and disposal of NORM. This replaces a 1991 publication titled LSA Scale in the Oil Industry. With regards to their radioactivity, most sands can be disposed of into the sea, following appropriate registration and authorization by government requirements. Hard scale if present can have higher levels of radioactivity and may require to be brought ashore to licensed sites.

Produced sands may also be coated or mixed with barium sulphate and other scale compounds. These scale compounds can contain traces of naturally occurring radioactive elements, sufficient to require special precautions in handling and disposal. Solids from limestone reservoirs would have a carbonate composition.

18.2 Sampling and Analysis of Solids

As far as oil in water emulsions are concerned, the most important properties of produced water solids are the size distribution, the overall quantity of solids, and the surface wetting characteristics.

18.2.1 Total Suspended Solids (TSS)

The sample point is flushed prior to pulling a sample in a jar or running a sample on-line (two different procedures, both contained herein below). The volume of filtered water and the weight of the collected solids are recorded. A well-mixed sample is filtered through a micro-fiber filter and the residue retained on the filter is dried and then weighed. The TSS in the sample is determined by the difference between the pre-weighed filter disc and the weight of the filter disc plus the filter cake (solids) after filtration has been carried out.

Assemble the apparatus in the figure below. Begin vacuum suction. Place a 0.45 micron Millipore filter in the filter disc holder. Wet the filter with a small volume of clean water to seat it. Stir the sample with a magnetic stirrer at a speed sufficient to shear larger particles, and if practical, to obtain a more uniform (preferably homogeneous) particle size. Carefully pour the sample into the reservoir over the filter, while continuing to stir the sample by hand.

Wash out the sample container with three successive 10-mL volumes of distilled water and pour through the filter, allowing complete drainage between washings. Continue suction for about 3 minutes after filtration is complete.

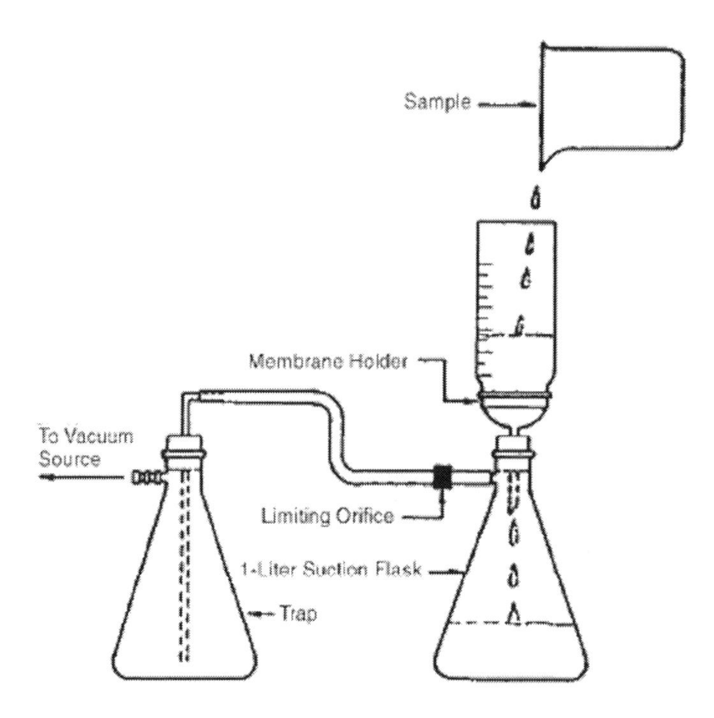

Figure 18.1. Suspended Solids test apparatus for laboratory use.

Often, when taking samples in the field, it will be desirable to get a particle size distribution of the solids. This requires using a series of the blue filter holders. The largest pore filter should be in the first position receiving flow, with successive smaller pore filters in each holder, until the smallest pore filter is used in the last filter holder. Attach the filter assembly to the sample point. To flush the filter cup, ensure the cup is facing upwards and crack open the ball valve A until the filter cup is full of water. Flush for one minute. Install the filter membrane using tweezers to carefully place the filter into the filter cup. Take note of the pore size of the membrane being used. Secure the filter by screwing the inner cup into the outer cup of the filter, ensuring that no bubbles are trapped, and that the two halves are securely sealed. Ensure that the filter is centralized on the holder to prevent water from escaping around the edges of the filter.

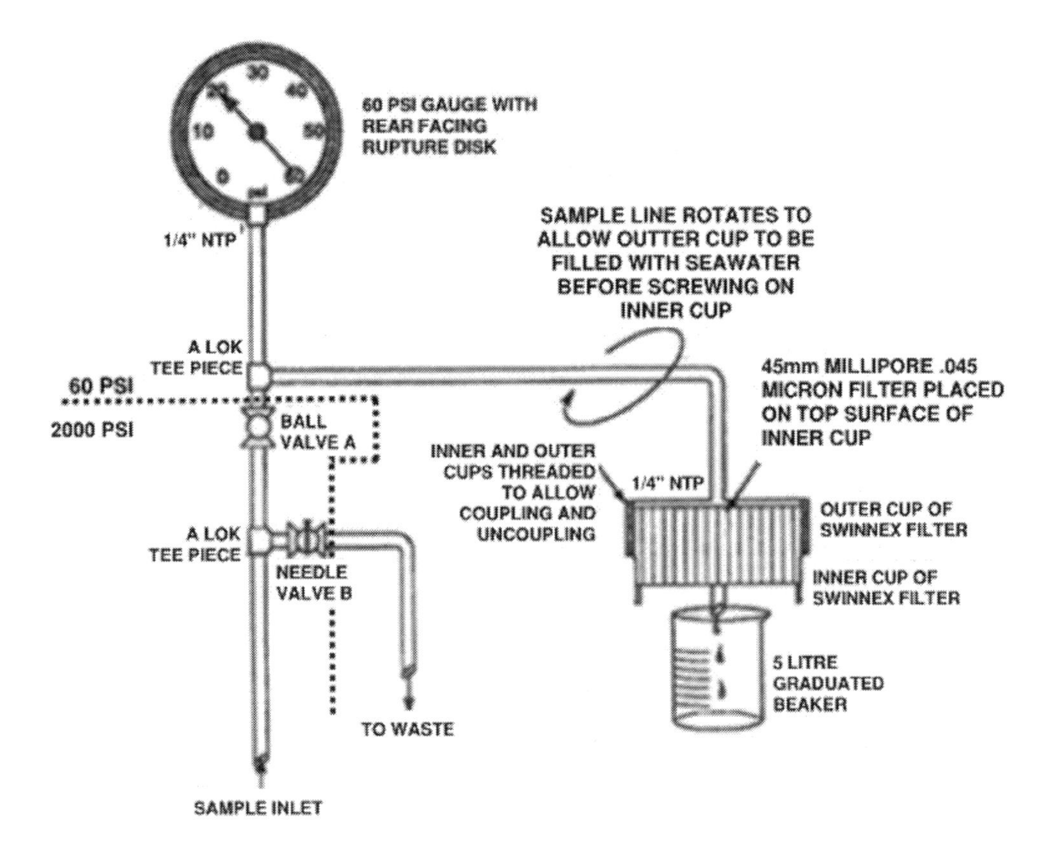

Figure 18.2. On-line Suspended Solids test apparatus for direct connection to process stream.

18.2.2 NACE Chemical Solvent Test

Using the NACE MR-173 – 2005 [12] test, solids were collected on a deepwater platform in the Gulf of Mexico in October 2006. These solids were collected from the produced water discharge of the wet oil tank. This discharge line was fed into a centrifugal pump and the stream was then recycled into the Bulk Oil Treater. A solids size analysis was not carried out but based on settling tests, the solids were very small diameter (less than a few microns) and had oil attached making them roughly neutrally buoyant in the produced water. At the time, the platform had several subsea production systems from significantly different reservoirs and the TDS of the combined produced waters was high (above 200,000 mg/L).

Some of the filter solids were subjected a series of solvent washes as outline in Table 1.xx.xx below. These sequential washes identified the presence of halite salts, oil compounds soluble in an aromatic solvent (waxes or asphaltenes), carbonate solids, iron-based solids, and formation fines. Table 1.xx.xx describes the results of this testing.

Table 18.1 Solids Analysis for solids collected on the Bullwinkle Platform (Oct 2006)

Gravimetric Wash Test	weight % of dry sample	Comments
Deionized water wash	19.7	Includes substances soluble in water such as salts
Xylene wash	9.6	Includes substances soluble in xylene such as paraffin, oil, and organics
Acetic acid wash	27.7	Includes substances soluble in weak acetic acid such as carbonate mineral scales
Hydrochloric acid wash	27.4	Includes substances soluble in 15 % HCl acid such as iron sulfide, and iron oxide.
Acid Insolubles	15.6	Includes substances insoluble in 15 % HCl acid such as sulfate scale, sand, silica fines
Total	**100**	

Another portion of the filtered solids were washed using deionized water followed by xylene and then subjected to inorganic solids composition analysis using EDAX, XRF and XRD. The XRD results were:

Positive for calcium carbonate scale

Positive for barium sulfate scale

Positive for silica fines

Positive for iron compounds

As given in the table above, we can conclude that roughly 70 wt % of the sample is composed of inorganic solids. Another portion of the sample was washed with deionized water, followed by xylene, followed by HCl. This leaves the Acid Insolubles fraction and this sample was analyzed using EDAX and XRD which gave a positive indication of sulfate scale and silica fines. The acid wash did not evolve noticeable levels of H_2S which rules out the presence of iron sulfide compounds. Optical microscopy verified the presence of cubic crystals which are typical of halide precipitates.

Organic solids analysis was performed using DSC/TGA, GC, and H NMR. The results are positive for both waxes and asphaltenes.

Results such as these must be interpreted by a qualified chemist since they involve a combination of qualitative and quantitative techniques. However, the final result in this case turned out to be close to that given in the above table of gravimetric wash results.

Table 18.2 Final results of solids sample analysis

Substance	Weight %
Organic material	10
Halide salt from high salinity	20
Calcium carbonate precipitate	28
Iron oxide solids	27
Barium sulfate precipitate and silica fines	15
Total	**100**

The conclusions from these analyses and observations were:

- mineral solids were being formed from a combination of processes (incompatibility leading to barium sulfate precipitation, pressure drop leading to carbonate precipitation);
- the solids were a oil-wet solid of inorganic precipitates and oil making them roughly neutrally buoyant in the high salinity produced water;
- high iron oxide solids content indicated corrosion processes which was verified by the frequent requirement to repair and replace section of the produced water piping;
- high solids content was a major contributing factor in poor water quality, and was a complicating factor in implementing clean-up technologies which suffer performance degradation in the presence of solids.

Over a period of several months, the problems were solved by a combination of chemical treatment, better corrosion prevention, segregation of fluids, and the re-routing of recycle streams.

18.2.3 Particle Size Distribution (PSD)

Particle Size Distribution is in many cases really particle diameter distribution. The word size is used because in some cases the particle is not spherical and the measured quantity is some type of average of the major axes. The measurement of particle size was discussed already in Chapter xx. That discussion will not be repeated here. Instead various field results will be presented.

A wide variety of sand particle size distributions are given in the figure below. There is significant variation from one geographic location to another. Variations within a given geographical area can also be considerable as well. Nevertheless, there is a clustering of curves which can be used as an initial starting point for process design.

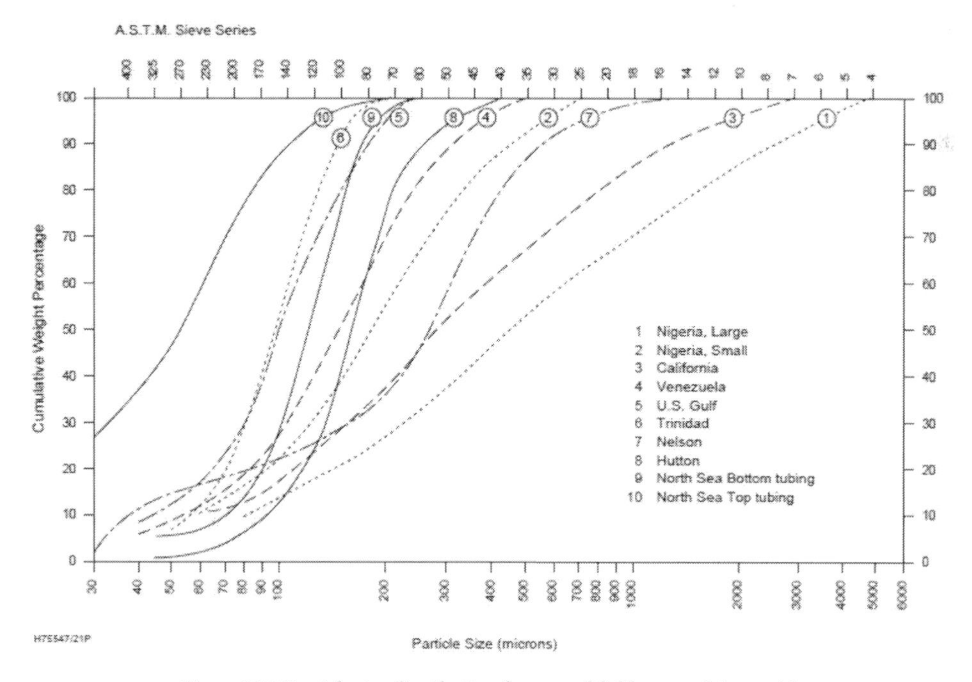

Figure 18.3 Particle size distribution for several fields around the world.

Some distributions have a high proportion of fine particles. Most reservoir solids finer than about 15 micron tend to be composed of clay rather than silica (sand) particles. It is also generally true that normal production of small quantities of sand tends to be composed of fine particles that get swept into the producing fluids. When a significant region of the formation around a well bore fails, the particles tend to have a broader distribution including larger particles that are not normally swept in the produced fluids. This highlights the difference between transient sand production and catastrophic sand failure mechanisms. Very fine particles can promote stabilization of emulsions and hence make oil / water separation more difficult. The larger grains of sand accumulate in separators causing a number of production problems.

The size of scale particles depends on how and where the scale is formed. Formation within the bulk liquid usually results in fine particles in the micron or even submicron range initially. Particles resulting from release of surface deposits could have a wide variety of sizes and shapes depending on detailed equipment configuration and the potential for further breakage during subsequent passage through pipework and equipment. For example, scale particles in samples of solids from bulk storage tanks at Forcados Terminal (Shell Nigeria) ranged in size up to about 15mm major dimension and 2-4 mm thickness. Carbonate scale particles in primary separators on the El Fateh field (Dubai Pet. Co.) were more granular in shape, mainly sized around 1mm or more. Proppants, if back produced, will by design, have narrower size distributions than produced sands, except for fragments formed by crushing. Commonly used sieve size bands for proppants range from 6/12 (3,350 to 1,700 μm) to 70/140 (212 to 106 μm).

As with the concentration, obtaining an accurate PSD is not easy. Representative sampling can be difficult and the sand production rate varies with time. Taking measurements from separators is not recommended, as the smaller particles might not deposit, giving an unrepresentative PSD. One recommended method which is not used much is to take a sample from the core, as available. It is important to take different samples vertically, as the grain size will vary with depth. Solids may also be produced if there is failure of a gravel pack. Its size can range from 500μm to greater than 1000 μm, depending on that used for the particular well activities. The subject is discussed in [9, 13].

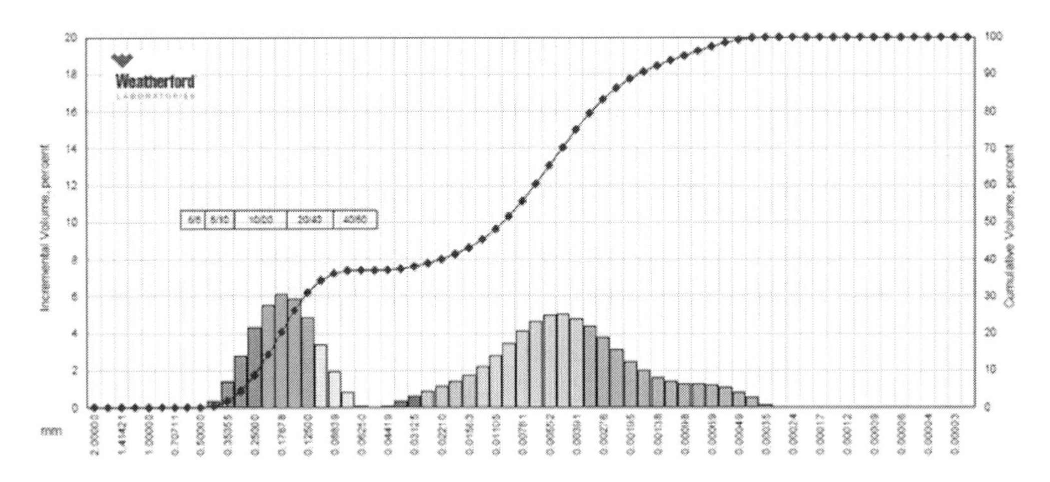

Figure 18.4 Particle size distribution for a particular field. Two distributions are shown. The distribution associated with larger particle sizes is composed of sand. The smaller size distribution is composed of asphaltene particles. Often bimodal distributions have significant overlap making the existence of sand and other particles less obvious.

Chemical treatment of conglomerate has been effective in many cases but chemicals for this purpose must be applied with care. The injection of acid is intended to displace the oil from the solid particle surface and thus return the solid particles to a water-wet state which then facilitates gravity separation and eventual discharge. However, acid will almost certainly cause corrosion problems.

The fluids in the deepwater GoM are particularly suitable to forming these conglomerate stabilized emulsions. They have relatively high asphaltene content from a few to several percent. Resins are also somewhat high varying in the range of a few percent to 15 % resin. Due to relatively high aromaticity and resin content the asphaltenes tend to be stable to marginally stable, with none of the fluids in the unstable region. While this means that asphaltene precipitation is not a problem in the reservoir, near well bore or tubing locations, asphaltenes do nevertheless precipitate to some extent in the topsides and contribute to the stability of both water in oil and oil in water emulsions. They also stick to solids particles causing solids stabilized emulsions. This, together with the relatively short residence times found in the deepwater, leads to a relatively challenging situation regarding oil and water separation. As already discussed, sophisticated electrostatic treater design and well managed chemical programs are required to separate drops of water from oil that are stabilized by asphaltene particles. Thus, it is not surprising that horizontal floatation would be of such great use.

As a specific example of processes that can lead to water treating problems due to solids, we consider an example from Shell Brazil. When a reservoir is water flooded with any water other than source water, there is the potential for scale formation as the two water chemistries intermingle. In a typical seawater breakthrough scenario, barium sulfate particles will form in the produced fluids. The particles can be very small (<1 micron to 2 micron), particularly in the presence of a scale inhibitor. As the barium sulfate particles form, the asphaltenes may also be precipitating. As asphaltenes precipitate, they may adsorb onto available surfaces including the barium sulfate particles. Such fine particles, with an asphaltenic coating are excellent water-in-oil emulsifying agents.

Chemical treatment of conglomerate has been effective in many cases but chemicals for this purpose must be applied with care. The injection of acid is intended to displace the oil from the solid particle surface and thus return the solid particles to a water-wet state which then facilitates gravity separation and eventual discharge. However, acid will almost certainly cause corrosion problems.

From a water treating standpoint, the production of fine solids can be difficult to deal with. Fine solids can be produced by either sand particles or by the precipitation of scale forming minerals such as ordinary salt, sulfate scales, and carbonate scales. In the best case the fine particles remain water wet and are discharged in the overboard water stream. Often though, a fraction of the particles will become oil wet by attachment of sticky crude oil components such as asphaltenes, waxes, naphthenates and naphthenic acids. As described elsewhere, fine oil-wet particles can stabilize emulsions. Such emulsions are then difficult to resolve because the presence of solids increases the density. A fraction of these solids stabilized emulsions will be neutrally buoyant in water thus making then impossible to be separated by gravity settling equipment and flotation units which rely on density differences to achieve separation.

The fluids in the deepwater GoM are particularly prone to forming these oil-wet solids stabilized emulsions. They have relatively high asphaltene content from a few to several or even 10 percent. Resins are also somewhat high varying in the range of a few percent to 15 % resin. Due to relatively high aromaticity and resin content the asphaltenes tend to be stable to marginally stable, with none of the fluids in the unstable region. While this means that asphaltene precipitation is not a problem in the reservoir, near well bore or tubing locations, asphaltenes do nevertheless precipitate to some extent in the topsides and contribute to the stability of both water in oil and oil in water emulsions. They also stick to solids particles causing solids stabilized emulsions. This, together with the relatively short residence times found in the deepwater, leads to a relatively challenging situation regarding oil and water separation. As already discussed, sophisticated electrostatic treater design and well managed chemical programs are required to separate drops of water from oil that are stabilized by asphaltene particles. Thus, it is not surprising that horizontal flotation would be of such great use.

Flotation is somewhat effective in separating neutrally buoyant solids stabilized emulsions. The gas bubbles stick to the solids themselves and to the oil-wet surface of the solids and carry them to the top of the water where they are floated over the spillover weir. This is an effective means to perform the separation of oil and water. However, it leads to problems downstream when the oil-wet solids are recycled through the system. This is described in greater detail below in the discussion of the Bullwinkle platform.

Further, should production in the GoM change over time, the use of horizontal IGFU may need to be challenged and revised as well since large vessels or multiple vessels may be required

There are many design details that affect the performance of a flotation system. However, there are two fundamental parameters that must be compatible:

- the flow rate of oil and the oil droplet size distribution,
- the flow rate of gas and the gas bubble size distribution.

The oil droplet size distribution is usually expressed as the percentage distribution of drops as a function of drop diameter. The distribution is based on either the volume-number or diameter-number. Either is acceptable, but whichever is chosen must be specified. The sum total of this distribution is by definition 100 %. Therefore, the volumetric flow rate of oil must be known (or oil concentration together with water flow rate). All of this is fairly well understood and accepted in the oil industry.

What is generally not so well understood is the importance of specifying similar characteristics for the gas. In general, the performance of a gas flotation unit will depend to a great extent on the characteristics of the gas bubbles, their size distribution and the volumetric flow rate of gas compared to that of water. This is critical to the proper selection of gas flotation equipment.

While an IGF will typically deliver a greater volume of gas than a DGF, there are many factors which affect the volume of gas delivered in an IGF. They are rotor diameter, rotor length, rotor design, rotor speed, and rotor submergence (depth of water above the top of the rotor at its non-energized state). The configuration of the rotor makes it difficult at best to measure volume of fluids pumped by a specific rotor, so manufacturers of submerged rotor designed IGF units have achieved optimum rotor operation designs based on years of experience.

18.3 Solids Management

As discussed in the introduction to this chapter, solids separation, handling and management are three distinct activities. Solids management is the more general activity which encompasses both solids separation and solids handling. It encompasses the entire lifecycle of dealing with solids starting with separation, and including handling, transportation, treatment, regulatory testing and analysis, application for a disposal permit and the final step of actually disposing of the solids. It also must include asset integrity issues related to valves, chokes, pumps and other equipment which may need to be upgraded in order to withstand the erosive nature of fluids containing solids. Simply stated, solids separation and handling are important but they are only the beginning of the solids management journey.

Good management of solids starts with the simple recognition that the production of solids is inevitable in the production of oil and gas yet few facilities are designed for it. Sand production is not unusual. Yet many production systems are designed without any solids separation, handling or management systems whatsoever. This fact makes it seem within the industry that solids production is some kind of mistake. Granted, high levels of solid production are hazardous (due to erosion), and can be precursors or warnings of an imminent production well failure. In order to prevent this occurrence, there should be a response plan already in place since rapid response may be required in order to prevent well failure of flow line plugging. But some quantity of solids production is truly inevitable and a management system should be put in place at the design stage or shortly thereafter in order to reduce solids management costs later. As the facility life continues, the solids management plan must also become more robust and detailed.

The most important variables in solids handling and management are source and volume (or mass) production rate. It is essential to know how much solids, and what type of solids are being produced in order to consider which separation and handling practices are economical. Low production rates allow a greater variety of handling options. High rates demand automated technical solutions. In order to find the right and best economical option, the engineer must know how much solids there are. It is also valuable to know where the solids are coming from so that options for reducing their production rate can be considered. For this purpose, solids composition and particle size distribution are useful.

It is important to do the math. Questions that need to be answered include:

- what composition
- what fluid flow rate
- what concentration
- what specific gravity
- what is the weight and volume of solids that must be handled and disposed

For many operators, the amount of sand or other solids they need to handle can be a unwelcome surprise. Answering these questions should be the first step. Often facilities engineers focus on how to separate the sand from the production fluids. Sand separation is important but not if there is no economic way to handle, transport and dispose of the sand. Looking at the bigger picture should be the first step.

What are the practical limits of putting solids on a boat. Look at the economics of the different options. What is the viability of the different options. A lot of options are not viable. Sand washing for example was expensive. So it had to be hauled away. Cleaning to reduce the liquid by-product. Put in drums and haul it away. Very manual.

Cost of other options is now lower and better known. It has to go somewhere. It will not disappear. It has to be put into a container. How can this be done. What size container. How strong does the container need to be. How much will it hold. How will the solids be transferred from the tank to the container. Are pulleys, levers, or other lifting and transfer machines available? How heavy is the sand, how much space will it occupy.

Offshore there are even fewer options than onshore. Can be a costly logistical issue. Can be a big task. Typically there is not enough staff offshore to handle the manual workload. The office staff are typically not experienced in dealing with solids and sand. Few companies have specialists with critical knowledge. Like produced water, solids are considered a nuisance. Most operating companies do not plan for it. It is not seen as a normal way of doing business.

Criteria that needs to be evaluated such as:

- Range of flow rates for each phase (oil/water/gas)
- Particle size distribution
- Total suspended solids
- Oil content of the solids
- Pressure losses
- Replacement parts
- Liquid loss (leachate)
- Space requirements
- Uptime / downtime
- Maintenance cost
- Initial CAPEX seems to be the focus
- But there is also ongoing CAPEX and OPEX

Equipment Tailored for Sand Production: Various valves and chokes have been designed with erosion resistant materials of construction. Also, various pieces of equipment have come on the market that can be used to separate sand from fluids, can remove sand from vessels without a shutdown, and that can be used to transport sand to locations where it can be disposed. They key in selecting this equipment is to ensure that the selected items match the equipment needs of the facility. This can only be done by competent fluid characterization.

18.4 Solids Separation from Fluids

In this section, various technologies for the removal of sand form multiphase fluids and produced water are discussed. The technologies include flotation, settling, and cyclonic Desander.

The preferred location for facilities sand removal is before the choke with a wellhead desander [4]; however, in cases where this technology is not currently used, the produced solids will collect in the primary separating vessel. Produced solids may erode chokes and flowlines, but the first place of accumulation in large quantities is the production separator [6 - 8]. Solids settle in production vessels and piping when the transport velocity drops below the limiting deposition velocity. Accumulated solids in gravity=-settling vessels result in loss of residence time, corrosion-enhancement zones, increased sediment in oil, increased oil content in produced water, and degraded injectivity of produced water [3]. Sand and solids are removed from production separators either off line (shut down for physical removal) or on line by use of jetting system.

18.4.1 Flotation Solids Removal

Flotation can be quite effective in separating neutrally buoyant, solids-stabilized reverse emulsions from produced water, through the attachment of bubbles which then create the density difference needed for separation. The gas bubbles stick to the solids themselves and to the oil-wet surfaces of the solids, carrying them to the surface of the water where they are floated over a spillover weir. It should be noted that while flotation can be highly efficient, it's application is generally limited to relatively low levels of oil-wet solid (e.g., less than 100 to 300 ppm). As a caveat, it should be noted that the recycle of contaminates collected as skim fluids from a flotation cell can be problematic and lead to a degradation of discharged produced water quality. Flocculating agents are also used with success to treat solids. When a floc forms, the larger diameter of the floc magnifies the effect of any density difference with water and facilitates separation by Stokes law.

However, flocculating agents must be applied with care. Flocculating agents must always be applied at a location where minimal shearing will occur downstream. Shearing does not improve mixing of the chemical. Instead it will break any floc that has formed and irreversibly render the chemical useless. If the system is over-treated with a flocculating agent, a sticky polymer floc can form from the combination of oil, solids and polymer. This combination becomes particularly troublesome in recycle streams. When this floc is rejected back into the process stream, it can accumulate more solids and oil. This can result in more emulsion, thicker pads in the separator vessels, and more viscous pads which in turn can lead to fouling of vessel internals, including plugging of inlet headers, collection headers, distribution screens, treater grids, and level detectors.

Solids Handling Experience with Horizontal IGF: Production of large grains of sand is minimal across the Shell deepwater. However, this is not because the reservoirs are themselves consolidated. Indeed, the geology consists of unconsolidated sandstone. Even without applied stress, a typical core sample will fall apart and form a heap of sand when taken out of its container. This fact was recognized very early in the development of deepwater and the well completions in the deepwater reflect a concerted and successful effort to control sand. Typical well completions involve frac and pack, 40/20 gravel pack and relatively tight sand screens. Further, sand monitoring programs are in place and well control is exercised to prevent sand failure. Nevertheless, the economics of the deepwater environment require high rate wells. These high rates tend to pull very fine grains of sand (sub-micron to roughly a few microns) through the formation and the production of very fine grains of sand and clay are common. These very fine grains of sand can cause significant water treating problems.

18.4.2 Conical Separator Solids Removal

Desanding Vessel: A Desanding Vessel is a separator that is tailored for the separation of sand from a multiphase fluid (oil, water, gas, sand). The two most important design features of a Desanding Vessel are residence time and the design of the conical bottom discharge. The required residence time can be determined based on particle size and hindered settling rates. Simulation tools are available for determining this parameter. Bottom discharge of sand is critical to the proper functioning of these separators in order to avoid the need for periodic manual cleanout.

The Desanding Vessel should be a vertical design. This is required for a number of reasons. First, it is critical that essentially all of the deposited sand be collected and discharged on a regular basis. Horizontal separators, with cylindrical bottoms, present a difficult geometry to ensure complete collection and discharge. The sand can collect in ridges and along the cylindrical wall such that sand jetting devices cannot reach all of the deposited sand. This is the main reason why some sand jetting devices do not work well. If there is a pocket of deposited sand that is not removed, the sand will eventually form a cement. Production platforms typically have small vibrations throughout the facility. When these vibrations are transmitted to a separator vessel containing sand, the sand particles move around each other and eventually form a low porosity, highly compacted sand bed. With further time, mineral scale may deposit and bacteria begin to multiply within the sand bed. These processes literally result in a cemented material that can only be removed by manual cleanout. The process of sand cementing can be avoided altogether by using a steep conical bottom geometry where all of the sand can be removed by application of jetting nozzles on a regular basis.

The second reason why a vertical and not horizontal separator should be used is to minimize deck space. Vertical vessels provide the minimum deck space for a given residence time. The height of the vessel is maximized to provide residence time at minimum cross-sectional area. A vertical configuration has the drawback that an oil/water pad can form and not be resolved effectively. Most pads require some degree of cross-sectional area for resolution. However, it is not the purpose of the Desanding Separator to resolve an oil/water pad since such a pad is more effectively resolved after gas and sand are removed. An advantage of the vertical design is that it will have a relatively narrow diameter which will result in a steep angle for the conical bottom. This is particularly important. An oil bucket based on an overflow weir can be placed along the circumference near the top of the vessel to capture oil for discharge. Degassing can be carried out using either a separate upstream degassing vessel or can be carried out in the Desanding Vessel.

In designing the jetting system, one of the primary considerations is to ensure that all of the sand can be removed periodically. Sand that is left behind will eventually form a cemented sand bed, as discussed above. One of the more effective designs is to use a ring jetting manifold with nozzles that point tangentially to the wall and slightly downward. Two or three rows of such nozzles can be used. The profile of the jet can be either flat (between 40 and 80 degree spread), or cylindrical (less than a 40 degree spread). It is recommended that a cylindrical jet be placed near the bottom of the conical section. It is recommended that the upper ring of nozzles have a flat profile. The placement of nozzles (number of nozzles in each ring) should be such that the jet of each nozzle strikes the next nozzle in the ring. This will ensure that sand does not accumulate behind the nozzles. In operation, the bottom ring of nozzles would be activated first, followed by the next upward ring, and so on. This ensures the complete removal of sand from the collection zone of the vessel. It also minimizes the volume of flush water. Experience has demonstrated that frequent washing with less water is preferable to less frequent washing using greater wash volumes. The former mode prevents the formation of a cemented sand bed.

Wash water composition / source is an important consideration. Characteristics such as salinity, pH, temperature, and oil content are mostly irrelevant, within reason. The most important characteristic is that the water is free of oxygen. This is imperative. When oxygen is present, erosion-corrosion can occur which is extremely detrimental to asset integrity. Corrosion rates in the range of hundreds of millimeters/year and higher have been known to occur when a slurry of solid particles is present together with oxygen. The sand slurry that is generated during discharge may be relatively clean and free of oil. When this is the case, then relatively clean water should be used for jetting. When the sand is associated with oil, then there is no point to use clean water. The water should be relatively free of solids particles in order to prevent nozzle plugging. In some Desanding Vessel designs, the flushing water is combined with platform gas in order to provide a powerful scouring mechanism. This results in a more effective flushing process, with less flush water required. However it involves a more complex design which must be weighed against the cost and schedule of project. The concentrated sand slurry that is generated by vessel jetting should be routed separately from the produced water treatment process.

As discussed, there are a number of parameters that be selected in order to design a functioning Desanding Separator. CETCO has design models, field experience from decades of operation within a super-major operating company, field experience of solids handling from a major consulting company that serves the minerals / mining industry, field experience from within the parent company Minerals Technologies Inc., and access to sand handling specialists who are currently employed in super-majors or recently retired.

18.4.3 Cyclonic Desander

Sand Cyclone: A sand cyclone is illustrated in the figure below. Multiphase fluid enters through one or more tangential nozzles (feed inlet). The tangential nozzle and the cylindrical entrance chamber ensures that the linear momentum of the incoming fluid is converted to rotational momentum. The vortex finder (small cylindrical tube at the overflow outlet) acts as a shield between the incoming fluid (sand containing) and the treated effluent fluid for which most of the sand has been separated. Sand particles migrate to the outer edge of the swirling flow and are discharged through the underflow nozzle along with some water which helps to maintain a slurry of solids suspended in water. All other fluids (oil, water, gas) are discharged through the overflow.

In much of the old literature on sand cyclones there is a tendency to refer to the device as a hydrocyclone. This is misleading. In this discussion, the terminology "sand cyclone" will be used for a cylindrical device that removes sand from multiphase (oil, water, gas) fluids. The terminology "hydrocyclone" will be used exclusively for a device that removes oil from water. These two devices have almost nothing in common except for their outward appearance as conical devices. The types of fluids that they handle, the flow and pressure operating envelopes, and they way they work are completely different. These two types of devices should not be confused with one another. It is well known that hydrocyclones are not effective in the Gulf of Mexico while they work very well in the North Sea and other regions. The reasons for this are well understood and have nothing to do with the performance of sand cyclones.

In this book, two types of cyclonic separation devices are discussed. The first is the deoiler hydrocyclone. It separates dispersed oil from water and is referred to as a hydrocyclone. In that device, centrifugal motion is used to drive the lighter oil phase into the center of a tapered cylinder and eventually out of the reject port. Chapter xxx provides an extensive discussion of hydrocyclones. The second type of device is the desander. In that device, centrifugal motion is used to drive suspended solids particles to the outer perimeter (wall) of the device. As shown in the figure, a fluid mixture (gas, solids, water) enters the unit by way of a tangential nozzle. This imparts centrifugal motion to the fluid. Gas and water are discharged through one end of the device. Solids are discharged through the opposite end.

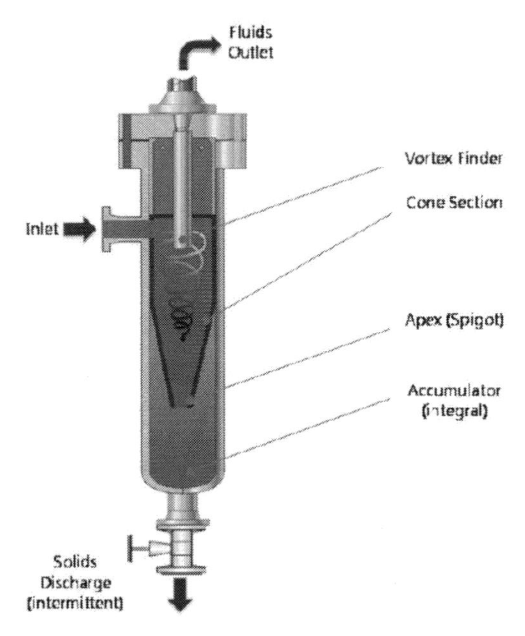

Figure 18.5. Cyclonic desander.

Unlike well-head desanders, produced water desanders are relatively well established technology. Sand and produced water densities and produced water viscosity are well defined, therefore, the separation performance of the desanding hydrocyclone can be predicted whereas the range on uncertainties in reservoir fluids makes this prediction difficult for well-head desanders.

The following limitations should be observed:

- The minimum operating pressure drop for a desander should be 5 psig (35 kPa)
- The practical limit for sand separation from water by a hydrocyclone is 10 microns

The following separation efficiency curves, supplied by Axsia Mozley Ltd, can be used to give an indication of the size of sand particles which will be removed in either a 2 or 3 inch produced water desanding cyclone. These curves should not be used for design purposes.

Shown in the figure is a single Cone device. The desander cyclone is itself a pressure vessel. Multi-Cone devices are available where several desander cyclones are housed within a pressure vessel.

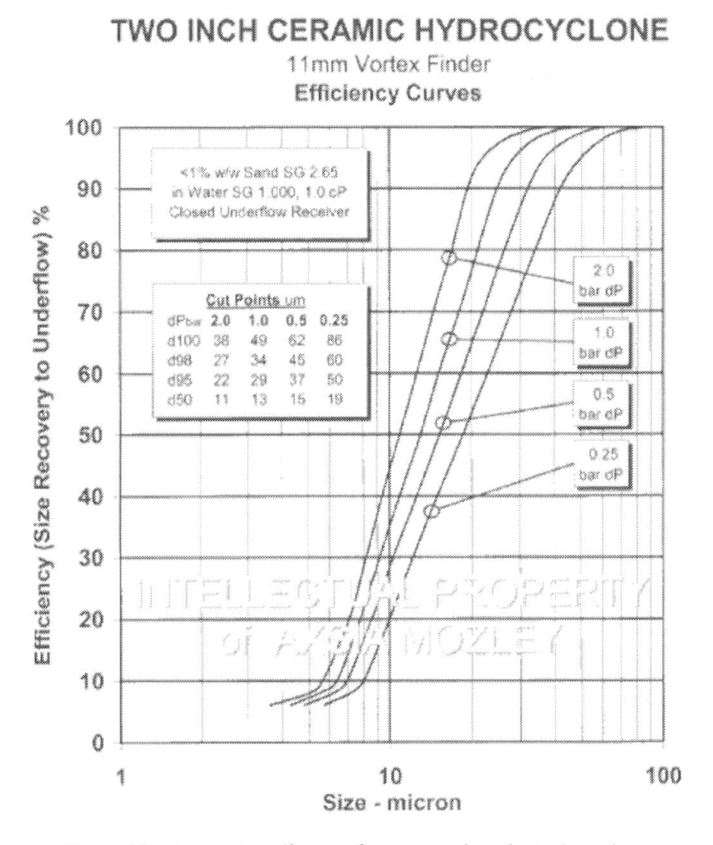

Figure 18.6. Separation efficiency for a two inch cyclonic desander.

18.5 Solids Removal from Separators

Settled solids within a production separator become harder to remove as time passes. Consolidation and binding in effect solidify the sand mass over time. Consolidation is a mechanical effect whereby the sand rearranges to a more-compact mass. Time, gravity, fluid lubrication and vibrations from platform and equipment all contribute to consolidate the sand. Binding occurs when foreign materials bridge between or cement together the sand grains. Asphaltene, wax, mineral precipitates, bitumen, scale, iron sulfide, bacteria, corrosion products and production chemicals can all contribute to binding. The decrease in temperature and pressure in the production separator, compared with tubular and wellhead conditions may also accelerate the binding effect. Consolidated or bound sand requires thermal, mechanical or chemical action to return to a free-flowing state. An efficient jetting-sand-removal system must therefore move or remove the solids while they are still capable of being fluidized.

Offline Sand Removal: Sand and solids removal from production separators can be done either online or offline. As an offline operation, the vessel is typically shutdown and isolated [4, 7]. This method has low capital expenditure but results in loss of production. Water hoses are attached to available nozzles and the slurry of solids is discharged through the vessel drain nozzles. The slurry must be routed to a suitable decanting system and ultimately disposed of. In situations where the solids have built-up or become solidified, personnel may have to enter the vessel and loosen and remove the solids. This is a time consuming task.

Online Sand Removal: It is much preferable to remove solids when the vessel is online. Various separator internal devices can provide online sand removal, which negate the need for equipment isolation and vessel entrance. These devices include spray jets and pans [6], conveyance sprays [14] vortex desanders [15], and eductors [16]. These devices typically fall under the general moniker of 'sand jetting' even though they may not use specific jet-spray devices. Each of these devices should be designed to remove sand accumulated from the water zone with minimal interface at the oil/ water interface.

The most common approach for online separator sand removal is the traditional jet and pan system. Spray water is introduced through an internal piping header aligned axially along the separating vessel. Spray nozzles are spaced along the header to introduce a jet of water that fluidizes and pushes the sand toward the bottom middle of the vessel. A sand pan (inverted V-trough with triangular slots) or sand cap (flat circular plate) direct the sand toward the outlet nozzle and prevent vortex formation.

Some success has been achieved using jetting systems [10]. Traditional jetting systems use spray nozzles to fluidize and push the sand toward one or more discharge nozzles. The intention is to sweep the solids from one end of the vessel to the other end where the discharge nozzle is located. In many cases, the normal water outlet nozzle is used in which case the produced water flow must be temporarily diverted to handle the solids slurry. Alternatively, the objective of cyclonic-jetting is to create a vortex that fluidizes sand in a circular zone near the bottom of the vessel such that the oil/ water interface is not disturbed. With these devices, the extent of solids removal is a function of the height, spray flow rate, and spacing of the vortex units. Single units can be optimized at a height of 10cm (4in), with spray pressures of 0.7 barg (11psig) to provide an area of influence of 1.1m2 (12 ft2) with 28 cm (11in) of sand bed depth. Placing two units in parallel with overlap of their affected zones reduces the 'egg-carton' effect associated with this technology. The egg-carton effect refers to a pattern of cleaning where sand is removed effectively in several circular areas without sand removal between the circles. This effect can be reduced when the circular areas overlap. A sand slurry with up to 60 wt% solids is generated and must be transported from the jetting system to the handling equipment. Slurry transport requires makeup water such that the flow rate is less than the erosion velocity (upper limit) but greater enough to maintain the flow of the slurry (minimum particle-transport velocity).

Many anecdotal papers have been published showing general jet and pan layout, but only Priestman et al. (1996) have published a detailed analysis for this type of system. The reader is referred to their excellent paper for full technical details on jet design and spacing, drain spacing, fluidization factor, and solids-evacuation procedure. Of key note from their recommendations is partitioning of the vessel into discrete wash look zones, with each zone having a length at three times the vessel diameter. The jet and pan internals are designed for a single zone and repeated along the length of the vessel.

18.6 Solids Handling

As discussed in the introduction to this chapter, solids separation, handling and management are distinct activities. The former involves separating solids from oil and water. This is typically done using equipment described above such as cyclones installed in a flow line, and jetting systems installed in vessels. Solids handling involves transportation, treatment, and ending with final disposal. Simply stated, once the solids are separated, solids handling is the task of moving it from point "a" to point "b" and getting rid of it.

Sand Handling and Disposal: As discussed above, dealing with sand can be broken down into two phases: sand separation and sand handling. There are two practical options for sand separation. Whether a Desanding Separator or Cyclonic Desander is selected, both separators will provide a sand slurry of roughly 5 to 10 wt % sand with the remainder water. Sand handling involves the transport of this sand slurry to cuttings boxes, removing the water from the slurry, offloading the cuttings boxes onto a boat for transportation to shore and final disposal.

Onshore disposal is the only option since the NPDES (National Pollutant Discharge Elimination System) permit for the EPA Region 6 states that there shall be no discharge of produced sand into open waters. Produced sand is defined in the permit. The definition includes desander discharge from produced water.

Discharge of a slurry form either the Desanding Vessel or the Cyclonic Desander makes the transportation relatively straightforward. However the slurry does require further concentration before practical disposal can take place. Back-washable filters are an option. The sand from the filters can be discharged directly into cuttings boxes. Loading the slurry into cuttings boxes and removing the decanted water using a vacuum system is another option. In either case, the solids will ultimately be loaded into cuttings boxes (25 bbl or 50 bbl) and transported to shore for disposal. For both options, a calculation of transport slurry volume and cuttings box wet sand volume and weight are required.

Staging: Choose a course filter to remove 80 % of the solids in a compact and inexpensive system. Then remove the rest of the solids in a finer filter. This two-stage approach will help prevent blinding of the filter which often occurs in a one-stage system. Blinding will dramatically reduce the dirt holding capacity of a filter. In the industry, there is too much emphasis on the use of a single stage unit. Staging can actually reduce the cost and weight compared to a single stage system. In seawater filtration, staged filtration is always used.

Filtering on a side stream: concentrated, lower flow rate. Total volume. Technology selected. One inch off of a 24 inch line. So side stream may not be representative. Off WOT or Slops Tank. Taking it off the right point.

18.6.1 Disposal in a Landfill Site

Most oil contaminated sand or solids that have been removed from produced water are disposed in a landfill site. There are various types of landfill sites. Regulations for such sites and disposal costs vary from one state to another. The U.S. Federal regulation is based on Subtitle C of the Resource Conservation and Recovery Act (RCRA). In 1978 the U.S. EPA developed a set of hazardous waste management standard that regulate the disposal of In 1988

U.S. RCRA: In 1976 the U.S. Resource Conservation and Recovery Act (RCRA) was passed. The primary objective of this legislation is to protect human health and the environment from the potential hazards of waste disposal. The legislation was codified in Title 40 of the CFR (Code of Federal Regulations) parts 239 through 282. It gives the EPA the authority to develop regulations for waste disposal. Subtitle C of the act focuses on hazardous solid waste. Subtitle D on non-hazardous solid waste.

In 1988, the US EPA determined that disposal of E&P wastes shall not be controlled under the hazardous waste restrictions of the RCRA (Subtitle C). This determination is often referred to as the Subtitle C exemption. The exempted wastes include produced water, produced sand, drill cuttings, drilling fluids, spent filters, filter media, and backwash, and production brines that are uniquely associated with E&P activities. These wastes are sometimes referred to as Normal Oilfield Waste (NOW) in order to emphasize that they are exempted from Subtitle C of the RCRA. The 1988 determination also exempts residual salts derived from evaporation and demineralization of produced waters.

However state regulations and permitting programs that are more stringent than the RCRA have been adopted. For example, California law does not exempt E&P wastes from its hazardous waste program. State and local radioactivity restrictions for solid waste disposal would apply.

Paint Filter Liquids Test: According to 40 CFR 264.314 and 264.315, solid waste intended for transportation and disposal is required to pass the Paint Filter Liquids Test (SW-846 Method 9095). The purpose of the test is to determine the presence of free liquid in a sample of waste. A conical paint filter (mesh number 60) is used to line the inside of a glass funnel. A minimum sample size of 100 gram or 100 mL is placed inside the paint filter. If no liquid is seen to pass through the filter within a five minute time period, the sample is considered to pass the test.

Toxicity Characteristic Leaching Procedure: This test is intended to determine if a sample of waste intended for landfill disposal is hazardous. The test procedure determines the concentration of regulated compounds in a liquid sample or liquid extract of a solid sample. A liquid sample is filtered in order to obtain a liquid extract. A solid sample is extracted using a formulated liquid extraction solvent. The liquid is then analyzed for the concentration of hazardous substances, as defined by the TCLP D-List. Each substance on the D-List has a MCL (Maximum Contamination Level).

References to Chapter 18

1. G.K. Wong, P.S. Fair, K.F. Bland, R.S. Sherwood, "Balancing act: Gulf of Mexico sand control completions, peak rate versus risk of sand control failure," SPE – 84497, paper presented at the SPE Annual Meeting and Exhibition, Denver (2003).

2. F. Ahmad, M. Ward, A. Fisher, "Grecko Wells; Bringing Sand to Surface, A Change in Well Design Philosophy," paper presented at the IADC/SPE Asia Pacific Drilling Technology Conference and Exhibition, Kuala Lumpur, Malaysia, 13-15 September. SPE-87956-MS. (2004).

3. J.S. Andrews, H. Kjorholt, H. Joranson, "Production enhancement from sand management philosophy. A case study from Statfjord and Gullfaks," paper presented at the SPE European Formation Damage Conference, Sheveningen, The Netherlands, 25-27 May. SPE-94511-MS. (2005)

4. C.H. Rawlins, S.E. Staten, I.I. Wang, I. I. "Design and Installation of a Sand Separation and Handling System for a Gulf of Mexico Oil Production Facility," paper presented at the SPE Annual Technical Conference and Exhibition. Dallas, Texas, 1-4 October. SPE-63041-MS. (2000).

5. "Naphthenate Deposits, Emulsions Highlighted in Technology Workshop," Journal of Petroleum Technology (July 2008).

6. R.W. Chin, "Oil and gas separators," Chapter 2 in Petroleum Engineering Handbook, Volume III, Facilities and Construction Engineering, ed. K. Arnold, Chapter 2, 33-35. Richardson, Texas: SPE (2007).

7. I. McKay, P.R. Russ, J.W. Mohr, "A sand management system for mature offshore production facilities," Presented at the International Petroleum Technology Conference, Kuala Lumpur, Malaysia, 3-5 December. IPTC-12784-MS (2008).

8. J. Tronvoll, M.B. Dusseault, F. Sanfilippo, "The Tools of Sand Management," paper presented at the SPE Annual Technical Conference and Exhibition, New Orleans, Louisiana, 30 September – 3 October, SPE-71673-MS. (2001).

9. A.G. Ostroff, Introduction to Oilfield Water Technology, National Association of Corrosion Engineers, Houston, TX (1979).

10. C.H. Rawlins, "Design of a cyclonic jetting and slurry transport system for separators," Oil & Gas Facilities, February (2016).

11. C.H. Rawlins, J. Costin, "Study on the interaction of a flooded core hydrocyclone (desander) and accumulation chamber for separation of solids from produced water," paper presented at the 2014 Produced Water Society Annual Seminar, Houston (2014).

12. NACE Standard TM0173-2005, Standard Test Method. Methods for Determining Quality of Subsurface Injection Water Using Membrane Filters, published by NACE International (2005).

13. C.C. Patton, Oilfield Water Systems, Campbell Petroleum Series, Norman, OK (1974).

14. R. Fantoft, T. Hendriks, R. Chin, "Compact subsea separation system with integrated sand handling," paper presented at the Offshore Technology Conference, Houston, 3-6 May. OTC-16412-MS. (2004).

15. M.S. Jasmani, E.C. Geronimo, L. Chan, "Installation of on-line vessel desander manifold," paper presented at the SPE Annual Technical Conference and Exhibition, San Antonio, Texas, USA, 24-27 September. SPE-101575-MS. (2006).

16. S.D. Coffee, "New approach to sand removal," paper presented at the Offshore Technology Conference, Houston, 5-8 May. OTC-19465-MS. (2008).

CHAPTER NINETEEN

Deepwater Best Practices

Chapter 19 Table of Contents

19.0 Introduction 459

19.1 Challenges in the Design of Deepwater Platforms 459

- 19.1.1 Solids in Produced Water and Fluid Compatibility 461
- 19.1.2 Oil Droplet Sizes in Produced Water 462
- 19.1.3 Deoiling Hydrocyclones on Deepwater Platforms 463
- 19.1.4 Induced Gas Flotation on Deepwater Platforms 466
- 19.1.5 Disc-Stack Centrifuges 467

19.2 Best Practices 469

- 19.2.1 Minimize Inlet Shear 471
- 19.2.2 Apply Heat Upstream 471
- 19.2.3 Solids Prevention, Separation, Removal 471
- 19.2.4 Minimize Hydrate Inhibitor 472
- 19.2.5 Minimize the use of Corrosion Inhibitor (Design for Lifecycle Integrity) 472
- 19.2.6 Separate Water Early in the Process 472
- 19.2.7 Provide an Effective Reject / Recycle Handling System 473
- 19.2.8 Chemical Injection Points & Metering 473
- 19.2.9 Provide an Effective Monitoring and Control System 474

Summary 474

References to Chapter 19 476

19.0 Introduction

Deepwater platforms in the Gulf of Mexico and elsewhere around the world are transitioning from dry oil or low water cut production to higher water cut production. These platforms face unique challenges resulting from relatively cold production fluids, platform motion, addition of various inhibitor chemicals (corrosion, hydrate, etc), co-mingled production from multiple reservoirs, process recycle streams, high shear in many cases, slugging, and a general lack of space and weight for water treatment equipment. The lessons learned by deepwater operators can provide better designs for future facilities and guidance for debottlenecking existing facilities. Equipment and practices which have worked well or not worked well for treating produced water on floating platforms are described and the issues pertinent to the deepwater operation of this equipment are identified.

19.1 Challenges in the Design of Deepwater Platforms

On floating deepwater facilities, separator vessel residence time is typically short due to the high cost of space and weight. During the design phase of such facilities, there is usually little knowledge of the produced fluid properties. Even in the best case where water and oil samples have been taken and carefully analyzed, such samples will not have contaminants with representative particle size distributions which exist in the produced fluids during actual production. Venting of gas and CO2 will alter the composition of dissolved components and may cause precipitation of minerals. Migration of constituents form the bulk phases to the interface occurs in the first 10 or 20 or so minutes after initiation of production and alters the stability of dispersions and other forms of emulsion. Thus, sampling and analysis is of limited use in design. In addition, as oil and/or water production rates increase during the life of the field, or as new fields are brought on, the fluid properties may change dramatically as a result of fluids from new fields being brought onto the platform via subsea tie-backs. Thus, designing oil/water separation systems for deepwater facilities is challenging.

It would be impractical to design a system that provides for all contingencies in future rates and fluid properties. Therefore, what is needed, and the lessons learned that are discussed here, involve judicious design decisions that build in flexibility with minimal space and weight requirements. In other words, the original system design must have sufficient built-in flexibility to allow debottlenecking and capacity increase with minimal facility modifications.

This chapter relies on the experiences of platforms in the Gulf of Mexico, offshore Brazil, the north Atlantic, and the North Sea [1]. However, the lessons are generic and should apply to deepwater facilities in most geographic locations. Some specific experiences from Shell E & P operations in the Gulf of Mexico (GoM) are discussed by way of example. Bullwinkle was the first deep water platform to begin production (1989) and this platform now processes fluids from a number of subsea and primary fields. Auger came on line in 1994 and currently handles production from a total of six fields. Mars receives production from ten sands.

The major issues relating to water treatment discussed here are: fluid compatibility, equipment design, process design, and technology selection. A Process Flow Diagram which is a composite of the systems on the various GoM platforms is shown, for illustrative purposes, in the figure below.

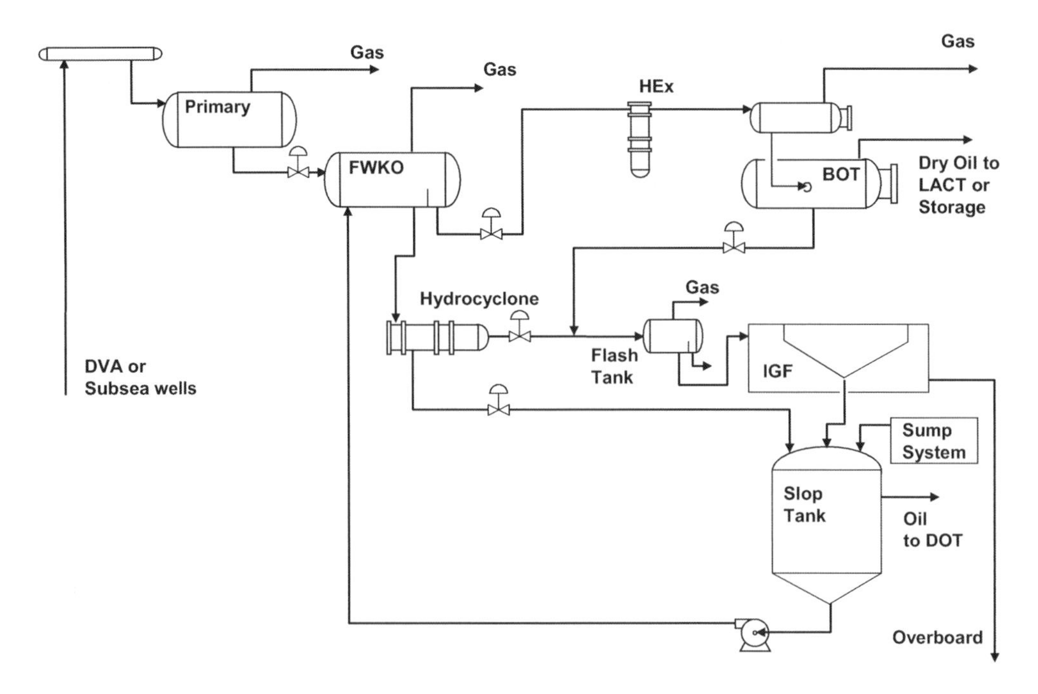

Figure 19.1 A composite Process Flow Diagram is shown which represents the essentials of the process systems installed on the Shell platforms in the GOM.

Driving Mechanisms in Deepwater Reservoirs: Many of the reservoir fluids in deepwater are driven by strong compaction forces, by moderate aquifer support, or a combination of both. For the most part, natural water drive is not significant in the deepwater GoM. In deepwater Nigeria, there is little driving force and waterflood is required to recover hydrocarbons. Thus, as shown in the figure above, most facilities are designed with two-phase primary separators. The producing sands are mostly unconsolidated turbidite sheets with a range of channel and fault densities from zero to a moderate degree of compartmentalization. The sands are highly compressible, in the range of 50 to 80 micro sips, so many reservoirs maintain pressure through compaction. Avoiding sand production is a major issue in well completion design and is for the most part successful [2]. Nevertheless, production of a small amount of very fine sand (1 micron or less) is typical. The fine sand has an impact on produced water quality as discussed below.

Water Cut: While individual wells have attained high water cuts, overall water production on any given platform has remained relatively low. As shown in **Table 1**, current water cuts are in the range of 10 to 30 %. But water cuts have been increasing steadily, and are likely to continue to increase for reasons such as the implementation of waterfloods, the implementation of improved hydrate flow assurance strategies, and artificial lift that will allow the continued production of subsea systems to relatively high water cut.

Table 19.1 Summary of Shell GoM Deepwater Platforms and their Current Production

Platform	Location	First Oil (date)	Oil Production 2004 (BOPD)	Water Production 2004 (BWPD)
Bullwinkle	Green Canyon 65	1989	64,000	24,000
Auger	Garden Banks 426	1994	56,000	30,000
Mars	Mississippi Canyon 807	1996	155,000	20,000
Ram-Powell	Viosca Knoll 956	1997	80,000	20,000
Ursa	Mississippi Canyon 810	1999	96,000	23,000
Brutus	Green Canyon 158	2001	40,000	8,000

19.1.1 Solids in Produced Water and Fluid Compatibility

A key component for designing a water treatment system and/or for understanding the manner in which a system does or does not effectively remove contaminants, is having an understanding of the physical and chemical character of those contaminants. Produced water solid contaminants come from three sources: scale mineral precipitates (often resulting from the mixing of incompatible waters, or a change in process conditions, such as a pressure or temperature change), asphaltene precipitation (from the mixing of incompatible crudes or from mixing condensates with crude), and formation fines.

Water compatibility can generally be predicted by the use of commercially available software [3]. The input to these programs includes the geochemical analysis of the water, the produced gas analysis, and process conditions (temperature and pressure). In determining whether or not any given water or mix of waters will have a scaling tendency, it is important to saturate the water with the gas composition at the conditions of specific points in the process, e.g., the FWKO, the point of mixing, etc. If this gas saturation is not taken into account, then the scaling tendency of the water cannot be correctly predicted. Predicting scale tendencies based solely on the laboratory analysis of a produced water can be misleading.

The figure below shows how the barium content varies for several water sources in the deepwater GoM. Based upon this variability, it can be expected that a well tailored scale inhibitor program will be required. Noting that scale inhibitors do not prevent the precipitation of scale, only its deposition onto surfaces, one can still expect that the cleaning of incompatible waters will require the use of induced gas flotation (IGF), media filtration, or both to capture the oleophilic scale mineral solids.

The Mars platform receives fluids from ten sands, as summarized in **Table 2**. For the most part, the aromatics content of the crudes are high, so the expectation is that the probability for asphaltene precipitation when mixing crudes will be low. Two exceptions would be the presence of either the Terra Cotta or the Ultra Blue crudes as the aromatics to asphaltenes ratio are lower for these crudes. Also, there is evidence that the Pink crude is bio-degraded, indicating the possible presence of organic acids which could impact water treatment.

Despite the expectation (and experience) that asphaltene precipitation from the subject crudes is unlikely to be an issue, the potential remains for the precipitation of asphaltenes when a condensate is mixed with the crude. For this reason, the mixing of a crude and a condensate is best delayed until both are fully dehydrated.

The third source of solids in produced water is fines which migrate from the formation. These fines are more often than not hydrophilic in nature and thus not a source for residual oil in the water. They can, however, impact total suspended solids (TSS) and turbidity. Also, they can have a tendency to become partially oil wet from the presence of waxes and asphaltenes. Oil-wet fine sand particles tend to stabilize interface emulsions in separators which then results in a degradation of produced water quality. They can also preclude the application of coalescing equipment or media due to the tendency to plug such equipment. Typically the solids content is not sufficient to justify installation of desanding cyclones which, because of the small particle sizes involved, may be of marginal value. Removing these solids generally requires the use of a high MW flocculent polymer for flotation. When these solids are present, then the regular draining of interface emulsion to a separate slop treatment system may be required if water quality is to be maintained.

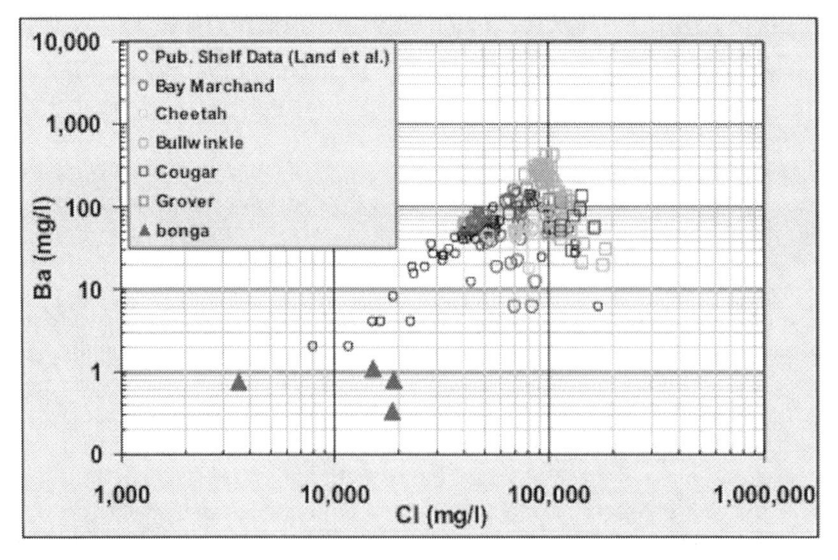

Figure 19.2 The Barium content of water from multiple deepwater GoM sources varies by nearly two orders of magnitude, indicating the potential for solids precipitation from mixing waters.

Table 19.2. Fluid Properties for the Mars Reservoirs

RESERVOIR	Depth	API	Saturates	Aromatics	Resins	Asphaltenes	Atomic Sulfur (wt %)	ACID No.
Lower Pink	11,821	19.6	23.1	58.4	15.2	3.3	2.5	
Pink	13,036	17.3	21.0	61.6	14.0	3.4	2.7	4.35
Lower Green	16,287	31.5	40.9	47.6	9.5	2.1	1.8	0.34
Ultra Blue	16,301	24.6	41.3	37.5	13.9	7.3	2.6	
Orange	16,550	24.4	27.4	54.5	14.5	3.6	2.4	1.00
Upper Green	16,910	23.5	24.3	55.6	14.9	5.2	2.6	
Magenta	17,610	24.2	28.1	54.1	13.5	4.3	2.6	0.61
Violet Ic	18,419	22.5	25.0	59.3	12.6	3.1	2.7	1.20
Lower Yellow	18,476	27.6	30.8	51.8	15.1	2.2	2.2	
Terra Cotta	18,476	22.1	25.2	50.2	13.4	11.2	2.8	0.95

19.1.2 Oil Droplet Sizes in Produced Water

Because of the low water cut on most deepwater platforms, making the primary gas/liquid separation in 2-phase vessels is logical. These vessels can be relatively small, with only 1 – 3 minutes liquid residence time, and gas is flashed at a relatively high pressure, thus minimizing compression costs. The problem for water treatment comes as the oil/water mixtures are transferred with a significant pressure drop to a lower pressure separator or FWKO. These pressure drops can be in the range of 200 to 600 PSI which shears water into small droplets in the oil and shears oil into small droplets in the water. As shown in **Table 3**, the oil droplet size generated by a 400 PSI pressure drop from an HP separator to a FWKO is small with a median on the order of 20 microns and a minimum size as small as 2 microns. An on-line survey of the oil droplet size distributions at various locations downstream of the 400 PSI pressure drop on one platform showed that the droplet size distribution remained essentially unchanged throughout the water treatment system. Thus the challenge was to select technology capable of recovering these small oil droplets from the produced water.

Table 19.3 The oil droplet size distribution after a 400 PSI pressure drop had a median size of 20 microns, with drops being as small as 2 microns. Total oil increased as the largest droplet size increased, but the median drop size did not change with the oil content.

Sample	Droplet Size (Microns)			Oil Conc.
Time	Min.	Max.	Mean	(mg/liter)
11:14	5	52	20	280
11:21	2	75	22	380
11:50	5	35	20	140
16:00	5	37	19	115
16:49	5	63	24	520
07:21	5	31	17	50
08:03	5	50	20	185

Other factors which impact oil droplet size and the stability of the distribution include the presence corrosion inhibitors and/or methanol in the produced water. Methanol is commonly utilized for hydrate control and may be injected continuously or periodically. Methanol in the produced water decreases the oil/water interfacial tension, which in turn helps to stabilize dispersed oil as smaller droplets. Methanol has other effects as well. One effect is the precipitation of otherwise soluble di-valent mineral salts (calcium, magnesium, barium sulfates and/or carbonates). These minerals tend to be hydrophobic and thus become neutrally buoyant oily solids which are not readily recovered by hydrocyclones or other gravity-based water treating equipment.

It should also be noted that methanol will tend to increase the solubility of some hydrocarbons in the produced water and that methanol is partially recovered by the solvent extraction method used for an EPA 1664 analysis. A direct IR analysis of the extraction solvent that contains methanol will register a higher TOG value than will an instrumental analysis which more closely approximates the evaporative methodology that is part of the EPA 1664 procedure. Thus using a wet IR instrument on a deepwater platform may yield higher TOG values that are somewhat sporadic since they will reflect when methanol is or is not present in the produced water.

Given that high pressure drops or the presence of chemicals that are related to deepwater oil/gas production may be present on a deepwater platform, the process designer would need to contemplate, for example, the need for higher efficiency oil/water hydrocyclones and the need to remove neutrally buoyant oily particulate from the water.

19.1.3 Deoiling Hydrocyclones on Deepwater Platforms

The initial experience with hydrocyclones in the GoM was somewhat unfavorable [4]. As discussed by Khatib, hydrocyclones were used successfully in the north Sea but not as much in the deepwater GoM. This was due to a lack of understanding of how the hydrocyclone fits into the typical water treatment process, their envelop of operation (pressure and flow rate), and process control of the unit. Recent experience with hydrocyclones has been somewhat improved in the deepwater GoM but there are still unique feature of deepwater platforms that reduce the effectiveness of hydrocyclones. This is discuss in some detail in Cpater x.x.x.

Hydrocyclones installed on deepwater facilities are typically composed of 25 or 35 mm diameter liners. Feed is from the water discharge of the FWKO. The pressure drop across and flow through the hydrocyclone depends on the FWKO level control system and can fluctuate significantly with time. The PDR (pressure differential ratio) is typically maintained in the range of 1.6 to 2.0, depending upon the manufacturer. In this setting, the hydrocyclone is not intended to produce overboard discharge quality water. Instead, it is intended to shave the peaks in oil content and provide an overall pre-conditioning for downstream flotation.

As shown in the figure below, hydrocyclone performance on one platform was very good from December 2003 through October 2004. During this time period, the operating procedures for the hydrocyclone were being carefully followed. This included daily backwashing, routine cleaning of the liners, adjustment of the number of active versus blank liners, and maintaining the PDR in the manufacturer's recommended range of 1.6 to 2.0. Also, the chemical vendor had an effective treatment program in place to prevent scale deposition throughout the facility, and, in particular, across the hydrocyclones. Starting in October 2004, those procedures were not followed as diligently and hydrocyclone performance began to deteriorate. Various factors contributed to this deterioration including staff turnover and the fact that much of the routine maintenance of the unit was cumbersome. Therefore, new installations of hydrocyclones incorporate improved operability features as listed below.

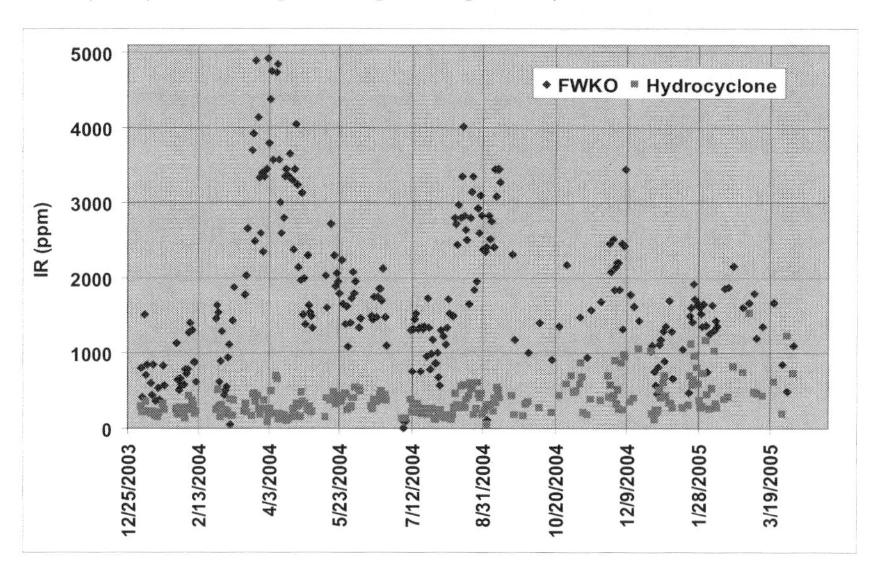

Figure 19.3 The hydrocyclone performance on one deepwater platform is shown over time. The deterioration in performance was found to be related to inadequate unit maintenance.

A summary of Best Practices formulated by Shell for hydrocyclone operation includes:

- Select hydrocyclone liners for removal of oil drops greater than 10 micron
- Regularly adjust the number of liners to match the flow requirements.
- Avoid rapid flow fluctuations
- Minimize solids to hydrocyclones and anticipate scale deposition in the reject orifices
- Minimize entrained gas in feed since it will reduce the true volume of reject liquid
- Clean and backflush regularly

- Increase reject port size to prevent blocking with solids and to accommodate break out gas volume
- Any new installation should have the following operability features:
    - Automated backwash
    - Quick-open enclosure on the containment vessel
    - Spare liners readily available for quick change-out and offline cleaning

The design of piping for transporting rejects to a slop vessel is particularly important for maintaining effective hydrocyclone performance. The designer needs to understand that due to the pressure drop involved, this fluid will have a substantial gas volume fraction and the liquids may be laden with fine, emulsion-stabilizing solids (typically <20 microns). Also, the piping should contemplate a reject ratio that is in the range of 5% of the design capacity of the hydrocyclone unit, again with the expectation that the fluid will be multiphase. If there is CO_2 in the evolved gas, then the precipitation of scale in the reject orifice of the hydrocyclone can be expected and easy access to the hydrocyclone liners for maintenance purposes is essential. Without adequate design, it will be difficult to maintain the required Differential Pressure Drop Ratio and the fluid reject rate required for effective hydrocyclone operation.

As mentioned above, if the process anticipates that the produced water will experience a substantial pressure drop (> 50 PSI) upstream of the hydrocyclone, then the selection of a high efficiency liner designed with a D_{50} droplet size removal capability on the order of 10 to 15 microns is essential.

Hydrocyclone performance is summarized for a number of platforms in **Table 4**. Any unit downstream of hydrocyclones must be designed to safely release gas arising from the pressure drop through the hydrocyclone. A well designed flash drum (degasser vessel) can contribute significantly to deoiling,and to the suppression of surges which can lead to a produced water sheen. Therefore it is considered a best practice to install oil skim facilities in the flash drum. This allows the discharge of the Bulk Oil Treater, for example, to be tied into a downstream vessel, rather than requiring that the stream be pumped into an upstream location. The presence of a flash tank helps to eliminate of a major source of shear and thus the generation of small droplets.

Table 19.4 Hydrocyclone performance experience from various Shell facilities [5]

Location	Hydrocyclone Inlet (ppm)	Hydrocyclone Outlet (ppm)	Degasser Outlet (ppm)	Hydrocyclone System
N. Sea	400	17		35 mm liners No degasser
N. Sea	600	25		60 mm liners No degasser
N. Sea	400	45	30	35 mm liners
US	520	39	20	K-Liner Pumped Feed
US	2310	160	6	35 mm liners Pumped Feed
US – Auger	1500	200	100	K-Liner FWKO Feed
Dubai	327	88	39	35 mm liners
N. Sea	102	30	17	G-Liner Production system

19.1.4 Induced Gas Flotation on Deepwater Platforms

While the importance of horizontal Induced Gas Flotation Units (IGFU) on Shell facilities in the deepwater GoM is taken for granted, such equipment is not necessarily part of the essential water treating equipment in other parts of the world. For example, at least a few high volume water producing Shell assets in the North Sea do not have flotation units. Shell deepwater Brazil (Bijupira-Salema) has a vertical IGFU. Horn Mountain (BP), Holstein, and Thunderhorse in the deepwater GoM also use vertical IGF units with varying degrees of success.

A typical horizontal IGFU consists of a tank with six chambers in series. The first chamber is an inlet section that provides momentum reduction, a quiet zone for chemical residence time, and an opportunity for most of the dissolved gas breakout to be vented. Without this, the liquid level in the first flotation cell would be higher than that in the downstream cells, making it difficult to consistently remove floating contaminants from all the cells. The next four chambers consist of induced gas flotation with rejects flowing over a spillover weir and water passing under a partition into the next chamber. The final chamber is provided for the final evolution of gas bubbles in order to minimize gas carry-under to the discharge piping.

Vertical Column flotation units were originally designed to address concerns over the effect of wave motion on deep water floating structures. Baffles and other devices have effectively minimized the effects of wave motion on the performance of horizontal IGF units. Column flotation units were adapted for use in the North Sea and on FPSO applications where atmospheric tanks and horizontal vessels are typically not accepted. Although column flotation units have performed well, single vessel applications can be prone to upsets while offering no redundancy or protection against mechanical failure. When column flotation units are utilized, the use of two or more vessels should be considered.

Much of the design and operating experience regarding horizontal IGFU for the deepwater was transferred directly from decades of use in the shallow water GoM. C.A. Leech [6] describes the design capacities and operating parameters for horizontal IGFU at several shallow water GoM locations. Flotation units at all of these facilities are similar in many respects to flotation units on deepwater facilities. The consensus opinion among Shell operations staff in the deepwater GoM is that induced gas flotation provides efficient separation of oil to discharge quality on a consistent basis and with a minimum of operations adjustment and maintenance. To some extent, the success of gas flotation units in the deepwater can be attributed to the fact that many deepwater operators have many years experience operating the shallow water assets where they gained relevant operating experience.

The character and quality of the skim fluids from an IGFU will vary depending upon the presence of oily solids and the character of the water clarifier or polymer utilized to support the flotation performance. Some types of water clarifiers such as dithiocarbamates, are very effective, but tend to generate a heavy, sticky floc that is not easily recovered downstream. Other times, a high molecular weight polymer may be utilized which, if not fully hydrated, can tend to accumulate in and downstream of the IGFU, again causing difficulties with contaminant handling. Also, the presence of water clarifiers and flocculants in fluids which are recycled to an upstream separator will often result in the formation of stable interface emulsions that upset free water knock outs and LP separators. Since it is impossible to know the character of the water clarifier which will be required on any given platform, it is particularly important to avoid the recycle of contaminants recovered from a flotation unit into any upstream separator or tank.

19.1.5 Disc-Stack Centrifuges

The overall experience with disc stack centrifuges on Shell deepwater facilities has been poor. This experience is summarized in the Table below. In general, the units did a good job of separating oil and water, when they were in operation. However, a combination of problems caused the uptime to be very low. In the case of the Auger platform, several weeks of troubleshooting were required to startup the unit. Once in operation, the unit ran for roughly a day before the shaft connecting the motor to the centrifuge bowl broke in two pieces. Given the time and effort that had gone into the unit it was decided to scrap any further work. The problems with the Auger centrifuge were characterized as rotating equipment problems with the final failure also being a rotating equipment failure.

Table 19.5 Summary of Operating Experience with Disc Stack Centrifuges

Location	Water Treating Effectiveness	Mechanical Reliability	Status
MP-252	Good	Difficult to maintain	Working
Auger	Not Tested	Failed immediately	Removed
Mars	Good	Failed	Removed
NaKika	Good	Failed	Removed
Holstein	Good	Failed	Removed

The centrifuge on Mars suffered some of the same problems as the unit on Auger plus control system problems, and fouling of the disc stack by sticky solids. The rotating equipment problems together with the control problems were sufficiently difficult that maintenance staff spent an inordinate amount of time working on the unit. A build up solids in the disc stack would cause an imbalance fault which required that the unit be disassembled for cleaning. This only increased the maintenance requirements to the point that the platform could not keep the centrifuge running with the available staff.

<u>**Mars:**</u>

- Model: Alfa Laval OFPX 517 (OF = Oil Field, PX = intermittent discharge / sludge space unit)
- Capacity: 6,000 BPD, usually operated at around 4,000 to 5,000 BPD.
- Specifications: 4,140 rpm; g-force at periphery of the disc = 5,000 g.
- Feed from Wet Oil Tank
- Objective: intermittent to continuous small volume use to clean up methanol containing produced water streams from the Europa subsea system.
- Operating experience: At the time that this one was installed, it was one of the biggest / highest speed units that Shell GoM had in service. Today it would be considered to be medium / small capacity. They had many problems initially. The original feed line was from the bottom of the Wet Oil Tank. They had a barium sulfate problem. Also they had a tendency to over-treat with DTC (dithio carbamate water clarifier). Long story short, they gummed up the disc stack over and over again. Alfa Laval was frustrated at the way Mars operated the unit. Shell was frustrated that Alfa Laval would not come out and "fix" it. Eventually they moved the Wet Oil Tank discharge / centrifuge feed to a higher point in the WOT vessel. This allowed the larger solids to settle below the WOT discharge and not go into the centrifuge. Scale inhibitor was applied for the barium sulfate problem and DTC injection was put on a trend with tighter control.

It turned out that Europa did not need continuous methanol injection. Due to high salinity, and low water content, the methanol volumes on startup were also low. Therefore Mars used the centrifuge to help clean up oily water that would otherwise be recycled through the other parts of the water treating system. It really was not a critical need and eventually they took it off the platform to provide space for other equipment. The mechanics didn't like it because they had to spend a fair amount of time tearing it down and cleaning or replacing the disc stack, even after they made the above modifications.

Auger:

- Model: Alfa Laval LEO-100
- Capacity: 15,000 BWPD
- Feed from Wet Oil Tank
- Objective: intermittent to continuous small volume use to clean up oily water having small oil drops.
- Operating experience: This unit never really got running. It had originally been purchased for use on MP-289C. It was not used there and it sat idle for two years. Auger was experiencing produced water problems and frequent sheens. When the unit was installed on Auger, the shaft broke within a month or so of operation. The shaft breakage caused other damage so the unit was sold for scrap.

MP-252:

- Model: Two small Westfalia units
- Capacity: 300 BPD and 750 BPD
- Objective: continuous small volume use to clean up methanol containing oily water from Tahoe.
- Operating experience: These small units seem to have far less mechanical problems than the large units described ablve. Although they are still a maintenance headache, the offshore staff have learned how to run and maintain them. The Tahoe water could not be treated without them. Only a slipstream of produced water is sent to the centrifuges, not the entire platform produced water volume.

Summary: The Shell experience with the use of Disc Stack Centrifuges can be summarized as follows:

- Units were tested in the laboratory and pilot field settings successfully
- At some locations there is a potential need for treating production with small oil droplets, although this water can probably be treated with induced gas flotation if an appropriate flotation polymer is utilized
- Insufficient on-board personnel were available for regular maintenance on the centrifuges
- Larger units (higher forces) failed in actual use
- Smaller units (1300 BWPD) clean water successfully, but require significant maintenance and operator training

Utilizing Low Shear Pumps

An oil droplet size distribution survey was conducted on the Mars platform to determine the impact of pumps moving oily water from the Wet Oil Storage Tank and from the Slop "Wet Oil" tank. The results are shown in the figure below. In both cases, the pumps were operating far off their peak efficiencies and the recycle of the very small oil droplets made it more difficult to clean the produced water. Based upon this and other experience in the Gulf of Mexico, Shell's guidelines for pump selection for produced water handling are summarized as follows:

- Use low shaft speed pumps (< 1800 rpm)
- Select pumps with high hydraulic pump efficiency (> 60 %)
- Specify large impellor diameter (goes along with slow speed for given gpm)
- Use an oversized discharge nozzle (slow discharge speed)
- Maintain a limited pressure boost per stage (< 50 psi)
- Operate the pump with a low specific speed Ns < 700

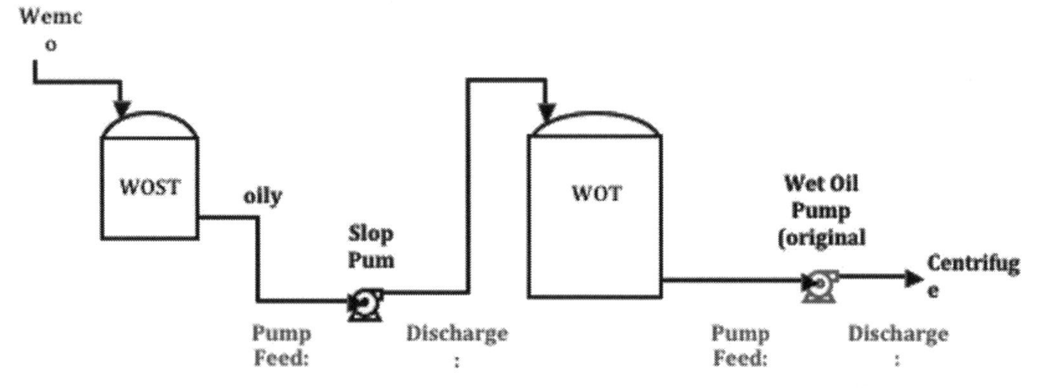

Figure 19.4 An on-line survey of oil droplet size distributions on Mars shows that low efficiency pumps were generating very small oil-in-water droplets that were difficult to remove in the water treatment system.

19.2 Best Practices

A composite, somewhat generic process flow diagram was presented in Figure 1 for a separation and water treating system on deepwater platforms. Based upon the above discussion of successful operational practices on deepwater platforms, several upgrades to the PFD of Figure 1 can be recommended. These include:

- Use of a PDR controller to the deoiling hydrocyclone system to maintain PDR above 1.6.
- The routing of all reject and skim fluids to a slop tank
- The limited blending of oil recovered from the slop tank with sales oil as long as the BS&W of the sales oil does not approach contract limits
- The return of clarified water to the produced water treating system, not to an upstream separator

- All pumps utilized are of the low shear variety and selected to operate near the peak efficiency on the pump curve
- Water clarifier is injected, as appropriate, immediately upstream of the cleaning technology that the clarifier is intended to support.
    - For example, any clarifier injected ahead of the flash tank would not be expected to retain activity that would support flotation in the IGF unit.

In developing the list of Best Practices, we use the "Dirty Dozen – Major Contributors to Poor Water Quality" developed by Ted Frankiewicz:

1. Upstream separators filled with solids
2. Separators operating outside of their design envelope
3. Recycle streams feeding contaminant laden emulsions to separators
4. Gas slugging below the water line in skimmers and separators
5. Neutrally buoyant, oil coated solids
6. Water soluble, film-forming corrosion inhibitors, especially in fresh water
7. High shear dispersion of oil droplet, especially as API Gravity increases
8. Excessive or incorrect chemical treatment
9. Ultra-fine scale mineral solids ($CaCO_3$, $CaSO_4$, $FeCO_3$)
10. High concentrations of organic acids in water phase
11. Iron sulfides
12. Highly variable flow rates (e.g. snap acting valves)

Many of the Best Practices discussed below are discussed in terms of eliminating these "Dirty Dozen Major Contributors to Poor Water Quality." Best practices for removing oil from water are given here in the case where there are no other constraints on the design of the system. The Best Practices are defined in terms of the following categories. These are described in detail below.

1. minimize inlet shear
2. apply heat upstream
3. prevent solids production, separate and remove solids
4. minimize the use of hydrate inhibitor
5. minimize the use of corrosion inhibitor
6. separate water early in the process
7. provide an effective rejects handling system
8. provide an effective chemical treating system
9. provide an effective monitoring and control system

19.2.1 Minimize Inlet Shear

A fact which is often overlooked in water treating is that system performance is often determined by what happens to the fluids upstream of the inlet separators. The inlet drop size and the stability of oil drops are two important factors that determine the performance of the separation equipment. Inlet drop size is determined by energy dissipation (shear intensity) that the fluid experiences. The oil drop stability, or conversely oil drop coalesce tendency, is a function of drop size, oil drop concentration, and presence of production chemicals such as spent acid, and hydrate and corrosion inhibitor. These chemical factors are discussed further below.

The extent of inlet shear depends on the pressure drop of the fluids through the well head choke, and the platform boarding choke. The greater the pressure drop across these valves, the smaller the average drop size.

19.2.2 Apply Heat Upstream

Most fluid processing facilities have some form of heat input. Heat is typically added in order to meet the vapor pressure specification of the export crude oil. Since the vapor pressure requirement is applied to the export crude, heat can be added as far downstream as the dry oil tank. Typically however, heat is usually added upstream of this point in order to realize benefit from decreased viscosity and improved oil and water separation.

The point at which heat is added is a tradeoff between several factors. If heat is added upstream, then heating will be required for high pressure system with a greater volume of fluid and gas. When they are placed upstream, they must not only heat the oil but also the water. In addition to the added fluid volume, water has a heat capacity that is a least a factor of two higher than that of most typical crude oil. Thus, the heat generating capacity of the platform must be larger. This adds to the weight, space and capital expense of a project. However, the benefit of adding heat upstream is that more of the gas will break out at higher pressure, thus reducing the required capacity of the gas compressors. Also, if heat is added upstream then oil and water separation processes realize the benefit of lower viscosity fluids, hence achieving better water quality. If heat is added downstream, then less heat is required but there is no other process benefit.

19.2.3 Solids Prevention, Separation, Removal

The presence of solids has several detrimental effects on oil/water separation. Solids tend to become partially oil wet and migrate to the oil/water interface where they stabilize an oil/water emulsion. Most of the solids are reservoir related sand/clay production.

Handling of solids involves three phases. The first phase is potentially the most effective and that is to prevent solids from entering the production fluids. Depending on the source of solids (discussed below), this can be accomplished in various ways. Effective well bore sand screens are essential in high rate wells. The second phase of solids handling is to separate them from the production fluids at a point relatively upstream of the oil/water separation equipment. This typically involves the use of sand cyclones, screens or filters. The third phase of handling solids is to remove any accumulation of solids in the production separation and water treating equipment. This typically involves jetting with external nozzles, or the operation of an installed sand jetting system. Both solids separation and removal are somewhat difficult to carry out. By far the most effective means of dealing with solids is to stop them at their source, namely downhole in the wells.

19.2.4 Minimize Hydrate Inhibitor

Effective water treating starts upstream of the facilities. One of the most important decisions regarding water treating is the hydrate strategy and other flow assurance strategies, including Asphaltene and corrosion control of the subsea flowlines. Many of the chemicals that are needed for flow assurance applications have an impact in the oil water separation process. Some are worse than others. Many of the chemicals required will lower the interfacial tension thus making the oil/water mixture more likely to form small drops of oil as a consequence of valve shearing.

19.2.5 Minimize the use of Corrosion Inhibitor (Design for Lifecycle Integrity)

Asset integrity for the lifecycle of a project can be a costly issue. Materials selection involves two

kinds of information. The first is the CAPEX for the materials of construction. This is relatively straightforward in terms of making cost estimates on the basis of material selection for flow lines, risers, wells head and so on. Often there is pressure on a project to reduce CAPEX in order to meet project value requirements for investment.

The second is based on the net present value of operational costs over the lifecycle. The cost of corrosion monitoring, use of corrosion inhibitor, and the cost of shut downs and repairs are taken into account. This is a much more difficult calculation since it depends on a number of unknown factors. It is often difficult to assess accurately the corrosion rate prior to operation. For example, it is unlikely that the production fluid composition for 5 or 10 years will be known with any degree of certainty. Thus, it is sometimes the case that corrosion inhibitor is required later in field life.

Like so many other decisions in field development, it would be prohibitively expensive to design for the worst case scenario. Nevertheless, the use of robust and corrosion resistant materials which minimize the necessity for injection of corrosion inhibitor is a practice which will help ensure water treating effectiveness.

19.2.6 Separate Water Early in the Process

While the topic of oil/water separation is a broad topic, the definition of Best Practices for water treating can be easily stated: remove the large drops of oil before they become sheared either by valves or by pumps.

The required residence time for separating oil and water in a three-phase primary separator depends on the fluid characteristics such as oil density, water density, oil drop size, and water viscosity (related to temperature). The greater the driving force for separation, the more fluid can be processed in a given vessel volume per unit time. Conversely, if the driving force is diminished then the vessel must be larger to achieve the same level of separation. Stokes Factor, as described below is a good measure of driving force for separation.

Three phase separation and the use of a hydrocyclone on the water discharge is an effective means of separating oil from water before the detrimental effects of valve shear. The interface level control valve for the separator is downstream of the hydrocyclone.

Fluid characteristics can vary significantly from one reservoir to another. If for example, a light hydrocarbon is mixed with a hydrocarbon having unstable or marginally stable asphaltenes, the light hydrocarbon will act as an anti-solvent and cause the asphaltenes to precipitate. The precipitated asphaltenes will migrate to the oil/water interface and cause stable emulsions. Incompatible water streams are also problematic. Thus, having a number of primary separators rather than a single large separator is better form a water treating standpoint. Once the fluids are separated, then comingling has much lesser impact.

19.2.7 Provide an Effective Reject / Recycle Handling System

All water treating equipment generates a reject stream. In selecting equipment and designing an oil/ water separation system, it is important to consider the flow rate and oil-in-water concentration of all reject streams. A material and flow balance of the overall process is required. If the primary separation equipment is not adequately sized to handle the reject flow, then the system will be bottlenecked. The operators will be forced to reduce either the oil retention time, the water retention time or the reject flow rate. Any one of these steps will result in poor oil and / or water quality.

While most water treating equipment provides better separation at higher reject flow rates, increasing the flow rate of the reject streams has serious negative consequences on the overall system design. Equipment selection should consider all relevant flows: the inlet flow and concentration, the effluent flow and concentration, and the reject flow and concentration. While this seems obvious, many equipment suppliers given only the inlet and effluent flows without specifying the separation efficiency as a function of the reject flow rate.

In many cases it is desirable to route the reject stream through the primary separation system. This greatly enhances the probability of coalescence and capture of the oil in the main oil stream destined for oil export. Realistic values of reject rate (percentage or flow rate) must be used in sizing the vessels which will handle both the reject and the primary production stream. Serious complications arise if vessel are not adequately sized for cross-sectional area, retention time, and discharge flow rate.

As discussed, it is necessary to minimize the volume of rejects while maintaining the performance of the water treating equipment. While it is imperative to do this in the design phase, it is also of great benefit to do this in real time during operations. This is achieved by measuring and controlling the reject flow rate in order to eliminate fluctuations. Control schemes for water treating reject streams are relatively simple but rare. Where they have been applied they result in a tremendous improvement of the overall operation of the oil and water separation system.

The design of facilities for offshore platforms is typically constrained in weight and space. Thus, it is often difficult to justify the addition of a slops or wet oil tank. Nevertheless, the presence of such a tank makes the handling and treating of reject streams much easier. Such a tank is used to provide settling time and another opportunity for chemical treating. Typically a solvent can be added which will prevent the formation of deposits due to recycling of water treating chemicals.

19.2.8 Chemical Injection Points & Metering

Water treating systems are typically designed with the assumption that water treating and deoiling chemical will be used. Separation efficiency of separators and water treating equipment is almost always stated with the assumption that chemical will be used. In many cases, the performance of water treating equipment in the absence of chemical is actually not well known in the industry.

A well designed water treating system will include adequate chemical storage, premixing, mixing with process fluids, sufficient number and location of injection points, metering, and delivery.

19.2.9 Provide an Effective Monitoring and Control System

The successful operation of a water treating facility depends to a large extent on operator adjustments. Most separators and water treating equipment require adjustment of levels and interfaces, as well as the periodic operation of an interface bleed. Some separators and water treating equipment have the facility for online cleaning which must be periodically operated. Control systems must be periodically tuned to ensure smooth operation of the level control valves. Chemical injection rates must be adjusted to accommodate changing well lineups and production rates.

Summary

The general design considerations for a produced water treating system on a deepwater platform include the following:

1. Upstream separators
    a. Design to accept slug flow so that flow rates to the downstream water treating system can be more temporally uniform
    b. Include cyclonic inlets or similar technology to avoid shattering oil into small droplets which are difficult to remove in downstream treating
2. 1st stage of water treating – deoiling hydrocyclones
    a. Favor the installation of high efficiency hydrocyclones on the expectation that pressure drops from HP and IP separators will be sufficient for generating oil droplets in water with a size range of 5 to 50 microns
    b. Design and install the hydrocyclones such that maintenance is convenient.
    c. Plan for a minimum 5% hydrocyclone reject volume which must flow multiphase to a slop or wet oil tank.
    d. Install a Differential Pressure Ratio controller on the hydrocyclone
    e. Expect the reject stream from the hydrocyclone to be solids-laden and not suitable for being returned to an upstream separator
3. Induced Gas Flotation
    a. Plan for a 5 to 10% skim liquid volume from the IGF unit(s)
    b. Route continuously skimmed contaminants to a slop or wet oil tank
    c. Consider a degassing unit upstream of the IGF only if excessive gas evolution is expected ahead of or in the IGF Otherwise, use this dissolved gas breakout to augment flotation gas by installing the LCV for the upstream separator(s) near to the IGF unit
    d. When the use of a high MW polymer is required, continuously and on-line pre-mix the polymer into the water stream using a venturi injector.
    e. The injection rate for water clarifiers should be controlled by a flow meter which sends a signal to the chemical injection pump. Continuous modulation of the chemical injection rate is preferred.

4. Eliminate recycle streams

   a. The slop or wet oil tank should have facilities to send oil to the LACT unit. Typically the oily emulsion recovered from the produced water will contribute <0.1% to the BS&W of the sales oil

   b. Clarified water from the slop or wet oil tank should be returned to the inlet of the water treatment system and not to an upstream separator where residual chemicals and solids can contribute to the formation of stable interface emulsions

   c. Solids recovered from the slop or wet oil tank should be sent for disposal, preferably by injection into a well dedicated to and permitted for this purpose

5. Utilize Low Shear Pumps

   a. Proper process design can minimize but not eliminate the need for pumping contaminant-laden process streams. Pumps should be selected such that they can operate near their maximum efficiency, preferably above 60%.

   b. Avoid the use of PD pumps when feeding a hydrocyclone as the pulsed flow will degrade hydrocyclone performance.

6. General

   a. Be cognizant that upstream chemical use, e.g., methanol, corrosion inhibitors, scale inhibitors, oil demulsifiers, etc., can negatively impact a water treatment system.

References to Chapter 19

1. J.M. Walsh, T.C. Frankiewicz, "Treating produced water on deepwater platforms: Developing effective practices based upon lessons learned," SPE – 134505, paper presented at the SPE ATCE Florence, Italy, (2010).

2. G.K. Wong, P.S. Fair, K. F. Bland, R. S. Sherwood, "Balancing Act: Gulf of Mexico Sand Control Completions, Peak Rate versus Risk of Sand Control Failure," SPE 84497, SPE ATCE Denver Co, (5-8 October 2003).

3. ScaleChem software from OLI Systems, Inc., New Jersey.

4. Z. I. Khatib, "Handling, treatment and disposal of produced water in the offshore oil industry," SPE 48992, SPE ATCE, New Orleans, La (27-30 September 1998).

5. M.F. Schubert, F. Skilbeck, H.J. Walker, "Liquid Hydrocyclone Separation Systems," p. 275 from Hydrocyclones – Analysis and Applications, ed: L. Svarovsky, M.T. Thew, Kluer Academic Publishers (1992)

6. C.A. Leech, S. Radhakrishnan, M.J. Hillyer, V.R. Degner, "Performance Evaluation of Induced Gas Flotation Machine through Mathematical Modeling," J. Petroleum Tech., p. 48 (Jan 1980).

CHAPTER TWENTY

Applications: GoM vs. North Sea

Chapter 20 Table of Contents

20.0 Introduction..481

20.1 Examples of GoM and North Sea Water Treatment Process Lineups........................486

20.1.1 North Sea Field Examples..487

20.1.2 Gulf of Mexico Field Examples..489

20.2 Field Data & Modeling..490

20.2.1 Oil Drop Diameter Distribution..491

20.2.2 Separation Equipment..492

20.2.3 Modeling Validation and Training Sets:..498

20.2.4 Effect of Shear and Temperature..500

20.2.5 Process Performance Diagram:..501

20.2.6 Results..502

20.3 Discussion:..504

20.4 Summary and Conclusions:..505

References to Chapter 20..507

20.0 Introduction

In general, water treating systems in the North Sea differ from those in the deepwater Gulf of Mexico. The two most apparent differences are the extensive use of hydrocyclones in the North Sea, and the use of large, multistage horizontal flotation units in the deepwater Gulf of Mexico (GoM). Deepwater GoM platforms employ hydrocyclones but not nearly to the extent as a typical North Sea platform. Typically in the North Sea, if flotation is used at all, it is a vertical compact unit. While there are exceptions to these generalizations, the majority of platforms in these locations follow these generalities. The objective of this chapter is to provide an understanding of the reasons for these differences.

Field data together with modeling results are presented to explain the differences in facilities design. The models accurately correlate the measured drop size and oil in water concentration observed in the two regions. In addition, the modeling tools are used to answer hypothetical "what if" questions. This allows isolation of individual variables such as fluid temperature, shear, separator residence time, and fluid density. Thus the modeling provides a detailed understanding of the relative importance of these variables. It also provides a direct comparison of the performance of North Sea versus Gulf of Mexico process configurations.

While the qualitative conclusions are well known (i.e. deepwater separation systems are designed to minimize weight and space), the detailed understanding provided here gives insight into the design of water treating systems in general. It also emphasizes in a quantitative way, the importance of carrying out effective water treating early in the process and the necessity of using large end-of-pipe equipment when this is not possible.

The differences in water treating systems between the two regions have been discussed qualitatively [1 – 3], as well as quantitatively [4]. As discussed by Bothamley [3], there are several differences between facilities in the North Sea versus those in the deepwater Gulf of Mexico (GoM). FPSOs (Floating Production Storage and Offloading) and fixed leg platforms are commonly used in the North Sea. These topsides facilities are typically much larger and heavier than those employed in the deepwater GoM. FPSOs have only recently been employed in the deepwater GoM,, and Bullwinkle is the only fixed leg deepwater platform. Floating Spars and Tension Leg Platforms are typically used in the deepwater GoM. These floating facilities are smaller and weight less than typical facilities in the North Sea. Other factors which account for the differences are capital availability and operating costs, extraction techniques, reservoir characteristics, the properties of the fluids being treated, and regulations.

In terms of extraction strategies, there are fundamental differences between the two regions as well. Almost all North Sea oil fields are developed, at least in early life, using pressure maintenance relying on water and/or gas injection. Fluid production in many of these fields has reached high water cut. In the deepwater Gulf of Mexico, most production is relatively dry and there are only a few fields that have applied water injection. These development strategies have a strong effect on the oil/water ratio over the life of a field.

As shown in Figure 1, a typical North Sea oil/water treating system consists of three-phase separation, together with hydrocyclones on the water discharge, throughout the facility. On many installations, heat is added upstream of the inlet separators (not shown in Figure 1). The effluent (product) from the hydrocyclones is typically routed to a degassing vessel, and the final stage is a compact flotation unit. In some cases, flotation is not required to achieve the overboard discharge target.

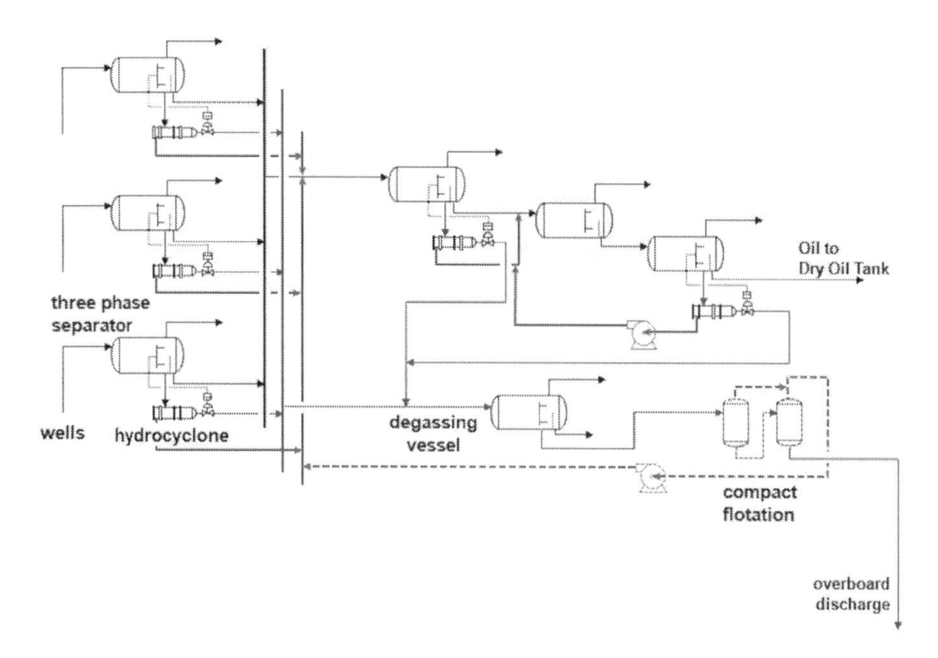

Figure 20.1 Typical North Sea oil / water separation system

The main features of North Sea systems can be summarized as follows:

- relatively more weight and space available than in the deepwater GoM;
- relatively high arrival temperature or heat added upstream;
- several primary separators to segregate incompatible fluids;
- almost all primary separators are three phase;
- hydrocyclones on every horizontal separator;
- hydrocyclones utilized upstream (on aforementioned three phase separators) which provides greater driving force for separation;
- hydrate inhibitor is not commonly used during steady state operation;
- slightly more corrosive fluids (higher CO_2 and H_2S) than in the deepwater GoM, and hence greater use of corrosion inhibitor;
- flotation flocculant added upstream of the degassing vessel and downstream of valves;
- flotation, if present, is a compact vertical unit.

As shown in Figure 2, and as discussed in [2, 4], typical deepwater Gulf of Mexico oil / water systems consist of one or two stages of two-phase separation (gas / liquid), followed by a three-phase separator such as a Free Water Knockout (FWKO). Depending on the hydrate prevention strategy, the fluids may be cooled somewhat by the long riser between the sea floor and the platform topsides. Where insulated piping is used, the cooling is less severe. Heat exchange to increase the temperature of the incoming fluids is not often utilized. Heat is typically added downstream, primarily to assist in achieving the target oil vapor pressure, and the oil dehydration target.

In the deepwater GoM, oil / water separation tends to be focused toward the end of the gas / oil / water separation process. The produced water may be processed through a hydrocyclone. However, in some facilities a hydrocyclone is installed but not used. The reason for this is discussed below. Large horizontal multistage flotation is commonly employed. However, in some facilities two stage vertical flotation is used. When this is the case, retention time in the vertical flotation units is comparable to that in a horizontal multistage unit and cannot therefore be accurately referred to as compact flotation. Compact flotation, having short residence time, is not effective enough for practical use in the deepwater GoM.

As discussed by Bothamley [3], offshore platforms in both regions produce dehydrated gas, without dew point control, and crude oil that has had its vapor pressure reduced and has been dehydrated. Thus, both regions employ a staged (cascaded) pressure reduction system which is designed to minimize the overall load on the gas compression system. This is more or less where the similarity stops.

In the deepwater GoM, spec crude oil is produced offshore which is suitable for direct routing to the refineries located on the coast of the GoM. The crude oil has low Reid Vapor Pressure (< 11 psi), and low BS&W (< 1% by volume). This is almost without exception since it is a requirement of utilizing the extensive pipeline infrastructure that exists. In the North Sea, typical crude oil vapor pressure from offshore has a vapor pressure in the range of 150 psia, and a water content of 2% by volume. The relatively recently deployed FPSOs are an exception where spec crude is produced for tanker loading.

Also, of relevance to the present study, primary separators in the North Sea tend to have a narrower range of pressures (from 150 to 750 psia) compared to those in the deepwater GoM (from 150 to 1,800 psia). Depending on the well productivity, and sand control requirements, wells in the deepwater GoM might be routed to high pressure (1,500 to 1,800 psia), intermediate pressure (750 psia) or low pressure (150 psia). The intermediate pressure and high pressure separators are two-phase as just discussed. Thus oil and water are discharged together, through the same nozzle, and are sheared intensely through the level control valve. This results in a high concentration of oil dispersed in the water phase and in small oil drops which are difficult to separate. In the North Sea, most wells fall into the low pressure range with some wells in the intermediate pressure range. Essentially all inlet separators in the North Sea are three phase which provides segregation of oil and water before the fluids are passed through the level control valve.

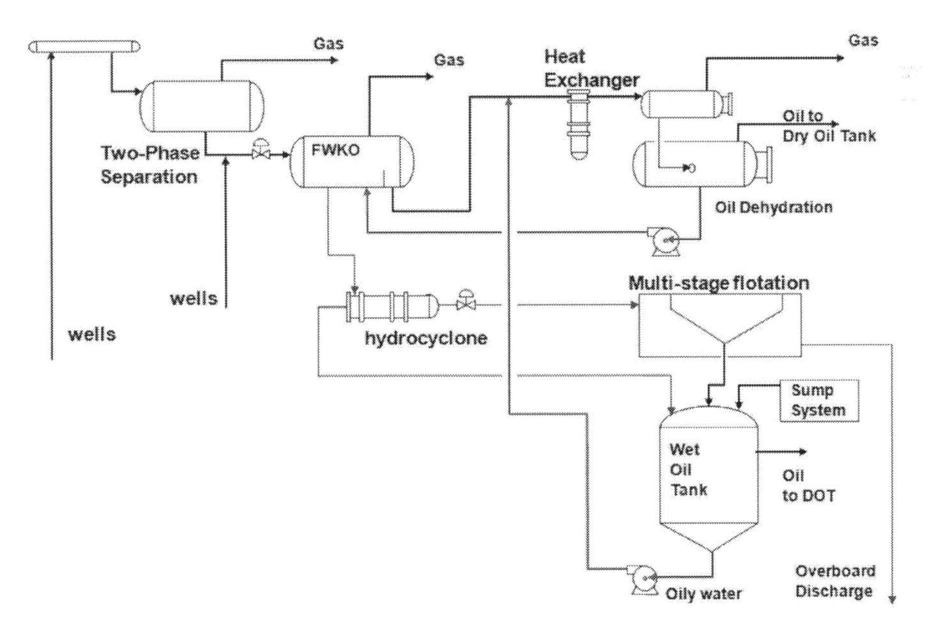

Figure 20.2 Typical Deepwater Gulf of Mexico oil / water separation system

The main features of deepwater GoM systems can be summarized as follows:

- relatively less weight and space available than in the North Sea;
- heat added downstream of the produced water separation vessels;
- few primary separators – limited segregation of crude oils and water;
- primary separators are small and two phase (sized for degassing) and thus have no hydrocyclones on water discharge;
- hydrocyclone units are operated at lower driving pressure;
- treated produced water contains a higher concentration of smaller oil droplets of less than 10 microns, as a result of more shearing and pumping in some case;
- moderately heavy use of hydrate inhibitor and other flow assurance chemicals;
- large multi-stage horizontal flotation is required.

As discussed, there are several differences between the typical deepwater GoM system and the typical North Sea system. However, from a water treating standpoint, there is one difference that has greater impact than most others. The process flow in the North Sea involves three-phase separators fitted with hydrocyclones. These separators provide two stages of oil / water separation upstream of the shearing effect of the interface control valve, gravity separation in the vessel followed by intense centripetal separation in the hydrocyclone. The configuration is shown in Figure 3 below.

As shown, the three phase separator has a spill-over weir and an oil discharge nozzle. Water accumulates upstream of the spill-over weir and is discharged. Other configurations are often used such as an oil bucket together with an underflow / overflow arrangement for the water. Not shown in Figure 3 is the inlet device, nor any perforated plates or gas demisting devices. All of these internal elements are crucial for achieving good separation. This detail is not shown at this point. The important issue at this stage is the difference between two-phase and three phase separation.

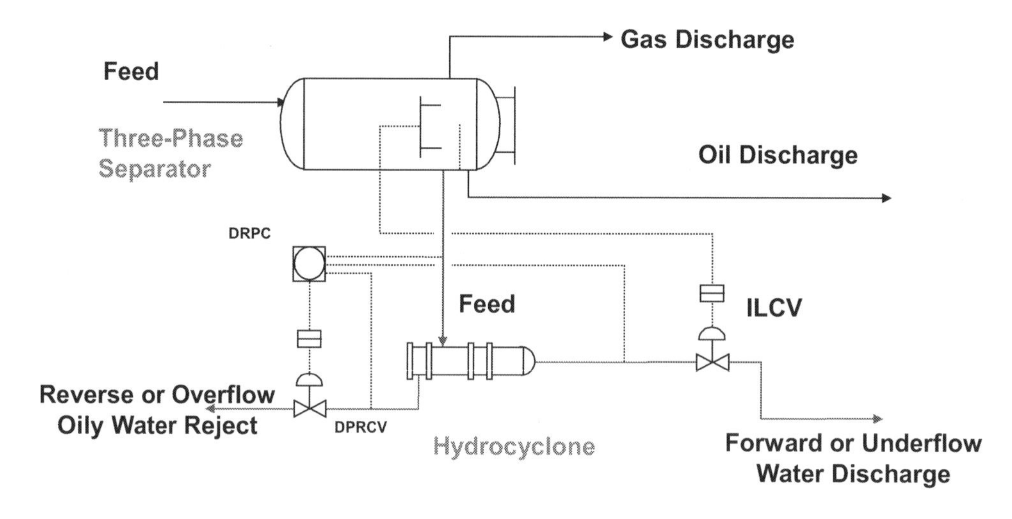

Figure 20.3 Typical North Sea three phase separator fitted with hydrocyclone. The interface level control valve is downstream of the hydrocyclone. The hydrocyclone control system is shown with a DP Pressure Controller, and an Interface Level Control Valve (ILCV)

The process flow in the deepwater GoM involves a combination of two-phase separators and three-phase separators. These separators provide initial oil / water separation upstream of the interface control valve. The configuration is shown in Figure 4 below.

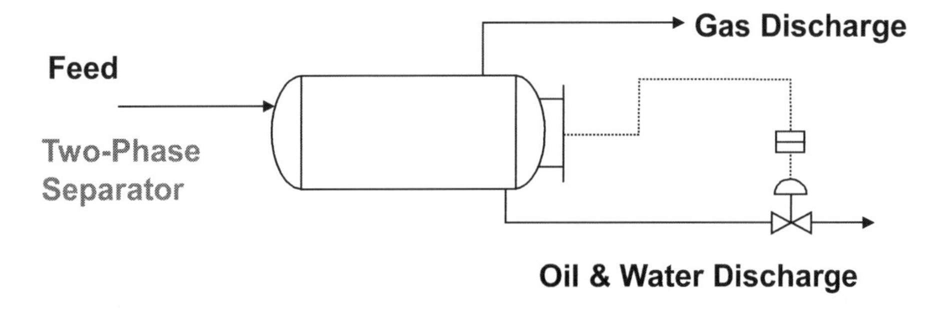

Figure 20.4 Typical deepwater GoM two-phase separator. Oil and water are discharged through a common nozzle and flow line. Both fluids are sheared due to the Level Control Valve (LCV)

As shown, the two-phase separator has a level indicator and a level control valve. Both oil and water are discharged together and flow through the separator discharge nozzle and through the level control valve where they are sheared. Not shown in Figure 4 are any internal devices which must be installed for good flow distribution and demisting of the gas.

In general, vessel internals of the separators in North Sea systems are well assessed by computational fluid dynamics modeling and by full scale pilot testing before final selection. In the GoM, vessel internal design very often is standard and not necessarily optimum for the application.

As shown in Table 1a and 1b, the typical weight of oil producing platforms in the North Sea and the deepwater GoM are significantly different. In many cases, for roughly the same oil production capacity, the platforms in the North Sea are two to three times heavier than those in the deepwater GoM. Another way to view this information is to say that weight availability is much higher in the North Sea due to the platform structure that is used in the relatively shallower water, compared to the deepwater GoM. This has been highlighted in the text below Figure 1 and Figure 2.

Not all platforms in the two regions are included. Only those platforms in each region that have moderate API gravity crude oil (25 to 30), and highly saline produced water (100 to 250 k mg/L TDS) were included. Most platforms in each region have fluid properties that fall in this range. However, there are also a few platforms in each region with fluid properties that do not. They were excluded from this analysis. Most of the excluded platforms have higher API gravity. For example the Ram Powell (deepwater GoM) gas processing facility was excluded. We could have included such platforms in this analysis because it turns out that they exhibit the same water treating line-up differences with their counterparts as seen with the moderate API gravity crude oils. For simplicity however, they were excluded.

Table 20.1 Weight of representative North Sea platforms

Platform	Year of first oil	Structure	Peak Oil Rate (kBOPD)	Topsides Weight (mt)
Gullfaks A	1986	Gravity-Based Concrete	180	48,590
Gullfaks B	1988	Gravity-Based Concrete	180	27,395
Gullfaks C	1990	Gravity-Based Concrete	180	52,935
Statfjord A	1979	Gravity-Based Concrete	200	41,300
Statfjord B	1982	Gravity-Based Concrete	200	43,000
Statfjord C	1985	Gravity-Based Concrete	200	48,000
Troll A	1996	Gravity-Based Concrete	140	25,400
Troll B	1995	Floating Concrete	140	22,400
Troll C	1999	Floating Steel	140	20,000
Heidrun	1995	Floating Concrete	200	65,000
Draugen	1993	Gravity-Based Concrete	100	28,000
Eider	1988	Fixed Steel	20	11,200

Note that the Peak Oil Rate given in this table was estimated from very granular data and as such is probably accurate to +/- 20%, and should not be used for any application requiring high accuracy. Also note that many of these platforms have experienced up to 40% water cut during the period of high oil production such that BOPD is not entirely representative of fluid handling capacity.

mt = metric ton (1,000 kg).

Table 20.2 Weight of representative deepwater Gulf of Mexico platforms.

<table>
<tr><th>Platform</th><th>Structure</th><th>Component</th><th>Peak Oil Rate (kBOPD)</th><th>Weight (mt)</th></tr>
<tr><td rowspan="2">Ursa</td><td rowspan="2">TLP</td><td>deck</td><td rowspan="2">220</td><td>5,700</td></tr>
<tr><td>topsides including the rig</td><td>10,180</td></tr>
<tr><td rowspan="3">Mars</td><td rowspan="3">TLP</td><td>hull</td><td rowspan="3">220</td><td>7,100</td></tr>
<tr><td>deck</td><td>3,300</td></tr>
<tr><td>topsides</td><td>6,200</td></tr>
</table>

mt = metric ton (1,000 kg).

TLP = Tension Leg Platform

Note that the peak oil rate for Ursa and Mars is roughly 220,000 BOPD, with an approximate water production of 10 to 20% during the period of peak oil production.

20.1 Examples of GoM and North Sea Water Treatment Process Lineups

In order to discuss the differences in produced water process lineups, it is necessary to have actual cases upon which to make observations. The elements discussed are: inlet fluid characteristics, equipment used, process lineup of the equipment, production chemicals added, and any unique operating practices. The equipment selected and the process lineup has a profound effect on the final effluent water quality. This is particularly true for offshore installations where space and weight are constrained. In addition to process lineup, the inlet fluid properties, chemical treating program and operations strategy have important effects on final effluent produced water quality. These later factors are discussed briefly below but are not the focus of this paper.

When properly defined, process lineups for water treating are complex. Each lineup is composed of oil / gas / water separation equipment and water treating equipment. The feed and discharge streams of that equipment are connected in a particular piping configuration. The details of that configuration are important and can have a major impact on the performance of a system. Reject handling, for example, is an important part of produced water treating. In a properly defined process lineup, all significant reject and recycling streams are shown. Since shearing of produced water leads to small oil drops and the potential for water treating problems, all significant valves and pumps are shown. Likewise, since the addition of heat can have an important impact on produced water treating, all significant heaters and heat exchangers are shown. Any process description that is missing these elements is incomplete.

20.1.1 North Sea Field Examples

North Sea Field A: High pressure wells (21 barg arrival pressure) are processed in a two-stage separator train comprised of three phase separators. Oily water is discharged from the first stage separator and processed through hydrocyclones. Low pressure wells (11 barg arrival pressure) are processed in a single three-phase separator. Oily water is discharged from the separator and processed through hydrocyclones.

Thus, all production is routed through three phase separators with hydrocyclones on the water discharge. The effluent from all hydrocyclones is routed to a produced water flash tank, and then discharged to the sea. Typical oil discharge quality is 2% BS&W. Typical water discharge quality is 20 to 40 mg/L oil content.

North Sea Field B: Oil is processed in two parallel separator trains, both with three stages, and comprised exclusively with three phase separators. First stage separators are operated at 56 barg and second stage at 15 barg. Third party oil production is also produced in the same production trains, entering at the second stage. Produced water is taken from all first- and second- stage separators and is treated in hydrocyclones, produced water flash tanks and flotation cells before discharge to sea. Typical oil quality to storage cells is 0.7 % BS&W. Typical discharge produced water quality is 10-20 ppm oil content.

North Sea Field C: The oil is "medium heavy oil" and is heated between first and second stage separator in order to achieve desired separation (from approx. 50°C to 70°C). The oil is processed in one separation train with four stages, with an electrostatic coalescer as the last stage. All separators are three-phase with hydrocyclones on the water discharge. The first stage separator is operated at 14 barg and second stage at 10 barg. Produced water is separated from the first- and second-stage separators and is treated in hydrocyclones, produced water flash tank and an Epcon Compact Flotation Unit before discharge to sea. Typical oil export quality is 0.5% BS&W. Typical discharge produced water quality is less than 10 ppm oil content.

North Sea Field D: The operator of this field has over time experienced challenges with naphthenate precipitations. This oil also needs heating between first and second stage separator in order to achieve the desired separation (from approx. 67°C to 77°C).

The oil is processed in one separation train with three stages. First stage separator is operated at 14 barg and second stage at 10 barg. Produced water is separated from the first- and second- stage separators and is treated in hydrocyclones, Epcon and produced water flash tank before produced water re-injection (PWRI).

Typical oil export quality is 0.3% BS&W, and typical produced water quality is 15-30 ppm oil content (PWRI).

North Sea Field E: The oil is easy separable oil produced from sandstone reservoir. The oil is processed in two parallel first stage separators and one second stage separator. The first stage separators are operated at 7 barg. Produced water is separated from the first stage separators and is treated in hydrocyclones and produced water flash tank before discharge to sea. Typical oil export quality is 1-2% BS&W, and typical discharge produced water quality is 20-30 ppm oil content.

North Sea Field F: The process lineup for produced water for this platform is given in Figure 1. As shown, fluids enter the platform from either direct platform wells or from various subsea systems. The typical hydrate control strategy is based on pipe-in-pipe or insulated pipe-in-pipe flow lines. This reduces the amount of hydrate inhibitor needed. This has had a beneficial effect on water treating.

There are five inlet separators. Two inlet separators are vertical and high API fluids (40 degrees or so) are processed. The three horizontal inlet separators are shown. One is lined up to a light fluid, and the other two are lined up to heavier fluids. This allows segregation of asphaltenic crude from light crude oils which might act like an anti-solvent for the asphaltenes. The three primary separators were designed for moderate flux rate.

All primary separators are three phase. The gas condensate separators are vertical without a hydrocyclone. All of the horizontal separators have a hydrocyclone on the water discharge line of the separators and upstream of the level control valve. The hydrocyclone reject is routed to the feed oil line of the downstream separator. This routing does not require a pump. It does introduce a water stream into an oil stream with a risk of forming a complex emulsion.

Originally a flash drum was installed between the hydrocyclone effluent and the flotation unit. A flash drum (or degassing vessel) provides greater separation and an opportunity to allow chemicals to mix and react.

There is no slops vessel (or Wet Oil Tank). Since this vessel is not present, much of the water reject from the downstream hydrocyclone and the flotation unit must be routed upstream without a chance to settle and coalesce.

The overall performance of the Platform water treating system has historically been poor. This is a pity because the upstream lineup is effective from a water treating standpoint. The upstream separators are three phase and the water discharge of those separators have hydrocyclones installed.. The three factors which affect the system performance are:

- Use of relatively high concentrations of corrosion inhibitor (100 to 150 mg/L),
- Excessive sand production (2 tons per day),
- Poorly performing flotation unit (20% separation efficiency).

Features beneficial to produced water treating:

- use of pipe-in-pipe insulation, thus moderate use of hydrate inhibitor,
- heat added upstream,
- several primary separators to segregate incompatible fluids,

- primary separators are three phase,
- hydrocyclones on every separator.

Features detrimental to produced water treating:

- no flash drum (degassing vessel) between hydrocyclone outlet and flotation feed,
- no wet oil tank for skimming and chemical treating.

<u>**North Sea Field G:**</u> The process lineup for produced water has the option of overboard discharge or injection in disposal wells. Three phase fluids (four phase including sand) enter the platform and are initially separated in the primary production separators. Water from the separators enters sand cyclones. There is an option to route water from the discharge of the sand cyclones to the water injection wells. This option was installed as part of a field trial to study produced water injectivity as a function of water quality. Normally, this option was not used. Normally the discharge of the sand cyclones is routed to the hydrocyclones. The hydrocyclone effluent water goes to a Produced Water Flash Drum (PWFD). The effluent water from the PWFD can either be discharged overboard or routed for further treatment ultimately disposed in an injection well. Further treatment involves filtration through 10 micron cartridge filters, and cooling through heat exchangers. It is then sent to the injection pumps for subsurface disposal.

Features beneficial to produced water treating for Reinjection:

- oil and solids separation equipment and process routing are suitable for injection targets,
- water cooling prior to reinjection is beneficial for maintaining injectivity.

Features detrimental to produced water treating:

- operability of cartridge filters is a problem. They require frequent change-out.

20.1.2 Gulf of Mexico Field Examples

<u>**GoM Field A:**</u> This field was one of the first deepwater platforms in the world. Due to capital constraints, it was deployed without water treating equipment. All inlet separators were relatively small two-phase vessels with high flux rates and short residence time. Incoming fluids were relatively cool (30 to 50 C), and heat was only added downstream for crude oil vapor pressure control.

For the first few years of operation, all water producing wells were shut-in. This was a reasonable strategy for several years since initial production remained essentially dry. However, when water production did commence, it rose rapidly to about 20%. During an 18 month period, roughly 16,000 BOPD was shut in due to an inability of the facilities to treat the water. This equated to roughly 8 MM bbl of oil deferred. During this time, a project was executed to install, and shortly thereafter to expand the water treating system.

The water treating process lineup eventually included the conversion of one of the two-phase primary separators to a three-phase vessel with the addition of hydrocyclones. CFD was used to for vessel internal element optimization. A produced water flash tank was installed downstream of the hydrocyclones. A multi-stage horizontal flotation unit was installed.

<u>**GoM Field B:**</u> The process lineup for this field is typical of deepwater GoM and as such is given in Figure 2. As a general design strategy for this field, the high pressure and intermediate pressure separators are two-phase with short residence times (and therefore high flux rates). The Free Water

Knockout is three phase but is also designed with short residence time as well. Hydrocyclones were only installed downstream, and heat was only added downstream. Thus, the topsides design was extremely compact, like its predecessor platform "A."

Reject from the hydrocyclones is fed to the Wet Oil Tank. Reject from the flotation unit is fed to a Slops Oil Tank and then to the WOT. Reject form the slops tank is fed to the Dry Oil Tank where it is then exported along with the platform produced oil.

The Slops Oil Tank was used to degas the flotation reject stream, and to separate oil which could then be pumped to the Dry Oil Tank. Oily water from the Slops Tank was pumped to the Wet Oil Tank.

Due to tight capital constraints, minimal flow line insulation was used. The typical hydrate control strategy was to use methanol on startup. This had an adverse effect on water quality.

Chemical treatment on the platform included injection of dithiocarbamate (DTC) upstream of the flotation unit. DTC is in general a fast acting and highly reactive chemical. This property makes it effective as a deoiling compound since it sticks well to oil drops. But it will also stick to metal surfaces forming a gum-like substance that hardens over time. This is typically referred to as polymer floc, or polymer scale. To prevent polymer floc when using DTC, the practice was to mix a hydrocarbon solvent into the Wet Oil Slop Tank. The hydrocarbon would react with the unreacted DTC in the flotation reject stream and prevent scale formation in the recycle. Initially the practice was to use HAN but later crude oil was used.

Historically, the average OiW concentration is in the range of 20 mg/L, and only a few sheens occur per year. The hydrocyclones operated at 70% efficiency and the flotation unit operated at 90 % efficiency.

Features beneficial to produced water treating for Reinjection:

- effective sand control,
- low corrosivity of produced fluids, hence limited use of corrosion inhibitor,
- water cooling prior to reinjection is beneficial for maintaining injectivity,
- use of multi-stage high efficiency flotation.

Features detrimental to produced water treating:

- HP & IP inlet separators are two-phase without hydrocyclones,
- high flux rates through the inlet separators,
- extensive use of un-insulated flow lines thus hydrate inhibitor is required on startup,
- heat is only added downstream.

20.2 Field Data & Modeling

In this section, modeling tools are applied, together with field data, which demonstrate the differences in a quantitative manner. The models have been developed using fundamental principles discussed in various parts of this book (see chapter 4, 6, 12). The modeling tools provide a quantitative estimate of the relative importance of various factors which differentiate the systems in the two regions (such as inlet fluid shear, temperature, water density, oil density, separator flux rate, residence time, application of hydrocyclones, flotation, etc).

The objective of the modeling is to quantify the differences in water treating systems between the two regions. The most straightforward way to carry out this analysis is to use a single representative water treating system from each region and to vary the fluid temperature and process lineup.

20.2.1 Oil Drop Diameter Distribution

Oil drop diameter is a key parameter in modeling water treating equipment. There are many ways of measuring oil drop diameter in the field and in the laboratory. The original method, based on photo-microscopy and manual counting of drops has been in practice for nearly 30 years. It remains one of the more accurate methods, if properly executed. Modern methods utilize particle imaging. Recently the quality of particle imaging instruments and software has achieved a level of accuracy that is comparable to the manual photo-microscopy method. Of course, the instruments are much faster.

One of the well-established parameters in modeling is the maximum oil drop diameter. It is loosely defined as the maximum drop diameter that will exist under a given set of conditions. For example, in determining the effect of valve shear, Dmax is the maximum drop diameter in the discharge of the valve. It can be estimated based on simple (Hinze-Kolmogorov) formulas which take into account the pressure drop, water viscosity, interfacial tension, etc..

Many separation devices are characterized by the Dmax, i.e. the smallest oil drop diameter that will be completely separated. By applying Stokes Law to gravity and enhanced gravity based separation, such as occurs in separators and hydrocyclones, Dmax and the oil in water concentration can be predicted. From a modeling standpoint, Dmax is a simple and useful parameter.

In reality though, oil drop diameter is a distribution function. The field measurement methods discussed above provide a histogram of oil drop diameter which can be readily converted into a distribution function. There is a significant and growing body of literature which provides modeling insight into the drop diameter distribution and how it changes with shearing, coalescence, and separation processes. Further, the formulas based on Dmax are inaccurate where significant drop shearing occurs, as will be discussed shortly. Thus, the modeling work presented here is based on oil drop diameter distribution, rather than Dmax.

In this book, drop diameter distribution is modeled using either a Log-Normal distribution or a Rosin-Rammler distribution. These distribution functions are discussed in Section 3.3 (Particle and Droplet Size Distributions). Numerous data sets from both field measurements and laboratory measurements have established that the Rosin-Rammler distribution is effective in modeling the small scale droplets that are generated in intense shearing. The Log-Normal distribution is more effective at modeling the shape of the oil droplet distribution near the mean.

One of the interesting findings is that drop diameter distributions tend to be bi-modal, as shown in Figure 5 below, particularly where significant drop shearing occurs. There is currently debate in the literature as to the fundamental mechanisms that cause this bi-modal distribution. One theory suggests that under high shear conditions, very small drops are formed by the violent smashing of larger drops. Another theory suggests that small drops are less likely to coalesce than larger drops and they get "left behind" so to speak as coalescence occurs downstream of valve shear. In any case, bi-modal distributions are often seen throughout the deepwater GoM, where valve shear is more pronounced, and to a lesser extent in the North Sea where shearing is less of an issue. The impact of this bi-modal drop diameter distribution, on oil in water separation, is discussed further below.

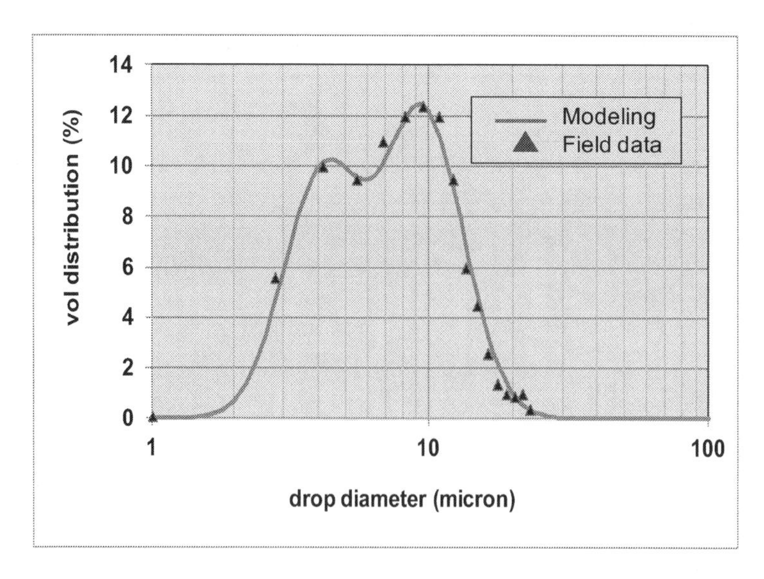

Figure 20.10 A pronounced example of a bi-modal drop diameter distribution from the deepwater Gulf of Mexico. The bimodal distribution is typical of systems where drop shear occurs. Points are field data measured using a drop size analyzer. The line is a curve-fit using two Log Normal distribution functions.

20.2.2 Separation Equipment

The fluids in the liquid settling zone of the vessel can be thought of as having three distinct zones. The upper-most zone is composed of an oil continuous liquid containing dispersed water drops. The water drops settle according to Stokes Law. The bottom-most zone is composed of a water continuous liquid containing dispersed oil drops. The oil drops cream (rise to the top) according to Stokes Law. Between these two layers, there is typically a dispersion zone [5] where the transition between oil-continuous and water-continuous fluids takes place. Stokes Law is applied to model oil separation from water in the water leg of gravity based separators. Stokes Law gives the rate of rise of oil drops in water. Stokes Law was discussed in Section 3.2.3 (Stokes Law Settling).

If the calculated $S(d)$ is greater than unity, then $S(d)$ is set equal to unity. An example of a Separation Efficiency Curve is given in Figure 6 (see the green line). According to equations 4 and 5, the separation efficiency curve for a gravity separator depends on the oil and water density, the water viscosity, the water residence time, and the volume and height of the water leg. In the equations above, the separation efficiency is expressed as a fraction which varies between 0 and 1. Typically however, it is expressed as a percentage, as shown in Figure 6.

At this point, it is important to point out that there are a number of approximations implied in equations 4 and 5 when applied to modeling gravity separators. These equations are only valid for an ideal set of conditions which almost never occur in practice. To account for real-world effects, the residence time is multiplied by a Hydraulic Efficiency Factor which has been assigned a value of 0.7. This factor accounts for the following set of non-ideal conditions. First, the actual residence time is a distribution function and not a single value. Part of the fluid follows a streamline from the inlet to the discharge. This fraction of fluid will have a short residence time. Another part of the fluid gets caught in swirls and eddies, with longer residence time. Internal devices such as perforated plates are intended to reduce such effects but they typically do not eliminate them. Most separator vessels have formation sand in the bottom which reduces the volume available for residence time. Also, drop-drop coalescence is not treated explicitly. It has been observed that a Hydraulic Efficiency Factor of 0.7 is reasonable.

The drop diameter distribution, and the oil concentration in the water phase discharged from the vessel are calculated as:

$$F(d)_E = F(d)_F(1 - S(d)) \quad \text{Eqn 20.1}$$

$F(d)_E$ oil drop diameter distribution of the effluent

$F(d)_F$ oil drop diameter distribution of the feed

In Figure 6, the Separation Efficiency Curve is based on Stokes Law settling for oily water in a primary separator. The Separation Efficiency Curve for Stokes Law settling has a characteristic maximum drop diameter. In the figure this value is 100 microns. In general this value depends on the residence time, fluid viscosity, and density difference between the oil and water, i.e. all of the Stokes Law variables. Note that, as shown in the figure, the maximum drop diameter is 100 micron (green curve) and the outlet drop diameter distribution (blue curve) has no oil drops of 100 micron or larger.

Other separation equipment can be modeled using equation 6, such as hydrocyclones, and flotation units. As discussed below, a modified version of Stokes Law is applied to model hydrocyclones. In those cases, the Separation Efficiency Curve derived for the particular equipment would be used. More detail is given below.

Using the above approach, it is possible to calculate the oil drop diameter distribution at various points within the water treating system. Combining these tools allows a complete model for the water treating system. The final modeling tool combines information from these individual tools into a map of the performance of the system.

Inlet Fluid Condition: The performance of a water treating system is greatly influenced by the processes upstream of the inlet separators. The condition of the inlet fluids (oil in water concentration, and oil drop size) has a major effect on the oil drop size distribution, and hence on the performance of the oil / water separation equipment. Upstream of the inlet separators, the fluids are subjected to shearing and coalescing processes due to artificial lift, slugging in pipeline and riser flow, and flow through chokes and valves. Both shearing and coalescence occur to a greater or lesser extent depending on the equipment, flow rate, temperature, and ultimately on the intensity of the turbulence.

As discussed in Section 3.4 (Turbulence) and 3.5 (Shearing of Oil Droplets), the turbulence intensity (turbulence energy dissipation rate) is a well defined quantity. It is expressed in units of watt / kg of fluid. A fluid pipeline has low turbulence intensity. When the turbulence intensity is low, the fluid motion promotes drop-drop collisions which lead to coalescence. This is why in many cases the shearing effect of an electrical submersible pump is minimized by the time the fluids reach the inlet separator. If the pipeline from the well head to the separator is relatively long (greater than ¼ mile), then significant coalescence can occur. Arnold [6] gives a general rule of 300 pipe diameters for the effect of oil drop coalescence to occur. A more detailed model of droplet size, which takes into account coalescence, is given in Section 4.5 (Coalescence of Oil Droplets).

Therefore, in the present work, the effect of an essentially unknown inlet fluid condition was minimized using the following approach. Data was gathered for the fluid discharge of the inlet separators. This included oil concentration in the water phase and the oil drop diameter distribution in the water phase. In other words, the data gathering and modeling employed here was focused on the oily water discharged from the inlet separators, rather than the inlet to the separator. The oily water discharge was the starting point for the analysis. A knowledge of residence time together with Stokes Law was then used to back-calculate the inlet fluid condition. This was a far more practical approach than to attempt to predict the inlet fluid condition. As discussed previously (see text following equations 4 and 5), various approximations are involved in this modeling approach.

Stokes Factor: In this work, the Stokes Factor is used to account for the properties of the fluids. The Stokes Factor was discussed in Section 3.2.5 (Stokes Factor). The effect of fluid properties, including temperature, on separation performance can be captured through calculation of the Stokes Factor. In Figure 7, the inverse of water viscosity ($1 / m_w$) is given as a function of temperature. Also given is the Stokes Factor as a function of temperature for an example crude oil and water combination. For this illustration, the oil is assumed to have an API of 27 degrees (density 871 kg / m^3 at 60 F, 15.5 C), and the produced water is assumed to have a density of 1,100 kg/m3. For this calculation, the density of the oil and water are assumed to not change significantly with temperature. Only the viscosity of the water is assumed to vary with temperature. This is an illustrative calculation.

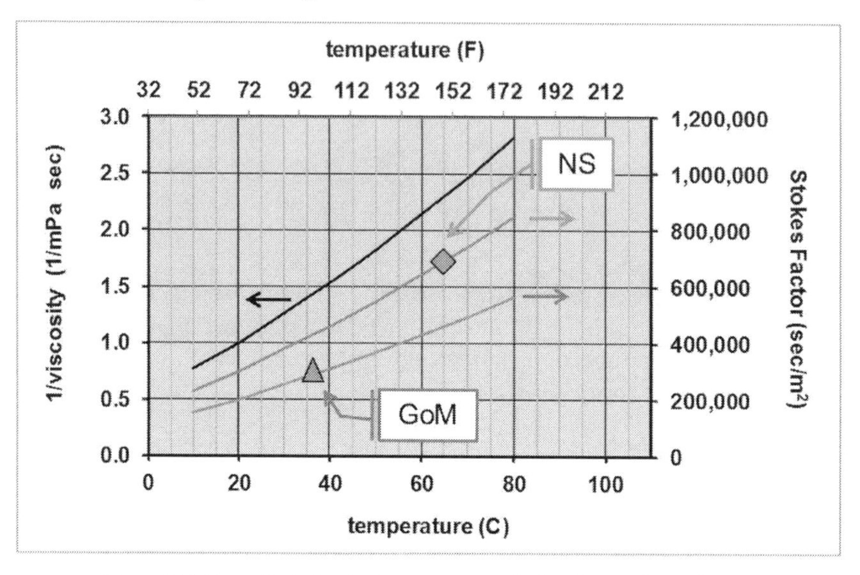

Figure 20.11 The effect of water temperature on Stokes Law settling velocity, as given by 1/viscosity as a function of temperature

Water Residence Time vs Stokes Factor: In designing a separator for oil / water separation, the geometrical dimensions of the vessel are determined from the flux rate required to resolve the dispersion band [5] and the residence time required to perform initial oil dehydration and water deoiling. Together these parameters determine the volume of the liquid section, the height of the oil/water interface and the height of the oil/gas interface, and the cross sectional area of the liquid section of the vessel.

Within the dispersion band, there is a honeycomb pattern of relatively large drops of oil and water separated from each other by thin films of the other liquid. Thus, within the dispersion band, the separation of oil and water is driven by the mechanics of film drainage and buoyancy forces that cannot be modeled by Stokes Law due to the non-isolated nature of the phases. From a macroscopic or continuum mechanics standpoint this process is characterized by flux (flow per unit area), rather than settling time or residence time. The most important property of the liquid is the oil viscosity since it controls the drainage rate of the liquid films between the large drops of water within the dispersion band. Without going into any more detail, separator cross section area determines the capacity of the separator to resolve the dispersion band.

The water section below the dispersion band is available for drops of oil to cream out (rise out) of the water phase. This later process is most appropriately modeled by Stokes Law. The residence time and vessel geometry (height to NIL and NLL) determines the capacity of the separator to provide initial deoiling of the water phase.

Figure 8 gives the residence time versus the Stokes Factor for a number of North Sea and deepwater GoM platforms. In all cases the residence time is for the water phase in the first (most upstream) three phase separator in the system. In the case of the North Sea systems, this is the primary inlet separator. In the case of the deepwater GoM systems, this is the Free Water Knock Out (FWKO), which is downstream of the HP and IP separators. Where there is an inlet heat exchanger, the discharge temperature is used. The residence time shown here is the so-called theoretical residence time defined by the volume of the water section divided by the volumetric flow rate of water. This is not an accurate measure of residence time. As discussed in the literature, the actual residence time is a distribution function which depends on the hydrodynamics of the vessel. In the modeling below, we use an hydraulic efficiency factor to account for this.

Each point represents a facility at a given period in time. In Some cases, a single platform will have more than one point in the figure in order to represent different flow rates and water cuts during the life of the platform. As water cut rises, which is a significant and consistent process in the North Sea, the water residence time has a tendency to decrease and the fluids become hotter. The facilities in the deepwater GoM have not experienced a significant increase in the water cut, at least not to the extent of the facilities in the North Sea.

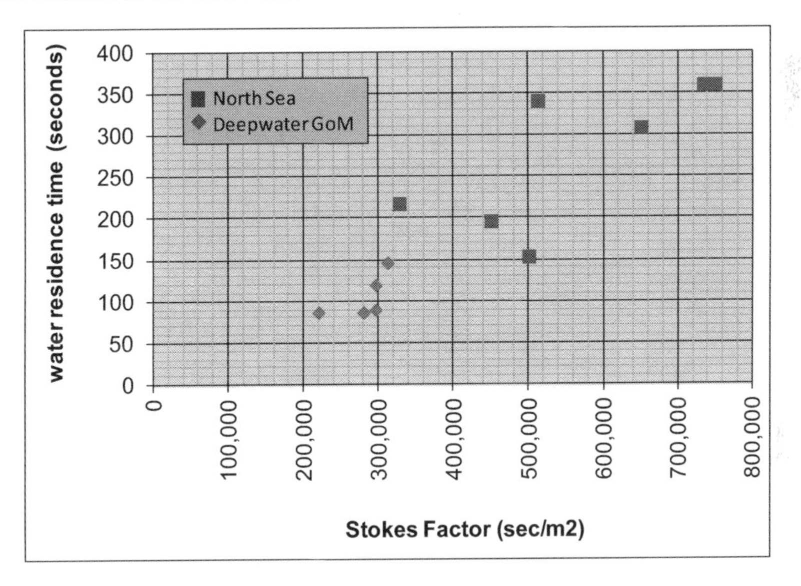

Figure 20.12 Water phase residence time versus Stokes Factor for North Sea and deepwater GoM platforms

It is evident from this survey data that water residence time is generally two to three times greater in the North Sea compared to deepwater GoM. Also, the Stokes Factor has a wide range in the North Sea but that it is generally greater in the North Sea than in the deepwater GoM. An interesting point to note is that the lowest value of Stokes Factor reported in the North Sea (320,000 sec / m2) is for the Gannet D Field which required multistage horizontal induced gas flotation. This point will be further discussed below. These field data for Stokes Factor and water residence time will be used below in modeling the differences between the North Sea systems versus the deepwater GoM systems.

Hydrocyclone Modeling: Modeling of hydrocyclones was discussed in Section 10.2 (Deoiling Hydrocyclone – Theory, Design and Modeling). Reitema [7] derived a quantity that has become known as the Hydrocyclone Number [8 – 10]:

$$N_{Hy} = \frac{Q_f \Delta\rho (d_{75})^2}{\mu D^3}$$ Eqn 20.2

N_{Hy}	=	Hydrocyclone Number (dimensionless)
Q	=	volumetric flow rate (m^3 / sec)
Dr	=	density difference between water and oil (kg / m^3)
d_{75}	=	drop diameter for which 75% of drops are separated (m)
m	=	viscosity of the water (Pa sec => kg / m sec)
D	=	involve diameter of the cyclone (m)

The fact that the Hydrocyclone Number is dimensionless can be verified. The parameter d_{75} is key to this relationship. It specifies a particular drop diameter. More precisely, it is defined as the drop diameter for which 75% of the oil volume will be separated. The other 25% of oil volume is discharged in the effluent (product) water.

There was no particular reason why the value of 75 percent was chosen. Another value, of say 50 percent, could equally well have been chosen. In that case, the value of the Hydrocyclone Number would be different. Mathematically any convenient value in the range of about 20 to 80 percent could have been chosen. It is a feature of hydrocyclones however that the 100 percent separation value cannot be chosen. Without going into details, the s-shape of the separation curve precludes the use of a d_{100} value.

Physically, the value of 75 percent relates to the inner 75% of the cross sectional area. Within this area, drops of this size (d_{75}) are close enough to the core of the cyclone such that they have enough time to migrate to the core and be separated from the bulk of the water stream which is discharged as effluent. The other 25% of drops of this diameter happen to enter the cyclone too far away from the core to migrate and be separated.

As demonstrated by Reitema, and verified by Colman and Thew and coworkers, The Hydrocyclone Number is only a weak function of the hydrocyclone Reynolds Number. Without going into details, this suggests that it is primarily a function of hydrocyclone geometry. Thus, for a given hydrocyclone liner design, it is relatively constant and can therefore be used to predict performance over a range of operating conditions.

The particular equations used in this paper are based on the measurements carried out at the Orkney Water Treating Center [8]. The Hydrocyclone Number is calculated using equation 9, for a given value of d_{75}. Separation efficiency is then calculated from the following empirical equation:

$$S(d) = 1 - \exp[c_1(\tilde{d}_{75} - c_2)^{c_3}]$$ Eqn 20.3

$S(d)$	=	separation efficiency curve, as a function of drop diameter (fraction)
$\tilde{d}_{75}$	=	d/d_{75} = reduced drop diameter (dimensionless)
c_1, c_2, c_3	=	adjustable parameters (dimensionless)

The value of d and hence $S(d)$ is calculated from equation 10. A plot of $S(d)$ versus d is given in Figure 9 below. The performance (separation efficiency) given in the figure is characteristics of a commercial hydrocyclone liner such as the Krebs K-liner.

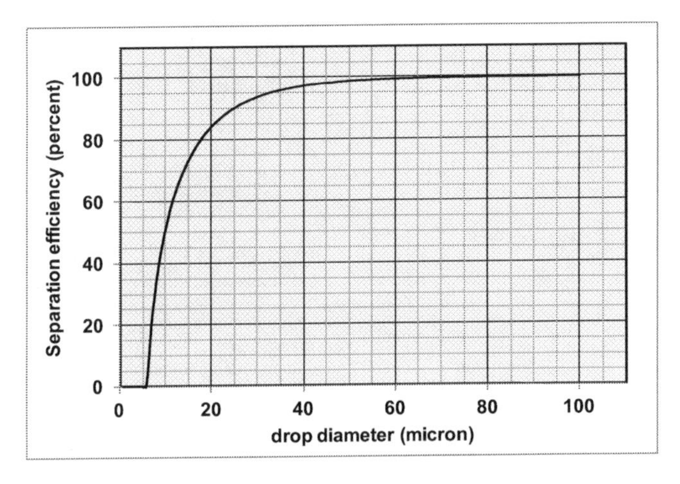

Figure 20.13 Example of a Separator Efficiency Curve for a hydrocyclone. Note that there is no sharp cutoff at large drop sizes (cannot define a d100 due to the s-shape of the curve)

To summarize, several variables are taken into account in the model including the forward flow rate as a function of the forward pressure drop. The magnitude of the centrifugal force is calculated using dimensional and similarity analysis. Stokes Law is used to calculate the speed at which an oil drop moves toward the core. Fluid properties such as temperature impact the separation in two ways, one it impacts the Stokes Law calculation of drop migration speed, and two it impacts the development of swirl motion for a given pressure drop. This later effect is crucial for an accurate understanding of hydrocyclone performance.

Flotation Modeling: Since flotation is required in the deepwater GoM in order to meet water quality requirements, a flotation model is required. The author has developed a quantitative model for flotation performance as a function of the flotation type, gas rate, and bubble and oil drop diameter. However, such detail is not necessary here. Instead, a comparison of horizontal multistage flotation versus single stage vertical compact flotation is given in Figure 10. As shown, for multistage flotation, separation efficiency is 85% for inlet oil drops in the range of 10 to 14 micron diameter. The single stage compact flotation achieves a separation efficiency of about 50%. However, as shown, the inlet drop oil concentration for compact flotation, as deployed in the North Sea is considerably lower than that in the GoM. Thus, low separation is not only adequate, it is considered prudent for this application.

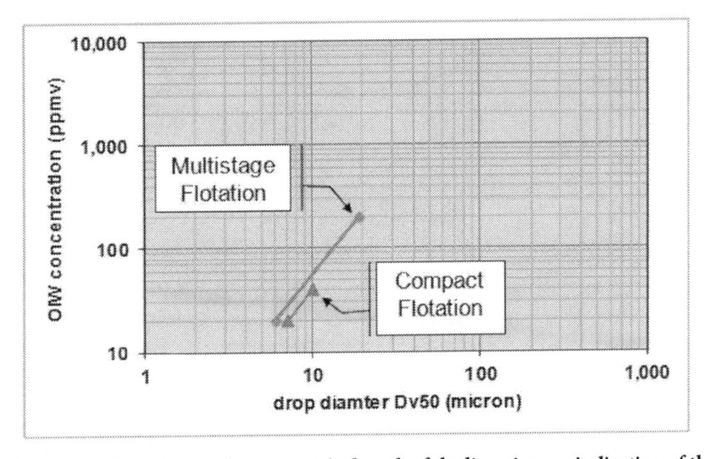

Figure 20.14 Example of flotation performance. The length of the line gives an indication of the separation efficiency. Note that the inlet drop size distribution for the two systems is different. Both flotation units are typically installed downstream of a hydrocyclone. Multistage horizontal flotation is typically used in the deepwater GoM. Single or double stage vertical compact flotation is sufficient for many applications in the North Sea.

20.2.3 Modeling Validation and Training Sets:

Literature and field data were used to validate the modeling tools, and to provide some additional understanding of the differences in the North Sea versus deep water systems, as shown in Figures 11 - 13. Validation of the hydrocyclone model has been carried out in association with the original Orkney Water Treating Center work. In that work, pilot systems were set up with four models of hydrocyclones. Oil concentration was measured in the feed, the effluent and in the reject. The data were then used to curve-fit the model parameters.

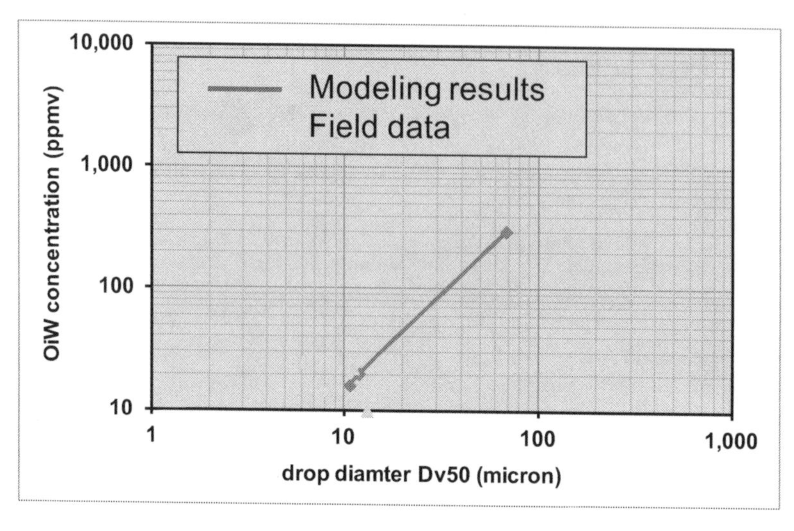

Figure 20.15 Performance diagram for hydrocyclone (and degassing vessel) operating at North Sea conditions, together with data from four North Sea platforms. The field data (yellow circle) were measured on the discharge from the degassing vessel. The modeling (blue points and line) are for hydrocyclone feed (upper right), hydrocyclone effluent, and degassing vessel effluent (lower left).

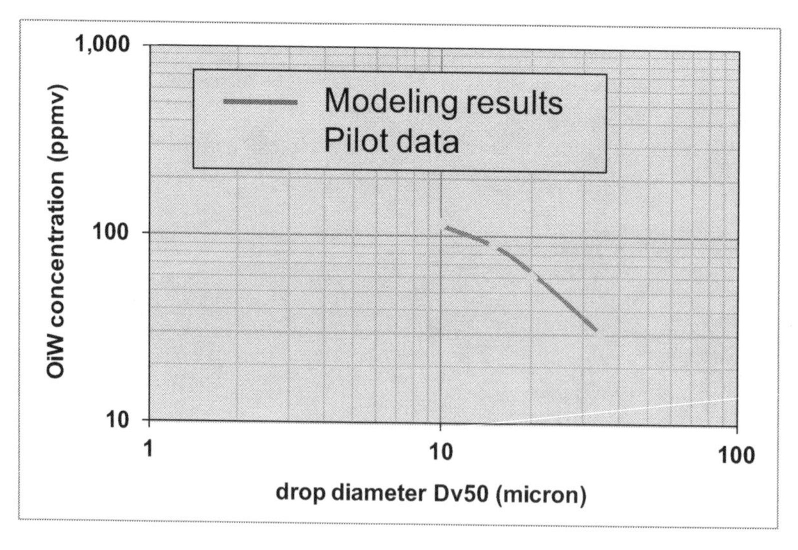

Figure 20.16 Modeling and experimental data. In this case, the inlet drop diameter was not measured. It was treated as a single adjustable parameter in the model. Thus, this data set does not provide a rigorous test of the model. However, since only one value of inlet drop diameter was used for all four measured data points, and all four measured outlet values were accurately reproduced, it does provide at least some sense that the model is accurate.

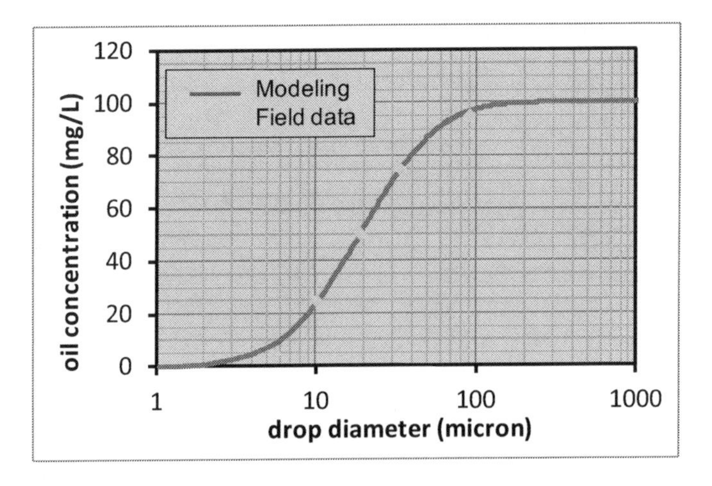

Figure 20.17 Hydrocyclone inlet measured drop diameter distribution (yellow points) for the Gannet platform, together with a Log-Normal distribution model (blue line). Good reproduction of the data is obtained by adjusting parameters in the Log-Normal distribution.

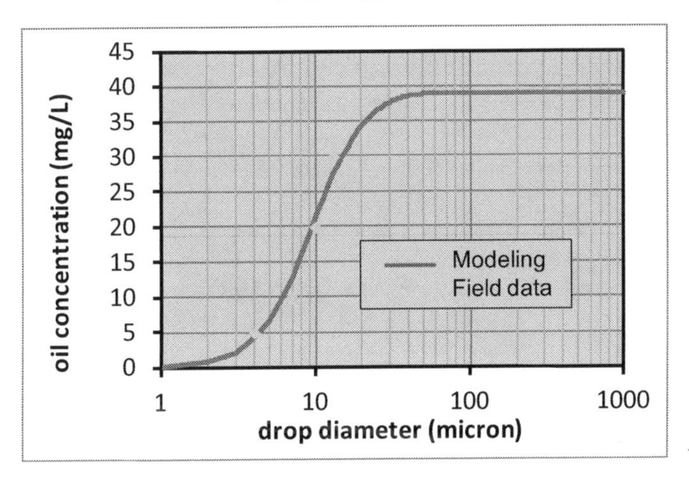

Figure 20.18 Outlet measured drop diameter distribution (yellow points) for the Gannet platform, together with model results for hydrocyclone performance (blue line).

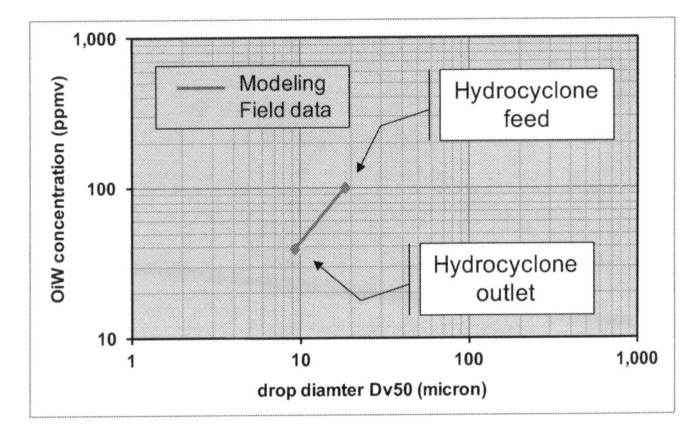

Figure 20.19 The same hydrocyclone inlet and outlet data and modeling results as shown in Figures 13a and 13b. The hydrocyclone feed characteristics are input to the model. The model then predicts the outlet characteristics. Good agreement between the model and measured data is obtained.

20.2.4 Effect of Shear and Temperature

Two important differences between the North Sea and deepwater GoM systems are temperature of the fluids and shearing through the system. North Sea fluid temperatures are higher than those of the deepwater GoM. This is due in part to different reservoir temperatures. It is also due in part to the longer riser lengths in the deepwater GoM. Fluid temperature strongly affects primary separation. The higher the temperature, the lower the water viscosity. Lower viscosity gives faster oil drop settling. The separation performance of both vessels and hydrocyclones benefits from higher temperature. But the important question is, how much of an effect does temperature have compared to other factors? Asked another way, could temperature alone be responsible for the fact that most North Sea platforms do not require extensive flotation?

Shown in Figure 14a are modeling results for the effect of hydrocyclone inlet fluid temperature. The temperature range considered is rather large (36 C to 72 C), but representative. In the modeling, the same drop diameter distribution for the feed was used for both the hot fluid and the cold fluid, The only difference was the fluid temperature. As shown by the length of the line, and by the final oil concentration, temperature has a dramatic effect.

Another important variable is shearing. Shearing itself is not difficult to model. There are many variations of the Hinze model that give satisfactory estimates of drop diameter through valves. However, shearing alone is not enough. Together with shearing, drop-drop coalescence is an important associated mechanism which occurs both in the valve itself and in the piping downstream of the valve. Rather than model these effects explicitly, the net result of these mechanisms is considered. In other words, a range of drop diameters was considered as input to a hydrocyclone.

Shown in Figures 14b are modeling results for the effect of smaller drop diameter which represent the effect of shear. In this figure, only the cold fluid (T = 36 C) representative of the GoM was studied. The blue line in Figure 14b is the same as the blue line in Figure 14a (same temperature and same inlet drop diameter).

A range of three inlet average drop diameters is shown (30, 42, 58 micron). This is an entirely reasonable range of hydrocyclone inlet drop diameters for the deepwater GoM. In this case, a dramatic shift in separation occurs as a function of inlet drop diameter. As shown, smaller drop size due to shearing, together with colder temperatures, can be responsible for significant performance deterioration.

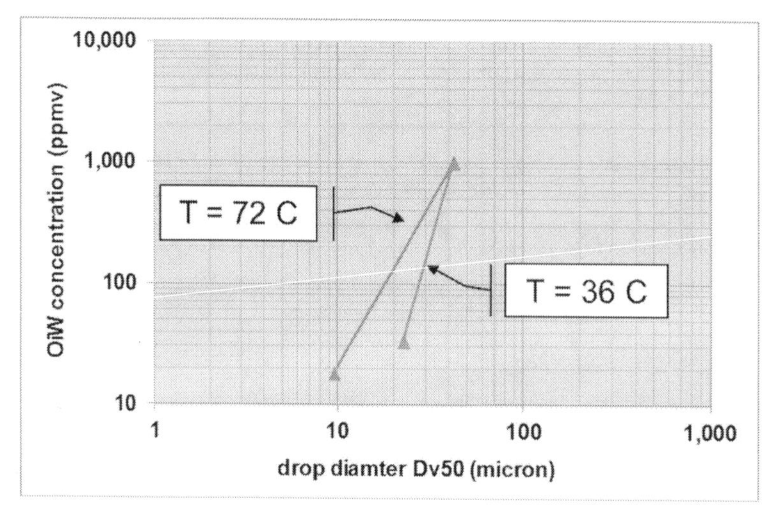

Figure 14a. Modeling results for the effect of temperature on the performance of a hydrocyclone.

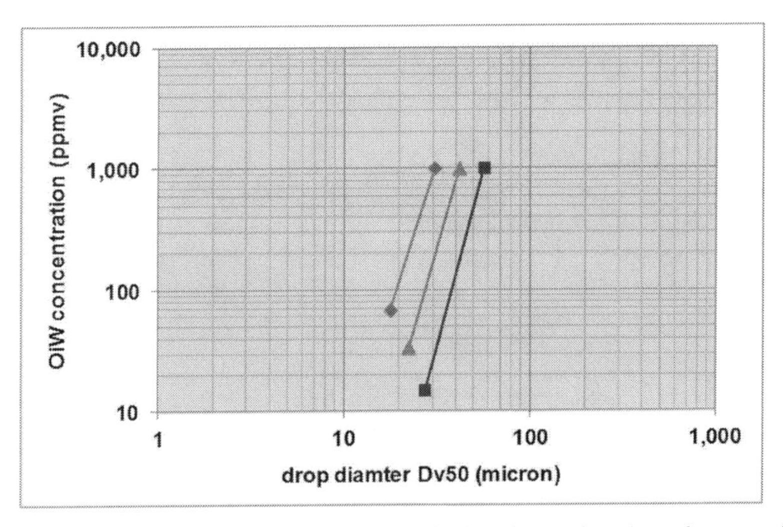

Figure 20.4 Modeling results for the effect of shear (smaller inlet drop diameter) on the performance of a hydrocyclone. All three lines are for fluid temperature of 36 C

Hypothetical modeling has thus been used to compare the effect of temperature and drop shear. Admittedly, only a limited number of temperature and drop size combinations have been presented. For example, warm fluid with small drops is not shown. This was done in order to keep the figures as simple as possible. Also it must be noted that a relatively high value of inlet oil concentration was selected. This was done for illustration purposes and to make the results more dramatic.

While it may appear that the conclusion of this modeling is that drop shear has a greater effect than temperature, this is not necessarily the case when a greater number of combinations is considered. For example, high temperature is capable of overcoming small drops to a fairly large extent. The conclusion of the authors is that the two effects (temperature and shear) are of similar consequence. This is an important result. It suggests that the differences between North Sea water separation processes are as much due to process configuration as to fluid temperature. This is discussed in greater detail in the next section.

20.2.5 Process Performance Diagram:

Two of the most important diagrams that characterize a water treating system are the Process Flow Diagram (PFD) and the Process Performance Diagram (PPD). Whereas the PFD gives the process schematic, the PPD gives the process performance. In other words, the PFD shows the equipment selected, the process line-up, the reject routing and all equipment and routing relevant to the system design. PFD examples are given in Figures 1 and 2. The PPD gives the inlet OiW concentration and oil drop diameter (D_{V50}) for each important location within the process line-up. An illustrative PPD is shown in Figure 15 below.

As discussed above, there are various measures of oil drop diameter that could be used in this figure. We have chosen here to use the D_{V50} value (volume average 50% value). The D_{V50} value is typically used whenever the entire drop diameter distribution is calculated.

In some cases however it is more appropriate to use the maximum drop diameter and not the entire drop diameter distribution. In that case, the x-axis would be the D_{V95} value or Dmax. In other cases, it is useful to use the D_{V10} drop diameter. This is used in cases where a pronounced bi-modal distribution is observed. In that case, it is important to track the small drops through the system.

The separation efficiency of each piece of equipment can be seen directly on the figure. The first point in the system represents the condition of the fluids upon entering the inlet separator. It is in the upper left hand side. As the fluids proceed through the system both the D_{V50} oil drop diameter and the OiW concentration decrease. The final drop diameter and OiW concentration is seen on the lower left hand side. The length of the line from one point to another is a measure of the separation efficiency of each piece of equipment. The inlet separator efficiency is given by the distance between the first two points (upper right hand side, and next point proceeding to the lower left hand side). The distance between the second and third points is a measure of the hydrocyclone separation efficiency.

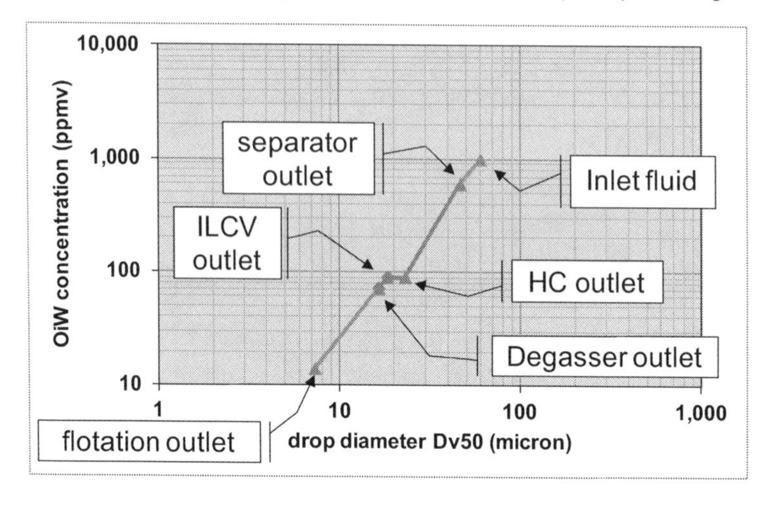

Figure 20.5. Process Performance Diagram (PPD). This diagram is for a typical platform and is for illustrative purposes. The PFD has a three phase separator, hydrocyclone, Interface Level Control Valve (ILCV) downstream of the hydrocyclone, degasser vessel and flotation unit

20.2.6 Results

The modeling tools discussed above were applied to a general set of fluid conditions and properties. The input variables used are given in the table below. It must be emphasized that these are generalized or model properties. Not all North Sea platforms are accurately described by these parameters. These parameters most accurately describe a moderate API oil, in the Norwegian sector of the North Sea where the inlet fluids are relatively hot.

Table 20.4

Parameter	NS	Deepwater GoM	Units
Stokes Factor	600,000	300,000	sec/m^2
Separator Residence time	300	120	sec
Temperature	70	35	C
Inlet pressure drop	2	10	bar
Water cut	25	25	%

Both temperature and Stokes Factor are given. For Stokes Law calculations, Stokes Factor alone is sufficient. However, for the hydrocyclone calculations, both Stokes Factor and water viscosity are required.

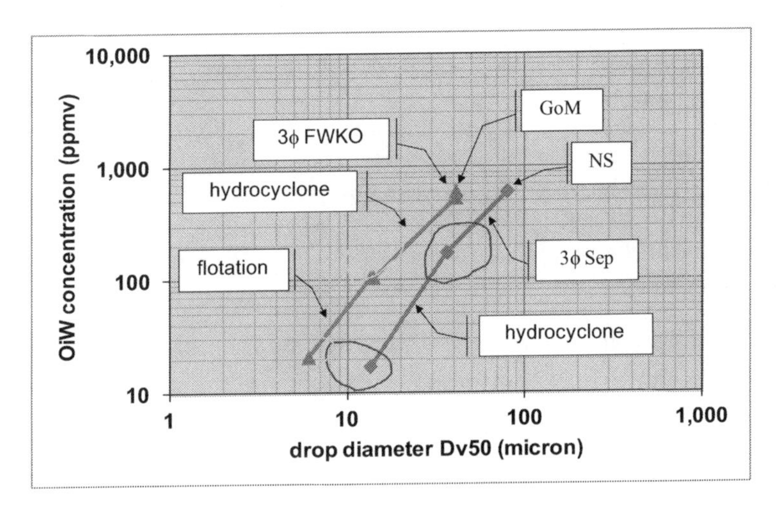

Figure 20.6 Process Performance Diagram (PPD) for platforms in the North Sea (NS) (green and red) and GoM (blue and yellow). The green points and line are from modeling of North Sea platforms. The red circles are from platform measurements for four North Sea platforms. The blue points and line are from modeling of deepwater GoM platforms. The yellow circles are from platform measurements for Mars, Auger and Ursa platforms from roughly 2001 to 2007

As shown in the figure, the inlet fluids in the deepwater GoM model have smaller drops and contain more oil than the North Sea model. This is due to the higher pressure drop upstream of the primary separator.

Field data are represented in the figure by circles which cover a range of drop diameter and oil concentration. These field data were collected from several surveys of four platforms in the North Sea and three platforms in the deepwater GoM. These platforms were admittedly selected for the present paper on the basis that the equipment and process was operating within reasonable ranges of the design. In other words, problematic platforms were not selected for this study. In fact, Figure 16 has been used by the author to help identify problematic platforms.

The separation efficiency of each piece of equipment, as calculated from the model, can be seen directly on the figure. The first point in the system represents the condition of the fluids upon entering the inlet separator (upper right hand side). As the fluids proceed through the system both the D_{v50} oil drop diameter and the OiW concentration decrease. The final drop diameter and OiW concentration is seen on the lower left hand side. The vertical drop from one point to another is a measure of the separation efficiency of each piece of equipment.

As discussed previously, the separation efficiency is given by the length of the line from one point to the next. As shown in the figure, the separation efficiency of the primary separator in the deepwater GoM is negligible. The two points are practically on top of each other. This is due to limited residence time and colder fluids which reduces the Stokes Factor.

The efficiency of primary separation and hydrocyclones for the North Sea model is much higher than that for the deepwater GoM model. The hydrocyclone effluent for the North Sea model hydrocyclone has a Dv50 of 10 micron. This is similar to that for the deepwater GoM model hydrocyclone, which is 12 micron. This difference is due to the higher temperature fluids in the North Sea which allows the hydrocyclone to separate smaller drops. Physically, the higher temperature reduces the viscosity of the water which improves the speed of drop migration to the overflow core, and which increases the swirl motion (driving force) at a given pressure drop through the hydrocyclone which increases the driving force for separation.

As discussed, the separation efficiency of the model flotation unit is given by empirical data from the deepwater GoM. As shown, flotation is required for the deepwater GoM model fluid but not for the North Sea model fluid.

Further, it is interesting to note that the slope of the lines is almost always found to be roughly in the range of 1.8 to 2.0. This reflects the importance of drop diameter in the three processes (primary separation, hydrocyclones, and flotation). For primary separation, Stokes Law governs where the slope would be expected to be close to 2.0. In this case, deviation from Stokes Law is due to hindered settling in a fluid containing a large number of drops.

Altogether the modeling reproduces the observations made in the field. That is, in those North Sea facilities where there is minimum drop shear in the boarding system, and where the primary separators have hydrocyclones, drop size is maintained and good separation efficiency is achieved without the requirement of a large flotation unit.

20.3 Discussion:

Water treating systems in the North Sea are different from those in the deepwater Gulf of Mexico. Factors which account for this deviation include capital and operating costs, extraction techniques, reservoir characteristics, the properties of the fluids being treated, the target specifications, and the obvious differences in platform type (fixed structure in shallow water versus floating structure in deep water).

Two approaches are used to understand why water treating systems in the North Sea differ from those in the deepwater Gulf of Mexico. The first approach is to compare the systems to Best Practices in water treating system design and to consider the reasons for deviation from the Best Practices. The second approach presented in this paper is to use a combination of field data and modeling to understand the impact of oil droplet size, temperature, and the use of three-phase versus two-phase gravity separators.

As discussed, the use of two-phase (gas/liquid) high pressure and intermediate pressure separators, as is common in deepwater, has the consequence that a hydrocyclone cannot be employed on the discharge of those vessels. A conventional hydrocyclone can really only be used when water is the predominant phase and the oil is present as droplets dispersed in the water. This requires three-phase separation with oil and water each having their own discharge nozzle. When both oil and water are discharged through the same nozzle, there are two phases (oil and water) and each occurs in percentage concentrations. A conventional hydrocyclone does not work well in this situation so they are typically not installed on the oil/water discharge of a two-phase vessel.

The other impact of two-phase separators is the shearing that occurs to the bulk oil/water mixture. Due to the pressure drop from one vessel to another, the oil and water mixture is extensively sheared. From the modeling results, it appears that such shearing is partly responsible for the need to employ large multistage horizontal flotation units. Thus, the use of two-phase inlet separators, which are significantly smaller and weigh much less than their corresponding three-phase counterparts, comes at a price. The price is the requirement to use a large multistage flotation unit for water treatment. The question is whether or not there is a net space and weight saving. To answer this question, the overall weight of the components can be compared.

A typical two-phase high pressure separator in the deepwater GoM weighs roughly 60 to 80 tons (operating weight), depending on the capacity. For example, the two-phase high pressure separator on the Mars platform has four inch thick steel walls and a liquid residence time of about 54 seconds. It is 72 inch diameter and is 28 feet long (seam to seam). It has an operating weight of 65 tons (60 metric tons). Since it is a two-phase separator, it is only required to separate gas and liquid. Given the density difference between liquid and gas, a short residence time is sufficient. If it were designed as a two-phase vessel with say 3 minutes residence time for each liquid phase, it would weigh at least three times more than the typical HP deepwater GoM vessel. Further, to effectively utilize a large three phase vessel, hydrocyclones should be added to the water discharge. This would add even more weight and require even more space.

A Wemco is a six chamber horizontal flotation unit with four active flotation cells. A Wemco unit that can process about 20,000 BWPD weighs about 30 tons wet weight, and operates with about six minutes retention time. It does occupy considerable space but since it operates at a low pressure, the wall thickness is low and hence the steel weight is only about half the weight of the unit. The other weight component is the water itself.

Thus, a tradeoff is made in the design of the deepwater facilities. By installing two-phase high pressure and intermediate pressure separators considerable weight (and space) is saved. But such systems shear the oil and water phases and make it all the more difficult to achieve oil and water quality. As a consequence, a high performance flotation unit must be used. Typical high-performance flotation requires a large multistage design. The net tradeoff is however still effective. The weight savings in using two-phase inlet separators is greater than the weight gain of having to use high performance flotation.

20.4 Summary and Conclusions:

In this chapter water treatment differences between North Sea and deepwater GoM platforms are analyzed. The scope of the analysis covers flow assurance strategies, chemical treatment programs, available separation technologies (gravity settling, hydrocyclones, and flotation), and process configuration. Both field data and modeling results are presented to explain why three-phase separator and hydrocyclones are widely used in the North Sea, while two-phase separator and multistage flotation units are applied in the deepwater GoM.

The present paper focuses mostly on fluid characteristics (such as temperature), equipment (two-phase versus three-phase separators, hydrocyclones, and flotation units), and process configuration (as characterized by the Process Performance Diagram). The differences can be summarized as follows:

- North Sea:
    - warmer fluids (lower water viscosity)
    - heat added upstream
    - less costly weight and space (shallow, no hurricanes)
    - three-phase primary separation (w/ hydrocyclone on each water discharge)
    - **hydrocyclones on all primary separators**
    - **no flotation required or just compact flotation required**
- Deepwater Gulf of Mexico:
    - cooler fluids (higher viscosity)
    - heat added downstream

- weight & space expensive (deep water, hurricanes)
- two-phase / short residence time primary separation
- **hydrocyclones used wherever possible (FWKO)**
- **large horizontal 4-stage flotation required**

Both regions appear to utilize Best Available Technology within the regional constraint of the cost of weight and space and fluid characteristics. The designers of North Sea platforms chose to employ three-phase separators together with hydrocyclones on the water discharge of each separator. This was a judicious design choice, given the availability of space and weight.

On the other hand, the designers of deepwater GoM platforms made the choice to use two-phase inlet separators due to the very high cost of weight and space. In doing so, a judicious tradeoff is made. Oil/water separation does not occur until late in the separation process, which subjects the oil and water to intense shearing. This requires more robust water treatment at the back end of the facility, such as horizontal multistage flotation. Given that flotation is a low pressure process, the weight of added steel due to such large flotation unit is small compared to the weight savings of two-phase primary separation. Thus, the typical deepwater GoM water treating system design appears to be justified from a holistic or whole-process perspective.

Modeling tools appear to correlate accurately the lab and pilot data. High quality field data on drop size distribution together with relevant process information is scarce. Benchmark data sets are needed. The modeling tools show that inlet fluid shearing is at least as important as fluid temperature in determining the required separation equipment. Inlet fluid shearing is greater in the GoM due to inlet two-phase separators. Thus, the fluids entering the platform, as well as the fluids entering the first three-phase separator (FWKO) have higher oil concentration and smaller oil drop diameter. The separator residence time is shorter than in the North Sea. The fluid temperature is lower as well. All these factors combine to give a relatively lower performance in the hydrocyclones and thus multistage flotation is required even though multistage flotation is heavier and occupies more space than the compact vertical flotation used in the North Sea.

It is only through a whole-process analysis, as demonstrated here that the justification for such an approach can be quantified. The choice of two-phase inlet separators in deepwater GoM is an economic decision which, as discussed in the paper, is judicious given the relative dryness of the fluids and the high cost of installing three-phase inlet separators in deepwater systems.

Not all of the systems evaluated by the author are shown in this paper. There are systems in the North Sea that have characteristics that are similar to those in the deepwater GoM. As in any survey of water treatment systems, there are exceptions to the general observations. Nevertheless, the general trends are seen more often than not.

Thus, the typical deepwater GoM water treating system design, with its use of large multistage horizontal flotation units appears to be justified from a holistic or whole-process perspective.

References to Chapter 20

1. J.M. Walsh, T.C. Frankiewicz, "Treating Produced Water on deepwater Platforms: Developing Effective Practices Based Upon Lessons Learned," SPE 134505 (2010).

2. J.M. Walsh, W.J. Georgie, "Produced water treating systems – comparison between North Sea and deepwater Gulf of Mexico," SPE 159713 (2012).

3. M. Bothamley," Offshore Processing Options for Oil Platforms," SPE 90325 (2004).

4. J.M. Walsh, "Produced water treating systems: Comparison of North Sea and deepwater Gulf of Mexico," SPE-159713-PA, *Oil and Gas Facilities*, (2015).

5. H.G. Polderman, J.S. Bouma, H. van der Poel, "Design Rules for Dehydration Tanks and Separator Vessels," SPE 38816 (1997).

6. K. Arnold, "Surface Facilities for Waterflooding and Saltwater Disposal," Chapter 15 of the Petroleum Engineers Handbook.

7. K. Rietema, "Performance and design of hydrocyclones – III. Separating power of the hydrocyclone," Chem. Eng. Sci., v. 15, p. 310 (19610.

8. S.J. Tulloch, "Comparative Testing of New Generation Deoiling Hydrocyclones by Orkney Water Test Center Ltd, December 1991 – April 1992," Project No.: OWTC/91/122, Document No.: R122-16B (1992).

9. D.A. Colman, M.T. Thew, "Correlation of Separation Results from Light Dispersion Hydrocyclones, "Chem. Eng. Res. Design, v. 61 (July 1963)

10. K. Nezhati, M.T. Thew, Aspects of the Performance and Scaling of Hydrocyclones for Use with Light Dispersions," Paper presented at the 3rd International Conference on Hydrocyclones, Oxford, England (1987).

CHAPTER TWENTY ONE

Applications: Dissolved & Water Soluble Organics

Chapter 21 Table of Contents

21.0 Introduction: 513

21.1 Chemistry and Analysis of WSO and Dissolved Organics: 514

- 21.1.1 Water Soluble Organics measured by EPA-1664: 514
- 21.1.2 Chemical and Phase Equilibria: 517
- 21.1.3 Organic Chemistry: 525
- 21.1.4 Dissolved, Dispersed, WSO and TOG: 529

21.2 Treatment Options for Removing Dissolved Organics: 531

21.3 Conclusions 535

References to Chapter 21 536

21.0 Introduction:

This chapter is based on an SPE paper published in 2014 [1], and on additional work carried out subsequently. In the paper, laboratory data, field data, and chemical and phase equilibria calculations are used to show that the so-called Water Soluble Organics (WSO), as measured using the EPA-1664 test are not equivalent to dissolved organics. Dissolved organics contribute to WSO. But WSO as measured using the EPA-1664 test are, for the most part composed of dispersed (suspended) oil. When dissolved organics contribute to WSO they are usually a minor component. Furthermore, very little of the dissolved material in a produced water sample is measured by the EPA 1664 test. This is an important point and a source of confusion in the industry. This is important because the removal of dispersed contaminants requires considerably different technologies than the removal of dissolved contaminants. Removal of dissolved organics is considerably more expensive. Removal of dispersed contaminants results in lowering of both TOG (Total Oil and Grease) and WSO. Thus, when high values of WSO are measured there is seldom a need to remove dissolved organics. Most WSO can be removed by removing the dispersed oil.

Produced water, whether it comes from primary production, gas condensate production, water flood, steam flood, or enhanced oil recovery, contains both dissolved and dispersed contaminants. The design of most oil and gas facilities includes units such as settling vessels and tanks, hydrocyclones, and flotation all of which are intended to remove dispersed oil and solids. These technologies are the subject of several chapters of this book. The technologies available to remove dissolved organics are well known but are typically more expensive and usually not required to meet regulatory compliance for surface discharge. Beneficial reuse as in feed to municipal water facilities, constructed wetlands, or agricultural irrigation will likely require reduction of dissolved organics. Considering these different requirements, it is critically important to understand whether the particular produced water in question has dissolved or dispersed contaminants. As mentioned, the WSO measured, as part of the EPA-1664 test cannot help in this regard.

The main objectives of this chapter are to provide guidance in determining when produced water really does contain dissolved organics as opposed to WSO, and to provide guidance in the selection of equipment and process lineups to remove dissolved organics.

The specific objectives of this discussion are to:

- develop a greater understanding of the relation between WSO and dissolved and dispersed components;
- develop a better understanding of what chemical components are measured and reported in the various monitoring and reporting analytical protocols;
- understand how process design choices influence WSO behaviour;
- develop a greater understanding of WSO behaviour through typical production systems;
- understand current extent of related work within the industry and anecdotal observations;
- consider the chemical and phase distribution of water soluble organics from a process engineering perspective;
- develop tools and techniques that guide process design toward optimal solutions for specific assets.

This is an Applications chapter. Up to this point in the book, the material has been fundamental and topics of chemistry have been separate from sampling and analysis as well as technologies. This segmentation of material was necessary in order to provide deep coverage of a broad number of topics. The purpose of this and all of the Applications chapters is to bring together all of these subjects into a single coherent approach to problem solving. In particular, this chapter brings together the chemistry of dissolved versus WSO components, how that chemistry impacts the EPA test, and the various technologies available for removing dissolved components.

Organization of this Chapter: This chapter is organized as follows. First, the EPA-1664 test is reviewed with emphasis on the reasons why that test cannot by itself be relied upon to determine the concentration of dissolved components. A short section on chemical and phase equilibrium is presented to help in this explanation. Second, the organic chemistry of oil and produced water are reviewed with emphasis on which compounds contribute to WSO and which contribute to dissolved organics. This subject was discussed in broad terms already in Chapter 2 (The Chemistry of Produced Water). Here, the connection between the chemistry of the components and the regulatory test (EPA-1664) is made explicit. Finally, various technologies for removing dissolved components from produced water are discussed. Technologies to remove WSO are not discussed because, as mentioned, most of the rest of the book is devoted to that subject.

21.1 Chemistry and Analysis of WSO and Dissolved Organics:

21.1.1 Water Soluble Organics measured by EPA-1664:

Total Oil & Grease (TOG) and Water Soluble Organics (WSO): In this book, as well as in most of the US-based oil and gas industry, TOG and WSO are defined in terms of the EPA 1664 analytical method [2]. That method is summarized here for reference.

Summary of EPA 1664 Method:

A 1-L sample of oily water is acidified to pH less than 2.

The sample is then serially extracted three times with n-hexane in a separatory funnel. The total quantity of hexane used in the extraction is 90 mL.

The extract is dried over sodium sulfate.

HEM: Hexane Extractable Material: the hexane extraction solvent is distilled from the extract at 85 C. The remaining material is referred to as HEM. The HEM is desiccated and weighed.

SGT-HEM: Silica Gel Treated HEM: the HEM generated in the above step is redissolved in n-hexane. An amount of silica gel proportionate to the amount of HEM is added. The intention of the silica gel is to remove polar components. After contact, the solution is filtered to remove the silica gel, the solvent is distilled from the extract at 85 C, and the SGT-HEM is desiccated and weighed.

SGA-HEM: Silica Gel Adsorbed HEM: the weight of material adsorbed by the silica gel. It is calculated as the difference: HEM – SGT-HEM.

In this book, TOG is equivalent to the n-hexane extractable material (HEM) of the EPA 1664 method. Likewise, water soluble organics (WSO) is equivalent to the fraction of the HEM that is adsorbed onto silica (SGA-HEM). The TPH components are those HEM components that do not adsorb on the silica gel. To summarize:

TOG = HEM (mg/L n-hexane extractable material)

TPH = SGT-HEM (mg/L silica gel treated, n-hexane extractable material)

WSO = SGA-HEM = TOG – SGT-HEM = (mg/L silica gel adsorbed, n-hexane extractable material)

The n-hexane extraction step is followed by evaporation of the n-hexane at 85 C. This results in vaporization of not only the n-hexane but also volatile components such as benzene and the protonated form of formic and acetic acid. Since the sample is acidified to pH < 2, almost all of the formic, acetic and in fact all organic acids will be protonated.

It is important to note that neither TOG, nor WSO nor TPH is equivalent to the dispersed oil content of an oily water sample. Further, WSO is not a direct or even indirect measure of dissolved organics in produced water. The term, Total Petroleum Hydrocarbon (TPH), is not entirely accurate either. Strictly speaking, a hydrocarbon compound is composed only of hydrogen and carbon. According to IUPAC there are four classifications of hydrocarbons (alkanes, unsaturated alkanes, cycloalkanes, and aromatics). As will be shown, the term TPH, as it is used in the oil and gas industry consists of much more than just hydrocarbons. It contains many polar organic molecules as well.

Dissolved/Dispersed versus WSO/TOG: There are two frameworks for analyzing organic compounds in produced water. One framework is based on the physical picture of dissolved versus dispersed organics. This is the phase partitioning framework. As discussed earlier, this framework has some complications due to difficulty in defining in defining the dividing line between dissolved and dispersed material, but conceptually the framework is self-evident. The other framework is the analytical framework. This framework distinguishes between organic compounds in water on the basis of the EPA 1664 analytical method which results in two main quantities, the Total Oil and Grease (TOG) and the Water Soluble Organics (WSO). There are other quantities that derive from this test such as HEM and SGTA-HEM and SGT-HEM. These are discussed in detail below. For now, the two frameworks are summarized here.

Table 21.1 Framework between Organic compounds in Water on the basis of the EPA 164 analytical method

Framework	Description
Phase Partition Framework Dispersed / Dissolved	The total organic concentration of a produced water sample is the sum of the dissolved organics plus the dispersed organics.
Analytical Framework TOG / WSO	Total Oil and Grease (TOG) is the total organic content that can be extracted using hexane, and which does not vaporize when the hexane is evaporated. Water Soluble Organics (WSO) are the polar (silica adsorbing) part of the evaporated hexane extract. It includes the dissolved organics and the polar organics in the dispersed oil. The remaining part of TOG, not included in the WSO, is the nonpolar part of the dispersed organics (TPH: Total Petroleum Hydrocarbons).

The difference between these two frameworks was illustrated in Chapter 7, Figure 7.5. Total oil and grease (TOG) is a combination of dissolved and dispersed organics. This same TOG value is also the sum of WSO (Water Soluble Organics) plus TPH (Total Petroleum Hydrocarbons - the nonpolar part of the dispersed organics).

There is confusion in the literature and in the industry over the difference between the dissolve/dispersed framework and the use of the TOG/WSO framework. It is sometimes assumed that WSO refers to the concentration of the dissolved organics in water. this is not the case. An additional component of WSO (polar dispersed component) comes from the dispersed oil.

The quantity of material in a sample that is measured as TPH or WSO is the result of a complex set of chemical and phase equilibria. The steps involved in the EPA-1664 test and the associated chemical and phase equilibria are shown in the table below.

Table 21.2 Steps in the EPA-1664 method and the compounds that are vented or not extracted.

Step	Chemical & Phase Equilibria	Compounds lost
Nitric acid	Protonate the acids which makes them more oil soluble	
Sample	Vent volatile compounds	BTEX Volatile organics Dispersed oil (if sample is not representative)
Hexane extraction	Leaves the highly water soluble compounds behind in the water	Low MWt acids & phenols
Evaporate the hexane	Flashes off the volatile compounds	BTEX Low MWt acids & phenols
Treat with silica	Adsorbs polar compounds	Any compound with a polar group

Impact of Evaporation Temperature: In order to demonstrate the temperature effect that occurs in the measurement of WSO the following test was performed. Samples of produced water were obtained from an offshore platform. The oil-in-water concentration was measured using an IR instrument. Two samples were used. One was acidized using mineral acid to pH = 2, as specified in the EPA-1664 test. The other sample was not acidized. Both samples were extracted using TCE (tetrachloroethylene) which is transparent in the IR region of the electromagnetic spectrum. As shown in the figure below, at room temperature, there is a large difference in the measured value of oil-in-water. At room temperature the protonated organic acids are detected in the acidized sample giving a total oil-in-water concentration of about 70 mg/L. Acidization caused the organic acids to become protonated which allows them to be extracted into the TCE solvent phase. In the unacidized sample the organic acids are ionized in water and there is no tendency to be extracted into TCE. They are not extracted and are not therefore detected in the IR reading. As the extract is heated to 60 C, the organic acids in the acidized sample are vaporized. At 60 C, only a low concentration of acid remained in the extracted sample, similar to that of the unacidized sample. This is essentially what happens to the light, low molecular weight organic acids in the EPA test when the hexane solvent is flashed off. Light acids such as formic and acetic acid have low boiling points and have a tendency to flash off during the EPA-1664 test when the hexane solvent is evaporated. The acetic acid concentration of the produced water was measured by Ion Chromatography. The acetic acid concentration was measured as 51 mg/L. If there is a small amount of formic and propionic acid, then the IC measurements would be in agreement with the results in the figure below.

There is little change in the unacidized sample since the organic acids remain ionized (unprotonated). They have no tendency to be extracted or vaporized. The slight decrease in the concentration of the unacidized sample is within the statistical error of the data. However, it is also possible that a small amount of hydrocarbon was vaporized as the temperature was increased.

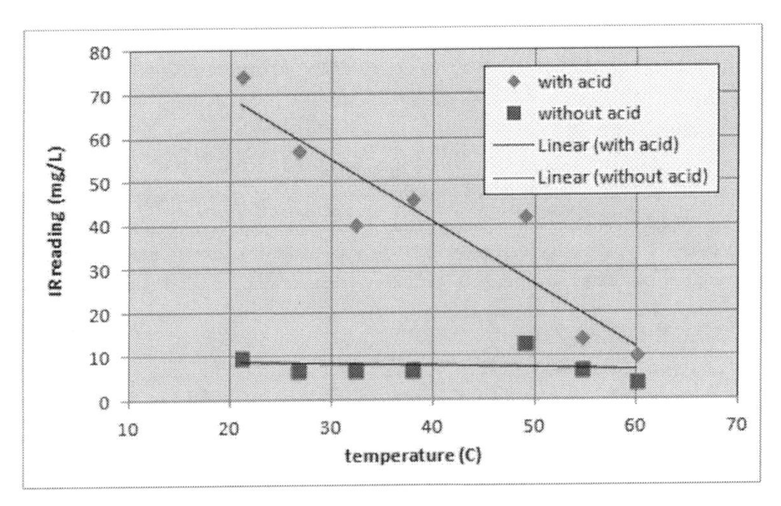

Figure 21.1 The effect of mineral acid and of temperature on the measurement of oil-in-water concentration. This produced water sample contained a fairly high concentration of light acids. Upon protonation with mineral acid, the light acids were extracted into the TCE (tetrachloroethylene – IR transparent) solvent and were measured as oil-in-water. Heating the sample vaporized these acids so that they no longer contribute to oin-in-water.

21.1.2 Chemical and Phase Equilibria:

In developing a treatment strategy for removing WSO from produced water, it is helpful to consider the chemical and phase equilibria involved. It has already been mentioned that there are a number of treatment options which differ significantly in their approach (chemical versus mechanical or some combination). A treatment that works well for BTEX may not work at all for acids. Thus, it is important to understand what class of compounds are present. The chemical and phase equilibria tells us why certain treatment options are effective for one class of compounds but not another.

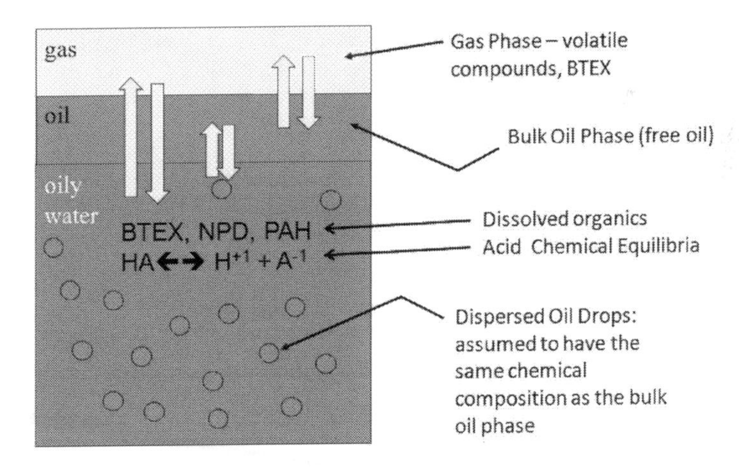

Figure 21.2 The chemical and phase equilibria involved in dissolved organics and WSO. All six classes of compounds are taken into account.

The important chemical equilibria is shown in the middle of the figure where the protonated acid (HA) is shown on the left hand side and the dissociation is shown on the right hand side of the equilibria. The dissociated acid is comprised of the anion (A^{-1}) and the proton H^{+}. The protonated acid (HA) partitions between the oil and the water phase. This partitioning is not equal. If the acid is a low molecular weight acid such as acetic acid, the partitioning strongly favors the water phase. In the case of formic and acetic acid, there is a tendency to form dimers which have a finite tendency to partition into the oil and gas phase, but the concentration is usually not significant for our purposes. Due to the very limited partitioning of these small molecular weight acids in an oil or hydrocarbon phase, these compounds do not contribute to WSO. If the acid is a higher molecular weight compounds such as decanoic acid, the partitioning favors the oil phase. The ions (right hand side) do not partition into the oil or gas phase. They are only soluble in the water phase and therefore do not contribute to WSO.

Carboxylic Acid Partition Coefficients: In the table below, partition coefficients are given for the small (low molecular weight) organic acids in water/hexane. The pH of the water was not adjusted and was therefore the pH value that resulted from the addition of the organic acid to the water.

Table 21.3 Partitioning of organic acids between water and hexane Framework between Organic compounds in Water on the basis of the EPA 164 analytical method

Acid	Formula	C_{Hexane} / C_{H2O}
Formic	HCO_2H	0.0002
Acetic	CH_3CO_2H	0.0008
Propanoic	$CH_3CH_2CO_2H$	0.003
n-Butyric	$CH_3(CH_2)_2CO_2H$	0.013
Valeric	$CH_3(CH_2)_3CO_2H$	0.053
n-Hexanoic	$CH_3(CH_2)_4CO_2H$	0.22
n-Heptanoic	$CH_3(CH_2)_5CO_2H$	0.89

As an example of the use of these partition coefficients, if the produced water sample contained 100 mg/L of acetic acid, then the hexane extract would contain 0.08 mg/L (prior to evaporation) – a negligible amount. This limited solubility in hexane prevents light acids from contributing to TOG and WSO. On the other hand if the produced water contained 100 mg/L of n-hexanoic acid, then the hexane extract would contain 22 mg/L of organic acid. This is a significant concentration. Note that the data in this table is not for pH of 2. The pH is that which naturally resulted from the addition of the organic acid to water and was not reported.

Benzene is relatively soluble in water but is more soluble in hydrocarbon or oil. It is also volatile. It does not contribute significantly to WSO because of its high vapor pressure. It flashes off at the elevated temperatures used to remove the hexane used as the extraction solvent. The other BTEX compounds have greater partitioning into the hydrocarbon phase and are less volatile. They can make a significant contribution to the WSO. NPD napth subset of the PHA these compounds partition between the disposed oil, free oil composition of the dispersed and free oil is the same.

Chemical and Phase Equilibria in the Production Process: Chemical and phase equilibria are relevant essentially in all stages of the production process. The chemical and phase equilibria that exists in the reservoir is disrupted by production when the pressure is reduced to induce flow. As the fluids emerge from the reservoir they experience shear, and temperature reduction. The shear provides mixing of phases which increases the surface area. Compounds that are prone to migrating to the oil/water interface are promoted to do so due to the large surface area created by shearing the oil and water phases into small droplets. Once the original chemical and phase equilibria established in the reservoir is disrupted by production activity, new equilibria are established rather quickly in most cases in the flow line and separators. These new equilibria result in volatile compounds entering the gas phase, this increases the pH in the water phase (due to carbon dioxide equilibria). An increase in pH deprotonates the organic acids and pushes some of them out of the oil phase and into the water phase.

Chemical and Phase Equilibria in the Sampling Process: Similar considerations can be made regarding the effect of sampling and analysis of the produced water. In the process of obtaining an atmospheric sample, the pressure is reduced and the temperature, generally, is reduced as well. As the produced water is vented to atmospheric pressure, vaporization of volatile compounds occurs including CO_2. This raises the pH, which ionizes the organic acids and pushes them from the oil phase into water phase. Thus, in some cases, the amount acids dissolved in the water phase will depend on the venting of CO2 in the sampling procedure. In order to standardize this effect, as well as preserve the sample, some sampling and analysis procedures require the addition of mineral acid to bring the pH down to a value of 2 or less. Reducing the pH with mineral acid will eliminate the sample to sample variation in pH due CO2 venting. But it will still result in a dependence between the concentration of acid dissolved in the water and the concentration of oil in the sample. This is discussed in greater detail below.

Chemical and Phase Equilibria in the Analysis Process: Once a sample is obtained, the analysis will typically involve additional modifications in the chemical and phase equilibria of the sample. For example, when hexane is added as an extractant, it increases the volume of the oil / hydrocarbon phase. This pushes oil soluble compounds into the oil phase. The technical name for these compounds is the Hexane Extractable Material (HEM). The oil and water phases are then separated. The water sample, after hexane extraction, is not longer required and it is typically disposed of. The oil phase containing the hexane is then heated so that the hexane can flash off. What is left is the HEM. The HEM is then poured into a column containing silica gel. The material that does not stick to the silica gel is collected and weighed and is referred to as the Total Petroleum Hydrocarbons (TPH). The difference between HEM and TPH is the WSO. All of these steps can and should be thought of in terms of the effect that they have on the chemical and phase equilibria of the four WSO components discussed above. For a more detailed analysis of this, see the SPE paper [1].

The figure below gives an example of the combined effect of chemical and phase equilibria as a function of pH for a particular sample of naphthenic acids. The y-axis is the total acid concentration in water. This concentration is the sum of the protonated acid (HA) and the ionized acid (H^+ and A^{-1}). The y-axis does not include the concentration of organic acid that has partitioned into the oil (hexane) phase. The figure can be considered from two vantage points. One is the effect of adding acid to a protocol sample of produced water. When acid is added, the pH goes down, to a value of 2 or less. This protonates a number of organics acids. The protonated form is more soluble in hexane than the ionized form so the protonated acid partitions into the hexane and the concentration of acid in the water phase goes down. The other vantage point is the effect of adding a mineral acid in a facility in order to reduce the concentration of organic acids in the water phase. Just as in the protocol analysis, the concentration of organic acid in the water goes down as the protonated acid partitions into the oil phase. This is an effective strategy for removing WSO from produced water but obviously it only works if the WSO is mostly composed of organic acids. Thus, the importance of recognizing the type of WSO present.

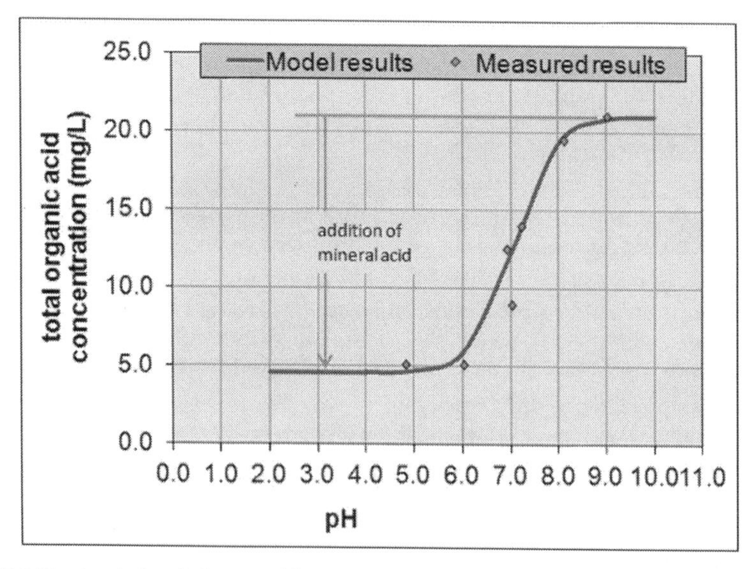

Figure 21.3 The chemical and phase equilibria involved in dissolved organics and WSO. All six classes of compounds are taken into account. y-axis: total acid (protonated and unprotonated) in the water phase [3].

These calculations are carried out for two systems. The first system is composed of water, a hydrocarbon compound, and an acid compound. This is an ideal system which can be modeled using well defined physical and chemical parameters. Most of the chemical and phase equilibria of much more complex field samples can be illustrated with this simple system. The modeling carried out here depends on measured data in order to adjust the equilibrium constants. Models based on first principles are available in [3 – 6, 11].

A plot of the anion concentration as a function of pH is given in the figure below. Two curves are given in the figure. In the blue curve (to the left), the organic acid is moderately strong (low pKa value). In the red curve (to the right), the acid strength is lower (higher pKa value). Recall that the pKa value is the pH at which half of the molar concentration of the acid is in the anion form and half is in the protonated (nonionic) form.

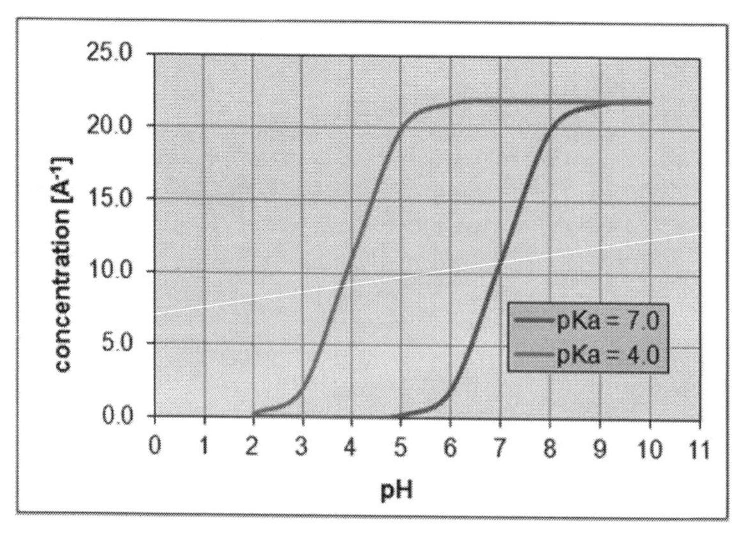

Figure 21.4 The concentration of acid anion as a function of the pH of the water. The pH is adjusted by addition of a mineral acid such as HCl.

The next figure shows the concentration of both species [HA] and [A^{-1}], dissolved in water, as a function of pH. Also plotted is their sum (red or top line), which remains constant. When no hydrocarbon solvent or oil phase is present, and when the acid does not vaporize, the total acid concentration (molar sum of HA and A^{-1}) in the water remains constant.

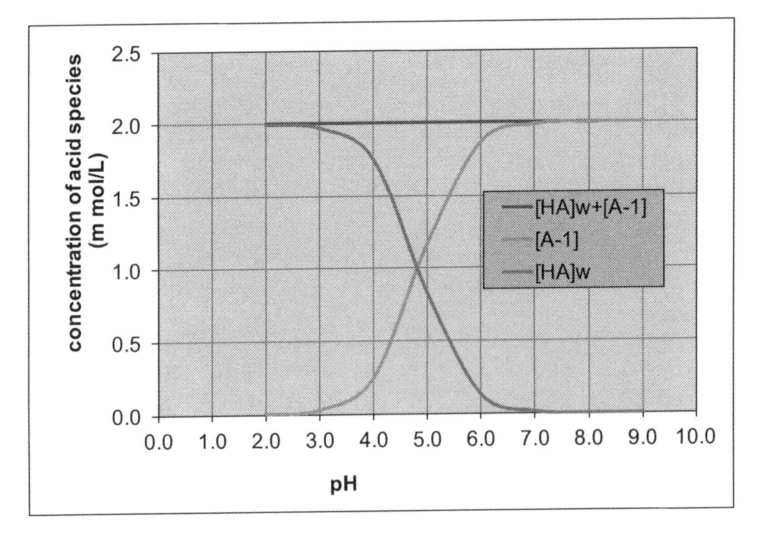

Figure 21.5 The concentration of acid anion as a function of the pH of the water. The pH is adjusted by addition of a mineral acid such as HCl.

The total molar concentration of acid for these calculations is 2 m mol/L (milli mol/liter). This value was chosen arbitrarily. If the acid of interest is acetic acid (molecular weight 59 mg/m mol), then the total acid concentration is 118 mg/L. If the acid of interest is hexanoic acid (molecular weight 115 mg/m mol) then the total acid concentration is 230 mg/L. The value of the equilibrium constant used for these calculations is pKa = 4.85 mol/L. Note that [HA]w = [A^{-1}] at the point where pH = pKa, as expected. This value is close to the typical value for C2 through C6 straight chain acids. The lowest pH chosen for these calculations is pH = 2. This corresponds roughly to the pH at which the EPA-1664 test is specified. As mentioned, these initial calculations are straightforward and can be found in an introductory textbook of organic or physical chemistry. Nevertheless, they are useful background to the next set of calculations which involve a non-aqueous phase.

Calculations were carried out for an organic acid in water in the presence of a hexane phase, as a function of pH. As shown in the figure below, the blue line (bottom curve) gives the concentration of the protonated acid in water. The green line gives the concentration of the ionized acid in water. All of the anion (A^{-1}) stays in the water phase. The red line gives the sum of the protonated and ionized acid dissolved in the water phase. Unlike the previous figure, the total concentration of acid in the water (red line) is not constant. Instead, some of the protonated acid $[HA]_o$ partitions into the hexane phase, as shown by the brown line. Due to this partitioning, the concentration of organic acid in the water phase decreases.

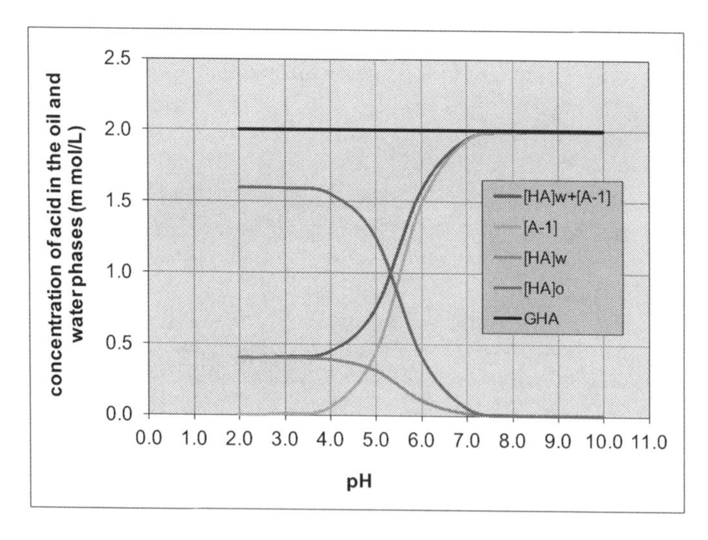

Figure 21.6 The concentration of all relevant species in the hydrocarbon/acid system as a function of pH. The pH is adjusted by addition of mineral acid such as HCl, or alkali such as NaOH.

The total molar concentration of acid for these calculations is again 2 m mol/L (milli mol/liter) and the pKa = 4.85, as in Figure 21.6. For the calculation in Figure 21.6, an equal volume of water and hexane was used. The hexane/water partition coefficient was chosen as 4. This is typical of the C8+ acids.

The preceding calculations, while simple, provide important insight into the behavior of organic acids in both real process systems, and in sampling and analysis procedures that are used. In a process system, the pH is governed by the presence of organic acids, the CO_2 concentration in the water (which is dominated by temperature and pressure), related produced water chemistry (e.g. alkalinity), and production chemicals (e.g. scale inhibitors, H2S scavengers). If the contribution of the production chemicals is not significant, then the pH of the system increases as the fluid traverses through the system. This is due to the degassing of the fluids which vaporizes the CO_2.

If the hydrocarbon phase in Figure 21.6 is assumed to be crude oil, then Figure 21.6 gives insight into the fate of organic acids in the production system. Referring to Figure 21.6, if the produced water pH is 6, then roughly 75 % of the acid will be dissolved in the water (red line). The other 25 % will be dissolved in the oil phase (brown line). If the produced water in the Free Water Knockout is pH = 6, then 25 % of the acid would partition into the oil phase and not the water phase. This is one mechanism by which organic acids are removed from the water. If, on the other hand, the produced water in the Free Water Knockout has a pH = 4.5, then roughly 75 % of the acid will be partitioned into the oil phase. Thus, the pH of the process has a dramatic impact on the fate of the dissolved organic acids.

When a sample is acidized, the organic acids that are present in the produced water are completely converted to the protonated form (right hand side of the reaction, left hand side of the figures). These molecules are still polar, but less so than the ionic form, and have a much greater tendency to partition into the hexane (or other organic) extraction solvent.

If a sample is not acidized, the organic acids that are present in the produced water are mostly in the form of the anion (left hand side of the reaction, right hand side of the figures). These molecules are completely insoluble in hexane or similar organic solvent. Thus, unacidized samples have a lower test reading regardless of the method (IR, UV, EPA-1664), provided that an organic solvent is used for extraction.

In fact, a simple test of whether or not organic acids are present and contributing to the TOG and WSO reading is to carry out a measurement of both the acidized and unacidized samples. If there is a difference of more than 20 % or so, then organic acids are contributing.

The Two-Component Model:

In the present Illustrative Case, field data are analyzed. Only two aggregate species are included in the model. One component is polar (adsorbs on silica gel) and partitions between the oil and water phases. The second component is nonpolar and does not partition. Both of these components are considered here to be aggregate components of many hundreds of chemical species.

The two components are defined in this model as:

C = partially soluble organics that contribute to both dissolved and dispersed oil as well as WSO

D = nonsoluble hydrocarbons that contribute to dispersed oil but not to WSO

Field data are plotted in the figures below.

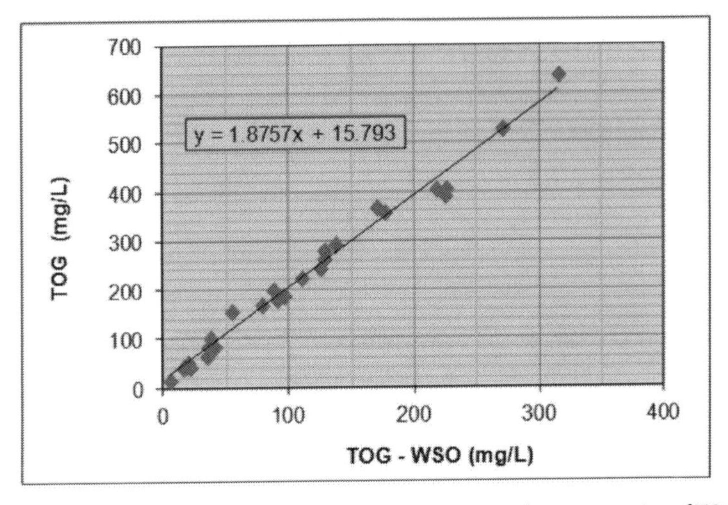

Figure 21.7 Measured field data for the concentration of TOG versus the concentration of TOG – WSO.

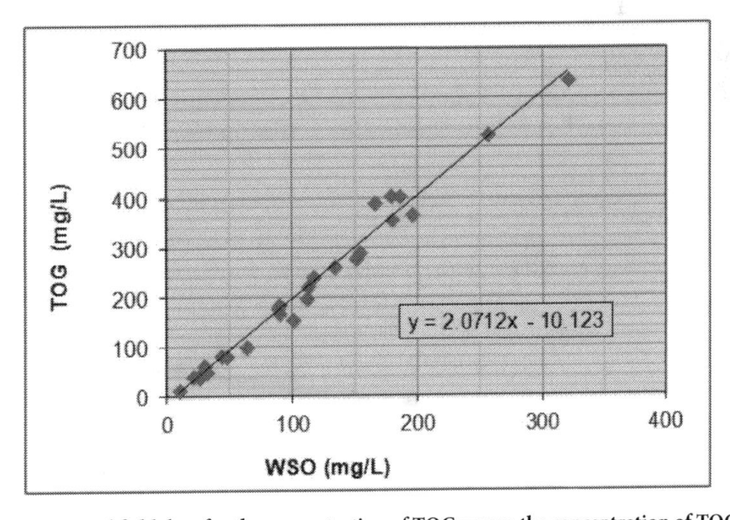

Figure 21.8 Measured field data for the concentration of TOG versus the concentration of TOG – WSO.

Based on the two-component model, the below figure was constructed.

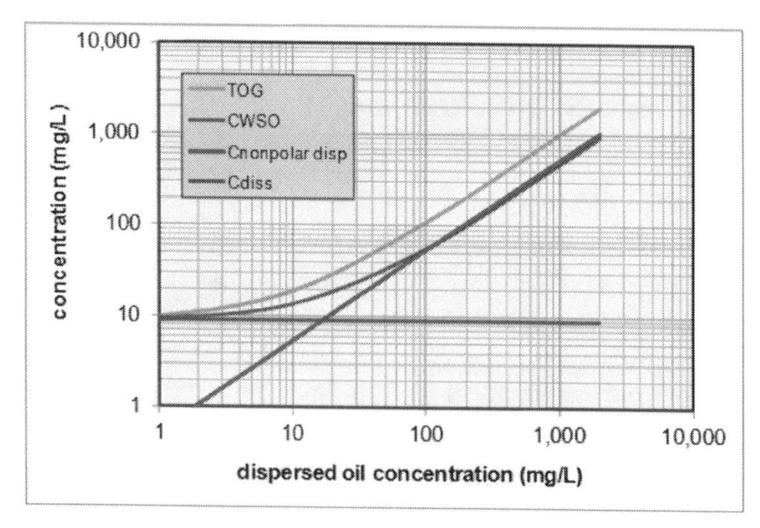

Figure 21.9 Calculations using the Two-Component Model using parameters based on field data.

The two component model can be used to show explicitly the difference between the two frameworks for analyzing the systems. As discussed in the Introduction, there are two frameworks for understanding organics in produced water. The most readily understood framework is the phase partitioning framework, which distinguishes between dissolved organics and dispersed oil. This framework has some complications due to difficulty in defining the dividing line between dissolved and dispersed material, but conceptually the framework is self-evident. The other framework is the TOG/WSO framework. This is the regulatory framework. It is more difficult to understand but using the two-component model, the contributions from each framework are readily understood. Having a model that allows calculation of all of the quantities in both frameworks is useful because it makes explicit and precise the difference.

As shown in the figure above, as the concentration of dispersed oil is reduced, the concentration of both TOG and WSO goes down. This is demonstrated by starting at the upper right hand side of the figure and following the lines toward the lower left hand side. The only curve that stays constant as the concentration of dispersed oil is reduced is the dissolved organics (purple horizontal line). It is logical that if dispersed and not dissolved oil is being removed, the model should predict a horizontal line, as shown.

From the above field data and modeling one of the data points was selected for further discussion. Throughout this book, a set of bar graphs have been used to explicitly point out the difference between dissolved organics and WSO. For illustration, one data point was plotted on the bar graph as shown in the figure below. The highest value of TOG was selected (TOG -= 320 mg/L). Based on the field data and the model, this TOG is broken down into WSO = 154 mg/L, of which 145 mg/L comes from the dispersed phase, and only 9 mg/L is dissolved in the water phase. This is a striking demonstration that if all of the dispersed oil is removed, most of the WSO is removed as well. Once the dispersed oil is removed, by a technology such as CETCO Hi-Flow, then only 9 mg/L is left. This final truly dissolved fraction can be removed by a number of media technologies.

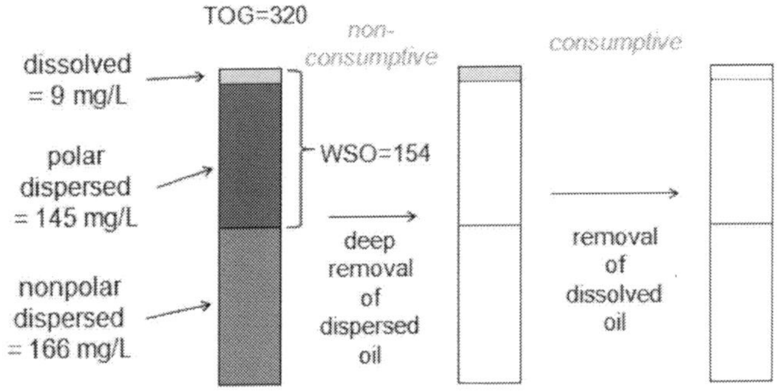

TOG=320
dissolved = 9 mg/L
polar dispersed = 145 mg/L
nonpolar dispersed = 166 mg/L
WSO=154
non-consumptive
deep removal of dispersed oil
consumptive
removal of dissolved oil

Figure 21.10 Calculations using the Two-Component Model using parameters based on field data.

Deep Removal of Dispersed Oil – CETCO HiFlow Coalescing Media: Hi-Flow is a cartridge based media that provides deep removal of dispersed oil. The media is designed to last without requiring a change-out for at least six months. In many installations, the media lasts much longer. The main factor that determines the life of the media is the presence of oily solids. When they are present, the media must be backwashed at least three of four times per day, and must be changed out at least every six months. In many installations where oily solids are present, a small cartridge filter system is installed upstream of the media. In order for this technology to be effective in removing WSO, most of the WSO must be in the dispersed oil. As discussed extensively above, this is often the case. However, it will not be the case if there is appreciable dissolved oil that is also extractable in hexane and non-volatile. Fortunately, that is often not the case.

For example, results from a deepwater Gulf of Mexico platform are given in the figure below. The TOG (Total Oil and Grease) was 320 mg/L. This sample was taken upstream of the flotation. Usually overboard produced water does not contain such high TOG. For the purpose of the evaluation of Hi-Flow, a high value of TOG was desired. Of that, 154 mg/L was composed of WSO. When the Hi-Flow media was applied it essentially removed all of the dispersed oil leaving only the dissolved oil behind. In this case, the dissolved oil was about 9 mg/L which is well below the overboard discharge limit. Thus, 146 mg/L of the WSO was dispersed. Hi-Flow provided deep removal of the dispersed oil thus removing what would have been a WSO problem. This example emphasizes the need to understand the chemistry of the WSO so that the proper treatment technology can be applied. In this case, the WSO were dispersed and could be removed using a non-consumptive media at very low cost.

21.1.3 Organic Chemistry:

If sampling and analysis of produced water from a facility indicates that there is a relatively high concentration of WSO, then some form of treatment will be required. What constitutes a high concentration varies from one region of the world to another. But generally speaking, the WSO concentration is near or exceeding the overboard discharge limit, then WSO must be reduced. There are various treatment options available for removing WSO from produced water. They vary significantly and include chemical treatments, mechanical treatments and combinations of chemical and mechanical treatment. The optimal treatment option for a given situation will depend on several factors starting with effectiveness, and including size, weight, capex versus opex, and staff time required to operate the equipment. All WSO treatment options have benefits and drawbacks. A good starting point for the selection of treatment options is to understand the chemistry of the WSO compounds present. There is a wide range of chemistries possible and it helps greatly to understand what kind of chemistry is present.

WSO Defined as a Solubility Class: As discussed extensively in Chapter 2 (The Chemistry of Produced Water), the number and type of individual molecules present in crude oil and produced water is enormous. Thus, it is not possible to discuss individual molecules, except in some special cases such as the low molecular weight acids, the BTEX (benzene, toluene, ethyl benzene, and xylene) compounds, and the low molecular weight phenols. All other compounds involved in WSO must be categorized into groups such as water-soluble acids, naphthenic acids, asphaltenes, aromatics, and saturates.

Water Soluble Organic compounds are defined as a solubility / volatility class. Individual chemical compounds can be identified, but the fundamental definition of WSO is given in the previous discussion as the fraction of the dissolved and dispersed hexane extractable material that does not flash off at 85 C and which sticks to silica gel. These characteristics of WSO come from the various steps used in the EPA-1664 test.

This may seem like a complex and unwieldy definition. In fact it is. However, it is not entirely unique to use this kind of definition in the oil and gas industry. A well-known similar case is asphaltene, which is a solubility class. Asphaltenes are those compounds that precipitate upon addition of pentane or hexane to crude oil, and which dissolve upon subsequent separation and contact with methylene chloride. In other words, the asphaltenes as a chemical aggregate group are only defined in terms of their solubility.

WSO Constituents in Produced Water: The table below lists the different classes of typical WSO. They are defined here in terms of their abundance, solubility, and volatility.

Table 21.4. The four major classes of compounds typically identified as water soluble organic compounds.

WSO Constituents	Characteristics
Phenol: C9+	significant partitioning into dispersed oil phase; soluble in extraction solvent; these compounds when present typically contribute to WSO
BTEX	benzene has high solubility in water but also soluble in extraction solvents; however, benzene is volatile and therefore does not typically contribute to WSO; TEX is less volatile and does contribute to WSO
Acids: C5+	somewhat water soluble depending on pH and carbon number extract into hexane at low pH (pH </= 2) phase partitioning (water/hydrocarbon) strongly dependent on pH high adsorption tendency onto silica gel main contributor to WSO in produced water with high WSO
NPD / PAH	low water solubility moderate solubility in hexane (but not completely soluble in hexane) most abundant components are the NPD compounds (see below)

NPD Compounds: The NPD compounds are naphthalene, phenanthrene, and dibenzothiophene. The first two of these compounds are members of the PAH class of compounds. Typically these three compounds are more abundant than the PAH compounds.

PAH (Polycyclic Aromatic Hydrocarbon) Compounds: The sixteen PAH compounds defined by the US EPA are listed in Section 2.5 (Hydrocarbons in Produced Water). These compounds are the most carcinogenic of produced water organics. They have a high persistency in seawater and thus can be traced through the food chain to the higher order species which are harvested for human consumption. Thus, in some parts of the world, such as Norway, there is strong emphasis on eliminating these compounds from overboard discharged produced water [7, 8].

Naphthenic Acids: In this book, the term naphthenic acid refers to any oil soluble acid. They were discussed in some detail in Section 2.6.2 (Organic and Naphthenic Acids). They typically make up the largest group of WSO.

Compounds that are not Typical WSO: It is helpful to identify those compounds that are not typically detected in the WSO. If the reader is able to understand why these compounds are not typically detected in the WSO, there is a good chance that he will understand the differences between dissolved organics and WSO.

Acids C1 to C4: These carboxylic acids are dissolved organics. They have high water solubility and high volatility. Even at pH = 2 they have low solubility in extraction solvents. These characteristics prevent them from being detected in the WSO test.

Phenol C6 to C8: These compounds have too high of water solubility, and hence too low of solubility in hexane or other extraction solvents to contribute to WSO. Like to low molecular weight acids, they are dissolved organics that do not contribute to WSO.

Examples of Dissolved Organics in Produced Water: Produced water varies significantly, from one reservoir to another, in the concentration and chemical species of dissolved organics. In the table below, an example of dissolved organics from four North Sea platforms is given [9]. These components were measured directly without using an extraction solvent. The acids were measured using isotachophoresis. As shown, there is a significant concentration of organic acids in each of the water samples.

Table 21.5 Dissolved Organics Measured in Produced Water from Four North Sea Platforms, Utvik [9]. All values are given in mg/L.

	North Sea Platforms			
Organic Compound Class	**Troll (mg/L)**	**Oseberg C (mg/L)**	**Oseberg F (mg/L)**	**Brage (mg/L)**
Organic acids	798	717	1,135	757
Phenols	0.6	11.0	11.5	6.1
BTEX	2.4	5.8	8.3	9.0
NPD	1.32	1.60	1.27	0.93
PAH (not NPD)	0.11	0.08	0.15	0.07
THC	33	60	44	58

At the bottom of the table, the measured value of THC (Total Hydrocarbon Content) determined using the old OSPAR (1997) protocol is given. The protocol for this measurement is: Freon extraction, quantitative IR measurement at the wave number of the aliphatic C-H stretch. Note that the sum of the concentration of the dissolved organics is much higher than the measured THC value. This is due to the fact that the Freon extraction step does not capture all, or even a large portion, of the dissolved organics. As discussed in the modeling section, solvent extraction involves phase partitioning of the dissolved organics between the produced water and the solvent. The more water soluble a component is, the less it will partition into the solvent and the lower the measured concentration in such tests as the old OSPAR protocol, and the EPA-1664 test, to name just two examples.

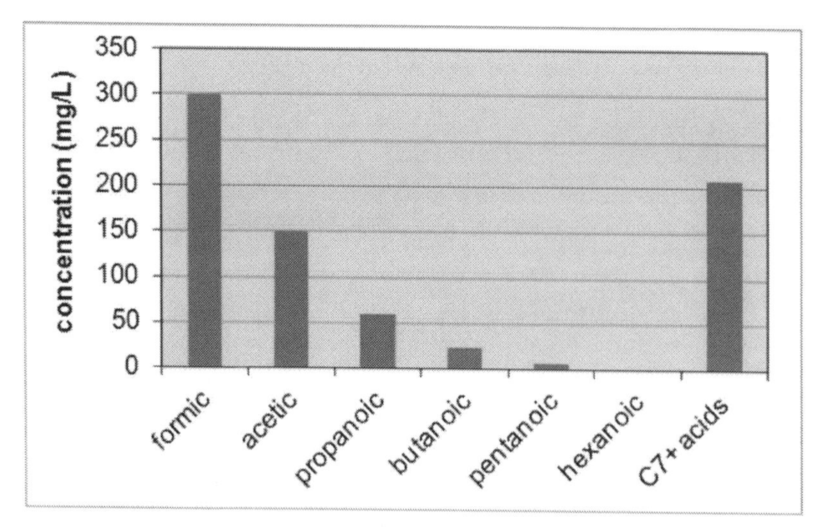

Figure 21.11 typical values for the concentration of specific acid compounds in produced water. Values for C1, C3-C7+ are from Utvik. The value for acetic acid was not reported by Utvik and was estimated independently.

For purposes of illustration, the relative concentrations of specific acids are given in Figure 21.11. Most of these values are reported by Utvik [9]. However, in the Utvik book there is no reported value for acetic acid. An estimate for acetic acid was made based on roughly a dozen samples from the North Sea that were gathered in relation to a separate study conducted by a major operator aimed at understanding the relationship between reservoir souring in water flood associated with the use of produced water containing organic acids. These data are given here for illustration only. As will be shown, almost none of the C1 to C4 acids contribute to the WSO value reported in the EPA-1664 method.

Shown in the figure below is an illustrative trend for the concentration of organic acids extracted from a number of produced water samples. These results are a composite of several data sources [9, 10, 12 - 14]. The intention of the Figure is to show the effect of solvent extraction, in this case, hexane, on the determination of WSO. As the carbon number increases (formic carbon number = 1, acetic acid = 2, etc), the concentration of the acid in produced water decreases, but the solubility in hexane increases. This gives rise to a slightly increasing, then decreasing trend with carbon number as shown in the Figure. As will be shown, almost none of the C1 to C3 acids contribute to the WSO value reported by the EPA-1664 method due to their high solubility in water [15 – 20] and high volatility at the solvent evaporation temperature. Sjoblom and co-workers [21] report the water/oil partitioning of C10 through C16 naphthenic acids. In that span of carbon number, the partition coefficient decreases by 4 orders of magnitude, from 0.01 to 0.00001. Thus, solubility in hexane (solvent for the EPA-1664 test) is assured for the higher molecular weight compounds.

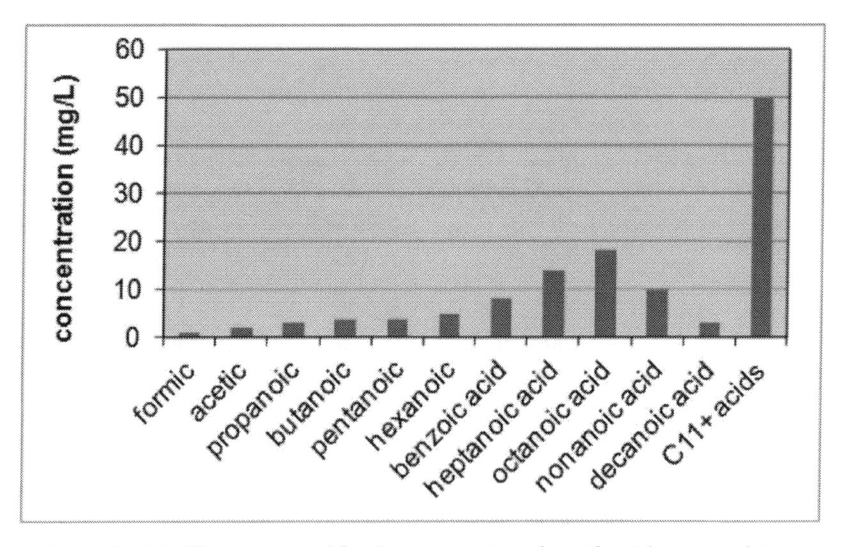

Figure 21.12 An illustrative trend for the concentration of specific acid compounds in a hexane extract of produced water. These data are a composite data from several sources [9 - 14].

In the above discussion, the various organic compounds involved in WSO analysis were identified. Also presented in an earlier discussion was the difference between the phase partitioning framework (dispersed/dissolved) versus the analytical framework (TOG/WSO) for distinguishing organic compounds in water. It is worthwhile to identify which organic compound classes are typically found in which phases, and which organic compounds are typically reported by each analytical method. This is summarized in the Figure below.

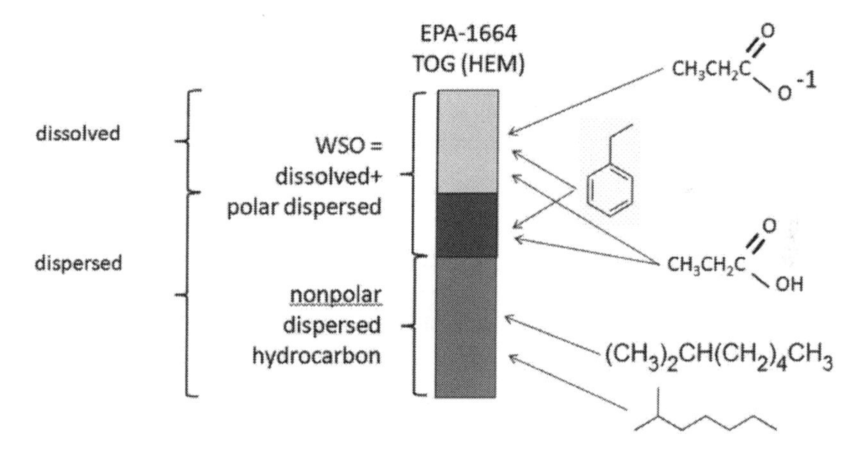

Figure 21.13 An illustration of the relationship between dissolved/dispersed organics and TOG/WSO, with examples of chemicals that contribute to each grouping.

21.1.4 Dissolved, Dispersed, WSO and TOG:

The name "water soluble organics" conjures the idea of organic compounds dissolved in produced water. However, given the industry use of the term WSO, and the association of the term with the EPA-1664 test method [2], there is little reason why reported values of WSO should be equivalent to dissolved organics. In some cases, the WSO content of a produced water sample is roughly equal to the content of dissolved organics. In other cases, WSO is not an accurate measure of dissolved organics. The reasons for this discrepancy are discussed below.

To understand what WSO are and how they differ from dissolved organics, an accurate definition is required of all of the related quantities of interest, i.e. dissolved organics, dispersed oil, Total Oil & Grease, and Total Petroleum Hydrocarbons. It is not possible to define one of these quantities without defining the others as well. Each of these quantities is related to the others and are defined as much by what they are, as by what they are not. Excellent discussions of these concepts are available in the literature [22 – 27]. There are several established and new technologies for treatment of dissolved organics [10, 28 – 36], many of which have specific applicability for compound types and concentrations. They are discussed below.

Dissolved versus Dispersed Organics: The concept of a dissolved organic molecule is well conceived. A dissolved organic molecule is simply an organic molecule that is surrounded by water. Likewise, the concept of dispersed oil is intuitive. Dispersed oil forms a distinct second phase, typically in the shape of drops that is distributed in the water phase. Due to buoyancy, the dispersed drops may cream (rise to the top) and form a free oil phase.

But not all of the organic material in the water is quite so simple. Many acids form dimers in water. These dimers too would be regarded as dissolved. Other polar molecules form small aggregates composed of say a few molecules to dozens of molecules. These might also be considered to be dissolved. Continuing the trend toward larger assemblies, produced water contains a wide range of concentrations and types of natural surfactant molecules. These can form micelles with polar groups at the oil/water interface and hydrocarbon molecules in the interior. At some point, the distinction between dissolved and dispersed organics becomes unclear.

While there is no universally accepted definition, in colloid and surface chemistry, organic molecules in water and assemblies of organic molecules having a diameter of less than 5 nanometers are generally considered to be dissolved [37]. Drops and particles having a diameter of greater than 50 microns are considered to be dispersed. These definitions leave a middle range of size that is not defined as either dissolved or dispersed. Assemblies of molecules having a diameter in between (5 nanometers to 50 micron) are considered to be colloidal. Colloidal material tends to be influenced by Brownian forces, and separation of these colloidal materials is dominated by surface chemistry, not simply settling or creaming. This is a rigorous definition, which may be helpful in some situations, but not here in the present application.

Faksness et al. [22, 24] use a practical approach to differentiate between dissolved and dispersed organics in produced water. They use 2 L jars to allow the dispersed organics to cream for a period of 48 hours. The organics that remain in the water are referred to as the Water Accommodated Fraction (WAF). Presumably they avoid using the term dissolved organics in recognition of the above mentioned ambiguity. In reality, they treat the Water Accommodated Fraction as dissolved organics. Using Stokes Law, and guessing at the physical properties of their system, it is likely that they remove all drops with a diameter larger then 2 to 3 micron.

Furthermore, as discussed below, produced water contains an enormous number and type of chemical compounds that have surfactant-like character. This includes the medium and long chain acids, resins and asphaltene classes of constituents. It is likely that in the produced water associated with moderate or high acid crude oils, that most of the dispersed material has these natural surfactant-like constituents residing at the organic/water interface. This detail is important in establishing the stability of the dispersed phase and whether or not it forms a free oil phase or remains dispersed. It is also important in establishing whether surfactant-like molecules reside in the dissolved phase or at the organic/water interface. Thus, at some future point in time, this detail might be incorporated in the model. This level of detail can be ignored by the use of "effective" partitioning coefficients. This is justified by the complexity of the model without considering additional detail.

Other methods to differentiate between dissolved and dispersed organics include the use of centrifugation or filtration. In either method, the result is a removal of dispersed organics with diameter in the range of 0.5 micron to 2 micron [38]. Any organics larger are generally considered to be dispersed. Those smaller are generally referred to as dissolved.

21.2 Treatment Options for Removing Dissolved Organics:

In this section treatment options for removing dissolved organic compounds from produced water are discussed. These options are only cost effective when the presence of dissolved organics have been confirmed. As discussed above, the so-called WSO reported as part of the EPA-1664 test are not necessarily dissolved organics. In the case where WSO are reported, but dissolved organics are not found, much less expensive treatments are available, as discussed above and in chapters related to the removal of dispersed oil droplets (Chapters 8 – 13). The material below is intended for the case where dissolved organics have been verified by other means than the EPA-1664 test. Chapter 7 (Sampling and Analysis) provides a discussion of methods that can be used to verify the presence of dissolved organic compounds in produced water. Much of the discussion below is focused on organic acids. They generally constitute the majority of dissolved organic compounds in produced water. As discussed in Chapter 2 (Chemistry) other dissolved components include phenol compounds and benzene.

Acidification followed by Oil/Water Separation: If mineral acid (or glycolic acid) is injected into the oil/water fluid mixture upstream of oil/water separation, the organic acids present in the produced water will become protonated and have a much greater tendency to partition into the oil phase. Thus acid treatment can be an effective treatment for acids dissolved in produced water. The total organic acid concentration in the water goes down as the organic acids partition into the oil phase.

However, acids are hazardous, corrosive, and they can reduce the effectiveness of other production chemistry treatments. Nevertheless, they are one of the least expensive and most effective options for dealing with dissolved organics. In addition to pushing the organic acids into the oil phase, the added acid has a tendency to reduce emulsion stability. The mechanism is due to neutralization of surface active compounds at the oil water interface.

As substitution for the mineral acids, Paul Hart [23, 26, 39] has identified an organic acid, glycolic acid, which is less hazardous to handle than the mineral acids and less corrosive. Glycolic acid is more acidic (greater tendency to ionize) than the typical organic acids found in produced water. Because glycolic acid is so extremely water soluble it can be added to the produced water without increasing the TOG concentration. It does not partition into the hexane phase during the solvent extraction used in the EPA-1664 test.

Liquid-Liquid Extraction – MPPE (Macro Porous Polymer Extraction): This is a liquid solvent extraction process. The liquid extractant is immobilized by capillary force in a porous media material. The media is made of small pellets that are themselves porous. The small pores trap the liquid extractant. As the produced water flows over and around the pellets, the liquid extractant comes into contact with the produced water. The chemical composition of the extractant is such that it does not mix with the produced water but it does adsorb aromatic hydrocarbons from the media. The media is packaged in cylindrical vessels as a packed bed. Each of the two beds is periodically taken out of service and regenerated in-situ. Low pressure steam is used to draw the hydrocarbons out of the extractant. The steam is then condensed which forms two phases, water and aromatic hydrocarbon. The two phases can be separated The aromatic hydrocarbon is sent to the Dry Oil Tank, or other suitable oil stream. The condensed steam is reused to generate steam for the next backwash cycle. A typical operating cycle is one hour of extraction followed by one hour of regeneration for each bed.

The treatment is effective particularly for aromatic compounds. However, overall the technology is a bit bulky for deepwater applications and has not been used there. It is however, suitable for locations nearer to shore, or onshore, where space and weight are not quite so expensive. In those locations, it has been found to be effective at removing BTEX, and some of the NPD and PAH compounds.

Gas Stripping using Flotation: Flotation can be used to provide gas stripping and removal of benzene and toluene. These aromatic compounds have a high vapor pressure. For this purpose, the available gas must not contain much benzene or toluene. In addition, the gas cannot be appreciably recycled. The objective is to use gas that has a low concentration of the aromatics as possible in order to enhance the partitioning of the aromatics in the gas phase.

The EPCON CFU (Compact Flotation Unit) has been evaluated for the removal of BTEX [40, 41]. The EPCON CFU is described in some detail in Section 12.2.6 (Vertical IGF). In this application a source of low BTEX gas was available for use in the flotation unit. The gas was not recycled. Instead, fresh gas was continuously fed to the unit. The concentration of BTEX in the influent water was 4 mg/L. The concentration in the treated effluent water was 0.5 mg/L. This equates to a separation efficiency of about 88 %.

Solvent Extraction using the CTour Process: The CTour process was was developed by the RF-Rogaland Research Institute, Stavanger, and the Norsk Hydro Research Center, Porsgrunn, Norway. It was commercialized by ProSep [42]. It's development is an example of joint industry / government / academia collaboration that occurs in Norway in the advancement of produced water technology. The technology is named after Charles Cagnaird de la Tour, a French engineer who is widely credited with the study and initial understanding of supercritical phase behavior and the existence of the critical point in fluid phase equilibria.

CTour is a solvent extraction process using light liquid condensate from the platform scrubber vessels. By contacting the condensate with the produced water, residual dispersed oil droplets and some of the dissolved components are extracted from the water into the condensate phase. In order for the process to be effective, the condensate be light, i.e. without BTEX or NPD components. If the condensate contains those components in appreciable concentration, then contact with the produced water under pressure will result in an increase in contamination of the water. Thus, the technology is not suitable for all platforms. Typically a HYSYS model is run, together with an evaluation of platform gas and condensate chromatography compositions. In the author's experience, most oil producing platforms in the deepwater GoM have condensate that is too rich for CTour to be effective. However, in the North Sea several installations of the technology have been made.

Voldum and Garpestad [43] report the successful implementation of a CTour system on Ekofisk. In field trial tests, the technology reduced the total oil to less than 2 mg/L, reduced naphthalenes and Polynuclear Aromatic Hydrocarbons (PAH) by 80 to 95 %, and showed an 80 % reduction in the Environmental Impact Factor (EIF). The CTour process is based on the use of available condensate from suction scrubbers as solvent. On Ekofisk this was not the case. Instead, Natural Gas Liquids (NGL) were used as the extraction fluid.

Descousse et al. [27] evaluated CTour for the removal of various compounds having a high EIF (Environmental Impact Factor). They found good removal efficiency for most compounds studied provided that the reference oil used in the extraction had low concentration of the contaminant compounds.

Oxidation Treatments: Partial or complete oxidation of the dissolved components both have the aim to chemically treat the organics and convert them to CO_2 or to compounds that are not detected by the EPA-1664 method. In the later case the compounds would be partially oxidized to low molecular weight organic acid compounds which are so highly soluble that they are not detected. The application of oxidation technologies is briefly discussed in Section 5.8 (Oxidation). Common oxidants that are used in the oil and gas industry:

1. sodium or calcium hypochlorite
2. chlorine dioxide
3. hydrogen peroxide
4. Fenton's reagent
5. ozone

Tests show wastewater containing about 160 mg/liter total phenols (including phenol, various methyl-phenols (a.k.a. cresols), some chloro-phenols, and other unspecified phenols), it was possible to drop concentrations to below detection limits using ClO_2. It was unclear what the phenols were being converted to, however, it was discovered that the major byproduct identified was benzoic acid, but that only represented about 10% of the initial concentration of total phenols.

Based on the flow rate and ClO2 production rate provided, where the dosage of about 3000 mg/liter, and the wastewater was run directly through the generator. BOD and COD are operationally defined parameters (i.e., consumption of oxygen or strong chemical oxidant under specific test conditions). Generally, biological processes are best for removing BOD, while phenols can be toxic. Therefore, pretreatment with ClO2 to remove the phenols could improve the effectiveness of biological treatment in this application.

Depending on the source of the wastewater, efficient BOD removal may also reduce the COD to acceptable levels. (BOD and COD were originally intended to be two measures of the same thing, but some compounds that contribute COD aren't very biodegradable. The relationship depends on the specific composition of the wastewater.)

When considering this application, it is necessary to have specific information on wastewater (e.g., concentrations of total and specific individual phenols, source of wastewater, BOD-to-COD ratio, other important constituents of wastewater, overall flow rate) and the current treatment operations and processes that are being used, but ClO2 shows some evidence of being effective.

The AquaPurge® system, manufactured by Clean Water Solutions, utilizes a combination of ozone and ultraviolet light to oxidize organic compounds dissolved in water. The amount of ozone and UV can be adjusted in order to selectively oxidize the dissolved organics, or to oxidize all of the dissolved compounds to CO_2 and H_2O. A benefit of this treatment is the oxidation of H_2S to sulfate and water, and the sterilization of the water. Also, dissolved iron will be oxidized and can be removed by coagulation followed by a number of different separation strategies such as clarification or flotation. According to the manufacturer's data, a 5,000 BWPD unit weighs 4 tons (8,000 lbs), and consumes 15 kW.

Consumptive Media - CrudeSorb: CrudeSorb is a modified clay material used for flow back treatment and produced water polishing treatment. It is provided by CETCO Energy Services. It is a consumptive technology. It has been found to have a higher capacity than activated carbon. It is packaged in either canisters which are easy to handle, or as bulk media which provies much greater capacity for a given occupied space. Descousse et al. [27] evaluated CrudeSorb for the removal of various compounds having a high EIF (Environmental Impact Factor). They found good removal efficiency for most compounds studied particularly the higher molecular weight compounds such as the PAH (Polycyclic Aromatic Hydrocarbons), substituted phenols, and the substituted naphthalene compounds. In other studies it has been found that BTEX compounds are also removed with good efficiency. Benzene removal of about 85 % can be achieved, while the other compounds (TEX) are removed with close to 90 % efficiency.

Consumptive Media - MyCelx: MyCelx is a consumptive technology. It is based on a specially treated cartridge filter that absorbs contaminants from produced water [44, 45]. A standard polypropylene cartridge filter (4 inch diameter x 30 inch long) is infused with polymerized triglyceride (linseed oil) with a binder and solvent. This mixture creates an amorphous layer of absorbent material on the cartridge filter fibers. Hydrocarbons dissolve in the amorphous polymer layer. Each cartridge has a capacity for oil of between 5 to 8 lbs (average 3 kg/cartridge). Contact time required is less than one second, which reduces footprint.

Example calculation of cartridge capacity:

Feed oil concentration: 100 mg/L

Effluent: 10 mg/L

Oily water flow rate of 10,000 BWPD (1,590 m3/day)

Daily oil removal (cartridge loading): 143 kg/day

Roughly 50 cartridges will be consumed per day (143 / 3 = 47.7 cartridges/day)

In reference [45] there is a report of a successful field implementation of the media for Anadarko in Vernal, Utah.

Consumptive Media – GAC (Granular Activated Carbon): Usually deployed in bulk. Packaged in a deep-bed configuration. GAC has already been extensively discussed in Section 13.6.4 (Activated Carbon Filtration).

Example calculation of bed capacity:

Capacity for hydrocarbon:	0.20 to 0.25 lbs HC/lb media
Typical bed weight:	3,000 lbs
Typical bed capacity:	600 to 700 lbs (270 to 320 kg)
Feed oil concentration:	100 mg/L
Effluent:	10 mg/L
Oily water flow rate:	10,000 BWPD (1,590 m3/day)
Daily oil removal (bed loading):	143 kg/day

Roughly 1 bed will be consumed every two days (143/300 = 0.5 bed/day)

For comparison, if this media were containerized as a cartridge, each cartridge would have a capacity of about 0.7 kg oil/cartridge. Compare this to the MyCelx technology which has a capacity of 3 kg/cartridge.

Consumptive Media – Resins: Resins for the selective removal of aromatics are available. These resins include BTEX-SORB, etc. They work best when there is deep removal of disposed oil and suspended oily solids upstream of the resin bed. They tend to be more expensive than GAC, on a pound for pound basis but they are more selective, have a higher capacity, and can be run at higher flux. All things considered, they can achieve lower cost than carbon in some selected applications.

21.3 Conclusions

The so-called WSO, as commonly identified with the silica adsorbed, hexane extractable material of the EPA-1664 test, is a combination of dissolved and dispersed components from produced water. To qualify as WSO, a chemical compound must be polar enough to adsorb on silica gel but nonpolar enough to partition into hexane. Another somewhat easier way to say this is the following. WSO are composed of dissolved organics from the water phase plus the polar part of dispersed oil. In this case, "polar" refers to the tendency to adsorb onto silica gel. Various classes of compounds were discussed which have this property.

A conclusion from this analysis is that for many produced water systems, WSO can be significantly reduced by deep removal of dispersed oil. Technologies that accomplish this are CETCO Hi-Flow, and some other media based systems. As dispersed oil is removed, both TOG and WSO are reduced. If all of the dispersed oil is removed, all that is left is the dissolved oil. Typically, very little of the dissolved fraction of produced water contributes to WSO (via EPA-1664).

References to Chapter 21

1. J.M. Walsh, J. Vanjo-Carnell, J. Hugonin, "Understanding water soluble organics in upstream production systems," SPE-170806, book presented at the SPE Annual Technical Conference and Exhibition held in Amsterdam, The Netherlands, (2014).

2. U.S. EPA, "Method 1664 Revision A: n-hexane Extractable Material and Silica Gel Treated n-Hexane Extractable Material by Extraction and Gravimetry," National Service Center for Environmental Publications EPA-821-R-98-002 (1999).

3. J. McFarlane, "Modeling of Water Soluble Organic Content in Produced Water," Oak Ridge National Lab (2004).

4. J. McFarlane, "Modeling of Water-Soluble Organic Content in Produced Water," ONRL Literature, May (2006).

5. J. McFarlane, D.T. Bostick, H. Luo, "Characterization and Modeling of Produced Water," Ground Water Protection Council, (2002).

6. D.T. Bostick, H. Luo, B. Hindmarsh, "Characterization of Soluble Organic in Produced Water," DOE Project Report (2002).

7. T.I.R. Utvik, J.R. Hasle, "Recent Knowledge about Produced Water Composition and the Contribution from Different Chemicals to Risk of Harmful Environmental Effects," JPT December (2002).

8. International Association of Oil & Gas Producers, "Aromatics in Produced Water: Occurrence, Fate & Effects, and Treatment," Report No. 1.20/324 January (2002).

9. T.I.R. Utvik, "Chemical Characterization of Produced Water from Four Offshore Oil Production Platforms in the North Sea," *Chemosphere,* Vol. 39, No. 15, pp. 2593-2606 (1999).

10. MyCelx Technologies Corporation, "Technical Report and Recommendations for Produced Water Treatment"

11. J. McFarlane, "Modeling of Water Soluble Organic Content in Produced Water," ORNL, February (2004).

12. 12 W.K. Seifert, "Carboxylic acids in petroleum and sediments," Fortschritte d. Chem. Org. Naturst., v. 32 (1975).

13. J.G. Speight, High Acid Crudes, Elsevier, Amsterdam (2014).

14. J.S. Brown, T.C. Sauer, M.J. Wade, J.M. Neff, "Chemical and toxicological characterization of produced water freon extracts," chapter from Produced Water, Edited by J.P. Ray and F.R. Engelhart, Plenum Press, NY (1992).

15. T.E. Havre, J. Sjoblom, J.E. Vindstad, "Oil/water partitioning and interfacial behavior of naphthenic acids," Journal of Dispersion Science and Technology v. 24, N. 6, p 789 (2003).

16. S. Kato, "Partition coefficients of fatty acids between water and n-hexane," Biochemical Engineering (1992).

17. R. Smith, C. Tanford. "Hydrophobicity of long chain n-alkyl carboxylic acids, as measured by the distribution between heptane and aqueous solutions," Proc. Nat. Acad. Sci., v. 70 p. 289 (1973).

18. M.A. Reinsel, J.J. Borkowski, J.T. Sears, "Partition coefficients for acetic, propionic, and butyric acids in a crude oil/water system," J. Chem. Eng. Data, v. 39, p. 513 (1994).

19. A.G. Gilani, H.G. Ghanadzadeh, S.L.S. Saadat, M. Janbaz, "Ternary liquid-liquid equilibrium data for the water+butyric acid+n-hexane or n-hexanol systems at T= 298.2, 308.2, 318.2 K," J. Chem. Thermodynamics, v. 60, p. 63 (2013).

20. F.C. Rublo, V.B. Rodriguez, E.J. Alameda, "A contribution to the study of the distribution equilibrium of solme linear aliphatic acids between water and organic solvents," Ind. Eng. Chem. Fundam., v. 25, p. 142 (1986).

21. J. Sjoblom, S. Simon, Z. Xu, "The chemistry of tetrameric acids in petroleum," Adv. Coll. Interface Sci., v. 205, p. 319 (2014).

22. L.-G. Faksness, P.G. Grini, P.S. Daling, "Recent results from correlating dispersed oil and PAH content in produced water, and its impact on the selection of treatment technologies," TUVNEL Produced Water Workshop, Aberdeen (March 2003).

23. P.R. Hart, "Progress Made in Removing Water Soluble Organics From GOM Produced Water," World Oil, v. 227, Sept (2006).

24. L.-G. Faksness, P.G. Grini, P.S. Daling, "Partitioning of Semi-Soluble Organic Compounds Between the Water Phase and Oil Droplets in Produced Water," Marine Pollution Bulletin 48, pp. 731-742 (2004).

25. S.A. Ali, L.R. Henry, J.W. Darlington, J. Occhipinti, "Novel Filtration Process Removes Dissolved Organics from Produced Water and Meets Federal Oil and Grease Guidelines," PWS Meeting (1999).

26. P.R. Hart, "Removal of Water Soluble Organics from Produced Water in the Gulf of Mexico," Produced Water Society meeting, Houston (2005).

27. A. Descousse, K. Monig, K. Voldum, "Evaluation Study of Various Produced-Water Treatment Technologies to Remove Dissolved Aromatic Components," SPE 90103, paper presented at the ATCE, Houston (2004).

28. J.W. Darlington, "New Technology Achieves Zero Discharge of Harmful Alkyl Phenols and Polyaromatic Hydrocarbons from Produced Water" CETCO Offshore (August 2002).

29. K.T. Klasson, C. Tsouris, S.A. Jones, M.D. Dinsmore, D.W. DePaoli, A.B.Walker, S. Yiacoumi, V. Vithayaveroj, R. M. Counce, S.M. Robinson, "Ozone Treatment of Soluble Organics in Produced Water," Petroleum Environmental Research Forum Project 98-04, January (2002).

30. A.B. Hayns, "Removal by Filtration of Dissolved & Dispersed Hydrocarbons from Produced Water," PWS (1996).

31. A. Seybold, J. Cook, R. V. Rajan, R. Hickey, J. Miller, A. Lawrence, T. Hayes, "Demonstration of Dissolved Organics Removal from Produced Water," SPE 37420 (1997).

32. S. Kwon, E.J. Sullivan, L. Katz, K. Kinney, R. Bowman, "Pilot Scale Test of a Produced Water-Treatment System for Initial Removal of Organic Compounds," SPE 116209 (2008).

33. H. Parthasarathy, "Diminishing Environmental Footprint & Costs in Offshore Produced Water Treatment," OTC 21893 (2011).

34. P.L. Edmiston, J. Keener, S. Buckwald, B. Sloan, J. Temeus, "Flow Back Water Treatment Using Swellable Organosilica Media," SPE 148973 (2011).

35. R. Dores, A. Hussain, M. Katebah, S. Adham, "Using Advanced Water Treatment Technologies to Treat Produced Water from the Petroleum Industry," SPE 157108 (2012).

36. L.R. Moore, C. Cardoso, M. Costa, A. Mahmoudkhani, "Removal of Total Organics and Grease from Oil Production Effluents by an Adsorption Process," SPE 163272 (2012).

37. R.M. Pashley, M.E. Karaman, Applied Colloid and Surface Chemistry, Wiley, England (2004).

38. M. Yang, personal communication (May 2014).

39. P.R. Hart, "Removal of Water Soluble Organics from Produced Brine without Formation of Scale," SPE – 80250, paper presented at the SPE International Symposium on Oilfield Chemistry, Houston (2003).

40. L. Jahnsen, E.A. Vik, "Field trials with Epcon technology for produced water treatment," paper presented to the Produced Water Society, Houston, TX (2003).

41. E.A. Vik, L.B. Henninge, "The Epcon CFU zero discharge technology. Ca\se studies 2001 – 2005," Aquateam report No. 04-025 (2005).

42. J.B. Sabey, "CTour – simultaneous removal of dispersed and dissolved hydrocarbons in produced water," paper presented at the Produced Water Society, Houston (2006).

43. K. Voldum, E. Garpestad, "The CTour process: an option to comply with zero harmful discharge legislation in the Norwegian waters – experience of CTour installation on Ekofisk after startup 4th quarter," SPE – 118012, paper presented at the Abu Dhabi International Petroleum Exhibition and Conference, Abu Dhabi, UAE (2008).

44. U.S. Patents: US 5437793; US 5746925; US 5961823.

45. P. Harikrishnan, R. Schlicher, J. Yu, "Deoiling for discharge – quality water," World Oil, p. 73 (Dec 2009)

APPENDIX A

Units, Standard Conditions, Common Calculations

Appendix A Table of Contents

Units 543

Parts per Million 543

Common Calculations 544

Standard and Normal Conditions 545

Units

Units of Measure: Most calculations in this book are carried out in SI units with a final conversion to oilfield units where applicable. The units used in this book are not defined by a single so-called system of units such as the International System of Units (SI), the cgs (centimeter, gram, second), or oilfield units. Instead, the units used in this book are a hybrid of these systems. This is a consequence of the nature of water treatment in the oil and gas industry. The industry relies on volumetric units such as barrel in the US, and cubic meter overseas. Concentrations in water treatment are expressed in mg/L. Thus, the subject of water treatment in the oil and gas industry uses a hybrid combination of units.

Anyone who spends time in the oil field must be able to make rough calculations quickly without a calculator, unless you have one that is intrinsically safe. Such calculations provide a good understanding of the processes that occur in a facility. For this purpose, the SI system of units is most convenient. An explanation of the units used in this book, as well as the tricks of the trade for doing calculations rapidly without a calculator is given here. The common units used for water treatment in the oil and gas industry are:

Length:	meter, foot, micron
Volume:	barrel, cubic meter
Volumetric flow rate:	barrels per day, barrels per minute, cubic meters per hour
Concentration:	mg/L, ppm, gal/Mbbl, moles/L
Droplet or particle diameter:	micron ($1x10^{-6}$ m)
Droplet rise rate:	centimeters per minute

These are just examples. Several other units are also used depending on the situation.

Parts per Million

Chemical analysis of brines is often reported in mg/L, which refers to a specific compound or ion in milligrams per liter of brine. In common usage, this is also referred to as ppm (parts per million). In the upstream oil and gas industry this is incorrect. Strictly speaking, ppm refers to parts per million. Parts per million could mean weight per million units of weight, or volume per million units of volume. Examples are mg/kg (weight) or micro liters per liter (volume) or cubic feet per million cubic feet (volume). But strictly speaking it is incorrect to use mixed units such as mg/L.

In the industrial water treatment industry, ppm almost always means mg/L. When ppm is used in the industrial water industry, there is almost no ambiguity.

In the upstream oil and gas industry ambiguity creeps in due to the frequent necessity to address gas composition which is almost always specified as standard cubic feet per million standard cubic feet. Another less frequently used composition is moles per million moles. These units are, as mentioned above, a strict application of the ppm definition (something per million somethings). It turns out that a standard cubic foot per million standard cubic feet is equivalent to a mole per million moles. There is no ambiguity. Given the importance of gas composition in the upstream industry, ppm has the preemptive claim to gas composition and water composition is therefore usually expressed as mg/L. ppm might mean mg/L when dealing with water, or it could mean standard cubic feet per million standard cubic feet. This latter terminology is often used in discussion of gas composition.

ppmw:	mg of component / kg of solution
ppmv:	micro L of component / L of solution

In dealing with chemical reactions, minerals that dissolve in water, and any concepts that involve stoichiometry, it is useful to use the units of moles/L (moles of a substance per liter of brine solution). The conversion of mg/L to moles/L involves dividing by the molecular weight. Typically molecular weight is expressed as g/mol. But it can also be expressed as mg / milli mol (milli gram per milli mole). The two units are equivalent. The latter is easier to use.

mol / L = (mg / L) / (g / mol) / (1000 mg / g)

milli mol / L = (mg / L) / (mg / milli mol)

mol / L = (milli mol / L) / (1000 milli mol / mol)

Concentration is also sometimes expressed in terms of the number of equivalents per liter. An equivalent weight of a substance is the molecular weight divided by the valence charge. The number of equivalents is then the weight of the substance (usually in mg) divided by the equivalent weight.

Common Calculations

There are a few calculations that every water specialist must be able to perform. These calculations should be used often in order to characterize the system and the amount of contaminant that must be removed. here. These calculations help to explain the various units that are used and they provide a straightforward way to calculate certain quantities.

Example 1: Conversion of contaminant concentration to contaminant loading:

Concentration:	mg/L
Flow rate:	barrels per day
Loading:	lbs/day

Step 1: convert flow rate from barrels per day to cubic meters per day by dividing bbl/day by 6.29.

Step 2: multiply concentration in mg/L by flow rate in m3/day, divide by 1,000. This gives the contaminant loading in units of kg/day.

Step 3: The final step is to convert kg/day to lbs/day. Take the result of step 2 and divide by 2.2.

Concentration:	20 mg/L
Flow rate:	12,500 BWPD

Calculation:

$$\text{Loading} = \frac{(20\ \text{mg/L})(12{,}500\ \text{BWPD})(1000\ \text{L/m}^3)(2.2\ \text{lb/kg})}{(6.29\ \text{bbl/m}^3)(1{,}000{,}000\ \text{mg/kg})} = 87.4\ \text{lb/day}$$

Eqn (A.1)

This is an important calculation since it shows clearly the relationship between concentration, flow rate, and loading of contaminant. If the contaminant is sand, for example, this calculation shows that a small concentration such as 20 mg/L with a modest flow rate of 12,500 BWPD has a total solids loading of 87 lbs. Removing 87 lbs of material per day is a challenging task. Thus, it is important to have an intuitive understanding of the relationship between concentration, flow rate and loading.

<u>Example 1: Conversion of contaminant concentration to contaminant loading:</u>

Concentration:	mg/L
Flow rate:	barrels per day
Loading:	lbs/day

Step 1: convert flow rate from barrels per day to cubic meters per day by dividing bbl/day by 6.29.

Step 2: multiply concentration in mg/L by flow rate in m3/day, divide by 1,000. This gives the contaminant loading in units of kg/day.

Step 3: The final step is to convert kg/day to lbs/day. Take the result of step 2 and divide by 2.2.

Concentration:	20 mg/L
Flow rate:	12,500 BWPD

Calculation:

$$\text{Loading} = \frac{(20\ \text{mg/L})\,(12{,}500\ \text{BWPD})(1000\ \text{L/m}^3)(2.2\ \text{lb/kg})}{(6.29\ \text{bbl/m}^3)\,(1{,}000{,}000\ \text{mg/kg})} = 87.4\ \text{lb/day}$$

Eqn (A.1)

This is an important calculation since it shows clearly the relationship between concentration, flow rate, and loading of contaminant. If the contaminant is sand, for example, this calculation shows that a small concentration such as 20 mg/L with a modest flow rate of 12,500 BWPD has a total solids loading of 87 lbs. Removing 87 lbs of material per day is a challenging task. Thus, it is important to have an intuitive understanding of the relationship between concentration, flow rate and loading.

Standard and Normal Conditions

Stock tank barrels.

The CUBIC METRE under STANDARD CONDITIONS is:
1 m^3 (st) at 15 C (288.15 K) and 101 325 Pa (1.01325 bar)

The NORMAL CUBIC METRE is:
1 Nm^3 at 0 C (273.15 K) and 101 325 Pa (1.01325 bar)"

It then goes on to talk about the m^3 (st) as being adopted by ISO, ASTM, AGA, API, GPSA, IP, & IGU. It is recommended by SIPM.

It also notes that the Nm3 is used in a few European countries which have traditionally used metric units.

So these definitions are "universal" in that they are adopted by industrial and international standards bodies. But the "standard" to use is what is in the laws and contracts that apply to whatever you're working on. In Venezuela, they've adopted the same "standard" as ISO, etc.

Definitions of "Standard" volume do vary around the world and, in the U.S., from State to State.

An SI Standard cubic meter = 101.325 kPa & 0 C.

An Oil Industry Standard cubic meter is 101.325 kPa & 15.56 C (60 F)

In Venezuela, contracts are written with oil industry standard conditions, 101.325 kPa (14.696 psia) and 15.56 C (60 F), so use the conversion values listed in the GPSA data book.

I have a Shell publication that mentions:

"1 normal m^3 (0 C, 760 mm Hg) = 37.33 standard ft3 (60 F, 760 mm Hg)"

- a normal cubic meter is an "SI Standard" cubic meter
- a standard cubic meter is an "Oil Industry Standard" cubic meter
- an SI standard cubic meter, when heated from 0 C to 15.56 C will expand from 35.315 "Oil Industry Standard" cubic feet to 37.33 "Oil Industry Standard" cubic feet

    1. 'Standard' pressure & temperature = 100kPa & 15 deg C respectively, and;
    2. 'Normal' pressure & temperature = 101.325 kPa (i.e. 1 atm) & 0 deg C.

Index

The section number is given here for the major subjects covered in this book.

Activated Carbon Filtration: 13.6.4

Aging and weathering of Produced Water: 2.3.2

Alkalinity: 2.12.2

Analytical Results, Interpretation: 16.4

Anionic Polyelectrolytes: 5.5.3

Archaea (Extremeophiles): 6.1.3

Asphaltenes and Resins in Emulsions: 4.5.7

Bacteria: 6.1.2

Bacteria as a Constituent of Produced Water: 2.8

Bacterial Contamination & Solids: 18.1.6

Bacteria Control in the Oil Patch: 6.6

Bacterial Digestion in a Petroleum Reservoir: 6.2

Bacterial Growth in the Subsurface: 6.3

Bacteria Sampling and Analysis: 6.5

Bancroft Rule: 4.4.5

Bench-Top Oil-in-Water Measurement Methods: 7.4.7

Benefits and Drawbacks of Deoiling Hydrocyclones: 10.1.12

Benefits and Drawbacks of Electrocoagulation: 17.5.8

Benefits and Drawbacks of Flotation: 12.1.9

Best Practices for Deepwater Systems: 19.2

Biocides – Nonoxidizing: 6.6.1

Biocides – Oxidizing: 6.6.2

Biodegradation and Total Acid Number: 2.2.4

Biological Oxygen Demand (BOD) Measurement: 7.5.4

BOD, COD, TOC Chemistry of: 2.3.3

Boron Removal: 13.8.1

Bubble Coalescence: 12.2.2

Bubble Rise Velocity (V12): 12.3.3

Bubble / Water Interface: 12.3.4

Cake or Pre-Coat Filtration: 13.5

Calculating Oil Droplet Retention Efficiency for a Horizontal Separator: 8.3

Capillary Rise: 4.4

Capture Efficiency (Ecapt): 12.3.6

Capture Mechanisms and Spreading Coefficient: 12.3.5

Carbon Dioxide (CO2), 2.4.1

Cartridge Filtration: 13.4.2

Cationic Polyelectrolytes: 5.5.2

Centrifuge Mechanism of Separation: 11.1

Centrifuge Operation and Maintenance: 11.3

Centrifuge Representative Performance Data: 11.2

CETCO Hi-Flow media: ... 13.9

Challenges in the Design of Deepwater Platforms: 19.1

Characterizing Produced Water, How To: 16.2

Charged Surfaces in Ionic Solutions (Brine): 4.3

Chemical Injection Points & Metering: 19.2.8

Chemical Injection Systems: 5.17

Chemical Oxygen Demand (COD): 7.5.1

Chemical Treatment of Naphthenates, Dissolved Organics and Water Soluble Organics: 5.9

Chemicals - Chemistry: 2.9

Chemistry and Characteristics of Polymeric Flocculating Agents: 5.5.1

Chemistry of Carbamates and Chelating Agents: 5.5.4

Chlorine Dioxide (ClO2): 5.8

Classification of Designs (MIGF, HIGF, VIGF, VDGF, etc): 12.2.3

Clay Media Filtration: 13.6.3

ClO2 (Chlorine Dioxide): 5.8

Coagulation Chemistries: 5.4

Coagulation, Flocculation and Coalescence Mechanisms: 5.3

Coagulation, Flocculation, Coalescence Overview: 4.2

Coalescence Chemistries: 5.6

Coalescence Efficiency: 4.6.3

Coalescence Mechanism and Modeling: 4.6.1

Coalescence Model: 4.6.5

Coalescence of Oil Droplets: 4.6

COD, BOD, TOC: 2.3.3

COD / TOC Ratio: 7.5.3

Collision Efficiency (Ecoll): 12.3.4

Collision Force: 4.6.2

Collision Frequency (Fcoll): 12.3.2

Collision Frequency: 4.5

Compact Flotation Unit: 12.2.6

Composition of Produced Water – General Overview: 2.3

Condensed Produced Water: 2.1.3

Conditions in the Crust: 6.3.1

Conical Separator Solids Removal: 18.4.2

Continuum Mechanics: 3.1

Cook Book: 16.5

Correlating Performance of Continuous Flotation Cells: 12.3.9

Corrosion: 6.4.4

Corrosion Inhibitors: 2.9.1

Corrosion Inhibitors – Use and Impact on Water Treatment: 5.13

CPI – Fouling Mechanism and Model: 9.2.4

CPI – Operation and Maintenance: 9.1.6

CPI – Representative Performance Data: 9.1.5

Critical Coagulation Concentration & Schultz – Hardy Rule: 4.3.7

Cross-Flow Filtration: 13.7

Crude Oil Origin: 2.2.1

Crude Oil Types and Associated Produced Water: 2.3.4

Cyclonic Desander: 18.4.3

Debye Length: 4.3.1

Deep Bed Filtration: 13.6

Defining a Separator Size: 8.2.2

Demulsifiers – Impact on Water Treatment: 5.12

Deoiling Hydrocyclone – Practical Applications: 10.1

Deoiling Hydrocyclones on Deepwater Platforms: 19.1.3

Deposits and Biofilm: 6.4.1

Design Considerations: 12.2.1

Design Differences and Characteristics: 10.1.4

Design of Water Treatment Systems: 1.5

Designing Skim Tanks for Horizontal Flow: 8.4.1

Designing Skim Tanks for Vertical Flow: 8.4.2

Dirty Dozen – 12 Common Causes of Poor Water Quality: 15.2

Disc-Stack Centrifuges: 19.1.5

Dispersants: 5.7

Dispersed, Dissolved, TPH, TOG and WSO: 7.4.3

Disposal in a Landfill Site, Solids: 18.6.1

Disposal Wells: 1.5.4

Disposal Water Quality: 13.2.1

Dissolved Mineral Analytical Equipment: 7.3.1

Dissolved Mineral Content: 7.3

Dissolved Organic Compounds in Produced Water: 2.6.1

Dissolved Organics Interpretation: 16.5.5

Dissolved Organics, Treatment Options for Removing: 21.2

Dissolved versus Dispersed versus WSO/TOG: 21.1

Dissolved versus Suspended Contaminants: 2.3.1

Distributions of Particle and Droplet Size: 3.3

Drilling and Completion Fluids (including hydraulic fracturing flow back fluids): 2.10

Drilling and Completion Water Definition: 1.3

Drilling and Completions Flow Back Water: 2.1.5

Drop Break-Up and Coalescence in Hydrocyclones: 10.2.6

Droplet Rise Velocity, Stokes Law, and Related Quantities: 3.2

Duplicate Samples: 7.2.4

Dynamics of Continuous Flow Systems: 15.3

Electrocoagulation for Boron Removal: 17.5.5

Electrocoagulation for Fine Solids Removal: 17.5.7

Electrocoagulation, Mechanism of Separation: 17.5.1

Electrocoagulation Practical Applications: 17.5

Electrocoagulation for Softening and Silica Removal: 17.5.6

Elasticity and Viscosity of the Oil/Water Interface: 4.6.7

Electrical Double Layer and Zeta Potential: 4.3.2

Electrophoresis and Measurement of Zeta Potential: 4.3.3

Emulsion Formation – How, Where, Why: 4.1.2

Emulsion Types: 4.1.1

Enhanced Oil Recovery (IOR and EOR) - Chemistry: 2.1.4

Environmental Impact Factor (EIF): 1.5.4

EPA-1664 and other Gravimetric Methods: 7.4.2

EPCON Vertical Compact Flotation Unit: 12.2.6

Examples of GoM and North Sea Water Treatment Process Lineups: 20.1

Field Data: 16.3.6

Field Measurement Methods: 7.4.6

Field Visit Checklist: 15.3.4

Film Drainage & High Internal Phase Volume Emulsions: 4.7

Film Drainage Time: 4.6.4

Flocculation Chemistries: 5.5

Flotation Chemical Application: 12.1.3

Flotation Equipment: 12.2

Flotation Practical Applications: 12.1

Flotation Solids Removal: 18.4.1

Flotation Theory, Design and Modeling: 12.3

Flow Assurance Data: 16.3.4

Fluid Inlet Devices: 8.2.3

Fluid Phases in a Separator: 8.2.1

Flux Factor: 12.3.11

Formation and Reservoir Solids (Sand, Clay, Proppant, Organics): 18.1.1

Formation Water Definition: 2.1

Gas Breakout within the Hydrocyclone: 10.1.9

Gas Chromatography and Flame Ionization Detection (GC-FID): 7.4.4

Gas Flotation Tank: 12.2.8

Gas Production and Water Treatment: 2.1.3

Gasses (CO2, H2S, O2) in Produced Water: 2.4

Geologic Processes that Affect Formation Water: 2.2

Geological Data: 16.3.1

Gibbs Elasticity: 4.4.7

Gulf of Mexico Field Examples: 20.1.2

H2S Control – Impact on Water Treatment: 5.15

H2S Scavengers: 2.9

Hardness: 2.11.3

Heat, Apply Upstream: 19.2.2

Hi-Flow media, CETCO: 13.9

Historical Development of Technology: 1.4

Horizontal Multistage Hydraulic IGF (HIGF): 12.2.5

Horizontal Multistage Mechanical IGF (MIGF): 12.2.4

Hydrate Inhibitors: 2.9.2

Hydraulic Fracturing Flow Back Fluids: 2.10.2

Hydraulic Fracturing Fluids: 2.10.1

Hydrocarbons in Produced Water: 2.5

Hydrocarbon Recovery Strategy: 1.5.1

Hydrocyclone – Theory, Design and Modeling: 10.2

Hydrocyclone Number & Performance Curve: 10.2.5

Hydrocyclones – Troubleshooting Procedure: 10.1.11

Hydrogen Sulfide (H2S): 2.4.2

Hydrophilic – Lipophilic Balance (HLB): 4.4.6

Identifying Sweet Spots in Operations: 15.3.3

Importance and Role of Platform / Facilities Operators: 15.3.1

Induced Gas Flotation on Deepwater Platforms: 19.1.4

Injection Impairment: 13.2.1

Injection Water Quality, Permian Basin: 17.3

Injection Water Quality, Oil Reservoir: 13.2.1

Inorganic Constituents in Water: 2.7

Inorganic Scale Forming Minerals: 18.1.4

Inorganic Solids – Dissolved (mineral ions): 2.7.1

Inorganic Solids – Suspended: 2.7.2

Interface Concentration versus Bulk Concentration: 4.4.4

Interfaces – Introduction: 4.4.1

Interfacial Tension, Elasticity, Wetting, Capillarity: 4.4

Interfacial Tension: 4.4.2

Interstitial and Condensed Produced Water: 2.1.3

Interstitial and Connate Water Definition: 2.1.2

Ion Exchange: 13.8

IOR, EOR, cEOR: 1.5.1

Iron Compounds: 2.7.4

Iron Oxides: 18.1.2

Iron Sulfide: 6.4.3

Iron Sulfide Impact: 16.5.7

Iron Sulfide: 18.1.3

Ishikawa Diagram - Five Categories of Causes: 15.4

Isokinetic Sampling: 7.2.1

Laboratory Methods: 7.6.1

Laboratory methods for Measuring Droplet Size: 7.7.2

Langelier Scaling Index: 2.13.1

Log-Normal Distribution: 3.3.1

Macro-Porous Polymer Extraction (MPPE): 21.2

Mare's Tail: 13.9

Material and Flow Balance: 15.1

Material Balance – Reject Flow Rate and Concentration: 12.1.5

Material Balance – Reject Oil Concentration: 10.1.6

Mechanical versus Chemical Problems: 15.1

Mechanism of Separation – Performance Variables: 10.1.3

Mechanism of Separation & Performance Variables: 12.1.2

Media Coalescence: 13.9

Metallic Salt Coagulants: 5.4.1

Microbial Growth – Impact of Reservoir Conditions: 6.3.2

Microbiological Growth – Problems Caused: 6.4

Microfiltration & Ultrafiltration: 13.7.1

Middle East Regulations: 1.5.4

Mineral Precipitation Tendency: 16.5.4

Modeling of Flotation Performance: 12.3.1

Modeling the Batch Flotation Process: 12.3.8

Monitoring and Control Systems: 19.2.9

MPPE (Macro-Porous Polymer Extraction): 21.2

Multi-Phase Flow: 3.6

NACE Chemical Solvent Test: 18.2.2

Naphthenic Acids in Produced Water: 2.6.2

Naturally Occurring Radioactive Material (NORM): 18.1.7

NORM (Naturally Occurring Radioactive Material) Components: 2.7.3

North Sea Field Examples: 20.1.1

NPD Compounds: 2.5

Nut Shell Filtration: 13.6.2

Oil Density vs Temperature: 3.2.1

Oil Drop Diameter Distribution: 20.2.1

Oil Droplet and Solid Particle Size: 7.7

Oil Droplet Sizes in Produced Water: 19.1.2

Oil Flow Assurance Interpretation: 16.5.2

Oil in Produced Water: 7.4

Oil Spill Cleanup: 9.1.8

Online Methods: 7.6.2

Online Methods for Measuring Droplet Size: 7.7.3

Online Oil in Water Measurement Methods: 7.8.2

Online Oil in Water Monitors: 7.4.8

Operation and Maintenance: 10.1.10

Operation and Maintenance: 12.1.7

Operations Competency: 15.3

Organic Content: 7.5

Organic Solids: 18.1.5

Organics (besides hydrocarbons) in Produced Water: 2.6

Origin of Petroleum: 6.1.4

Overall Chemical Optimization Strategy: 5.11

Oxidation: 5.8

Oxygen (O2): 2.4.3

PAH (Polycyclic Aromatic Hydrocarbons) Compounds: 2.5

Particle Size Distribution (PSD): 18.2.3

Particle Size Measurements: 7.8.4

PDR: Pressure Differential Ratio: 10.1.5

PECT-F: 13.9

pH, Alkalinity, and Hardness: 2.12

pH: 2.12.1

Plate Interceptors: CPI, TPI & PPI – Practical Applications: 9.1

Plate Interceptors – Theory, Design and Modeling: 9.2

Polyaluminum Chloride (PACl) Coagulants: 5.4.2

Polyelectrolyte Coagulants: 5.4.3

Pourbaix Diagram: 17.5

Practical Devices for Coalescing Oil Droplets: 4.6.8

Precipitation and Scaling Tendency: 2.13

Precision of Analysis Methods: 7.8

Process Configuration: 15.4, p. 15-15

Process Configuration and Control: 10.1.7

Process Configuration, Rejects Handling, Control: 12.1.6

Process Configuration, Waste Handling, Control: 14.1.3

Process Control Philosophy and Instrumentation: 8.2.6

Process Engineering Data: 16.3.3

Process (FWKO, Skim or Wash) Tank Design Considerations: 8.4

Process Performance Diagram: 15.2

Process Performance Diagram: 20.2.5

Produced Water Characteristics and Crude Oil Types: 2.3.4

Produced Water Definition: 1.3, p. 7; 2.1

Pump Shearing: 3.5.3

Reference Methods for Oil-in-Water: 7.4.5

Reference Oil in Water Analysis Methods: 7.8.1

Regulations Gulf of Mexico: 1.5.4

Regulations Middle East: 1.5.4

Regulations North Sea, OSPAR: 1.5.4

Reject / Recycle Handling System: 19.2.7

Representative Performance Data: 10.1.8

Representative Performance Data: 14.1.4

Reservoir Engineering Data: 16.3.2

Reservoir Souring: 6.4.2

Resins and Asphaltenes (SARA Crude Oil Fractions): 2.6.4

Resource Recovery and Conservation Act (RCRA): 1.5.4

Reynolds Number: 3.2.2

Rosin-Rammler Distribution: 3.3.2

Routing of BOT Water Discharge to Upstream: 15.4.1

Sample Containers: 7.2.2

Sample Point Design: 7.1

Sample Point Location: 7.1.1

Sample Preservation and Storage: 7.2.3

Sample Probe: 7.1.2

Sampling and Analysis of Solids: 18.2

Sampling and Sample Handling: 7.2

Sampling for Droplet Size and Solid Particle Size Measurement: 7.7.1

Sand and Multi-Media Filters: 13.6.1

Sand Removal and Interface Drains: 8.2.5

Scale Inhibitors – Impact on Water Treatment: 5.14

Scaling Tendency: 2.13

Schultz – Hardy Rule: 4.3.7

Seawater Evaporites: 2.2.3

Selection of Flotation Gas: 12.1.4

Separate Water Early in the Process: 19.2.6

Separation Efficiency as a function of Particle (or Droplet) Size: 3.3.3

Separation Equipment: 20.2.2

Separator Design Considerations: 8.2

Shale Produced Water, Characteristics: 17.1

Shale Produced Water, Reuse Options: 17.2

Shale Produced Water, Disposal Injection: 17.3

Shearing: 3.5.4

Shear and Temperature, Effect of: 20.2.4

Shearing of Oil Droplets: 3.5

Shearing Through a Pump: 3.5.3

Shearing Through a Valve: 3.5.2

Silica and Silicon Compounds: 2.7.5

Sizing of Equipment based on Fluid Density Difference: 3.3.4

Small Droplets Impact: 16.5.9

Solids Impact: 16.5.6

Solids Content (TSS – Total Suspended Solids): 7.6

Solids Handling and Management: 18.3

Solids in Produced Water and Fluid Compatibility: 19.1.1

Solids in Produced Water: 18.0

Solids, Origin and Types of: 18.1

Solids Prevention, Separation, Removal: 19.2.3

Solids Separation from Liquid: 18.4

Solids Stabilized Emulsions: 2.13.2

Sources of Data (Data Mining): 16.3

Sparger Systems: 12.2.9

Spillover Effect in Operations: 15.3.2

SP-Pack media: 13.9

Stokes Factor: 16.5.1

Stokes Factor: 3.2.5

Stokes Law in Rotational Acceleration: 10.2.1

Stokes Law Settling (and Creaming) Rate: 3.2.3

Straining Filtration: 13.4

Subsea Produced Water Quality Measurement: 7.9

Surface Active Constituents in Crude Oil: 2.6.5

Surface Active Production Chemicals: 16.5.8

Surface and Interfacial Tension: 4.4.2

Surface Excess Concentration & Surface Activity: 4.4.3

Surface Wetting: 4.4.8

Sweep Factor: 12.3.10

System Performance Modeling: 20.2

Tangential Velocity & Vortex Analysis: 10.2.3

TDS, TOG, TOC, BOD, COD, TSS: 2.3.3

Tertiary Treatment, When is it required? : 13.2

Thermodynamic State of Produced Water: 2.3.2

Total Organic Carbon (TOC): 7.5.2

Total Suspended Solids (TSS): 18.2.1

Toxicity: 2.11

Treatment Options for Different Types of Emulsions: 4.1.3

Triazine H2S Scavengers: 2.9.3

Troubleshooting General Methodology: 15.5

Troubleshooting Procedure & Bench Top Flotation Testing: 12.1.8

Troubleshooting Procedure for CPI: 9.1.7

TSS Analysis Methods: 7.8.3

Turbulence: 3.4

Turbulence Energy Dissipation Rate: 3.4.2

Turbulent Flow – Maximum Droplet Size: 3.5.1

UIC (Underground Disposal Control): 13.2.1

Units of Measure: 1.2, p. 4; Appendix A1

Valve Shearing: 3.5.2

Vertical DGF (VDGF): 12.2.7

Vertical IGF (VIGF): 12.2.6

Vortex Breakers: 8.2.4

Water Analysis Data: 16.3.5

Water Cycle: 2.2.2

Water Flood – brief discussion: 1.5.1

Water Flood, Water Quality: 13.2

Water Quality Analysis (TDS, TOG, TOC, BOD, COD, TSS): 2.3.3

Water Soluble Organics Chemistry: 21.1.3

Water Soluble Organics, Chemical and Phase Equilibrium: 21.1.2

Water Soluble Organics measured by EPA-1664: 21.1.1

Water Soluble Organics (WSO), Chemistry: 2.6.3

Water Soluble Treatment Options: 21.2

Water Treatment Chemical Selection: 5.10

Water Treatment Chemicals, Terminology: 5.1

Weathering and Aging of Produced Water: 2.3.2

Weber Number: 3.4.3

Wedge-Wire Filtration: 13.4.1

Wetting of Solids in Produced Water: 4.4.10

Wetting of Solids Surfaces: 4.4

What is Oil in Water? 7.4.1

Zeta-Potential – Gas Bubbles in Water: 4.3.6

Zeta-Potential – Oil Droplets in Water: 4.3.4

Zeta-Potential – Solid Particles in Water: 4.3.5

Made in the USA
Columbia, SC
07 July 2024